LIQUID CRYSTALS
Fundamentals

LIQUID CRYSTALS
Fundamentals

Foreword by David A Dunmur

Shri Singh
Banaras Hindu University

World Scientific
New Jersey • London • Singapore • Hong Kong

Published by

World Scientific Publishing Co. Pte. Ltd.

P O Box 128, Farrer Road, Singapore 912805

USA office: Suite 1B, 1060 Main Street, River Edge, NJ 07661

UK office: 57 Shelton Street, Covent Garden, London WC2H 9HE

British Library Cataloguing-in-Publication Data
A catalogue record for this book is available from the British Library.

LIQUID CRYSTALS: FUNDAMENTALS

ISBN 981-02-4250-6

This book is printed on acid-free paper.

Printed in Singapore by World Scientific Printers (S) Pte Ltd

Foreword

Although liquid crystals were discovered as an interesting phenomenon at the end of the 19th century, it took nearly 30 years of research to establish their true identity. Georges Friedel in the 1920's was responsible for recognising that liquid crystals are indeed a new state of matter that is intermediate in structure and molecular organisation between the amorphous liquid state and the solid crystalline state. After a further 80 years of research, the number of known liquid crystal phases has increased from the original two, named by Friedel as nematic and smectic, to between 15 and 20 mesophases. The discovery of liquid crystals coincided with a period of rapid expansion in terms of the physical techniques available to study materials. This, together with great advances in the theory of condensed matter, made liquid crystals an exciting area for scientific research. It attracted experimental and theoretical physicists, applied mathematicians and synthetic chemists, all of whom have contributed to the foundations of the science of liquid crystals. A multi-disciplinary approach has been a characteristic of studies in liquid crystals, and those working in the area are expected to have some familiarity with the chemical constitution of the materials and their properties, together with a knowledge of the physics of the phenomena which they exhibit.

This book entitled "Liquid Crystals: Fundamentals", deals in a scholarly way with all aspects of liquid crystal science, including the chemical nature of the materials, their physical properties, and the complex phase behaviour which is shown by them. The author, Professor Shri Singh (Banaras Hindu University) has worked in the area of liquid crystals for many years and has made important contributions to the statistical theory of liquid crystal properties. He brings a special perspective to the subject, and readers will benefit from both the breadth of his treatment, and from the more detailed description of some selected topics. The study of liquid crystals is rewarding not only because of the exotic nature of their properties, but also because it cuts through conventional boundaries of scientific disciplines. Thus a liquid crystal scientist is a specialist on liquid crystals, but is also informed about many other aspects of chemistry, materials science and physics. Careful study of this volume on the fundamentals of liquid crystals will give the reader a sound basis from which to explore further the fascinating properties of liquid crystals.

David A Dunmur
Southampton, United Kingdom

December 2001

Preface

Liquid crystals are partially ordered systems without a rigid, long-range structure. They are intermediate in symmetry and structure between the solid crystalline state and the amorphous liquid state. The study of liquid crystalline materials covers a wide area : chemical structure, physical properties and technical applications. Due to their dual nature – anisotropic physical properties of solids and rheological behaviour of liquids – and easy response to externally applied electric, magnetic, optical and surface fields, liquid crystals are of greatest potential for scientific and technological applications. The subject has come of age and has achieved the status of being a very exciting interdisciplinary field of scientific and industrial research.

This book is an outcome of the advances made during the last three decades in both our understanding of liquid crystals and our ability to use them for various purposes. It deals with the various aspects of the science of liquid crystals, including the chemical nature of the materials, their properties, and the complex phase behaviour exhibited by them.

The book, containing 11 chapters, is meant for the benefit of both new and advanced researchers in the subject. The mode of presentation is simple to enable readers with a little or elementary knowledge of the subject to understand the basic and essential elements of the science of liquid crystals. For readers with considerable familiarity with the field some aspects have also been covered in depth. The book contains not only a systematic account of the existing literature but also tends to present some new findings. It is expected to be of real value to the graduates and scholars involved in higher research in condensed matter physics, chemical physics, materials science and engineering.

A book of this kind naturally owes a number of ideas to the reviews and books on this subject. The author expresses his thanks to all the authors who have written on the subject before, too many to acknowledge individually. It also owes much to all with whom the author has had an opportunity and advantage of collaborating or discussing the subject in depth. First of all, I express my debt to Prof. Yashwant Singh with whom I started my career and who initiated me into the study of condensed matter physics. It was in the early 1980's that Prof. Yashwant Singh ardently and energetically pursued theoretical work in soft matter physics at Banaras Hindu University and produced a small band of committed and sincere student-researchers in this department. I take this opportunity to pay my compliments to Prof. Singh both as a man and as a scholar in Physics.

Appreciation, advice and encouragement received from time to time from Prof. D.A. Dunmur (University of Southampton, U.K.) is gratefully acknowledged. It was very kind of him to closely go through the manuscript, some of the chapters of the volume, and to make comments and suggestions. I am especially grateful for the foreword to this book. I owe a deep debt of gratitude to him for his constant help and encouragement in my pursuit of studies in liquid crystals. I am thankful to Prof. N.V. Madhusudana (RRI, Bangalore), Prof. T.V. Ramakrishnan (I.I.Sc., Bangalore) and Prof. V.N. Rajasekhran Pillai (NAAC, Bangalore) for their encouragement, help and keen interest in my academic ventures. I also want to thank Dr. Ajit M. Srivastava (IOP, Bhubaneswar) with whom I discussed many aspects of the Physics of defects in condensed state.

It is my privilege to express thanks to my young colleagues and students who kept constantly reminding me to complete this book; in particular to Dr. Sanjay Kumar, Dr. T.K. Lahiri, Mr. Navin Singh and Mr. P.K. Mishra for their interest and help throughout the preparation of this book.

Special thanks go to Prof. Suresh Chandra who encouraged and helped me. I also want to mention Prof. D. Pandey, Prof. S.S. Kushwaha and Prof. R.P. Singh for their good wishes and help.

This book has been produced directly from camera-ready manuscript and its appearance owes much to the skill and patience of Mr. Vidya Shankar Singh. It is a pleasure to thank the editorial staff of World Scientific Publishing Co. Pte. Ltd. for inviting me to write the book and their patience to bear with me in the unavoidable delay in completing it.

Last, but not least, I wish to thank my wife Asha, children and Nati Rishabh for their cooperation, moral support and understanding during the course of this work.

Banaras Hindu University **Shri Singh**
June 2002

Contents

1

Liquid Crystals : Main Types and Classification

1.1 INTRODUCTION

Most of us trained in science, since our years in school, have been conditioned into recognising three states of matter – solid, liquid and gas. Our everyday experience strongly reinforces these concepts. A solid may be either crystalline or amorphous. The most important property which differentiates between solids and liquids is flow. The liquids flow and adopt the shape of container, whereas solids do not flow and tend to retain their shape. The optical properties of some solids and liquids can also be very different. For instance, some solids change the polarisation of light whereas liquids do not. Due to these concepts in mind, the early investigators could not think about the existence of other phases, although they found substances which did not neatly fall into these categories. They noted their observations with unresolved questions. At that time it was anticipated that if we take a crystalline solid and heat it then it will step through the phases solid-liquid-gas, unless it sublimes and evaporates before the gas phase is reached. The interconversion of phases, particularly, in case of organic compounds, usually takes place at well defined temperatures. Contrary to it an unusual sequence of phase transition was observed by an Austrian Botanist Reinitzer[1] in the year 1888. While investigating some esters of cholesterol he observed two melting points. At $145.5^{\circ}C$ cholesteryl benzoate melted from a solid to a cloudy liquid and at $178.5^{\circ}C$ it turned into a clear liquid. Some unusual colour behaviour was also observed upon cooling; first a pale blue colour appeared as the clear liquid turned cloudy and then a bright blue-violet colour as the cloudy liquid crystallised. Reinitzer sent the samples of this substance to a German Physicist Lehmann[2] who was studying the crystallisation properties of various substances. Lehmann observed the sample with his polarising microscope and noted its similarity to some of his own samples. He observed that they flow like liquids and exhibit optical properties like that of a crystal. The subsequent studies established that these observed intermediate phases represent a new thermodynamic state of matter that are quite distinct from the isotropic liquid. The mechanical and symmetry properties of these phases are intermediate between those of a crystalline solid and an isotropic liquid. Lehmann first referred to them as flowing crystals and later used the term "liquid crystals".

The molecules in a crystal are ordered whereas in a liquid they are not. The existing order in a crystal is usually both positional and orientational, i.e., the molecules are constrained both to occupy specific sites in a lattice and to point their molecular axes in specific directions. Contrary to it, the molecules in liquids diffuse randomly throughout the sample container with the molecular axes tumbling wildly. When a molecular material composed of anisotropic molecules is heated from the solid phase following possibilities exist at the melting point (see Fig. 1.1) :

(i) Both types of order (positional and orientational) disappear simultaneously and the resulting phase will be an "isotropic liquid (IL)" possessing T(3) × O(3) symmetry.

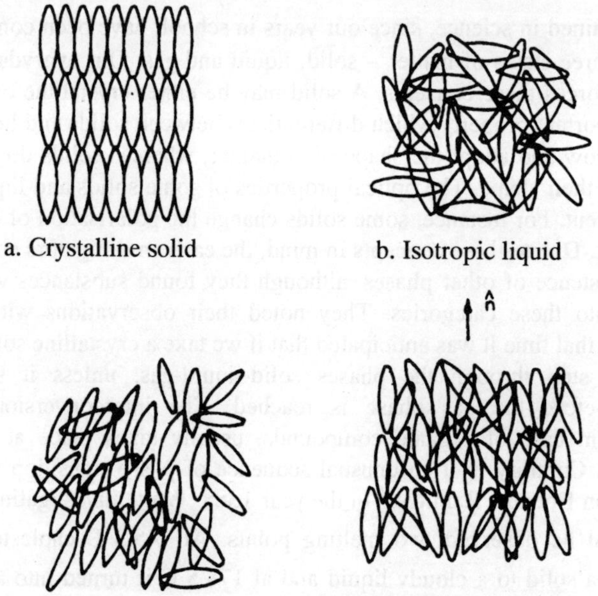

Fig. 1.1 Schematic arrangement of molecules in various phases.

(ii) Only orientational order disappears leaving the positional order intact and the corresponding phase is called a "plastic crystal (PC)". The materials in this phase are said to exhibit rotator phases; the molecules freely rotate along one or more of their molecular axes, whereas their centers of mass are fixed in a lattice.

(iii) The positional order either fully or partially disappears while some degree of orientational order is maintained. The phase thus derived is called as "liquid crystal (LC)". An other name in use is mesophase (meaning intermediate phase) or mesomorphic phase. In this phase the unique axes of the molecules remain, on average, parallel to each other, leading to a preferred direction in space. It is convenient to describe the local direction of alignment by a unit vector $\hat{\mathbf{n}}$, the director, which gives at each point in a sample the direction of a preferred axis. A compound that exhibits a mesophase is called a mesogenic compound. A range of books[3-58] are now available which review the liquid crystal phases and science of these materials.

The liquid crystalline phase possesses some of the characteristics of the order, evidenced by X-ray diffraction, found in crystalline solid and some of the disorder, evidenced by ease of flow, existing in liquids. The molecules in mesophases diffuse about much like the molecules of a liquid, but as they do so they maintain some degree of orientational order and sometimes some positional order also. The amount of order in a liquid crystal is quite small relative to a crystal. There remains only a slight tendency for the molecules to point more in one direction than others or to spend more time in various positions than others. The value of latent heat (around 250 J/g) indicates that most of the order of a crystal is lost when it transforms to a mesophase. In case of a liquid crystal to an isotropic liquid transition the latent heat is much smaller, typically about 5 J/g. Yet the small amount of order in a liquid crystal reveals itself by mechanical and electromagnetic properties typical of crystals.

This chapter is basically the introductory one which describes the molecular structure, known types and broad classification of liquid crystal phases.

1.2 GENERAL TYPES OF LIQUID CRYSTALS

Considering the geometrical structure of the mesogenic molecules, the liquid crystals can be grouped into several types. The liquid crystals derived from the rod-shaped molecules (i.e., one axis is much longer than the other two) are called "calamitics". This class of materials are well investigated and extremely useful for the practical applications. The mesophases formed from disc-like molecules (i.e., one molecular axis is much shorter than the other two) are referred to as "discotics". However, the often used term 'discotic' is not quite proper (see sec. 1.5, chapter 7). This class of mesogens were not known prior to 1977 until discovered independently by Chandrasekhar et al.[59] and Billard et al.[60] Intermediate between rod-like and disc-like molecules are the lath-like species.

Transitions to the mesophases may be brought about in two different ways; one by purely thermal processes and the other by the influence of solvents. The liquid crystals obtained by the first method are called "thermotropics" whereas those obtained by the second one are "lyotropics". Amphotropic materials are able to form thermotropic as well as lyotropic mesophases. The main subdivision of calamitic thermotropic liquid crystals is into nematic and smectic mesophases (see sec. 1.4). There exist many types of smectic phases, indicated as S_A, S_B, S_C,...,S_L which differ in the orientation of the preferred direction of the molecules with respect to the layer normal and the distribution of the centers of the molecules within the layer. In case the preferred direction is tilted with respect to the layer normal the smectic phase becomes biaxial. A detailed description is given in sec. 1.4 and chapter 6. On the basis of the structures clearly identified so far, the liquid crystals of disc-like molecules fall into two distinct categories, the discotic nematic and columnar (see sec. 1.5 and chapter 7).

The lyotropic liquid crystal phases are formed by the dissolution of amphiphilic molecules of a material in a suitable solvent. It is interesting to find that the lyotropic mesophases were observed in the myelin-water mixture before their thermotropic analogue but their significance was not realized. Considering the concentration of the material in the solvent and the temperature, there exist several different types of lyotropic liquid crystal phase structures (see chapter 10).

Liquid crystals are also derived from certain macromolecules (e.g. long-chain polymers) usually in solution but sometimes even in the pure state. They are known as "liquid crystal polymers (LCPs)" (see chapter 8). The polymers are long-chain molecules formed by the repetition of certain basic units or segments known as monomers. Polymers having identical, repeating monomer units are called "homopolymers" whereas those formed from more than one polymer type are called "copolymers". In a copolymer the monomers may be arranged in a random sequence, in an alternating sequence, or may be grouped in blocks.

The liquid crystalline order is also observed on the dissolution of a mesogenic material of long, rigid molecules in a solvent at high concentrations. Good examples are poly (γ-benzyl L-glutamate) (PBLG) and the tobacco mosaic virus (TMV). In order to differentiate them from liquid crystal polymers and lyotropic liquid crystals these phases are normally known as "polymer solution".

1.3 CLASSIFICATION OF LIQUID CRYSTALS : SYMMETRY AND STRUCTURE

The symmetry of liquid crystalline phases can be categorized in terms of their orientational and translational degrees of freedom. The nematic, smectic and columnar phase types (characteristic features are given below) possess, respectively, 3, 2 and 1 translational degrees of freedom, and within each type there can exist different phases depending on the orientational or point group symmetry. The point group symmetries of common liquid crystal phases are given in Table 1.1. Only those phases are included which have well-established phase structures; the so-called crystal smectic phases have been omitted. It can be seen from the Table 1.1 that most nematics, orthogonal smectics and columnar phases are uniaxial having two equal principal refractive indices, while the tilted smectic and columnar phases are biaxial, with all three different principal refractive indices.

Table 1.1 Symmetries of common liquid crystal phase types.

Liquid crystal phase	Point group and translational degrees of freedom	Optical symmetry : uniaxial (u(+), u(-)), biaxial (b), helicoidal (h)
I. Non-chiral phases :		
Calamitic, micellar, uniaxial nematic N_u	$D_{\infty h} \times T(3)$	u (+)
Nematic discotic (N_D), columnar nematic	$D_{\infty h} \times T(3)$	u (–)
Biaxial nematic (N_b)	$D_{2h} \times T(3)$	b
Calamitic orthogonal smectic or lamellar phases (S_A)	$D_{\infty h} \times T(2)$	u (+)
Tilted smectic phase (S_C)	$C_{2h} \times T(2)$	b
Orthogonal and lamellar hexatic phase (S_B)	$D_{6h} \times T(1)$ locally $D_{6h} \times T(2)$ globally	u (+)
Tilted and lamellar hexatic phases (S_F and S_I)	$C_{2h} \times T(1 \text{ or } 2)$	b
Discotic columnar : hexagonal order of columns, ordered or disordered within columns (D_{ho} or D_{hd})	$D_{6h} \times T(1)$	u (–)
rectangular array of columns (D_{ro} or D_{rd})	$D_{2h} \times T(1)$	b
molecules tilted within columns (D_{to} or D_{td} or $D_{ob,d}$)	$C_{2h} \times T(1)$	b
II. Chiral phases :		
Chiral nematic or cholesteric (N_t or N*)	$D_{\infty} \times T(3)$	b, h, locally biaxial but globally u (–)
Chiral smectic C phase (S_C^*)	$C_2 \times T(2)$	b, h
Tilted and lamellar hexatic phases (S_F^* and S_I^*)	$C_2 \times T(1 \text{ or } 2)$	b, h

Several analytical techniques are in use at present for the characterization and identification of phase structures. In some liquid crystal phases the classification is relatively simple and these phases can be identified by employing just one technique. Whenever minimal differences exist in the phase structures the precise classification often requires the use of several different techniques. The phase classification can be done on the basis of following different criteria :

- structures
- microscopic textures
- miscibility rules.

The structure is characterized by the arrangement and the conformation of the molecules and intermolecular interactions. The structural investigation is done mainly by X-rays supported by other methods like neutron scattering, nuclear magnetic resonance, infrared and Raman spectroscopy. The textures are pictures which are observed microscopically usually in polarised light and commonly in thin layers between glass plates. They are characterized by defects of the phase structure which are generated by the combined action of the phase structure and the surrounding glass plates. Depending upon the special experimental conditions for a given structure several different textures exist. This method allows a detailed classification of the mesophases. The textures are also known to exist[12,17,19,42] in several phase types and with blurred pictures. Such textures do not allow the detailed classification of the phases. The miscibility properties have been used extensively by Halle liquid crystal group to classify the mesophases. In the course of their work they developed, as their main tool, the "miscibility rule" which states that the liquid crystals are of the same type if they are miscible in all proportions. As the miscibility studies can only give the probability, but not the certainty, other two criterion can be used for the confirmation.

The optical polarising microscopy in combination with the investigation of the miscibility of binary mixtures has become the most powerful technique for the identification of mesophases. Microscopy reveals that a distinct optical texture results from each different liquid crystalline phase. Differential scanning calorimetry (DSC) is used as a complementary tool to optical microscopy. The DSC reveals the presence of mesophases and liquid crystal phases by detecting the enthalpy change associated with a phase transition. Although the calorimetry can not identify the type of phase, some useful information about the degree of molecular ordering within a mesophase can be derived on the basis of the level of enthalpy change. In the miscibility investigation a material with unknown mesophases is mixed with a known mesogen of already identified phase types. If the two materials belong to the same symmetry group, they are completely miscible and it can be concluded that the phases of each compound are identical and belong to the same miscibility group. Strictly speaking this criterion of selective miscibility only applies if the two compounds possess the same molecular symmetry. Hence, when the two different compounds do not mix, this does not necessarily show a difference between the two phases. Some conclusions can only be derived from the miscibility of the two phases, and not from an observed lack of miscibility.

The X-ray analysis is the final technique for the identification and classification of phase types. The X-ray diffraction provides direct information about the residual positional order in liquid crystals and hence determines the phase structure and classification to which the particular phase belongs. For the initial characterization the data on unoriented samples can be useful to distinguish unambiguously between a number of phases (e.g., S_A and N_u, S_E and S_B, etc.). However, this is not always the case (e.g. S_A and S_C may be confused). Powder data are very useful for deriving information about the intermolecular spacing and relative intensities. In general, to maximise the information it is necessary to study the aligned samples. Other techniques which can be used include neutron scattering studies usually on partially deuteriated samples, nuclear magnetic resonance (particularly useful for analysing lyotropic liquid crystal phases), electron paramagnetic resonance, and infrared and Raman spectroscopy. Each of these techniques provide specific information.

The following sections are devoted to the classification of various liquid crystal phases. The detailed description about them are given in the subsequent chapters.

1.4 CALAMITIC THERMOTROPIC LIQUID CRYSTALS

1.4.1 Non-chiral Calamitic Mesophases :

Following the nomenclature as proposed originally by Friedel[61], the liquid crystals of non-chiral calamitic molecules are generally divided into two types – nematics and smectics.

a. Nematic phases : The nematic phase of calamitics is the simplest liquid crystal phase. In this phase the molecules maintain a preferred orientational direction as they diffuse throughout the sample. There exists no positional order. Figure 1.2 shows the nematic phases occurring in different types of substance. An isotropic liquid (Fig. 1.1b) possesses full translational and orientational symmetry $T(3) \times O(3)$. In case of isotropic liquid-nematic transition the translational symmetry $T(3)$ remains as in isotropic liquid, but the rotational symmetry $O(3)$ is broken. In the simplest structure the group $O(3)$ is replaced by one of the uniaxial symmetry group D_∞ or $D_{\infty h}$ and the resulting phase is the uniaxial nematic phase (N_u) with symmetry $T(3) \times D_{\infty h}$. The molecules tend to align along the director $\hat{\mathbf{n}}$. A biaxial nematic (N_b) phase may result due to the further breaking of rotational symmetry of the system around the director $\hat{\mathbf{n}}$. Figure 1.2b illustrates the N_b phase existing in compounds with lath-like molecules ($T(3) \times D_{2h}$ symmetry). The existence of exotic nematic phases of higher symmetry such as hexagonal, cubic and icosahedral have also been conjectured[8b]. Optically, a hexagonal phase is uniaxial. The structures of the nematic phase of a comb-like liquid crystal polymer and the lyotropic nematic phase of a stiff polymer in solution are also shown in Fig. 1.2.

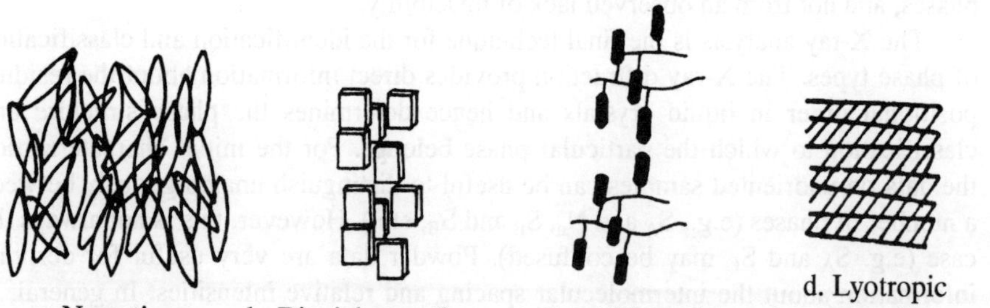

Fig. 1.2 Schematic arrangement of molecules in nematic phases of nonchiral calamitic mesogens.

b. Smectic phases : When the crystalline order is lost in two dimensions, one obtains stacks of two dimensional liquid. Such systems are called smectics. Smectics have stratified structures, with a well defined interlayer spacing. The molecules exhibit some correlations in their positions in addition to the orientational ordering. In most smectics the molecules are mobile in two directions and can rotate about one axis. The interlayer attractions are weak as compared to the lateral forces between the molecules and the layers are able to slide over one another relatively easily. This gives rise to the fluid property to the system having higher viscosity than nematics. The main features of nonchiral mesophases are given in Table 1.2.

A smectic can be characterized by its periodicity in one direction of space, and by its point group symmetry. A priori no point group is forbidden. As a result, an infinite number of smectic phases can be expected. The observed smectic phases differ from each other in the way of layer formation and the existing order inside the layers. The simplest is the smectic A (S_A) phase with symmetry T(2) × $D_{\infty h}$. In S_A phase the average molecular axis is normal to the smectic layers (Fig. 1.3a). Within each layer the centers of gravity of molecules are ordered at random in a liquid-like fashion and they have considerable freedom of translation and rotation around their long-axes. Thus the structure may be defined as orientationally ordered fluid on which is superimposed a one-dimensional density wave. The flexibility of layers leads to distortions which give rise to beautiful optical patterns known as focal–conic textures. When temperature is decreased, the S_A phase may transform into a phase possessing even lower symmetry. The breaking of $D_{\infty h}$ symmetry may lead to the appearance of tilting of molecules relative to the smectic layers. The phase thus derived is called smectic C(S_C) which possess the symmetry T(2) × C_{2h} (Fig. 1.3b).

Table 1.2 Phase types of nonchiral smectic liquid crystals.

Phase type	Orientational ordering	Positional ordering	Molecular orientation	Molecular packing
S_A	Long range	Short range	Orthogonal	Random
S_C	Long range	Short range	Tilted	Random
S_B	Long range	Short range	Orthogonal	Hexagonal
S_I	Long range	Short range	Tilt to apex of hexagon	Pseudo-hexagonal
S_F	Long range	Short range	Tilt to side of hexagon	Pseudo-hexagonal
S_L (S_{Bcry})	Long range	Long range	Orthogonal	Hexagonal
S_J	Long range	Long range	Tilt to apex of hexagon	Pseudo-hexagonal
S_G	Long range	Long range	Tilt to side of hexagon	Pseudo-hexagonal
S_E	Long range	Long range	Orthogonal	Orthorhombic
S_K	Long range	Long range	Tilted to side a	Monoclinic
S_H	Long range	Long range	Tilted to side b	Monoclinic

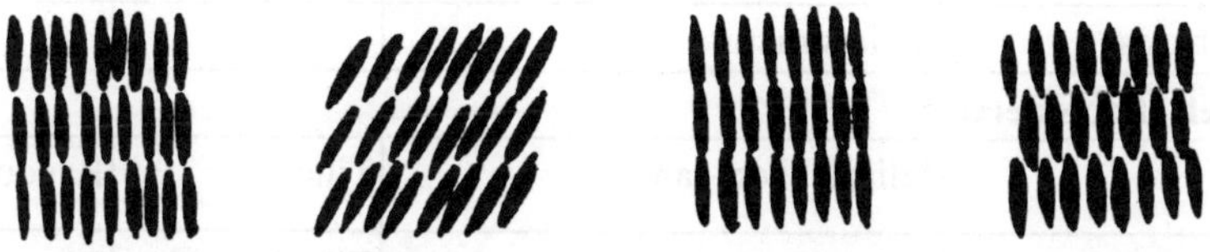

a. S_A b. S_C c. hex S_B d. S_E

Fig. 1.3 Schematic arrangement of molecules in smectic phases of nonchiral calamitic mesogens.

There exist several types of smectic phases with layered structures in which the molecules inside the layer possess effective rotational symmetry around their long axes and are arranged in a hexagonal (S_B) or pseudohexagonal (S_F, S_G, S_I, S_J) manner. In a S_B phase (Fig. 1.3c) the molecules are orthogonal with respect to the layer plane whereas in other phases they are tilted. The existence of S_B phase and other phase types of higher order has also been observed in polymeric liquid crystals. Several smectic phases in which the rotation around the long-molecular axis is strongly hindered have also been observed. The highly ordered molecules produce an orthorhombic lattice if the long-axes of the molecules are orthogonal with respect to the layer planes (S_E) and a monoclinic lattice in case of a tilted arrangement of the long-axes (S_H and S_K). Due to the three dimensional order in these phases they are also considered as crystals. The so called smectic D phase is a cubic phase and does not form layers. So the D phase is, in fact, not smectic. The S_B, S_I and S_F phases are 2D in character, i.e., inside the layers the molecules are oriented on a 2D lattice having no long-range correlations. The higher order smectic phases S_L (or S_{Bcry}), S_J, S_G, S_E, S_K, S_H possess 3D long-range order and therefore are also known as "crystal smectics". In S_L, S_J and S_G the rotation of molecules around the molecular long-axes is strongly hindered. In the S_E, S_K and S_H phases the rotational hindrance around the long-axes is so strong that only 180° jumps between two favoured positions are possible.

1.4.2 Chiral Calamitic Mesophases :

Chiral compounds are able to form liquid crystals with structures related to those of nonchiral materials but with different properties. An overview of most important phases occurring in chiral compounds is given in Table 1.3.

Table 1.3. Mesophases in chiral materials.

Phase type	Ordering	Molecular orientation	Molecular packing
Nonferroelectric types			
N_t (or N*)	Helical nematic structure	Uniaxial	-
Blue phases	Cubic structure	-	-
Ferroelectric smectics			
S_C^*	Helical, short-range	Tilted	Random
S_I^*	No layer correlation Short-range in-plane correlation	Tilted to side	Pseudo-hexagonal
S_F^*	Helical structure	Tilted to apex	Pseudo-hexagonal
S_J^*	Long-range layer correlation Long-range in-plane correlation	Tilted to apex	Pseudo-hexagonal
S_G^*	No helical structure	Tilted to side	Pseudo-hexagonal
S_K^*	Long-range layer correlation Long-range in-plane correlation	Tilted to side	Herring-bone
S_H^*	No helical structure	Tilted to apex	Herring-bone

a. Non-ferroelectric phases : When an optically active material forms a uniaxial nematic phase the preferred direction of the long molecular axis is not constant over the whole sample, as it would be in a uniaxial nematic, but rather it varies in direction throughout the medium in a regular way and displays a continuous twist along the optic axis leading to a helical structure (Fig. 1.4a). Hence the name twisted nematic, or chiral nematic, (N_t or N^*) is given to this phase. Since the derivatives of cholesterols were the first materials to exhibit this phase, it is often called "cholesteric". Due to the twisted or helical structure this phase possesses special optical properties which makes the material very useful in practical applications. This phase cannot be ferroelectric because of symmetry properties. Below their clearing point anomalous phases appear in many of them which are collectively known as "blue phases"[41,62,63]. Blue phases (BP_S) occur in a narrow temperature range between the cholesteric and isotropic phases. In many chiral compounds with sufficiently high twist upto three distinct blue phases appear. The two low-temperature phases, blue phase I (BPI) and blue phase II (BPII), have cubic symmetry, while the highest temperature phase, blue phase III (BPIII), appears to be amorphous.

b. Ferroelectric smectic phases : All of the chiral smectic phases with tilted structure exhibit ferroelectric properties. Due to their low symmetry they are able to exhibit spontaneous polarization and piezoelectric properties and are known as ferroelectric liquid crystals. It is well established that any tilted smectic phase derived from chiral molecules possesses a spontaneous polarization $\mathbf{P}_S$ that is oriented perpendicular to the director $\hat{\mathbf{n}}$ and parallel to the smectic layer plane. The presence of permanent dipoles fundamentally alters the nature of the interactions between the molecules themselves and with any cell wall or applied electric field.

The ferroelectric smectic C (S_C^*) phase (Fig. 1.4b) becomes optically active due to the formation of a helix. A macroscopic helical arrangement of molecules occurs as a result of precession of the molecular tilt about an axis perpendicular to the layer planes. The tilt direction of the molecules in a layer above or below an object layer is rotated through an azimuthal angle relative to the object layer. This rotation always occurs in the same direction for a particular material, and thus a helix is formed which is either left-or right-handed. The helical twist sense is determined by the nature and position of the chiral center with respect to the central core of the mesogenic material. One 360^0 rotation of the helix for the S_C^* phase usually extends over hundreds of layers. Thus, the azimuthal angle is relatively small and is usually of the order of one-tenth to one-hundredth of a degree.

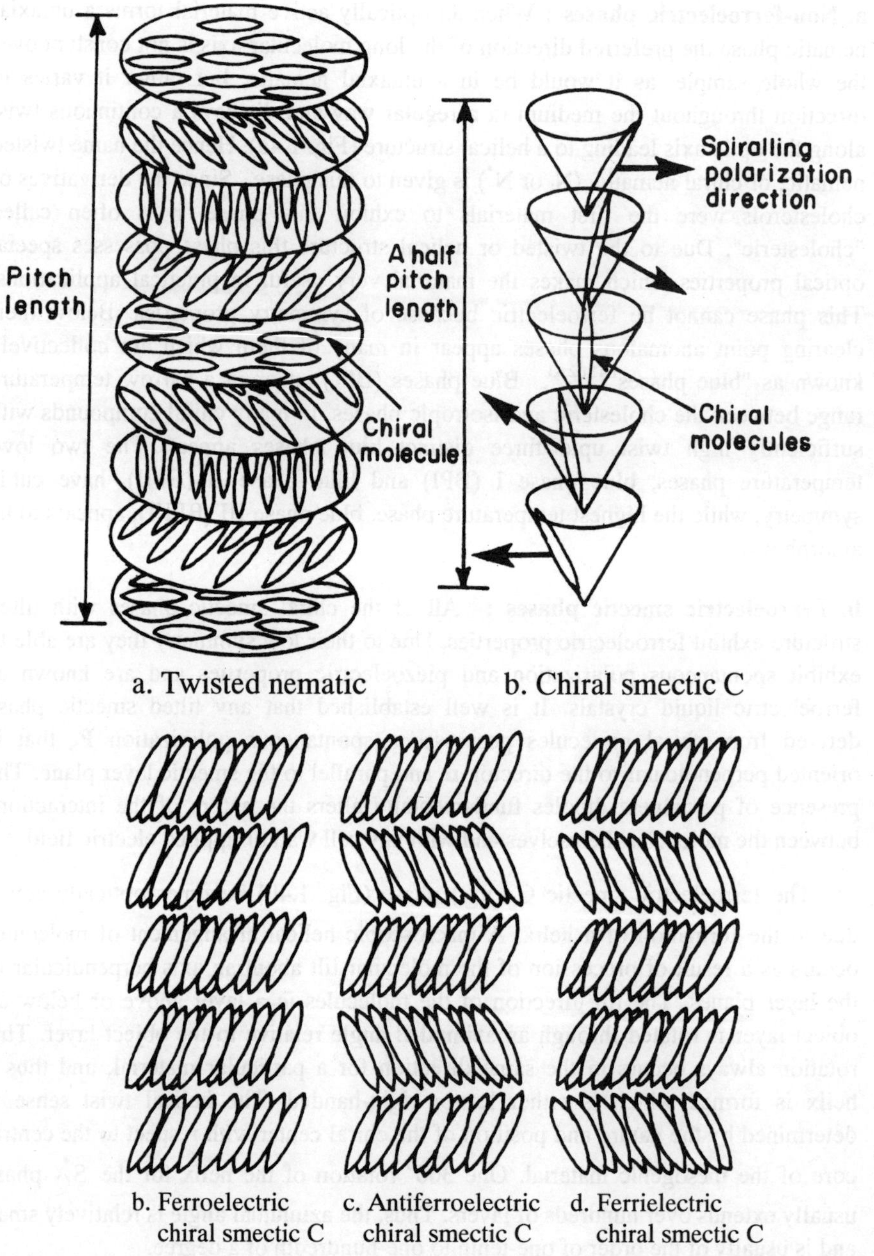

Fig. 1.4 Schematic arrangement of molecules in chiral calamitic mesogens.

c. Antiferroelectric and ferrielectric phases: Two new associated phenomena[64–66] are observed which show fundamental differences to the situations existing in S_C^* phase. This leads to two different arrangements of the molecules in their layer planes from that seen in S_C^* phase with associated variations in polarization directions. These new phases are known as antiferroelectric (Fig. 1.4c) and ferrielectric phases (Fig. 1.4d). From the suggested structure in an antiferroelectric smectic C* phase the molecular layers are arranged in such a way that the polarization directions in subsequent layers point in opposite directions which results in an average of the spontaneous polarization equal to zero. In a ferrielectric smectic phase, the layers are stacked in such a way that there is a net overall spontaneous polarization which is measurable.

1.4.3. Cubic Mesophases :

Optically isotropic cubic mesophases are known to occur in some thermotropic systems and more commonly in lyotropic systems[67]. In lyotropic materials the different chemical nature of the two parts of the molecules, the hydrophilic polar head groups and the hydrophobic alkyl tails, lead to the formation of micelles, which in turn can arrange themselves to generate cubic structure. In contrast to the rich information available on lyotropic materials, not much is known about cubic mesophases in thermotropic systems. Only very few cases of thermotropic cubic phases were known till few last years[19,68]. However, the finding of these mesophases in diverse thermotropic systems has led to a spurt of activity in this field[68,69], and the investigations on their structure and properties are now becoming available[70,71]. Such a symmetry has also been observed in compounds without optical activity. Classical examples are provided by substituted alkoxybiphenyl carboxylic acids :

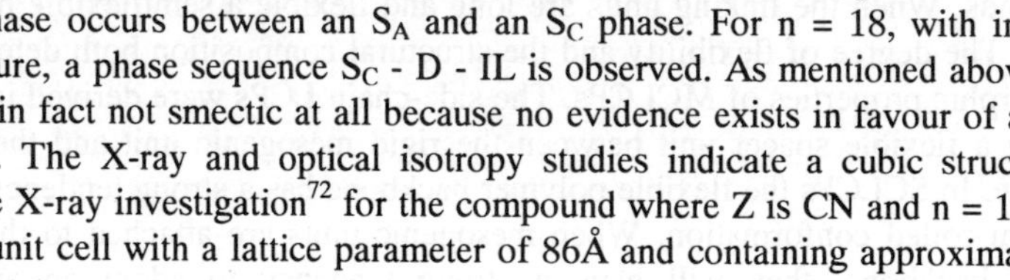

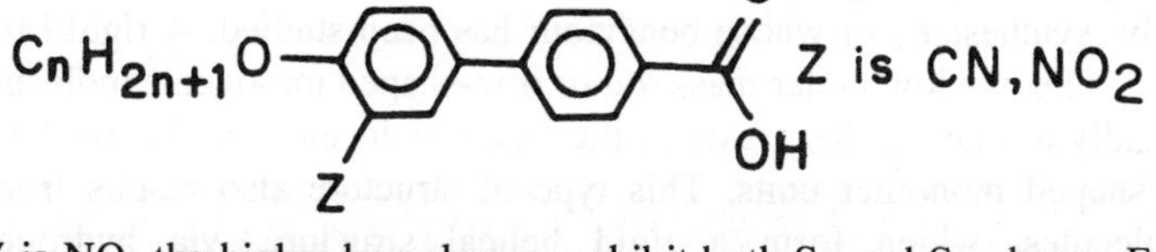

In case Z is NO_2 the higher homologues exhibit both S_A and S_C phases. For n = 16 the D phase occurs between an S_A and an S_C phase. For n = 18, with increasing temperature, a phase sequence S_C - D - IL is observed. As mentioned above, the D phase is in fact not smectic at all because no evidence exists in favour of a layered structure. The X-ray and optical isotropy studies indicate a cubic structure. An extensive X-ray investigation[72] for the compound where Z is CN and n = 18 lead to a cubic unit cell with a lattice parameter of 86Å and containing approximately 700 molecules.

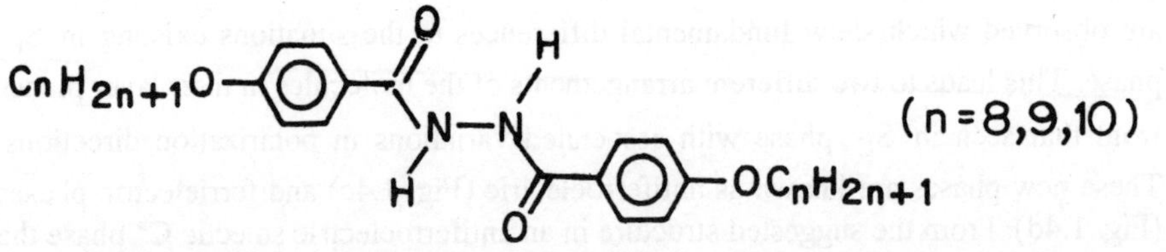

The cubic mesophases were observed[73] in the homologous series, with n = 8, 9, 10, at temperatures below an S_C phase. These phases probably are different from the D phase. For n = 8 the cubic structure has a lattice parameter of 45.7 Å and comprises about 115 molecules.

Another important example of cubic liquid crystals is the cholesteric blue phases.

1.4.4 Liquid Crystal Polymers :

The low molar mass mesogens can be used as monomers in the synthesis of liquid crystal polymers (LCPs). On the basis of molecular anisotropy two types of structure are possible (Fig. 1.5). In the first kind, the macromolecule as a whole has a more or less rigid form, whereas in the second class only the monomer unit possesses a mesogenic structure connected to the rest of molecule via a flexible spacer. In the first kind, the liquid crystallinity is due to rigid macromolecule as a whole or from several monomer units made up of rigid chain segments. The dominant factor in the second class is the structure of the individual monomer units which make up the flexible macromolecule.

On the basis of the location of mesogenic group LCPs are divided into two kinds : main chain liquid crystal polymers (MCLCPs) and side-chain liquid crystal polymers (SCLCPs). Figure 1.5 shows all types of LCPs which have been chemically synthesized or whose behaviour has been studied. A rigid MCLCP can be derived if known low molar mass rod or disc-shaped mesogenic molecules can be joined rigidly to one another. A calamitic macromolecule can also be derived from the disc shaped monomer units. This type of structure also results from flexible macromolecules, which form a rigid helical structure, via hydrogen-bonded compounds. When the linking units are long and flexible a semiflexible polymer is derived. The degree of flexibility and the structural composition both determine the mesomorphic properties of MCLCPs. The side-chain LCPs were derived in 1978 by inserting a flexible spacer unit between the rigid mesogenic unit and the polymer backbone. In SCLCPs the flexible polymer backbone has a strong tendency to adopt a random-coiled conformation. When mesogenic units are attached to the flexible polymer backbone, they will have a strong tendency to adopt an anisotropic arrangement. Clearly, these two features are completely antagonistic, and where the mesogenic groups are directly attached to the backbone the dynamics of the backbone usually dominate the tendency for the mesogenic groups to orient anisotropically; accordingly, mesomorphic behaviour is not usually exhibited. However, if a flexible spacer moiety is employed to separate the mesogenic units from the backbone, then the two different tendencies of the mesogen (anisotropic orientation) and backbone (random arrangement) can be accommodated within one polymeric system. According to this theory, the backbone should not influence the nature and thermal stability of the mesophases, but this is not true because the spacer unit does not totally decouple the mesogenic unit from the backbone. However, enhanced decoupling is generated as the spacer moiety is lengthened.

Calamitic	Discotic	
Rigid Main Chain ▭▭▭ ▯▯▯ Flexible Main Chain ~▭~▭~ ~▭~▭~	○○ ◐◐◐ ~○~○~ ~◐~◐~	LC Main Chain Polymers
⊤ ⊤ ▭ ▭	⊤ ◐	LC Side Chain Polymers
~▭⊤▭⊤~ ▭ ▭	~▭⊤▭⊤~ ◐ ◐	Combined LC Polymers

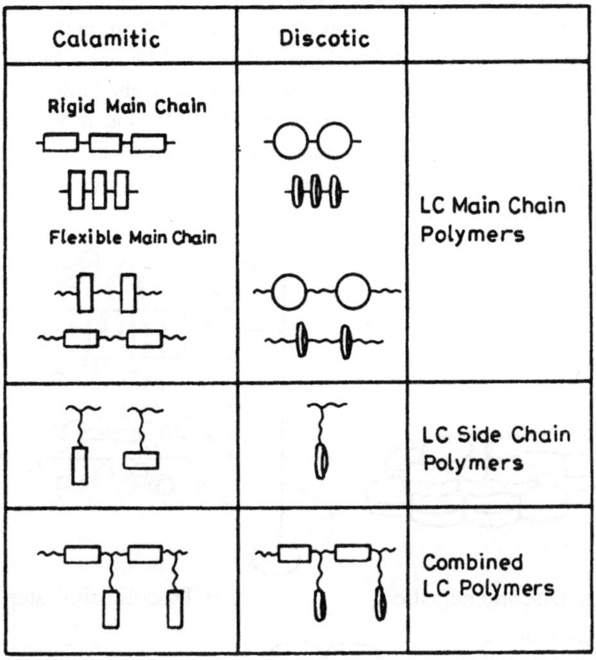

Fig. 1.5 Schematic representation of the different types of liquid crystal polymers.

The nature of the mesophase depends sensitively on the backbone, the mesogenic units and the spacers. Polymers invariably consist of a mixture of chain lengths, the size and distribution of which depend on the structural unit present and the synthesis procedure. Accordingly, there is no such thing as pure polymer and the average size of each chain is often referred to as the "degree of polymerization (DP)". A polydispersity of one (monodisperse) denotes that all of the polymer chains are identical in size.

1.5 MESOPHASES OF DISC-LIKE MOLECULES

Mesogens composed of disc-shape molecules exhibit two distinct classes of phases, the columnar (or canonic) and the discotic nematic phases (Fig. 1.6). When the crystalline order is lost in one direction (i.e. the system is melted in one dimension), one obtains aperiodic stack of discs in columns; such systems are called columnar phases. The different columns constitute a regular 2D array and hence the structure has translational periodicity in two dimensions, but not in three. A number of variants of this columnar structure have been identified and analysed systematically in terms of 2D (planar) crystallographic space groups. The fact that the system is a 1D liquid indicates only the absence of correlations in the arrangement of the centers of mass of the molecules along a straight line. The discotic nematic phase, like its calamitic analogue, is the least ordered mesophase and the least viscous. A smectic-like phase is also reported but the precise arrangement of the molecules in each layer is not yet fully understood. In the columnar structures the molecules are stacked upon each other, building columns which may be arranged in hexagonal, tetragonal and tilted variants. Along the axes of the columns, long-range order or disorder of the molecules can exist.

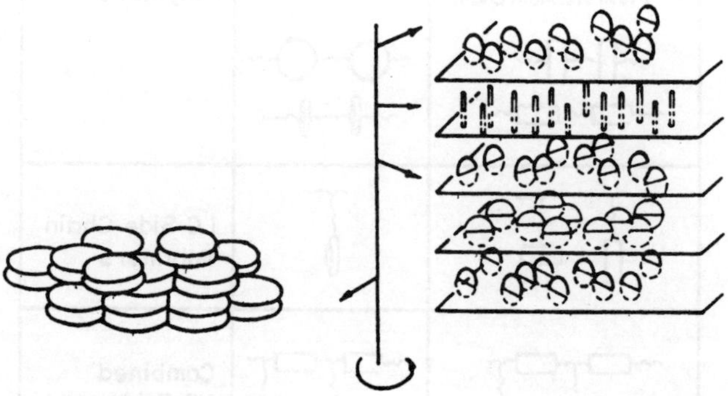

a. Discotic nematic b. Discotic cholesteric

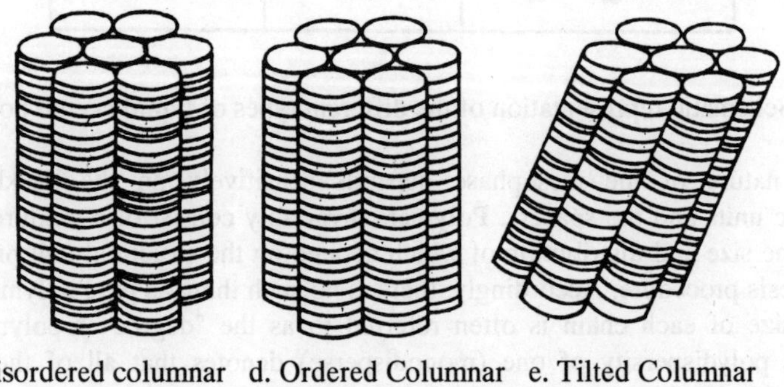

c. Disordered columnar d. Ordered Columnar e. Tilted Columnar

Fig. 1.6 Structures of disc-like mesogens

1.6 LYOTROPIC LIQUID CRYSTALS

Lyotropic liquid crystals are always mixtures (or solutions) of unlike molecules in which one is a nonmesogenic liquid. Generally, one of the components is formed by amphiphilic molecules and another is water. A typical example is the solutions of soap and water. As the water content is increased several mesophases are exhibited. On the basis of X-ray diffraction studies three different types of lyotropic liquid crystal phase structures have been identified - the lamellar, the hexagonal and the cubic phases (see chapter 10). The lamellar or neat phase is characterised by a layer arrangement of amphiphilic molecules. In this phase water is sandwiched between the polar heads of adjacent layers whereas the hydrocarbon tails are in a nonpolar environment. The hexagonal phase consists of micellar cylinders of infinite length packed in a hexagonal manner. There exist two types of hexagonal phases, the hexagonal phase (H_1) and the reversed hexagonal phase (H_2). In the cubic or viscous isotropic phase the layers form spherical unit with polar heads lying on the surface of the sphere and the hydrocarbon chains filling up the inside. The spherical units form a body-centred cubic arrangement.

1.7 MESOGENIC MATERIALS

As discussed above, a number of different structural classes of materials exhibit liquid crystalline phases of various different types. The mesogenic molecules must be highly geometrically anisotropic in shape, either elongated or disc-like. A broad class of calamitic molecules can be represented by the structure as shown in Fig. 1.7. The bridging group A (or A and B) together with the ring structures O_1, O_2 (O_1, O_2 and O_3) form the rigid core of the calamitic molecules. Many a times a linking group is not used to link the core units. The core rigidity is essential for the formation of liquid crystalline phases because it makes the interactions with other molecules anisotropic. The two terminal groups R_1 and R_2 can be linked to the core with groups X and Z but often the terminal groups are directly linked to the core. M and N are the lateral substituents which are usually used to modify the phase morphology and the physical properties of the material. The rigid core alone is not usually sufficient to generate liquid crystal phases and a certain flexibility is required to ensure reasonably low melting points and to stabilise the molecular alignment within the phase structure. The flexibility is provided by the terminal substituents R_1 and R_2 which may be polar or nonpolar. Many different types of synthetic methodology are available[40,41] which are employed to generate liquid crystal phases. Some typical examples of mesogenic compounds are given in Tables 1.4-1.7.

A number of different groups have been used as bridging groups, for example, the Schiff bases, diazo and azoxy compounds, nitrones, stilbenes, esters, biphenyls, etc., to form nematic liquid crystals. The ring structures are usually aromatic rings, but cyclohexane and other ring structures are also in use. A nematic liquid crystal with strong electrical properties can be synthesized by adding a polar terminal group to the end of the molecule.

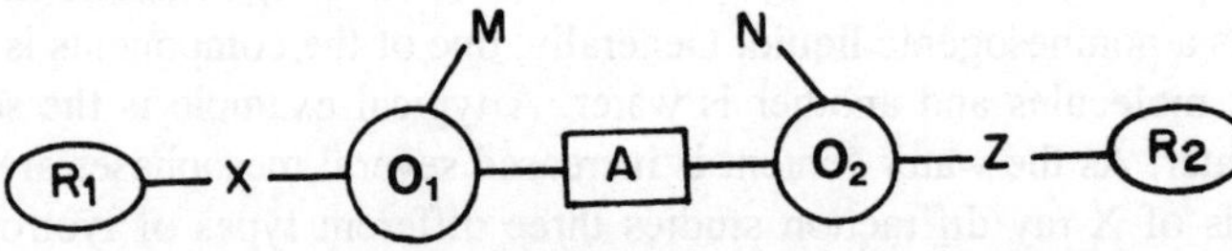

Fig. 1.7 Typical calamitic mesogenic molecule.

Table 1.4 Some typical examples of nematogenic compounds and observed phase sequences. Transition temperatures are given in ^{0}C, K stands for the crystalline state and other symbols are given in the text.

1 4,4′-dimethoxyazoxybenzene (p-azoxyanisole, PPA)

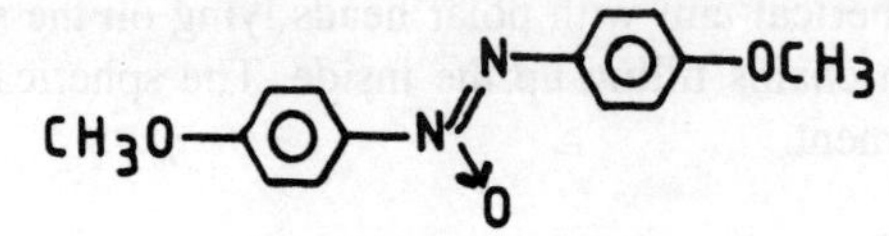

K 118.2 N_u 135.3 IL

2. N-(4-methoxybenzylidene-4′-butylaniline (MBBA)

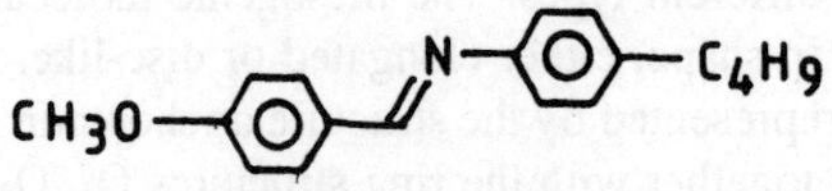

K 22 N_u 47 IL

3. 4-pentylphenyl-trans-4′pentylcyclohexylcarboxylate

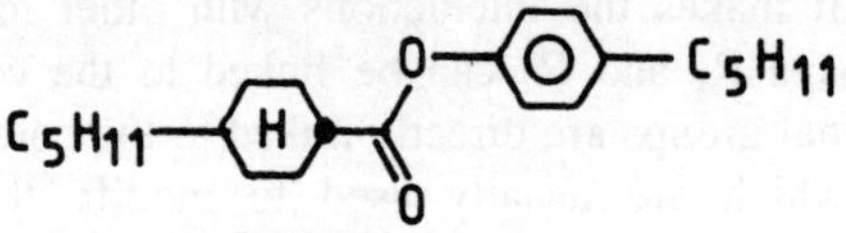

K 37 N_u 47 IL

4. p-quinquephenyl

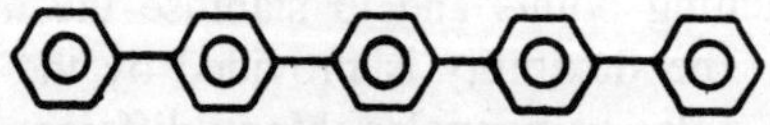

K 380 N_u 431 IL

5.

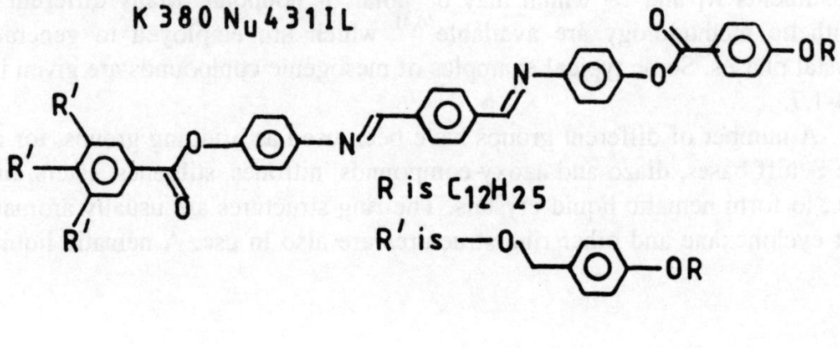

The smectogenic molecules are very similar to those exhibiting nematic phases but tend to be slightly longer. This can be achieved by increasing the length of the rigid core of the molecule with the addition of more rings or by making the end groups longer. Extensive investigations have been carried out to study the influence of the terminal groups on the mesomorphic behaviour which suggest certain general trends. When the terminal groups are very short, the material melts from the solid to the nematic phase and finally to the isotropic liquid phase. As the end groups become longer, the sequence of the phase transition changes; the material melts first from the solid to a smectic phase, then to a nematic phase and ultimately to an isotropic phase.

Table 1.5 Some typical examples of smectogenic compounds and observed phase sequences.

1. p,p′-dinonylazobenzene

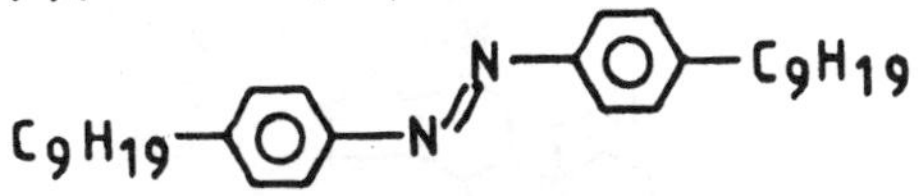

K 37 S_B40 S_A53 IL

2. p,p′-diheptyloxyazoxybenzene (HOAB)

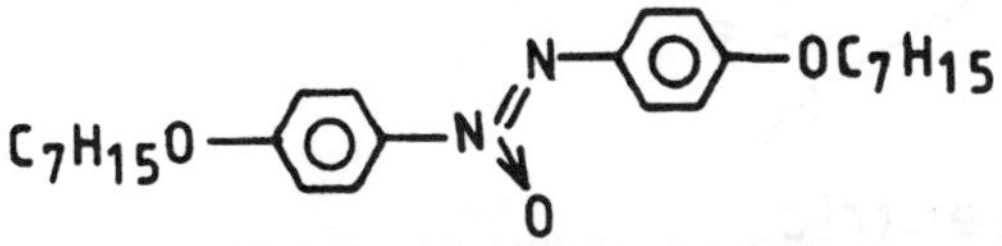

K 74.5 S_C 95.5 N_u 124 IL

3. Terephthalylidene-bis-(p-butylaniline) (TBBA)

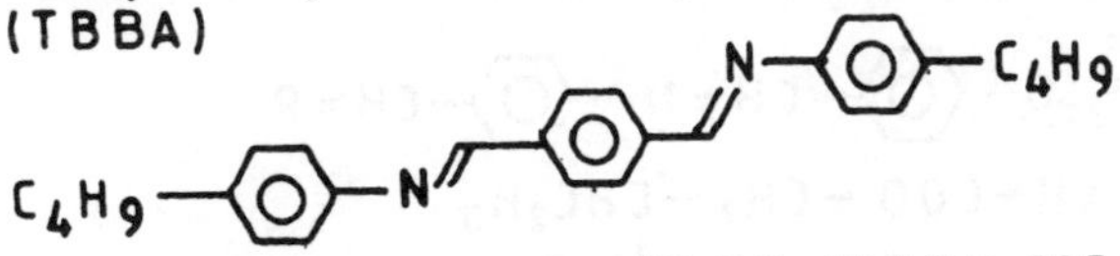

K 112.5 S_G144 S_C 172.5 S_A198.5 N_u235.5 IL

4. p-sexiphenyl

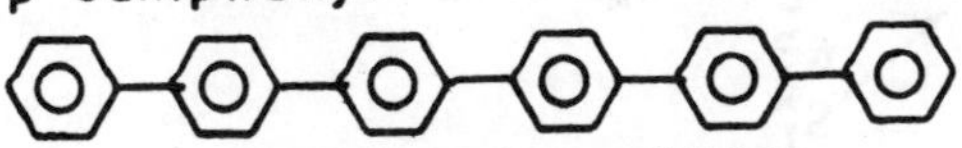

K 435 S_A470 N_u>500 IL

5. N-(4-n-pentyloxybenzylidene)-4′-n-hexylaniline

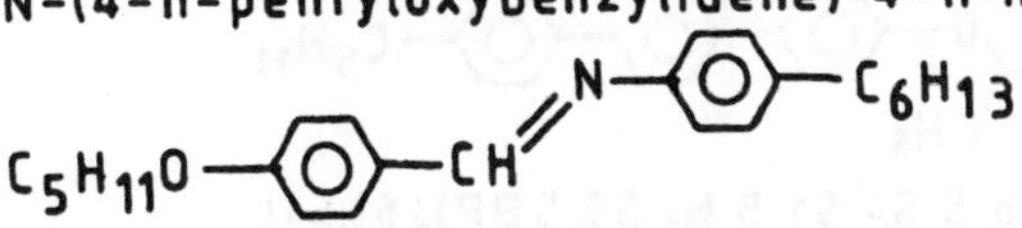

K 36 S_G38.4 S_F42.4 S_B50 S_C 51.8 S_A60.3 N_u72.8 IL

Table 1.6 Some typical examples of chiral mesogens and observed phase sequences.

A. Nonferroelectric

1. N-(p-ethoxybenzylidene)-p'-(β-methylbutyl)aniline

   $K_{15} N_t 60 IL$

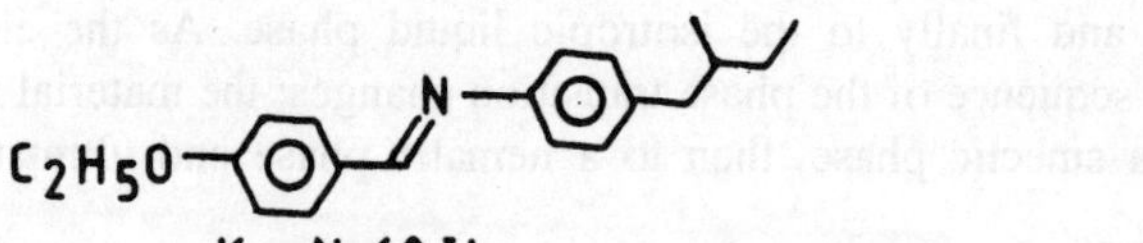

2. Cholesteryl nonanoate

   $K\ 78.6\ N_t\ 91\ BPI\ 91.3\ BPII\ 91.45\ IL$
   75.5
   S_A

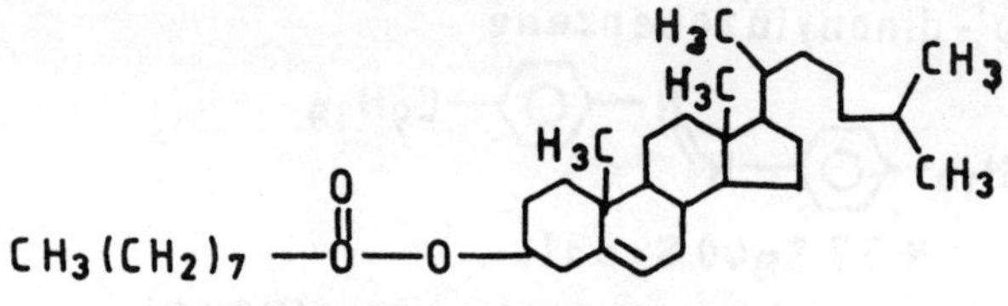

B. Ferroelectric

3. p-decyloxybenzylidene-p'-amino-2-methylbutylcinnamate (DOBAMBC)

   $C_{10}H_{21}O-\bigcirc-CH=N-\bigcirc-CH=R$
   R is $CH-COO-CH_2-CHC_2H_5$
                                        $|$
                                        CH_3

   $K\ 76\ S_C^*\ 95\ S_A\ 117\ IL$
   63
   S_I^*

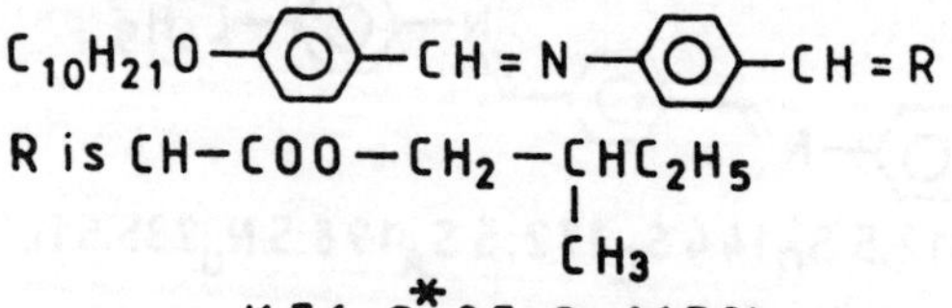

4. Difluoroterphenyl

   $K\ 38.5\ S_C^*\ 51.5\ N_t\ 59.5\ BPII\ 61.5\ IL$

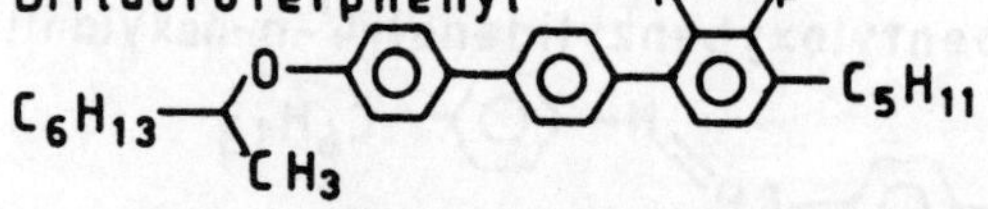

Table 1.7 Some typical examples of disc-like mesogens and observed phase sequences.

1.

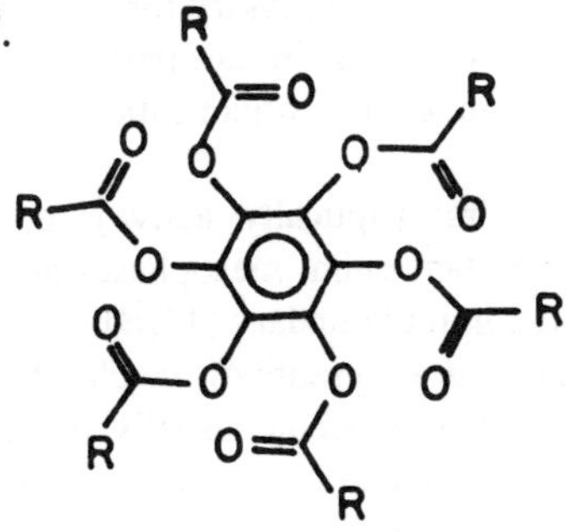

R is C_7H_{15}

K 80 D_{rd} 83 I L

2.

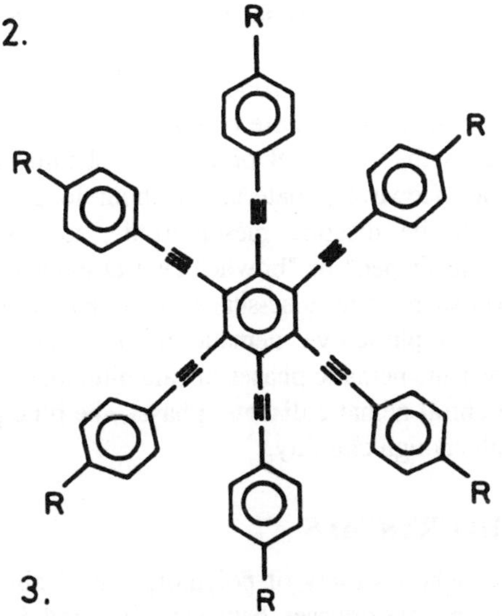

R is C_8H_{17}

K 80 N_D 96 I L

3.

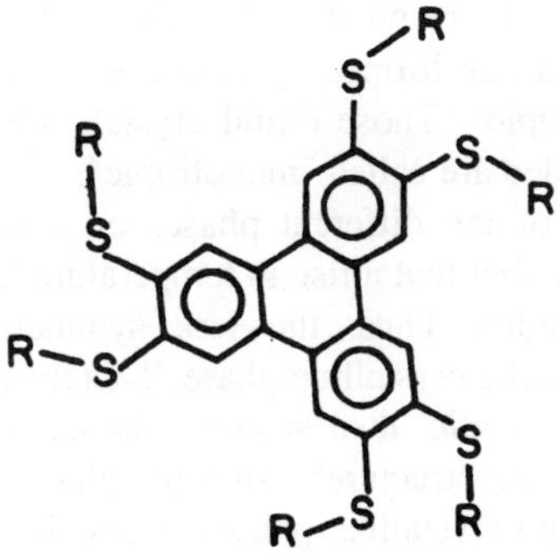

R is C_6H_{13}

K 65 D_{ho} 72 D_{hd} 93 I L

With even longer terminal groups the sequence of transition becomes solid ↔ smectic ↔ isotropic liquid. Many evidences are available which suggest that long terminal groups, especially hydrocarbon chains, are necessary for a material to form smectic liquid crystals. If the structures contain strong dipoles perpendicular to the long-axis of the molecule, side-by-side molecular organization, i.e., the formation of smectic phases, is favoured.

The mesogenic materials composed of chiral (optically active) molecules exhibit helical structure. Examples of some important chiral mesophases are chiral nematic, blue phases, chiral smectic phases, twist grain boundary (TGB) phases (see chapter 9). The magnitude of the pitch of the helix is a measure of the chirality. The shorter the pitch, the more chiral the phase. In a molecule usually chirality is introduced by incorporating a single chiral center. The chirality of phases composed of molecules with two chiral centers of the same handedness is greater than for molecules with one chiral center. In addition, the closer the chiral center is to the rigid core, higher is the chirality.

Typical examples of molecules exhibiting discotic nematic and columnar mesophases are given in Table 1.7. These molecules are comprised of a core based on benzene, triphenylene, or truxene with 3, 4, 6, 8 or even 9 substituents; the central core being more or less flat. However, materials without long flexible substituents have also been found to be discotic mesomorphic, for example, "pyramidic" compounds having a "cone-shaped" or "bowlic" central moiety. There exist many similarities between rod-shape and disc-shape mesophases. Longer terminal groups favour columnar discotic phase over nematic discotic phase just as they prefer to form smectic phases over the nematic phases in calamitic mesophases. Addition of a chiral center causes a chiral nematic discotic phase. The blue phases may be formed with molecules of high enough chirality.

1.8 POLYMORPHISM IN LIQUID CRYSTALS

The mesogenic materials exhibit the richest variety of polymorphism (Tables 1.4-1.7). They pass through more than one mesophases between solid and isotropic liquid. The transition between different phases corresponds to the breaking of some symmetry (Table 1.1). When a mesophase can be formed by both the cooling and heating processes, it is known as "enantiotropic". Those liquid crystals which are obtained only when the substance is supercooled are called "monotropic".

One can predict the order of stability of the different phases on a scale of increasing temperature simply by utilizing the fact that a rise in temperature leads to a progressive destruction of molecular order. Thus, the less symmetric the mesophase, the closer in temperature it lies to the crystalline phase. This means that upon cooling the isotropic liquid, first the nematic, then smectic phases "without order", smectic phases "with order-hexagonal structure", smectic phases "with order-herring bone structure", and finally solid/crystalline phases appear in a fixed sequence. If in a mesogen all of the above phases can exist, they are expected to appear according to the sequence rule, which, in general, may be written as

$$\text{IL–blue–N–}S_A\text{–}S_C\text{–}S_B\text{–}S_I\text{–}S_{Bcry}\text{–}S_F\text{–}S_J\text{–}S_G\text{–}S_E\text{–}S_K\text{–}S_H\text{–Solid/Crystal} \quad (1.1)$$

No single compound is known yet in which the complete sequence (1.1) has been observed. In most of the materials only a small part of different phases exist. Exceptions to the above sequence (1.1) have been observed[74,75] in which higher symmetric mesophase re-appears in-between two lower symmetric mesophases. This kind of phenomenon is known as "reentrant polymorphism".

REFERENCES

1. F. Reinitzer, Monatsch Chem. **9**, 421 (1888).
2. O. Lehmann, Z. Physikal Chem. **4**, 462 (1889).
3. G.W. Gray, Molecular Structure and the Properties of Liquid Crystals, (Academic Press Inc., London, 1962).
4. C. Robinson, Liquid Crystals (Gordon and Breach Science Publishers, London, 1966).
5. J.L. Ericksen, Liquid Crystals and Ordered Fluids (Plenum Press, N.Y., 1970).
6. G.H. Brown, and M.M. Labes, eds., Liquid Crystals 3 (Gordon and Breach Science Publishers, 1972).
7. R.S. Porter, and J.F. Johnson, eds., Liquid Crystals and Ordered Fluids (Plenum Press, N.Y., 1973).

8.(a) P.G. de Gennes, The Physics of Liquid Crystals (Clarendon Press, Oxford, 1974).

(b) P.G. de Gennes, and J. Prost, The Physics of Liquid Crystals (Clarendon Press, Oxford, 1993).

9. G.W. Gray, and P.A. Winsor, eds., Liquid Crystals and Plastic Crystals, Vol. 1 (Ellis Horwood, Chichester, England, 1974).
10. E.B. Priestley, P.J. Wojtowicz, and P. Sheng, eds., Introduction to Liquid Crystals (Plenum Press, New York, 1975).
11. S. Chandrasekhar, Liquid Crystals (Cambridge Univ. Press, 1977, 1992).
12. D. Demus, and L. Richter, Textures of Liquid Crystals (Verlag Chemie, New York, 1978).
13. F.D. Saeva, ed., Liquid Crystals : The Fourth State of Matter (Marcel Dekker Inc., 1979).

14. G.R. Luckhurst, and G.W. Gray, eds., The Molecular Physics of Liquid Crystals (Academic Press Inc., London, 1979).

15. H. Kelker, and R. Hatz, Handbook of Liquid Crystals (Verlag Chemie, Weinheim, 1980).

16. L. Bata, ed., Advances in Liquid Crystal Research and Applications (Pergamon, Oxford, 1980).

17. M. Kléman, Points, Lines and Walls (Wiley, New York, 1983).

18. L.M. Blinov, Electro-optical and Magneto-optical Principles of Liquid Crystals (Wiley, Chichester, 1983).

19. G.W. Gray, and J.W. Goodby, Smectic Liquid Crystals: Textures and structures (Leonard Hill, London, 1984).

20. A.C. Griffin, ed., Ordered Fluids and Liquid Crystals (Plenum Press, New York, 1984).

21. J.L. Ericksen, and D. Kinderlehrer, eds., Theory and Applications of Liquid Crystals (Springer-Verlag, New York, 1987).

22. G.W. Gray, ed., Thermotropic Liquid Crystals (Wiley, New York, 1987).

23. P.S. Pershan, Structure of Liquid Crystals (World Scientific, 1988).

24. G. Vertogen, and W.H. de Jeu, Thermotropic Liquid Crystals, Fundamentals (Springer-Verlag, New York, 1988).

25. V.A. Belyakov, and V.E. Dmitrienko, Optics of Chiral Liquid Crystals (Academic Publishers, Glasgow Harwood, 1989).

26. P.J. Collings, Liquid Crystals : Nature's Delicate Phase of Matter (Princeton University Press, 1990).

27. B. Bahadur, ed., Liquid Crystals : Applications and Uses (World Scientific, (a) Vol. **1** (1990), (b) Vol. **2** (1991), (c) Vol. **3** (1992)).

28. A.L. Tsykalo, Thermophysical Properties of Liquid Crystals (Gordon and Breach Science Publishers, New York, 1991).

29. S.A. Pikins, Structural Transformations in Liquid Crystals (Gordon and Breach Science Publishers, New York, 1991).

30. I.C. Khoo, and F. Simoni, eds., Physics of Liquid Crystalline Materials (Gordon and Breach Science Publishers, Amsterdam, 1991).

31. L. Lam, and J. Prost, eds., Solitons in Liquid Crystals (Springer-Verlag, New York, 1992).

32. S. Martelucci, and A.N. Chester, eds., Phase Transitions in Liquid Crystals (Plenum Press, New York, 1992).

33. I.C. Khoo, and S.T. Wu, Optics and Nonlinear Optics of Liquid Crystals (World Scientific, 1993).

34. G.R. Luckhurst, ed., The Molecular Dynamics of Liquid Crystals (Kluwer Academic, Dordrecht, 1994).

35. L.M. Blinov, and G. Chigrinov, Electro-optic Effects in Liquid Crystal Materials (Springer-Verlag, New York, 1994).

36. R.Y. Dong, Nuclear Magnetic Resonance of Liquid Crystals (Springer-Verlag, New York, 1994).

37. E.I. Kats, and V.V. Labedev, Fluctuational Effects in the Dynamics of Liquid Crystals (Springer-Verlag, New York, 1994).

38. S. Kumar, ed., Liquid Crystals in the Nineties and Beyond (World Scientific, 1995).

39. G.P. Crawford, and S. Zumer, eds.,Liquid Crystals Confined to Complex Geometries Formed by Polymer and Porous Networks (Taylor and Francis Ltd., London, 1996).

40. P.J. Collings, and M. Hird, Introduction to Liquid Crystals : Chemistry and Physics (Taylor and Francis Ltd., London, 1997).

41. P.J. Collings, and J.S. Patel, eds., Handbook of Liquid Crystal Research (Oxford Univ. Press, Oxford, 1997).

42. S. Elston, and R. Sambles, eds., The Optics of Thermotropic Liquid Crystals (Taylor and Francis Ltd., London, 1998).

43. D. Demus, J.W. Goodby, G.W. Gray, H.W. Spiess, and V. Vill, eds., Handbook of Liquid Crystals, Vol. **1** (Wiley-VCH, Weinheim, 1998).

44. G.W. Gray, ed., Physical Properties of Liquid Crystals (John Wiley and Sons Ltd., 1999).

45. A. Ciferri, W.R. Krigbaum, and R.B. Meyer, eds., Polymer Liquid Crystals (Academic Press, New York, 1982).

46. A. Blumstein, ed., Polymer Liquid Crystal (Plenum, New York, 1985).

47. L.L. Chapoy, ed., Recent Advances in Liquid Crystalline Polymers (Elsevier Applied Science Publishers, 1985).

48. C.B. McArdle, Side Chain Liquid Crystal Polymers (ed. Blackie, Glasgow, 1989; Chapman and Hall, New York, 1989).

49. A. Ciferri, ed., Liquid Crystallinity in Polymers : Principles and Fundamental Properties (VCH, New York, 1991).

50. A.M. White, and A.H. Windle, Liquid Crystal Polymers (Cambridge University Press, Cambridge, 1992).

51. A.A. Collyer, ed., Liquid Crystal Polymers: From Structures to Applications (Elsevier, Oxford, 1993).

52. N.A. Plate, ed., Liquid Crystal Polymers (Plenum, New York, 1993).

53. C. Carfagna, ed., Liquid Crystalline Polymers (Pergamon, Oxford, 1994).

54. L.A. Beresnev, Ferroelectric Liquid Crystals (Gordon and Breach Science Publishers, New York, 1988).

55. J.W. Goodby, R. Blinc, N.A. Clark, S.T. Lagerwall, M.A. Osipov, S.A. Pikins, T. Sakurai, K. Yoshino, and B. Žekš, eds., Ferroelectric Liquid Crystals: Principles, Properties and Applications (Gordon and Breach Science Publishers, 1991).

56. A. Buka, ed., Modern Topics in Liquid Crystals: From Neutron Scattering to Ferroelectricity (World Scientific, 1993).

57. J.W. Goodby, R. Blinc, N.A. Clark, S.T. Lagerwall, M.A. Osipov, S.A. Pikins, T. Sakurai, K. Yoshino, and B. Žekš, eds., Ferroelectric Liquid Crystals : Principles, Properties and Applications (Gordon and Breach Science Publishers, 1999).

58. H.S. Kitzerow, and Ch. Bahr, eds., Chirality in Liquid Crystals (Springer-Verlag, New York, 2001).

59. S. Chandrasekhar, B.K. Sadashiva, and K.A. Suresh, Pramana **9**, 471 (1977).

60. J. Billard, J.C. Dubois, H.T. Nguyen, and A. Zann, Nouveau J. Chim. **2**, 535 (1978).

61. G. Friedel, Ann. Physique **18**, 273 (1922).

62. H. Stegemeyer, Th. Blumel, K. Hiltrop, H. Onusseit, and F. Porsch, Liquid Crystals **1**, 3 (1986).

63. D.C. Wright, and N.D. Mermin, Rev. Mod. Phys. **61**, 385 (1989).

64. A.D.L. Chandani, Y. Ouchi, H. Takezoe, A. Fukuda, K. Terashima, K. Furukawa, and A. Kishi, Jpn. J. Appl. Phys. Lett **28**, 1261 (1989).

65. E. Gorecka, A.D.L. Chandani, Y. Ouchi, H. Takezoe, and A. Fukuda, Jpn. J. Appl. Phys. **29**, 131 (1990).

66. S. Inui, S. Kawano, M. Saito, H. Iwane, Y. Takanishi, K. Hiraoka, Y. Ouchi, H. Takezoe, and A. Fukuda, Jpn. J. Appl. Phys. Lett. **29**, 987 (1990).

67. For a review, see J.M. Seddon, and R.H. Templer, in, Handbook of Biological Physics, eds., R. Lipowsky, and E. Sackmann (Elsevier, Amsterdam, 1995) Vol. 1.

68. K. Borisch, S. Diele, P. Göring, H. Kresse, and T. Tschierske, J. Mater. Chem. **8**, 529 (1998).

69. S. Diele, and P. Göring, in, Handbook of Liquid Crystals, eds., D. Demus, J. Goodby, G.W. Gray, H.W. Spiess, and V. Vill (Wiley–VCH, Weinheim, 1998).

70. H. Kresse, H. Schmalfuss, B. Gestblom, K. Borisch, and C. Tschierske, Liquid Crystals **23**, 891 (1997).

71. D.S. Shankar Rao, S. Krishna Prasad, V. Prasad, and S. Kumar, Phys. Rev. **E59**, 5572 (1999).

72. G. Etherington, A.J. Leadbetter, X.J. Wang, G.W. Gray, and T. Tajbakhsh, Liquid Crystals **1**, 209 (1986).

73. D. Demus, A. Gloza, H. Hartung, A. Hauser, I. Rapthel, and A. Wiegeleben, Cryst. Res. Techn. **16**, 1445 (1981).

74. T. Narayanan, and A. Kumar, Phys. Rep. **249**, 135 (1994).

75. S. Singh, (a) Phys. Rep. **324**, 107 (2000); (b) Phase Transitions **72**, 183 (2000).

2

Distribution Functions and Order Parameters

As discussed in the previous chapter, the most fundamental characteristic of liquid crystals is the presence of orientational order of the anisotropic molecules, while the positional order of the center of mass is either absent (nematic phases) or limited (smectic and columnar phases). In this chapter, we shall discuss in a systematic manner how this ordering can be described. The background materials given here are essential for the understanding of the properties of liquid crystals. In the following section, we shall introduce distribution functions and other concepts which can be helpful in calculating order parameters and other relevant averages. We discuss, in particular, the singlet and pair distribution functions. In section 2.2, order parameters are introduced by expanding the singlet distribution in a complete basis set. The influence of the molecular and mesophase symmetries on the order parameters will be analysed. Finally, some of the important methods used for the measurement of microscopic order parameters are discussed.

2.1 DISTRIBUTION FUNCTIONS

We consider a system composed of N rigid molecules contained in volume V and temperature T. All equilibrium properties of the system can be evaluated, if the intermolecular interaction energy and the distribution functions are known. In the "rigid molecule" approximation the intermolecular potential energy, $U(\mathbf{r}^N, \boldsymbol{\Omega}^N)$, of the system depends only on the positions of the center of mass $\mathbf{r}^N (\equiv \mathbf{r}_1, \mathbf{r}_2, ..., \mathbf{r}_N)$ of the N molecules and their orientations $\boldsymbol{\Omega}^N (\equiv \boldsymbol{\Omega}_1, \boldsymbol{\Omega}_2, ..., \boldsymbol{\Omega}_N)$; any dependence on the vibrational or internal rotational coordinates is neglected. The orientation $\boldsymbol{\Omega}_i$ of the ith molecule is described by the Euler angles, θ_i, ϕ_i, ψ_i for nonlinear or θ_i, ϕ_i for linear molecules, respectively. For the molecular fluids, it is convenient to introduce several types of distribution functions, correlation functions and related quantities:

(i) The angular pair correlation function $g(\mathbf{x}_1, \mathbf{x}_2)$ where $\mathbf{x}_i (\equiv \mathbf{r}_i, \boldsymbol{\Omega}_i)$. It is proportional to the probability density of finding two molecules with specified positions and orientations. This provides complete information about the pair of molecules.

(ii) The site-site correlation function $g_{\alpha\beta}(\mathbf{r}_{\alpha\beta})$. It is proportional to the probability density that sites α and β on different molecules are separated by distance $\mathbf{r}_{\alpha\beta}$, regardless of where the other sites are. The set of $g_{\alpha\beta}$ always provides less information than $g(\mathbf{x}_1,\mathbf{x}_2)$; $g_{\alpha\beta}$ can be calculated from $g(\mathbf{x}_1,\mathbf{x}_2)$ but not vice versa.

(iii) The total and direct pair correlation functions. The total correlation function is defined as

$$h(\mathbf{x}_1,\mathbf{x}_2)=g(\mathbf{x}_1,\mathbf{x}_2)-1 \tag{2.1}$$

It measures the total effect of a molecule 1 on a molecule 2 and differs from $g(\mathbf{x}_1,\mathbf{x}_2)$ in that it approaches zero in the limit r_{12} (intermolecular distance) → ∞. h can be expressed as a sum of two parts : (a) a direct effect of 1 on 2; this is short-ranged and is characterized by direct pair correlation function, $C(\mathbf{x}_1,\mathbf{x}_2)$, and (b) an indirect effect, in which molecule 1 influences other molecules 3, 4 etc. which in turn affect 2. For nonspherical molecules

$$h(\mathbf{x}_1,\mathbf{x}_2)=C(\mathbf{x}_1,\mathbf{x}_2)+\rho_o\int d\mathbf{r}_3\,\langle C(\mathbf{x}_1,\mathbf{x}_3)\,h(\mathbf{x}_3,\mathbf{x}_2)\rangle_{\mathbf{\Omega}_3} \tag{2.2}$$

The second term on the right is the indirect contribution made up of direct effects between 1 and 3, and 3 and 2, and so on. The angular bracket $<\cdots\cdots>$ indicates the ensemble average. ρ_0 (= N/V) is the number density.

(iv) Spherical harmonic coefficients of $g(\mathbf{x}_1,\mathbf{x}_2)$. The function $g(\mathbf{x}_1,\mathbf{x}_2)$ can be expanded[1] in terms of spherical harmonic coefficients $g(\ell_1\ell_2\ell;n_1n_2;r_{12})$.

(v) Spherical harmonic coefficients of $C(\mathbf{x}_1,\mathbf{x}_2)$. It can be expanded in terms of spherical harmonic coefficients $C(\ell_1\ell_2\ell;n_1n_2;r_{12})$.

The rotationally invariant physical quantity $f(\mathbf{r}_{12},\mathbf{\Omega}_1,\mathbf{\Omega}_2)$ (e.g., pair potential, pair correlation function, direct pair correlation function, etc.) can be expanded[1,2] (in the convention of rose[2]) for nonlinear molecules as

$$f(\mathbf{r}_{12},\mathbf{\Omega}_1,\mathbf{\Omega}_2)=\sum_{\ell_1,\ell_2,\ell}\ \sum_{m_1,m_2,m}\ \sum_{n_1,n_2} f(\ell_1\ell_2\ell;n_1n_2;r_{12})$$
$$C_g(\ell_1\ell_2\ell;m_1m_2m)\,D^{\ell_1 *}_{m_1n_1}(\mathbf{\Omega}_1)\,D^{\ell_2 *}_{m_2n_2}(\mathbf{\Omega}_2)\,Y_{\ell m}{}^{*}(\mathbf{\Omega}) \tag{2.3}$$

where $f(\ell_1\ell_2\ell;n_1n_2;r_{12})$ are the m-independent f-harmonic coefficients. The function $D^{\ell_i}_{m_in_i}(\mathbf{\Omega}_i)$ are the Wigner rotation matrices, C_g the Clebsch-Gordon coefficients and $Y_{\ell m}(\mathbf{\Omega})$ are the spherical harmonics with $\mathbf{\Omega}$ describing the orientation of intermolecular vector. If one knows all the coefficients

$f(\ell_1\ell_2\ell;n_1n_2;r_{12})$ ($\ell_i = 0$ to ∞ and $n_i = -\ell_i$ to $+\ell_i$), it is equivalent to know $f(\mathbf{r}_{12},\mathbf{\Omega}_1,\mathbf{\Omega}_2)$.

For the linear molecules

$$f(\ell_1\ell_2\ell;n_1n_2;r_{12}) = f(\ell_1\ell_2\ell;00;r_{12})$$

$$= \left[\frac{(2\ell_1+1)(2\ell_2+1)}{(4\pi)^2}\right]^{\frac{1}{2}} f_{\ell_1\ell_2\ell}(r_{12}) \tag{2.4}$$

and

$$D^{\ell_i}_{m_i,0}(\mathbf{\Omega}_i) = (4\pi/(2\ell_i+1))^{1/2}\, Y_{\ell_i m_i}(\mathbf{\Omega}_i) \tag{2.5}$$

(vi) The configurational probability density $\rho(\mathbf{r}^N,\mathbf{\Omega}^N)$. This gives the probability that a specific molecule 1 is at $(\mathbf{r}_1,\mathbf{\Omega}_1)$, another specific molecule 2 is at $(\mathbf{r}_2,\mathbf{\Omega}_2)$, etc., and is, therefore, known as a specific probability density.

The evaluation of above mentioned functions for a system of anisotropic molecules is very difficult. This problem is very well treated by Gray and Gubbins[1]. So we shall be very brief and cover only that part of discussion which is relevant for defining order parameters.

2.1.1 n-Particle Density Distribution

There are many properties which can usually be expressed in terms of lower-order or reduced (2, 3 or 4 molecule) specific distribution function.

In the canonical (NVT) ensemble, the n-particle density distribution, defined as the probability of finding n particles out of the given N in the range $\mathbf{x}_1+d\mathbf{x}_1$, $\mathbf{x}_2+d\mathbf{x}_2$, ..., $\mathbf{x}_n+d\mathbf{x}_n$, can be written as

$$\rho^{(\mathbf{n})}(\mathbf{x}_1,\mathbf{x}_2,...,\mathbf{x}_n) = \left(\frac{N!}{(N-n)!\,Z_N}\right)\int d\mathbf{x}_{n+1}\cdots d\mathbf{x}_N \exp[-\beta U(\mathbf{x}_1,\mathbf{x}_2,\cdots,\mathbf{x}_N)] \tag{2.6}$$

where $U(\mathbf{x}_1,\mathbf{x}_2,\cdots,\mathbf{x}_N)$ is the potential energy of N particles and $\beta = 1/k_BT$ with k_B as Boltzmann constant. The volume element $d\mathbf{x}_i$ is equivalent to $d^3\mathbf{r}_i\, d\mathbf{\Omega}_i$ where $d^3\mathbf{r}_i = d\mathbf{x}_i\, dy_i\, dz_i$ and $d\mathbf{\Omega}_i = \sin\theta_i\, d\theta_i\, d\phi_i\, d\psi_i$, Z_N is the configurational partition function of the system defined as

$$Z_N = \int d\mathbf{x}_1\, d\mathbf{x}_2\cdots d\mathbf{x}_N \exp[-\beta U(\mathbf{x}_1,\mathbf{x}_2,\cdots,\mathbf{x}_N)] \tag{2.7}$$

In the absence of possible ambiguities, for the notational convenience, we use only one integration sign to indicate the, possibly multiple, integration over all the variables whose volume elements appear. Integration is extended to the sample volume V for positions and to the usual domains $0 \le \theta \le \pi$, $0 \le \phi \le 2\pi$ and $0 \le \psi \le$

2π for angles. Here $\rho^{(n)}$ is not normalized to one but to the number of n-plets that can be formed by choosing n variables out of N, i.e.,

$$\int \rho^{(n)}(\mathbf{x}_1,\mathbf{x}_2,\cdots,\mathbf{x}_n)\, d\mathbf{x}_1 d\mathbf{x}_2 \cdots d\mathbf{x}_n = N!/(N-n)! \tag{2.8}$$

We usually require one- and two-particle distributions often known as the singlet and pair distribution functions. These can be used to evaluate the ensemble average of any physical quantity which depends on the position and orientation of one or two particles;

$$< f(\mathbf{x}_1) > = \frac{1}{N} \int f(\mathbf{x}_1)\, \rho^{(1)}(\mathbf{x}_1)\, d\mathbf{x}_1 \tag{2.9}$$

and

$$< f(\mathbf{x}_1,\mathbf{x}_2) > = \frac{1}{N(N-1)} \int f(\mathbf{x}_1,\mathbf{x}_2)\, \rho^{(2)}(\mathbf{x}_1,\mathbf{x}_2)\, d\mathbf{x}_1\, d\mathbf{x}_2 \tag{2.10}$$

The singlet distribution $\rho^{(1)} (\equiv \rho(\mathbf{x}_1))$ gives the probability of finding a molecule at a particular position and orientation

$$\rho(\mathbf{r}_1,\Omega_1) = \left(\frac{N}{Z_N}\right) \int d\mathbf{x}_2 \cdots d\mathbf{x}_N \exp[-\beta U(\mathbf{x}_1,\mathbf{x}_2,\cdots,\mathbf{x}_N)] \tag{2.11}$$

Similarly, the pair distribution function, giving the probability of finding simultaneously a particle in a volume $d\mathbf{r}_1\, d\Omega_1$ centered at $(\mathbf{r}_1,\Omega_1)$ and a second one in a volume $d\mathbf{r}_2\, d\Omega_2$ centered at $(\mathbf{r}_2,\Omega_2)$, is

$$\rho(\mathbf{r}_1,\mathbf{r}_2,\Omega_1,\Omega_2) = \left(\frac{N(N-1)}{Z_N}\right) \int d\mathbf{x}_3 \cdots d\mathbf{x}_N \exp[-\beta U(\mathbf{x}_1,\mathbf{x}_2,\cdots,\mathbf{x}_N)] \tag{2.12}$$

These functions can also be defined alternatively as[3–5]

$$\rho(\mathbf{r}_1,\Omega_1) = N\langle \delta(\mathbf{r}_1-\mathbf{r}_1')\,\delta(\Omega_1-\Omega_1')\rangle \tag{2.13}$$

and

$$\rho(\mathbf{r}_1,\mathbf{r}_2,\Omega_1,\Omega_2) = N(N-1)\langle \delta(\mathbf{r}_1-\mathbf{r}_1')\,\delta(\Omega_1-\Omega_1')\,\delta(\mathbf{r}_2-\mathbf{r}_2')\,\delta(\Omega_2-\Omega_2')\rangle \tag{2.14}$$

where $\delta(a-b)$ is a Dirac delta function and the integration is over the primed variables.

Similarly, for the n-particle density distribution,

$$\rho(\mathbf{r}_1,\mathbf{r}_2,\cdots,\mathbf{r}_N,\Omega_1,\Omega_2,\cdots,\Omega_N) = \left(\frac{N!}{(N-n)!}\right) \langle \delta(\mathbf{r}_1-\mathbf{r}_1')\cdots\delta(\mathbf{r}_N-\mathbf{r}_N')\rangle \langle \delta(\Omega_1-\Omega_1')\cdots\delta(\Omega_N-\Omega_N')\rangle \tag{2.15}$$

In the limit of low densities, a limiting expression for the n-particle distribution can be derived. In fact, when the distance between particles is very large and their reciprocal influence is negligible, the probabilities of finding them in their respective volume elements becomes statistically independent. Therefore, for

$\left|\mathbf{r}_i - \mathbf{r}_j\right| \to \infty$, the n-particle joint probability can be written as the product of single-particle distributions,

$$\rho^{(n)}(\mathbf{x}_1, \mathbf{x}_2, \cdots, \mathbf{x}_n) = \left(\frac{N!}{(N-n)!\,N^n} \right) \rho(\mathbf{x}_1)\,\rho(\mathbf{x}_2) \cdots \rho(\mathbf{x}_n) \qquad (2.16)$$

This property can be used to define reduced n-particle distributions or correlations which tend to unity as the intermolecular distances tend to infinity, i.e., when these distances become many orders of magnitude larger than the typical intermolecular distances. Thus, we define

$$g^{(n)}(\mathbf{x}_1, \mathbf{x}_2, \cdots, \mathbf{x}_n) = \frac{\rho^{(n)}(\mathbf{x}_1, \mathbf{x}_2, \cdots, \mathbf{x}_n)}{\sum\limits_{i=1}^{n} \rho(\mathbf{x}_i)} \qquad (2.17)$$

The most important of this class of $g^{(n)}$ is the pair correlation function (PCF),

$$g(\mathbf{x}_1, \mathbf{x}_2) = \rho(\mathbf{x}_1, \mathbf{x}_2) / \rho(\mathbf{x}_1)\,\rho(\mathbf{x}_2) \qquad (2.18)$$

For a uniform system the physical properties are invariant under translation and the interaction energy $U(\mathbf{x}_1, \mathbf{x}_2, \cdots, \mathbf{x}_N)$ depends only on the relative distances. Thus for an ordinary isotropic fluid, or for a nematic, we can write

$$\rho(\mathbf{r}_1, \Omega_1) = \rho_o\, f(\Omega_1) \qquad (2.19)$$

and

$$\rho(\mathbf{r}_1, \Omega_1, \mathbf{r}_2, \Omega_2) = \rho_o^2\, f(\Omega_1)\, f(\Omega_2)\, g(\mathbf{r}_{12}, \Omega_1, \Omega_2) \qquad (2.20)$$

where $f(\Omega_1)$ is a purely orientational singlet distribution,

$$f(\Omega_1) = \left(\frac{V}{Z_N} \right) \int \exp[-\beta U(\mathbf{x}_1, \mathbf{x}_2, \cdots, \mathbf{x}_N)\, d\mathbf{x}_2\, d\mathbf{x}_3 \cdots d\mathbf{x}_N \qquad (2.21)$$

which is normalized to unity

$$\int f(\Omega_1)\, d\Omega_1 = 1 \qquad (2.22)$$

For an isotropic molecular fluid, we obtain

$$\rho(\mathbf{r}_1, \Omega_1) = \rho_o / 8\pi^2 \qquad (2.23)$$

and

$$\rho(\mathbf{r}_1, \Omega_1, \mathbf{r}_2, \Omega_2) = (\rho_o / 8\pi^2)^2\, g(\mathbf{r}_{12}, \Omega_1, \Omega_2) \qquad (2.24)$$

In case the particles are spherical (e.g. atoms) the pair correlation function depends only on the inter-particle separation alone and $g(r_{12})$ is known as the radial distribution function.

For very dilute fluid systems ($\rho_0 \to 0$)

$$g(\mathbf{r}_{12}, \Omega_1, \Omega_2) = \exp[-\beta u(\mathbf{r}_{12}, \Omega_1, \Omega_2)] \qquad (2.25)$$

and if no correlations exist at all,

$$\lim_{r_{12}\to\infty} g(\mathbf{r}_{12},\Omega_1,\Omega_2)=1 \qquad (2.26)$$

Here $u(\mathbf{r}_{12},\Omega_1,\Omega_2)$ is the pair potential energy. Another interesting limiting situation is the behaviour of $g(\mathbf{r}_{12})$ in an ideal solid. If this system is composed of spherical particles exactly positioned at the lattice sites, we obtain

$$g(\mathbf{r}_{12})=\left(\frac{1}{4\pi r_{12}^2 \rho_0}\right)\sum_i \gamma_i\, \delta(r-r_i) \qquad (2.27)$$

where γ_i represents the number of nearest neighbours at a distance r_i. Expressing the delta function as a gaussian of vanishing width, we can write

$$g(\mathbf{r}_{12})=\left(\frac{1}{4\pi r_{12}^2 \rho_0}\right)\lim_{\epsilon\to 0}\sum_i \gamma_i (4\pi\epsilon)^{-\frac{1}{2}} \exp[-(r-r_i)^2/4\epsilon] \qquad (2.28)$$

It follows that for an idealized solid $g(r_{12})$ consists of a series of peaks corresponding to the various shells of neighbours. In a real crystal, due to the possibility of thermal oscillations, the peak positions cannot be defined with absolute certainty and the peaks will obviously be smeared out. In a liquid, the peaks will be even more diffuse giving the indication of the existence of short-range order. In a fluid $g(r_{12})$ decays to one in the absence of correlations, while in a solid $g(r_{12})$ keeps oscillating even when the separation r_{12} is extremely large. Thus there exists long-range positional order in a solid which vanishes at the melting transition.

2.2 ORDER PARAMETERS

One mesophase differs from another with respect to its symmetry. The transition between different phases corresponds to the breaking of some symmetry and can be described in terms of the so-called order parameter (OP). It represents the extent to which the configuration of the molecules in the less symmetric (more ordered) phase differs from that in the more symmetric (less ordered) one. In general, an order parameter Q may be defined such that

(i) $Q=0$, in the more symmetric (less ordered) phase, and

(ii) $Q\neq 0$, in the less symmetric (more ordered) phase.

These requirements do not define an order parameter in a unique way. In spite of this arbitrariness, in many cases the choice follows in a quite natural way. In the case of liquid-vapour transition the order parameter is the difference in the density between liquid and vapour phases and is a scalar. In the case of ferromagnetic transitions without anisotropic forces, the order parameter is the magnetization which is a vector with three components. The anisotropy in some tensor property can serve the role of order parameter for the nematic-isotropic transition. Thus the choice of order parameters requires some careful considerations.

2.2.1 Microscopic Order Parameters

We describe the orientational and positional orders from the microscopic point of view. A central role is played here by the distribution functions which depend on the positions and orientations of the molecules. The transitions from one phase to another can be described in terms of the modifications that this produces in the distribution functions. When the distribution depends on the orientations alone, we will have orientational order parameters. In case this depends on both the positions and orientations, we obtain positional–, orientational–, and mixed (positional–orientational) order parameters. In practice, it is very difficult to determine the full distribution functions. For this reason, we write the expansion of this function such that the expansion coefficients form a set of microscopic order parameters. Only a few of them can be determined experimentally.

By definitions the microscopic order parameters may contain more information than just the symmetry of the phase. Various approaches have been adopted to define them.

A. Method based on singlet distribution

The single-particle distribution function (eq. (2.11)) is the best candidate to be used for defining the order parameters. Here we introduce the order parameters as the expansion coefficients of the singlet distribution in a suitable basis set[3,6,7].

Using the Fourier integral representation of the spatial delta function,

$$\delta(\mathbf{r}) = (1/V) \sum_{\mathbf{G}} e^{-i\mathbf{G}\cdot\mathbf{r}} \tag{2.29}$$

and the representation of angular delta function,

$$\delta(\Omega - \Omega') = \sum_{\ell,m,n} (2\ell+1)\, D^{\ell}_{m,n}(\Omega)\, D^{\ell *}_{m,n}(\Omega') \tag{2.30}$$

The singlet distribution (eq. (2.13)) can now be defined as

$$\rho(\mathbf{r},\Omega) = \frac{N}{V} \sum_{\mathbf{G}} \sum_{\ell,m,n} (2\ell+1)\, e^{-i\mathbf{G}\cdot\mathbf{r}}\, D^{\ell}_{m,n}(\Omega) < e^{-i\mathbf{G}\cdot\mathbf{r}'}\, D^{\ell *}_{m,n}(\Omega') > \tag{2.31}$$

Here $\mathbf{r}$ and $\mathbf{\Omega}$ are field variables (fixed points in space) and therefore, unlike the dynamical variables $\mathbf{r}'$ and $\mathbf{\Omega}'$ are not affected by the ensemble average.

Equation (2.31) can be written as

$$\rho(\mathbf{r},\Omega) = \rho_o \sum_{\mathbf{G}} \sum_{\ell,m,n} Q_{\ell mn}(\mathbf{G})\, e^{-i\mathbf{G}\cdot\mathbf{r}}\, D^{\ell}_{m,n}(\Omega) \tag{2.32}$$

where

$$\begin{aligned} Q_{\ell mn}(\mathbf{G}) &= (2\ell+1) < e^{-i\mathbf{G}\cdot\mathbf{r}'} D^{\ell *}_{m,n}(\Omega') > \\ &= \overline{Q^{\ell *}_{m,n}} \\ &= \left(\frac{2\ell+1}{N}\right) \int \rho(\mathbf{r},\Omega) e^{-i\mathbf{G}\cdot\mathbf{r}} D^{\ell *}_{m,n}(\Omega)\, d\mathbf{r}\, d\Omega \end{aligned} \tag{2.33}$$

are recognized as the order parameters and **G** the set of reciprocal lattice vectors of the crystalline phase.

From eq. (2.33) the following order parameters can be defined :

$$Q_{000}(0) = 1 \tag{2.34a}$$

$$Q_{000}(\mathbf{G}) = \overline{\mu}_G = \frac{1}{N} \int d\mathbf{r}\, d\Omega\, \rho(\mathbf{r},\Omega)\, e^{-i\mathbf{G}\cdot\mathbf{r}} \tag{2.34b}$$

$$Q_{\ell 00}(\mathbf{G}) = (2\ell+1)\overline{\tau}_{G\ell} = \left(\frac{2\ell+1}{N}\right) \int d\mathbf{r}\, d\Omega\, \rho(\mathbf{r},\Omega)\, e^{-i\mathbf{G}\cdot\mathbf{r}} P_\ell(\cos\theta) \tag{2.34c}$$

$$Q_{\ell mn}(0) = (2\ell+1)\overline{D}^{\ell *}_{m,n} = \left(\frac{2\ell+1}{N}\right) \int d\mathbf{r}\, d\Omega\, \rho(\mathbf{r},\Omega)\, \overline{D}^{\ell *}_{m,n}(\Omega) \tag{2.34d}$$

and so on.

Here $\overline{\mu}_G$ are the positional order parameters for monatomic lattices which is characterized by a set of reciprocal lattice vectors $\{\mathbf{G}\}$. $Q_{\ell mn}(0)$ (or $\overline{D}^{\ell *}_{m,n}$) are the orientational order parameters and $Q_{\ell 00}(\mathbf{G})$ (or $\overline{\tau}_{G\ell}$) the mixed order parameters. There can be up to $(2\ell+1)^2$ order parameters of rank ℓ. We can reduce this number by exploiting the symmetry properties of the mesophase and of its constituent molecules and applying the effects of all the operations of the relevant space groups of the symmetry.

a. Uniaxial mesophases

In the uniaxial mesophases (e.g. N_u and S_A) the singlet distribution must be invariant under rotation about the director. If this is chosen to be the Z-axis, it follows that m must be zero in $Q_{\ell mn}(o)$ (or $\overline{D}^{\ell *}_{mn}$). In addition, if the mesophase has a symmetry plane perpendicular to the director ($D_{\infty h}$ symmetry) only terms with even ℓ can appear in $Q_{\ell mn}(o)$.

The most important order parameters are the orientational ones defined as

$$\overline{D}^{\ell *}_{o,n} = \int d\Omega\, f(\Omega)\, D^{\ell *}_{o,n}(\Omega) \tag{2.35}$$

where the orientational singlet distribution function, $f(\boldsymbol{\Omega})$, defined by eq. (2.21) is normalized to unity.

Let us consider the simplest case of the uniaxial nematic phase composed of molecules of cylindrical symmetry. In this case the rotation about the molecular symmetry axis should not modify the distribution which means n = 0 and the $f(\Omega)$ has to depend only on the angle θ between the director and the molecular symmetry axis. Accordingly, we obtain

$$\overline{D}^{\ell *}_{0,0} (\equiv \overline{P}_\ell) = \int d\Omega\, f(\theta)\, P_\ell(\cos\theta) \tag{2.36}$$

The $\overline{P}_\ell$, the ensemble average of the even Legendre polynomial, are called the Legendre polynomial orientational order parameters. Thus from a knowledge of $f(\theta)$ all the orientational order parameters $\overline{P}_2$, $\overline{P}_4$, etc., can be calculated. Several experimental techniques are in use to measure these order parameters (sec. 2.3). Due to symmetry $\theta \rightarrow \pi-\theta$, all odd moments of singlet distribution function $f(\theta)$ vanish. So $f(\theta)$ can be calculated from a knowledge of all the even order moments.

For a uniaxial nematic composed of molecules of noncylindrical symmetry some of the important lower-order orientational order parameters are given by

$$\overline{D}^{2*}_{0,0} = \overline{P}_2 = \langle P_2(\cos\theta) \rangle \tag{2.37a}$$

$$\overline{D}^{2*}_{0,2} = \langle \frac{\sqrt{3}}{2} \sin^2\theta \cos 2\phi \rangle \tag{2.37b}$$

$$\overline{D}^{4*}_{0,0} = \overline{P}_4 = \langle P_4(\cos\theta) \rangle \tag{2.37c}$$

It should be mentioned that $\overline{P}_\ell$ measures the alignment of the molecular $\hat{e}_z$ axis along the space-fixed (SF) Z-axis (or the director). The order parameter $\overline{D}^{2*}_{0,2}$ is an indicator of the difference in the alignment of the molecular axes $\hat{e}_x$ and $\hat{e}_y$ along the director. When the mesogenic molecules possess axial symmetry, the molecular axes $\hat{e}_x$ and $\hat{e}_y$ are indistinguishable and the order parameters $\overline{D}^{2*}_{0,2}$, $\overline{D}^{4*}_{0,2}$, etc., vanish.

For the smectic phase with positional order in one dimension only, the singlet distribution simplifies to

$$\rho(\mathbf{r},\boldsymbol{\Omega}) = \rho_0 \sum_{\mathbf{G}} \sum_{\ell,m,n} (2\ell+1)\, \overline{D}^{\ell *}_{m,n}(\mathbf{G}) \exp[iG_z z]\, D^{\ell}_{m,n}(\Omega) \tag{2.38}$$

Here the Z-axis is parallel to the layer normal. Exploiting the symmetry of the mesophase and of the constituent molecules, eq. (2.38) can be again simplified. For the S_A phase which has symmetry $T(2) \times D_{\infty h}$ and is composed of cylindrically symmetric molecules,

$$\rho(\mathbf{r},\Omega) = \rho_0 \sum_{\ell} \sum_{q}{}' (2\ell+1)\, \overline{D}^{\ell *}_{0,0}(q) \cos[2\pi q z/\xi_0]\, P_\ell(\cos\theta) \tag{2.39}$$

where $|G| = 2\pi q / \xi_0$, with ξ_0 being the average interlayer spacing. The prime on the summation sign indicates the condition that ℓ is even. Here $\overline{D}_{0,0}^{0*}(q)(=\overline{\mu}_q)$ and $\overline{D}_{0,0}^{\ell*}(0)(=\overline{P}_\ell)$ represent, respectively, the positional-, and orientational order parameters. The $\overline{D}_{0,0}^{\ell*}(q)(=\overline{\tau}_{q\ell})$ is the mixed (positional–orientational) order parameter.

b. Biaxial mesophases

The order parameters for the biaxial mesophases (e.g., biaxial nematic, smectic C, etc.) can be defined from eq. (2.33). In defining order parameters for these phases it is usually assumed that the ordered phases have the same symmetry as the constituent molecules. Thus, to characterize a biaxial nematic (N_b) phase the following orientational order parameters can be identified,

$$\overline{P}_2 = \overline{D}_{0,0}^{2*} = \langle P_2(\cos\theta) \rangle \quad (2.40a)$$

$$\overline{\eta}_2 = \overline{D}_{0,2}^{2*} = \langle \tfrac{\sqrt{3}}{2} \sin^2\theta \cos 2\psi \rangle \quad (2.40b)$$

$$\overline{\mu}_2 = \overline{D}_{2,0}^{2*} = \langle \tfrac{\sqrt{3}}{2} \sin^2\theta \cos 2\phi \rangle \quad (2.40c)$$

and

$$\overline{\tau}_2 = \overline{D}_{2,2}^{2*} = \langle \tfrac{1}{2}(1+\cos^2\theta) \cos 2\phi \cos 2\psi - \cos\theta \sin^2\phi \cos 2\psi \rangle \quad (2.40d)$$

The order parameter $\overline{P}_2$ measures the alignment of the molecular $\hat{e}_z$ -axis along the director, and the $\overline{\eta}_2$ measures the departure in the alignment of the molecular $\hat{e}_x$ and $\hat{e}_y$ axes along the director. The other two order parameters $\overline{\mu}_2$ and $\overline{\tau}_2$ are a measure of the biaxial ordering existing in the system.

For the smectic-C phase with positional order in one dimension only, the singlet distribution is defined by eq. (2.38) and the corresponding order parameters identified are the positional $\overline{\mu}_q$, the orientational $\overline{D}_{0,0}^{2*}$, $\overline{D}_{0,2}^{2*}$, $\overline{D}_{2,0}^{2*}$, and $\overline{D}_{2,2}^{2*}$ and the mixed $\overline{\tau}_{q\ell}$, order parameters.

On the basis of above discussions, the various ordered phases can be characterized in the following way :

(i) $\overline{\mu}_q \neq 0$, $\overline{P}_\ell = 0$, $\overline{\tau}_{q\ell} = 0$, plastic phase

(ii) $\overline{\mu}_q = 0$, $\overline{P}_\ell \neq 0$, $\overline{\tau}_{q\ell} = 0$, N_u of axial molecules

(iii) $\overline{\mu}_q = 0$, $\overline{D}_{m,n}^{\ell*} \neq 0$, $\overline{\tau}_{q\ell} = 0$, N_b phase

(iv) $\overline{\mu}_q \neq 0$, $\overline{P}_\ell \neq 0$, $\overline{\tau}_{q\ell} \neq 0$, S_A phase of axial molecule

(v) $\overline{\mu}_q \neq 0, \quad \overline{D}^{\ell *}_{m,n} \neq 0, \quad \overline{\tau}_{q\ell} \neq 0$, S_C phase

(vi) $\overline{\mu}_q \neq 0, \quad \overline{D}^{\ell *}_{m,n} \neq 0, \quad \overline{\tau}_{q\ell} \neq 0$, crystalline phase

only a few of these microscopic order parameters can be measured experimentally.

B. Method based on pair distribution

For an anisotropic fluid the key quantity to be calculated is the pair distribution $\rho^{(2)}(\mathbf{x}_1, \mathbf{x}_2)$ (eq. (2.14)). We can write[3]

$$\rho^{(2)}(\mathbf{r}_{12}, \mathbf{\Omega}_1, \mathbf{\Omega}_2) = \rho_0^2 \, G(\mathbf{r}_{12}, \Omega_1, \Omega_2) \tag{2.41}$$

where $G(\mathbf{r}_{12}, \mathbf{\Omega}_1, \mathbf{\Omega}_2)$, termed as reduced distribution, is normalized as

$$\int d\mathbf{r}_{12} \, d\Omega_1 \, d\Omega_2 \, G(\mathbf{r}_{12}, \Omega_1, \Omega_2) = V(1 - 1/N) \tag{2.42}$$

and satisfies the relations

$$G(\mathbf{r}_{12}, \Omega_1, \Omega_2) = f(\Omega_1) f(\Omega_2); \qquad r_{12} \to \infty \tag{2.43}$$

and

$$G(\mathbf{r}_{12}, \Omega_1, \Omega_2) = f(\Omega_1) f(\Omega_2) \exp[-\beta u(\mathbf{r}_{12}, \Omega_1, \Omega_2)]; \quad \rho_0 \to 0 \tag{2.44}$$

For the sake of simplicity, we confine our discussion here to the uniaxial nematic phase and ignore the smectic phases. In a laboratory frame $G(\mathbf{r}_{12}, \Omega_1, \Omega_2)$ depends on the intermolecular separation r_{12}, the orientations Ω_1 and Ω_2 of molecules as well as on the orientation of the intermolecular vector Ω. We can write down the expansion[3]

$$G(\mathbf{r}_{12}, \Omega_1, \Omega_2) = \sum_{\ell_1, \ell_2, \ell} \; \sum_{m_1, m_2, m} \; \sum_{n_1, n_2} G(\ell_1 \ell_2 \ell; m_1 m_2 m; n_1 n_2)$$
$$D^{\ell_1}_{m_1, n_1}(\Omega_1) \, D^{\ell_2}_{m_2, n_2}(\Omega_2) \, D^{\ell}_{m,0}(\Omega) \tag{2.45}$$

This distribution has to be invariant under the symmetry operations of the molecules and of the phase. In addition, permutation of identical molecules should leave $G(\mathbf{r}_{12}, \Omega_1, \Omega_2)$ unaltered. If a further assumption is made that the intermolecular vector is spherically distributed, the rotation of $\mathbf{r}_{12}$ does not change $G(\mathbf{r}_{12}, \mathbf{\Omega}_1, \mathbf{\Omega}_2)$. Thus for a rotationally invariant fluid of axial molecules the reduced pair distribution, which does not depend on the orientation of the intermolecular vector, must be a function of relative orientations only and can be expressed as

$$G(\mathbf{r}_{12}, \Omega_1, \Omega_2) \equiv G(\mathbf{r}_{12}, \Omega_{12})$$
$$= \sum_{\ell} \left[(2\ell + 1)/64\pi^4\right] G^{00}_{\ell}(r_{12}) \, D^{\ell}_{0,0}(\Omega_{12}) \tag{2.46}$$

where $G^{00}_{\ell}(r_{12})$ is an lth rank angular correlation. This simple equation has some salient features. Since it is fully rotationally invariant it can be used to describe the orientational order of nematic phase in the absence of external fields. It is important to mention here that when we identify the order parameters as the expansion coefficients appearing in the expansion of singlet distribution we, more or less, assume implicitly the existence of an aligned mesophase with a uniform director. This, in turn, implies the existence of an external field to break the full rotational symmetry of the mesophase above the isotropic transition. Since the pair correlation can tell us about the extent of orientational correlations, this can be used to understand how long-range ordering is set up in a system exhibiting nematic phase in the absence of a field.

From the definition of $G(\mathbf{r}_{12}, \mathbf{\Omega}_{12})$ and the orthogonality of Legendre polynomials, $G_\ell(r_{12})$ can be expressed as

$$\begin{aligned} G_\ell(r_{12}) &= \int d\Omega_1\, d\Omega_2\, G(r_{12}, \Omega_{12})\, D^\ell_{00}(\Omega_{12}) \\ &= \overline{P_\ell[\cos\theta_{12}(r_{12})]}\; G^{00}_0(r_{12}) \\ &\equiv G^{00}_\ell(r_{12})\; G^{00}_0(r_{12}) \end{aligned} \tag{2.47}$$

Thus $G_\ell(\mathbf{r}_{12})$ is an ℓth rank angular correlation. For $\ell = 0$,

$$G^{00}_0(r_{12}) = \int d\Omega_1\, d\Omega_2\, G(r_{12}, \Omega_{12}) \tag{2.48}$$

defines the center of mass pair distribution. Any $G_\ell(r_{12})$ with $\ell \neq 0$, say $G_2(r_{12})$, measures the correlation in orientation between a molecule at the origin and another at a distance r_{12} from it. Thus $G_\ell(0) = 1$. For small values of intermolecular separation such that it can be taken to be the average nearest neighbour distance, $G_2(r_{12})$ can be used to define short-range order. For large separations $G_2(r_{12})$ is a measure of long-range order. At the nematic-isotropic transition the expected behaviour of $G_2(r_{12})$ changes dramatically. Above the clearing point $G_2(r_{12})$ should decay to zero with distance, while in the nematic phase it should tend to the square of the order parameter.

Another advantage of the pair correlation coefficients $G_\ell(r_{12})$ is that they can be conveniently used to explain the pretransitional phenomena, for example, magnetically induced birefringence and light scattering.

2.2.2 Macroscopic Order Parameters :

In most of the cases, the microscopic order parameters, as defined above, provide an adequate description of real mesogenic systems. However, in some cases, this microscopic description is no longer adequate and some other means must be found for specifying the degree of order. A significant difference between the high-temperature isotropic liquid and the liquid crystalline phase is observed in the measurements of macroscopic tensor properties. Thus the macroscopic properties, e.g., the diamagnetic susceptibility, the refractive index, the dielectric permittivity, etc., can be used to identify the macroscopic (tensor) order parameters. In this section, we describe a macroscopic approach in which an order parameter will be constructed independent of any assumption regarding the interactions of the constituent molecules.

A tensor order parameter can be defined in the following way. Let us apply some field **h** to the system. The resulting response **A** of the system will be given by

$$A_\alpha = T_{\alpha\beta}\, h_\beta \tag{2.49}$$

where $\mathbf{T}_{\alpha\beta}$ is a symmetric tensor, i.e., $\mathbf{T}_{\alpha\beta} = \mathbf{T}_{\beta\alpha}$, and h_β and A_α represent, respectively, the components of **h** and **A** in a given coordinate system. For example, if **h** and $\tilde{\mathbf{T}}$ denote, respectively, the external magnetic field **H** and the susceptibility tensor $\tilde{\boldsymbol{\chi}}$; $\mu_0^{-1}\mathbf{A}$ is the magnetisation **M** with μ_0 the permeability of the vacuum. In a properly chosen coordinate system $\tilde{\mathbf{T}}$ can be written in a diagonal form

$$\tilde{\mathbf{T}} = \begin{pmatrix} T_1 & 0 & 0 \\ 0 & T_2 & 0 \\ 0 & 0 & T_3 \end{pmatrix} \tag{2.50}$$

The diagonal elements are temperature dependent. The anisotropic part of $\tilde{\mathbf{T}}$ represents the tensor order parameter. Imposing the condition $\sum_{i=1}^{3} T_i = T$, the elements T_i can be expressed as

$$T_1 = \frac{1}{3} T(1 - Q_1 + Q_2) \tag{2.51a}$$

$$T_2 = \frac{1}{3} T(1 - Q_1 - Q_2) \tag{2.51b}$$

$$T_3 = \frac{1}{3} T(1 + 2Q_1) \tag{2.51c}$$

Accordingly the tensor $\tilde{\mathbf{T}}$ can be expressed as

$$T_{\alpha\beta} = T\left(Q_{\alpha\beta} + \frac{1}{3}\delta_{\alpha\beta}\right) \tag{2.52}$$

where the tensor $\tilde{\mathbf{Q}}$ having elements $Q_{\alpha\beta}$ is identified as the tensor order parameter,

$$\tilde{\mathbf{Q}} = \begin{pmatrix} -\frac{1}{3}(Q_1 - Q_2) & 0 & 0 \\ 0 & -\frac{1}{3}(Q_1 + Q_2) & 0 \\ 0 & 0 & \frac{2}{3}Q_1 \end{pmatrix} \tag{2.53}$$

The isotropic phase is described by $Q_1 = Q_2 = 0$. Only one order parameter $Q_1 \neq 0$ (and $Q_2 = 0$) is required to define an anisotropic phase of uniaxial symmetry.

Transition of uniaxial symmetry to a biaxial one requires a nonzero value of Q_2 also. Obviously both the order parameters depend on the temperature.

As an example, we consider the susceptibility tensor $\tilde{\boldsymbol{\chi}}$. The relationship between the magnetic moment **M** (due to the molecular diamagnetism) and the field **H** has the form

$$M_\alpha = \chi_{\alpha\beta} H_\beta \tag{2.54}$$

where $\chi_{\alpha\beta}$ is an element of the susceptibility tensor $\tilde{\boldsymbol{\chi}}$, α,β = x,y,z. When the field **H** is static, the tensor $\chi_{\alpha\beta}$ is symmetric ($\chi_{\alpha\beta} = \chi_{\beta\alpha}$). In the isotropic phase, it has the simple form

$$\chi_{\alpha\beta} = \chi\,\delta_{\alpha\beta}$$

where $\delta_{\alpha\beta}$ is the Kronecker delta.

For the uniaxial nematic phase, where the Z-axis is parallel to the nematic axis, $\tilde{\boldsymbol{\chi}}$ can be written in the diagonal form

$$\tilde{\boldsymbol{\chi}} = \begin{pmatrix} \chi_\perp & 0 & 0 \\ 0 & \chi_\perp & 0 \\ 0 & 0 & \chi_\parallel \end{pmatrix} \tag{2.55}$$

where $\chi_\parallel$ and $\chi_\perp$ are the susceptibilities parallel and perpendicular to the symmetry axis, respectively.
We can express $\tilde{\boldsymbol{\chi}}$ as

$$\chi_{\alpha\beta} = (\chi_\parallel + 2\chi_\perp)(Q_{\alpha\beta} + \frac{1}{3}\delta_{\alpha\beta}) \tag{2.56}$$

with $Q_1 = \Delta\chi/(\chi_\parallel + 2\chi_\perp)$ and $Q_2 = 0$, where $\Delta\chi = \chi_\parallel - \chi_\perp$ denotes the anisotropy in the susceptibility.

The general expression for the tensor order parameter can be obtained by an arbitrary rotation of the coordinate system. Let us consider a cartesian coordinate system with basis vectors $\bar{\mathbf{e}}_\alpha$. The elements of this diagonal representation are $\overline{Q}_{\alpha\beta}$. Next an arbitrary rotation is carried out leading to a coordinate system with basis vectors $\mathbf{e}_\alpha$. The elements $Q_{\alpha\beta}$ of the tensor $\tilde{\mathbf{Q}}$ with respect to new coordinate system are given by

$$Q_{\alpha\beta} = -\frac{1}{3}(Q_1 - Q_2)(\mathbf{e}_\alpha \cdot \bar{\mathbf{e}}_1)(\mathbf{e}_\beta \cdot \bar{\mathbf{e}}_1) - \frac{1}{3}(Q_1 + Q_2)$$

$$(\mathbf{e}_\alpha . \bar{\mathbf{e}}_2)(\mathbf{e}_\beta \cdot \bar{\mathbf{e}}_2) + \frac{2}{3}Q_1(\mathbf{e}_\alpha \cdot \bar{\mathbf{e}}_3)(\mathbf{e}_\beta \cdot \bar{\mathbf{e}}_3) \tag{2.57}$$

This is the general expression of the tensor order parameter for a biaxial phase. In case of uniaxial nematic the direction of unique axis given by the director $\hat{\mathbf{n}}$

coincides with one of the basis vectors of cartesian coordinate system in which $\tilde{\mathbf{Q}}$ is diagonal. Here $\hat{\mathbf{n}} = \bar{e}_3$ or $\hat{\mathbf{n}} = -\bar{\mathbf{e}}_3$; i.e. $Q_2 = 0$. Now eq. (2.57) reduces to

$$Q_{\alpha\beta} = -\frac{1}{3}Q_1(\mathbf{e}_\alpha \cdot \bar{\mathbf{e}}_\gamma)(\mathbf{e}_\beta \cdot \bar{\mathbf{e}}_\gamma) + Q_1(\mathbf{e}_\alpha \cdot \hat{\mathbf{n}})(\mathbf{e}_\beta \cdot \hat{\mathbf{n}})$$

$$= -\frac{1}{3}Q_1\delta_{\alpha\beta} + Q_1(\mathbf{e}_\alpha \cdot \hat{\mathbf{n}})(\mathbf{e}_\beta \cdot \hat{\mathbf{n}}) \tag{2.58}$$

Here we have used the relation

$$(\mathbf{e}_\alpha \cdot \bar{\mathbf{e}}_\gamma)(\mathbf{e}_\beta \cdot \bar{\mathbf{e}}_\gamma) = \delta_{\alpha\beta}$$

Thus the general expression for the tensor order parameter $\tilde{\mathbf{Q}}$ for a uniaxial nematic is given by

$$Q_{\alpha\beta} = Q_1(n_\alpha n_\beta - \frac{1}{3}\delta_{\alpha\beta}) \tag{2.59}$$

Here the order parameter Q_1 depends on the temperature.

If the values of $\tilde{\mathbf{T}}$ are known for the perfectly aligned phase we can also define an order parameter with values between 0 (isotropic phase) and 1 (perfectly aligned nematic). The susceptibility tensor (eq. (2.55)) can also be expressed as

$$\chi_{\alpha\beta} = \frac{1}{3}(\chi_\parallel + 2\chi_\perp)\delta_{\alpha\beta} + \Delta\chi_{\alpha\beta} \tag{2.60}$$

where

$$\Delta\tilde{\boldsymbol{\chi}} = \Delta\chi_{max}\, S \begin{pmatrix} -\frac{1}{3} & 0 & 0 \\ 0 & -\frac{1}{3} & 0 \\ 0 & 0 & \frac{2}{3} \end{pmatrix} \tag{2.61}$$

Here $\Delta\chi_{max}$ is the maximum anisotropy that would be observed for a perfectly ordered mesophase. Now the order parameter

$$S = \Delta\chi / \Delta\chi_{max} \tag{2.62}$$

is defined between 0 and 1 and the original tensor order parameter is related as

$$Q_{\alpha\beta} = \Delta\chi_{\alpha\beta} / (\chi_\parallel + 2\chi_\perp) \tag{2.63}$$

2.2.3 Relationship Between Microscopic and Macroscopic Order Parameters

When the molecules can be approximately taken as rigid, one may find a simple connection between the macroscopic tensor associated with $\tilde{\mathbf{Q}}$ and the microscopic quantities defining the microscopic order parameter $\overline{Q}^{\ell *}_{mn}(\equiv \mathbf{Q}^M)$. In fact, the level of knowledge is not the same for all the macroscopic tensor properties. Like the macroscopic order parameter $\mathbf{Q}$, the microscopic order parameter $\mathbf{Q}^M$ is the symmetric traceless, second rank tensor. We should mention that only in some simple cases there is a straightforward connection between the macroscopic and microscopic approaches.

A relationship between $\mathbf{Q}$ and $\mathbf{Q}^M$ can be written for the case of a rigid rod model by noting that the anisotropic part of the diamagnetic susceptibility is proportional to $Q^M_{\alpha\beta}$:

$$\Delta\chi_{\alpha\beta} = N\chi_a Q^M_{\alpha\beta} \tag{2.64}$$

where N is the particle density number and χ_a is the anisotropy of the molecular magnetic susceptibility. Since by definition $\Delta\chi_{max} = N\chi_a$, within the framework of the rigid rod model for the uniaxial nematic, one obtains

$$Q_{\alpha\beta} = Q^M_{\alpha\beta} \tag{2.65}$$

For the realistic models the relationship between $\mathbf{Q}$ and $\mathbf{Q}^M$ may be quite complicated. Thus the relation (2.65) obviously cannot be generalized.

2.3 MEASUREMENT OF ORDER PARAMETERS

The order existing in liquid crystalline phases can be measured by using the relationship between microscopic and macroscopic properties. However, no simple formalism is appropriate for all the mesophases and the difficulties are further compounded because of complex chemical structure and non-rigidity of the molecules. The order parameter can be directly related to certain experimentally measurable quantities, for example, diamagnetic anisotropy (susceptibility), dielectric anisotropy, optical anisotropy (birefringence and linear dichroism), etc. The measurements of these anisotropies can give the value of orientational order parameter $\overline{P}_2$. Other important techniques sensitive to local fluctuations are magnetic resonance spectroscopy, neutron scattering, X-ray scattering, electron paramagnetic resonance, infra-red and Raman spectroscopies, etc. Each of these gives specific information, for example, gyration radii for neutron experiments, populations of conformational states for infra-red and Raman spectroscopies, orientations of molecular tensors or vectors in external fields of Raman spectroscopy, nuclear magnetic and electron paramagnetic resonances. The order parameters are measured as a function of temperature. However, the results obtained by using different techniques often differ from each other. It is advisable to plot a smooth graph using these values and estimate the value of order parameter at a particular temperature from this graph. In this section, we describe some of the important methods which have been used extensively for the determination of the microscopic order parameters.

2.3.1 Magnetic Resonance Spectroscopy

Magnetic resonance spectroscopy[8-10] is a powerful technique which has been used extensively to investigate the macroscopic ordering and molecular motions in liquid crystals. The magnetic interactions responsible for the spectral profiles are anisotropic, i.e., their magnitude depends upon the relative orientation of molecules with respect to the direction of the applied magnetic field. So the direct information on the degree of orientational order of the molecules can be derived from the line (energy-level) positions and the spacings in the multiplet structures of the spectra. In addition, in the relaxation experiments when the equilibrium conditions are abruptly altered, the spectral profiles provide information on the characteristics of the molecular dynamics.

The basic ideas behind magnetic resonance are common to both the nuclear magnetic resonance (nmr) and electron spin resonance (ESR), but the magnitudes and signs of the magnetic interactions involved differ in these two experiments which principally lead to the divergences in the experimental techniques used.

When a steady magnetic field **H** is applied on the system an interaction between the field and magnetic moment takes place which may be represented by the total Hamiltonian. In a nmr experiment, for inducing transitions between the two nuclear spin levels an oscillating electromagnetic field is applied to the system. Absorption of energy occurs provided (i) the magnetic vector of the oscillating field is perpendicular to the steady field **H**, and (ii) the frequency ν of the oscillating field satisfies the resonance condition. One can observe a nuclear resonance absorption by varying either the magnetic field **H** or the frequency ν. In practice, one usually prefers to use a fixed frequency and obtain a spectrum by sweeping the magnetic field through the resonance value. The absorption of energy is plotted as a function of magnetic field strength. Alternatively, the absorption spectra can be obtained by keeping the magnetic field fixed and altering the radio frequency power through the resonance value. Because of the variations in the nuclear g factor and spin, different nuclei require different values of **H** and ν for the resonance to occur. In the ESR experiment the spectra results due to the interaction between the electron magnetic moment and the applied field **H**. An application of the oscillating field perpendicular to **H** induces transitions provided the frequency ν satisfies the resonance condition.

The general aspects of the applications of nmr and ESR techniques in the study of mesogenic systems can be treated as much as possible in a unitary way because their theoretical description is formally the same. Since the liquid crystalline molecules are diamagnetic, nmr spectroscopy can be carried out directly whereas doping with paramagnetic probes is required for the ESR study. In practice, commercial spectrometers work by sweeping the field at constant frequency to satisfy the resonance conditions, and in the ESR investigation the first derivative of the absortion curve is recorded, rather than the absorption itself.

The positions of the absorption lines in the spectra can be calculated from the quantum mechanical spin Hamiltonian describing the interactions of the magnetic moments with the external fields and among themselves. The interaction of the spins with the static magnetic field is called the Zeeman interaction. The exchange and dipolar interactions arise due to the interactions between different spins. In addition, typical interactions of the nuclear quadrupole charge distribution with the electric field gradient may arise. The relevant interactions involved in the nmr and ESR experiments are as follows[11] :

(i) Zeeman interactions

(a) electronic $\mu_B \mathbf{B}\cdot\mathbf{g}\cdot\mathbf{s}$,

(b) nuclear $-\mu_N \mathbf{B}\cdot\boldsymbol{\sigma}\cdot\mathbf{I}$,

where μ_N and μ_B are, respectively, the nuclear and Bohr magnetons. **g** and $\boldsymbol{\sigma}$ are called the g-tensor and chemical shielding tensor, respectively, **B** is the magnetic flux desity, **s** is the electron spin operator and **I** is the nuclear spin operator.

(ii) Anisotropic spin-spin couplings

(a) electron-electron, or zero-field splitting term

$$\mathbf{S}\cdot\mathbf{D}_0\cdot\mathbf{S},$$

(b) electron-nucleus, or hyperfine interaction

$$\mathbf{I}\cdot\mathbf{A}\cdot\mathbf{S},$$

(c) nucleus-nucleus

$$\mathbf{I}_1\cdot\mathbf{D}\cdot\mathbf{I}_2$$

Here $\mathbf{D}_0$ is the zero-field splitting tensor, **A** is the nuclear hyperfine tensor, and **D** is the spin-spin coupling tensor.

(iii) Nuclear quadrupole interaction

$$\mathbf{I}\cdot\mathbf{Q}\cdot\mathbf{I}$$

This term is present whenever the nuclear spin $I \geq 1$ (e.g. deuterium and nitrogen). The quadrupole tensor **Q** and the tensor $\mathbf{D}_0$ are traceless.

All of the above interactions are, in general, anisotropic and therefore have been represented by tensorial quantities. As is obvious, the spin Hamiltonian can be expressed as a sum of terms of the form

$$\mathbf{U}\cdot\mathbf{T}\cdot\mathbf{V}$$

where **U** and **V** are spin operators (or the magnetic field) and **T**, the interaction tensor, is a molecular property.

In order to make the point more clear, we consider, in brief, some specific cases of nmr investigation. The nmr is sensitive to the orientation of some molecular tensor with respect to the field and hence to the director. The protons present in the molecule are the more direct nmr probes to use but, in some cases, more precise investigations have been carried out through the use of other nuclei ^{13}C, ^{14}N, ...

Proton nmr :

Let us first consider an idealised case. The rigid rod contains two protons of spins $\mathbf{I}_1$ and $\mathbf{I}_2$ ($I_1 = I_2 = \frac{1}{2}$) with a separation d. An external magnetic field **H** is applied parallel to the Z-axis. The energy levels of the proton spins are determined by the spin Hamiltonian

$$\hat{H} = \hat{H}_Z + \hat{H}_D \tag{2.66}$$

where the Zeeman Hamiltonian representing the coupling of the magnetic moment of each proton with the external field is given by

$$\hat{H}_Z = -\hbar\gamma_H \mathbf{H}\cdot(\mathbf{I}_1 + \mathbf{I}_2) \tag{2.67}$$

and the dipolar Hamiltonian

$$\hat{H}_D = \mathbf{I}_1 \cdot \mathbf{D} \cdot \mathbf{I}_2 \tag{2.68}$$

describes the spin-spin dipolar coupling between the nuclear magnetic moments. The coupling tensor **D** depends on length d and its relative orientation to **H.** γ_H is the proton gyromagnetic ratio.

In our notation eq. (2.66) can be expressed as

$$\hat{H} = -\hbar\gamma_H H(I_{1Z} + I_{2Z}) - \frac{(\hbar\gamma_H)^2}{d^3}[3(\mathbf{I}_1\cdot\hat{\mathbf{e}})(\mathbf{I}_2\cdot\hat{\mathbf{e}}) - \mathbf{I}_1\cdot\mathbf{I}_2] \tag{2.69}$$

where $\hat{\mathbf{e}}$ is the unit vector along the rod. In liquids the direction of the long-molecular axis (the vector $\hat{\mathbf{e}}$) changes with time on a much faster scale (~ 10^{-9} s). Thus in the limit of this rapid motion $\hat{H}_D$ can be replaced by its average over the orientation of $\hat{\mathbf{e}}$. Taking into account the uniaxial symmetry around the Z-axis, the averaging procedure gives

$$<\hat{H}_D> = -\frac{(\hbar\gamma_H)^2}{d^3}(3I_{1Z}I_{2Z} - \mathbf{I}_1\cdot\mathbf{I}_2)\overline{P}_2 \tag{2.70}$$

Now introducing the total spin $\mathbf{I} = \mathbf{I}_1 + \mathbf{I}_2$ and using the fact that both moments have spin 1/2, we get

$$<\hat{H}> = -\hbar\gamma_H H I_Z - \frac{1}{2}\frac{(\hbar\gamma_H)^2}{d^3}(3I_Z^2 - I^2)\overline{P}_2 \tag{2.71}$$

where the constants have been ignored as they do not contribute to the energy intervals that determine the spectrum. For two protons I may take only two values I = 0 and I = 1, corresponding to a singlet and a triplet state, respectively. In a resonance experiment only transitions between neighbouring triplet levels can be observed. These levels are characterized by $I_Z = -1, 0, +1$. The term involving I^2 remains same for these three levels and hence may be ignored in the calculation of the relevant energy differences. So

$$<\hat{H}> \rightarrow -\hbar \gamma_H H I_Z - \frac{3}{2}\frac{(\hbar \gamma_H)^2}{d^3} I_Z^2 \overline{P}_2 \tag{2.72}$$

The corresponding energy levels and line shapes observed, in the absence and presence of $\hat{H}_D$, are shown in Fig. 2.1, and the transition frequencies are given by

$$h\nu_0 = \hbar \gamma_H H \tag{2.73a}$$

$$h\nu_{0\text{-}1} = h\nu_0 - \frac{3}{2}\delta \tag{2.73b}$$

$$h\nu_{0+1} = h\nu_0 + \frac{3}{2}\delta \tag{2.73c}$$

with

$$\delta = \frac{(\hbar \gamma_H)^2}{d^3} P_2(\cos\theta) \tag{2.74}$$

Here $\nu_0 = \gamma_H H/2\pi$, is the proton Larmor frequency (ν_0 is 42.5 MHz for a field of 1T). Thus the nmr spectrum consists of a single line in the isotropic phase where $\overline{P}_2 = 0$. In the presence of $\hat{H}_D$, this line splits into a doublet in the nematic phase with a spacing (in frequency unit)

$$\Delta\nu_D = \left(\frac{3\hbar \gamma_H^2}{2\pi d^3}\right) P_2(\cos\theta) \tag{2.75a}$$

This is the splitting observed for a static rod, but in a liquid crystal the molecular (or rod) orientation fluctuates. If this motion is very rapid, such that $2\pi\Delta\nu_D \tau < 1$ (τ is correlation time), it averages the splittings. So the measured splitting can be obtained as

$$\Delta\nu_D = \frac{3\hbar \gamma_H^2}{2\pi d^3} \overline{P}_2 \tag{2.75b}$$

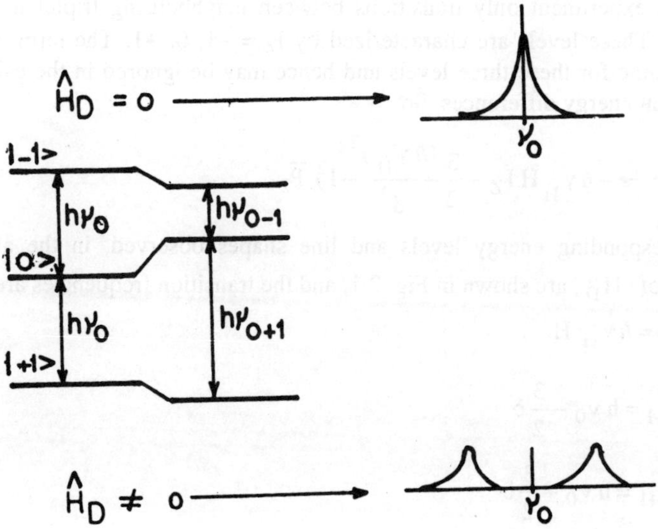

Fig. 2.1. The energy levels and line shapes in the absence and presence of $\hat{H}_D$.

If the distance d between the protons is fixed (rigid molecule) and is known, i.e., the dipolar field $(= \hbar\gamma_H / d^3)$ is known, $\overline{P}_2$ can be extracted from the experimental splitting. However, in practice, the situation is not so simple because many experimental difficulties in case of real mesogenic molecules may arise. The accuracy of the result depends on the accuracy of the value and the direction of relevant d. Both the magnitude and the direction of d are very sensitive to the bond angles which are determined from X-ray data. Since a large number of protons are usually present in a real molecule, normal nmr spectroscopy can not be applied directly. These protons couple to each other giving a very complicated spectra which cannot be analysed easily. In spite of this difficulty ^{1}H-nmr has been useful, often in combination with the analysis of the shape of the broad lines. The complication arising due to the presence of many protons can be dealt with by replacing all protons, except those in one specific part of the molecule, by deuterons. Since deuterons possess smaller magnetic moment, the D-nmr signals are distinct from the ^{1}H-nmr signals, and any coupling between them is small. This method has been widely used and reasonable values of $\overline{P}_2$ have been obtained.

^{13}C-nmr :

This nucleus is present in the skeleton of a mesogenic molecule with a natural abundance of 1.1%. Consequently the dipolar coupling between like spins can be neglected. There remains a dipolar coupling with unlike spins, such as those of the neighbouring protons. This coupling can be removed by irradiating the sample at the proton frequency. Under these conditions each ^{13}C behaves as an isolated nucleus and the total Hamiltonian can be expressed as

$$\hat{H} = \hat{H}_Z + \hat{H}_{CS} \tag{2.76}$$

where

$$\hat{H}_Z = -\hbar \gamma_C \, \mathbf{H} \cdot \mathbf{I} \tag{2.77a}$$

and

$$\hat{H}_{CS} = -\hbar \gamma_C \, \mathbf{H} \cdot \boldsymbol{\sigma} \cdot \mathbf{I} \tag{2.77b}$$

describes the coupling of the nucleus with the orbital diamagnetism of the bonding electrons around it. This term is responsible for the so called "chemical shift" of the resonance. Since the coupling tensor $\boldsymbol{\sigma}$ depends sensitively on the electronic structure of the chemical bond, chemically different carbons have distinct resonances.

Isotopic substitution of one specific nucleus (D or ^{14}N) can be used in combination with carbon-13 nmr to derive very specific information on the motions and the order parameters.

nmr quadrupole splitting :

Apart from the dipolar effects in nmr, other resonance methods are also in use for the determination of order parameters. These are studies of nmr quadrupole splittings, using nuclei with spin $I > 1$ (D or ^{14}N) and studies of the anisotropy of the Zeeman and hyperfine couplings in the electron spin resonance (ESR) of dissolved free radicals. The nmr quadrupolar splitting technique is now widely used because of its accuracy and ability to probe well defined parts of a molecule.
The Hamiltonian of a nucleus possessing an electric quadrupole can be written as

$$\hat{H} = \hat{H}_Z + \hat{H}_Q \tag{2.78}$$

where

$$\hat{H}_Z = -\hbar \gamma_D \, \mathbf{H} \cdot \mathbf{I}$$

and the quadrupolar Hamiltonian

$$\hat{H}_Q = \mathbf{I} \cdot \mathbf{Q} \cdot \mathbf{I} \tag{2.79}$$

is orientation dependent. γ_D is the deuteron gyromagnetic ratio. Since the dipolar coupling is generally weak for nuclei with $I > 1$, it has been neglected in writing eq. (2.78). The coupling tensor $\mathbf{Q}$ is proportional to the electric quadrupole moment eQ of the nucleus and to the electric field gradient tensor $\mathbf{V}$. In quadrupole resonance experiments the quadrupolar energy is a weak perturbation of the Zeeman term so that the corresponding energy levels can be simply estimated by perturbation theory.

A first order calculation gives the structure of levels and line shapes, under irradiation with a radiofrequency field, as shown in Fig. 2.2.

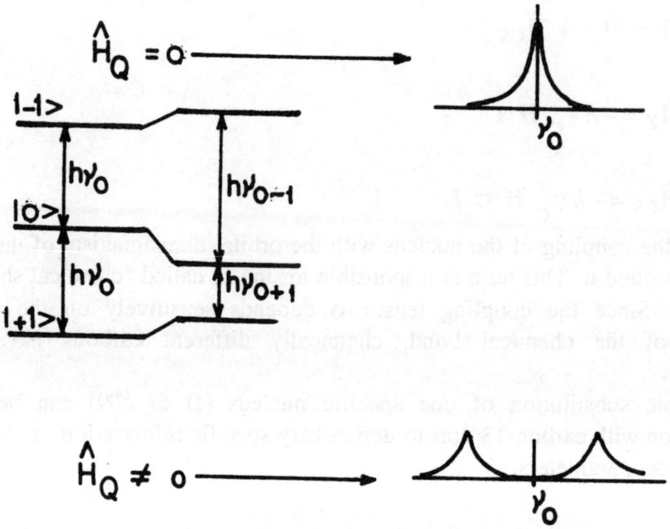

Fig. 2.2. Quadrupolar nmr splitting.

If there occurs a rapid uniaxial motion around the long-molecular axis, in the presence of $\hat{H}_Q$ the splitting of the lines is

$$\Delta\nu_Q = \frac{3e^2qQ}{2h}\,\overline{P}_2\,P_2(\cos\Theta) \qquad (2.80)$$

where Θ is the angle between the long-molecular axis and the CD bond. The static quadrupolar coupling constant, $\nu_Q = e^2qQ/h$, is of the order of 10^5 Hz for deuterium and 10^6 Hz for nitrogen; eQ and eq denote, respectively, the magnitude of the electric quadrupole moment and the electric field gradient. Thus the time scales of the experiments are 10^{-5} and 10^{-6} s. The determination of $\overline{P}_2$ with deuterons suffers because ν_Q (or eQ) is not known precisely and Θ is usually far from 0° or 90°.

2.3.2 Polarised Raman Scattering

Several techniques such as Raman scattering, elastic neutron scattering, and electron paramagnetic resonance line shape analysis are capable of determining both the usual nematic order parameter $\overline{P}_2$ and the higher order parameter $\overline{P}_4$. In principle, the Raman scattering technique[12,13] measures the average of the square of a molecular polarizability. This quantity depends on the spherical harmonics of 0, 2 and 4. The scattering of well chosen vibrational bands (corresponding to C≡N or C–C bonds) provides information on the orientational order of selective parts of mesogenic molecules.

In order to describe some basic aspects of Raman scattering in a simple way, let us consider the scattering of light by particles small compared with the incident wavelength. A time dependent electric moment **m** is induced due to the incident electric field **E** (= $E_0 \cos(\omega_0 t)$),

$$\mathbf{m} = \alpha \mathbf{E} = \alpha E_0 \cos(\omega_0 t) = m_0 \cos(\omega_0 t) \tag{2.81}$$

where α is the polarizability of the particle. According to classical electrodynamics this moment radiates light of frequency ω_0 (Rayleigh Scattering) and of intensity $I \sim m_0^2 \omega_0^4$. Because of the molecular vibrations $\boldsymbol{\alpha}$ itself is time-dependent. For a single vibration with normal coordinate q_i and eigenfrequency ω_i, i.e.,

$$q_i = Q_i \cos(\omega_i t) \tag{2.82}$$

α can be approximated for small amplitudes as

$$\alpha = \alpha_0 + \left(\frac{\partial \alpha}{\partial q_i}\right)_{q=0} q_i + \cdots \tag{2.83}$$

Now eq. (2.81) reads

$$m = \alpha_0 E_0 \cos(\omega_0 t) + \frac{1}{2}\left(\frac{\partial \alpha}{\partial q_i}\right)_0 Q_i E_0 \cos[(\omega_0 + \omega_i)t]$$

$$+\frac{1}{2}\left(\frac{\partial \alpha}{\partial q_i}\right)_0 Q_i E_0 \cos[(\omega_0 - \omega_i)t] \tag{2.84}$$

The first term leads to Rayleigh scattering. The remaining terms are associated with the oscillations at frequencies $\omega_0 \pm \omega_i$ giving rise to Raman scattering. According to a quantum mechanical treatment the Stokes Raman scattering at $\omega_0 - \omega_i$ originates from a lower energy level than the anti-Stokes Raman scattering at $\omega_0 + \omega_i$. As a result the Stokes Raman lines are much more intense than the latter ones. The molecular factor $(\partial \alpha / \partial q_i)_0^2 Q_i^2$ contributes to the intensity of the Raman scattering. For the anisotropic molecules eq. (2.84) can be used by considering the tensorial nature of the Raman polarizability

$$\boldsymbol{\alpha}^{\mathrm{eff}} = \left(\frac{\partial \bar{\boldsymbol{\alpha}}}{\partial q_i}\right)_0 Q_i \tag{2.85}$$

The scattering intensities will be proportional to the squared elements of $\boldsymbol{\alpha}^{\mathrm{eff}}$.

Suppose we choose a molecular coordinate system in which the Raman polarizability tensor associated with the selected vibration is diagonal

$$\boldsymbol{\alpha}^{\mathrm{eff}} = \alpha_0 \begin{pmatrix} a & 0 & 0 \\ 0 & b & 0 \\ 0 & 0 & 1 \end{pmatrix} \tag{2.86}$$

In order to get rid of α_0, one considers the intensity ratios. In the nematic phase, in the back scattering geometry, three ratios can be constructed

$$R_1 = \frac{I_{yz}}{I_{zz}}, \quad R_2 = \frac{I_{zy}}{I_{yy}}, \quad R_3 = \frac{I_{yx}}{I_{xx}}.$$

where $I_{\alpha\beta}$ denotes the scattered intensity; the first index α refers to the polarization of the incident radiation and the second β to the scattered one (Z is the optical axis), and α,β = x,y,z. The intensities are proportional to the squared elements of the Raman tensor. Transforming to the laboratory coordinate system, and averaging overall molecular orientations, one gets,

$$R_1 \sim \frac{<\alpha_{yz}^{\mathrm{eff}^2}>}{<\alpha_{zz}^{\mathrm{eff}^2}>}, \qquad R_2 \sim \frac{<\alpha_{zy}^{\mathrm{eff}^2}>}{<\alpha_{yy}^{\mathrm{eff}^2}>}, \qquad R_3 \sim \frac{<\alpha_{yx}^{\mathrm{eff}^2}>}{<\alpha_{xx}^{\mathrm{eff}^2}>}. \tag{2.87}$$

Both $<\cos^2\theta>$ and $<\cos^4\theta>$ are involved in these depolarization ratios. The measurement of the depolarization ratio R_{iso} in the isotropic phase gives a relation between the temperature independent elements a and b of the Raman tensor. Thus four measurements can be made to determine four unknowns; so a, b, $<\cos^2\theta>$ and $<\cos^4\theta>$ can be calculated, i.e., $\overline{P}_2$ as well as $\overline{P}_4$ can be obtained. More details on the measurements and the sensitivity of the results can be obtained elsewhere[14,15].

2.3.3 X-ray Scattering

X-ray or neutron scattering experiment provides access to the one-particle distribution function in the limit of large-angle scattering which in turn determines $\overline{P}_2$ and $\overline{P}_4$.

The X-ray diffraction techniques can be divided into two groups on the basis of the use of monodomain and powder samples. Using monodomain samples more information can be derived as compared to the powder samples. However, the monodomain samples are not always available. Monodomains of liquid crystals can be obtained either by orienting a powder sample with a magnetic field or by melting a single crystal.

The usual X-ray equipment for the study of monodomain samples is shown schematically in Fig. 2.3. A monochromatic X-ray beam is diffracted by the sample kept at constant temperature is an oven and oriented on a goniometer head. The diffracted beams are collected at a film placed behind the sample. Sometimes the sample is kept oscillating in order to obtain more information. Instead of film, an X-ray counter can also be used. For powder samples the equipment is similar to that for monodomains, except that the goniometer head is not needed.

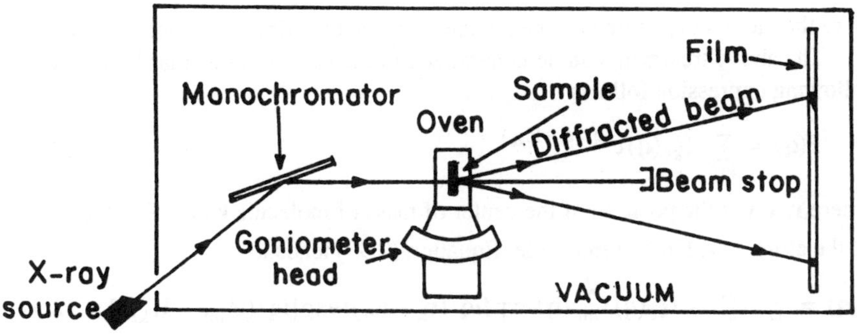

Fig. 2.3. A simple X-ray diffraction apparatus.

Referring to the scattering experiment as shown in Fig. 2.3 the incident and scattered wave vectors have a magnitude

$$k_i = k_s = \frac{2\pi}{\lambda}$$

The scattering wave vector is given by

$$\mathbf{q} = \mathbf{k}_s - \mathbf{k}_i, \qquad q = |\mathbf{q}| = 4\pi \sin\alpha / \lambda \tag{2.88}$$

where 2α is the angle between $\mathbf{k}_s$ and $\mathbf{k}_i$. For a scattering center at $\mathbf{r}$, the scattering of the incident radiation is described by the scattering amplitude

$$f\, e^{i\mathbf{q}\cdot\mathbf{r}}$$

with f known as the scattering power. Corresponding to N centers the scattering amplitude is defined as

$$F(\mathbf{q}) = \sum_{j=1}^{N} f_j\, e^{i\mathbf{q}\cdot\mathbf{r}_j} \tag{2.89}$$

where $\mathbf{r}_j$ defines the position of scattering center j. In case the scattering centers have a continuous distribution

$$F(\mathbf{q}) = \int \rho_e(\mathbf{r})\, e^{i\mathbf{q}\cdot\mathbf{r}}\, d\mathbf{r} \tag{2.90}$$

Here $\rho_e(\mathbf{r})$ denotes the time averaged electron density. If $f(\mathbf{q})$ denotes the atomic scattering amplitude, for a group of atoms, we can write

$$F(\mathbf{q}) = \sum_{j=1}^{N} f_j(\mathbf{q})\, e^{i\mathbf{q}\cdot\mathbf{r}_j} \tag{2.91}$$

It is seen that eqs. (2.89) and (2.91) are identical. This shows that the diffraction by a set of atoms can be treated in terms of a diffraction by a set of scattering points by taking the proper account of the variation of atomic scattering amplitude. Now the intensity can be expressed as

$$I(\mathbf{q}) = |F(\mathbf{q})|^2 \tag{2.92}$$

Here the factors depending on the geometry of the experiment are not included.

The above treatment can be generalized to the case of molecular liquids and the following expression follows

$$F(\mathbf{q}) = \sum_{k,\ell} f_{k\ell}(\mathbf{q}) e^{i\mathbf{q}\cdot(\mathbf{r}_k - \mathbf{R}_{k\ell})} \tag{2.93}$$

where $\mathbf{r}_k$ gives the position of the center of mass of molecule k and $\mathbf{R}_{k\ell}$ the position of the atom ℓ within that molecule. Equation (2.93) leads to

$$I(\mathbf{q}) = \sum_{k,\ell} \sum_{m,n} < f_{k\ell}(q) f^*_{mn}(\mathbf{q}) \exp[i\mathbf{q}\cdot(\mathbf{r}_k - \mathbf{r}_m)] \exp[i\mathbf{q}\cdot(\mathbf{R}_{mn} - \mathbf{R}_{k\ell})] > \tag{2.94}$$

The intensity $I(\mathbf{q})$ can be expressed as

$$I(\mathbf{q}) = I_m(\mathbf{q}) + D(\mathbf{q}) \tag{2.95}$$

where the molecular structure factor $I_m(\mathbf{q})$ which corresponds to the scattered intensity as observed from a very dilute gas of the same molecule is given by

$$I_m(\mathbf{q}) = \sum_k < \sum_{\ell,n} f_{k\ell}(\mathbf{q}) f^*_{kn}(\mathbf{q}) \exp[i\mathbf{q}\cdot(\mathbf{R}_{kn} - \mathbf{R}_{k\ell})] >$$

$$= N \left\langle \left| \sum_{\ell} f_{k\ell} \exp[-i\mathbf{q}.\mathbf{R}_{k\ell}) \right|^2 \right\rangle \tag{2.96}$$

and the so-called interference function $D(\mathbf{q})$ is written as

$$D(\mathbf{q}) = < \sum_{k \neq m} \exp(i\mathbf{q}\cdot\mathbf{r}_{km}) \sum_{\ell,n} f_{k\ell}(\mathbf{q}) f^*_{mn}(\mathbf{q}) \exp[i\mathbf{q}\cdot(\mathbf{R}_{mn} - \mathbf{R}_{k\ell})] > \tag{2.97}$$

Here $\mathbf{r}_{km} = |\mathbf{r}_k - \mathbf{r}_m|$. For a dilute gas $D(\mathbf{q})$ gives information only on components larger than the molecular diameter, i.e. $D(\mathbf{q}) \to 0$ at large values of $\mathbf{q}$, where the small distance contributions are dominant.

The diffraction pattern obtained for nematics is liquid-like but the presence of two characteristic distances, parallel and perpendicular to $\hat{\mathbf{n}}$, is evident. For the case $\mathbf{q}$ is parallel to $\hat{\mathbf{n}}$ the characteristic distance d, which is usually of the order of the molecular length ℓ, corresponds to the maximum of the scattering amplitude. The origin of this diffraction peak must be attributed to the correlations in the molecular arrangement along $\hat{\mathbf{n}}$. In fact, due to the orientational disorder of the molecules one finds[16] $\ell - d \sim 2$ Å for $\ell = 17$ to 25 Å. For the situation $\mathbf{q} \perp \hat{\mathbf{n}}$ a strong peak in the diffraction pattern is observed for a variety of mesogenic materials near $q \sim 1.4\ A^{-1}$. This corresponds to a distance of about 4.5 Å. From the intensity profiles of an aligned sample a detailed information about singlet orientational distribution function can be derived. In this profile a pronounced arc around the equatorial plane $\mathbf{q} \perp \hat{\mathbf{n}}$ appears which can be interpreted as arising from differently aligned clusters or coherence volumes. Assuming that there exists a perfect orientational order within each cluster $D(\mathbf{q})$ can be calculated. In this situation the evaluation of $I(\mathbf{q})$ reduces to a geometrical problem[17].

REFERENCES

1. C.G. Gray, and K.E. Gubbins, Theory of Molecular Fluids, Vol. I: Fundamentals (Clarendon Press, Oxford, 1984).
2. M.E. Rose, Elementary Theory of Angular Momentum (Wiley, New York, 1957).
3. C. Zannoni, Distribution Functions and Order Parameters, in, The Molecular Physics of Liquid Crystals, eds., G.R. Luckhurst, and G.W. Gray (Academic Press Inc., London, 1979) Chapt. 3, p. 51
4. R. Balascu, Equilibrium and Non-equilibrium Statistical Mechanics (Wiley, New York, 1975).
5. J.P. Hansen, and I.R. McDonald, Theory of Simple Liquids (Academic Press Inc., London, 2nd ed., 1986).
6. Y. Singh, Phys. Reports **207**, 351 (1991).
7. S. Singh, Phys. Reports **277**, 283 (1996); Phys. Reports **324**, 107 (2000).
8. A. Carrington, and A.D. McLachlan, Introduction to Magnetic Resonance (A Harper international edition, 1967).
9. R.Y. Dong, Nuclear Magnetic Resonance of Liquid Crystals (Springer-Verlag, New York, 1994).
10. (a) P. Diehl, and C.L. Khetrapal, NMR Basic Principles and Progress, Vol. 1 (Springer - Verlag, 1969).
    (b) J.W. Emsley, and J.C. Lindon, NMR Spectroscopy Using Liquid Crystals Solvents (Pergamon, New York, 1975).
11. (a) J. Charvolin, and B. Deloche, Nuclear Magnetic Resonance Studies of Molecular Behaviour, in, The Molecular Physics of Liquid Crystals, eds., G.R. Luckhurst, and G.W. Gray (Academic Press Inc., London, 1979) Chapt. 15, p. 343.
    (b) P.L. Nordio, and U. Segre, Magnetic Resonance Spectroscopy. Statatic Behaviour, in, The Molecular Physics of Liquid Crystals, eds., G.R. Luckhurst, and G.W. Gray (Academic Press Inc., London, 1979) Chapt. 16, p. 367.

12. P.G. de Gennes, and J. Prost, The Physics of Liquid Crystals (Clarendon Press, Oxford, 1993).

13. G. Vertogen, and W.H. de Jeu, Thermotropic Liquid Crystals, Fundamentals (Springer - Verlag, 1988).

14. S. Jen, N.A. Clark, P.S. Pershan, and E.B. Priestley, J. Chem. Phys. **66**, 4635 (1977).

15. L.G.P. Dalmolen, and W.H. de Jeu, J. Chem. Phys. **78**, 7353 (1983).

16. A.J. Leadbetter, Structural Studies of Nematic, Smectic A and Smectic C Phases, in, The Molecular Physics of Liquid Crystals, eds., G.R. Luckhurst, and G.W. Gray (Academic Press Inc., London, 1979) Chapt. 13, p. 285.

17. A.J. Leadbetter, and E.K.Norris, Mol. Phys. **38**, 669 (1979).

3

Physical Properties of Liquid Crystals

The order existing in the liquid crystalline phases destroys the isotropy (all directions are equivalent) and introduces anisotropy (all directions are not equivalent) in the system. This anisotropy manifests itself in the elastic, electric, magnetic and optical properties of a mesogenic material. The macroscopic anisotropy is observed because the molecular anisotropy responsible for this property does not average out to zero as is the case in an isotropic phase. Many applications of thermotropic liquid crystals rely on their physical properties and how they respond to the external perturbations. The physical properties can be distinguished into scalar and nonscalar properties. Typical scalar properties are the thermodynamic transition parameters (transition temperature, transition enthalpy and entropy changes, transition density and fractional density changes). The dielectric, diamagnetic, optical, elastic and viscous coefficients are the important nonscalar properties. In this chapter, we describe the various aspects of some of these properties and how they can be related with the molecular parameters.

The liquid crystalline states can be characterized through their orientational, positional and conformational orders. The orientational order is present in all the liquid crystal phases and is usually the most important ones. The positional order is important to characterize the various mesophases (nematic, smectic A, smectic C, etc.) and can be investigated by their textures (see chapter 11) or their X-ray diffraction patterns which allow a determination of the layer thickness. The conformational order is a measure of how the rings, fractional groups and end chains within a molecule are oriented and arises because of the short-range intramolecular and intermolecular forces. The relative orientation of the molecular segments can usually be measured by employing the nuclear magnetic resonance (nmr) technique. Since the conformational order quickly averages out over macroscopic distances, it is not of much concern as far as practical applications are concerned. Many details of orientational and positional order parameters are given in the previous chapter 2.

3.1 SCALAR PHYSICAL PROPERTIES

The transition temperatures and transition densities are the important quantities characterizing the materials. The difference in the transition temperatures between the melting and clearing points gives the range of stability of the liquid crystalline phases. For polymorphous substances the higher ordered phase exhibits the lower transition temperatures. Within a homologous series pronounced variations are observed for the respective phase transitions; with increasing alkyl or alkoxy chain length the clearing point decreases. The compounds with odd chain length have lower value of clearing points and thus an "even-odd effect" is observed. A rough estimate on the purity of a substance can be made from the temperature range over which the clearing takes place. High purity compounds have narrow clearing ranges of typically 0.3 to 0.5K. With increasing amounts of impurities this range will steadily increase to over 1K. The transition enthalpies and entropies between the solid and the liquid crystalline states, between the various liquid crystalline states and between the liquid crystalline state and the isotropic state are related to the degree of internal order present in the system. When a mesogenic material is heated from its crystalline state to above the clearing point a variety of transitions with accompanying enthalpy and entropy changes might occur.

A knowledge of both the temperature and the heat of transition is necessary if the principles of physical analysis are to be applied to study the mesophase transitions. From this information the transition entropy can be calculated which may play a pivotal role for evaluating the type and degree of order present in a system. When a material melts, a change of state occurs from a solid to a liquid and this melting process requires energy (endothermic) from the surroundings. Similarly the crystallization of a liquid is a exothermic process and the energy is released to the surroundings. The melting transition from a solid to a liquid is a relatively drastic phase transition in terms of the structural change and relatively high transition energy is involved. The relatively small enthalpy changes show that the liquid crystal phase transitions are associated with more subtle structural changes. Although the enthalpy changes at a transition cannot identify the types of phases associated with the transitions, the magnitude of the enthalpy change is proportional to the change in structural ordering of the phases involved. Typically, a melting transition from a crystalline solid to a liquid crystal phase or an isotropic liquid phase involves an enthalpy change of around 30 to 50 kJ/mol. This indicates that a considerable structural change is occurring. The liquid crystal to liquid crystal and liquid crystal to isotropic liquid transitions are associated with very much smaller enthalpy changes (~ 4-6 kJ/mol). Other smectic and crystal smectic mesophases are also characterized by values of similar order. The nematic-isotropic liquid transition usually gives a smaller enthalpy change (1-2 kJ/mol). The enthalpy changes of transition between various liquid crystal phases are also small. For example, the S_C

S_A transition is often difficult to detect because the enthalpy change is typically less than 300 J/mol. The enthalpy for S_AN_u transition is also fairly small (1 kJ/mol) and the S_CN_u transition has enthalpy less than 1 kJ/mol.

The transition from one phase to another is not necessarily sharp but streches over a certain range. The nematic-isotropic liquid transitions tend to be smaller than about 0.6 to 0.8K, whereas crystalline-smectic or smectic-isotropic liquid transitions are generally much wider. We define the melting (and clearing) point as the temperature at which the first change in optical transmission can be observed. Usually, the transition of the melting point is less sharp than the clearing point leading to an absolute error of ± 1K and ± 0.1K, respectively.

A number of techniques (e.g. optical microscopy, mettler oven, microscope hot stage, differential thermal analysis, calorimetry, Raman spectroscopy, etc.) are available for the measurement of transition quantities. Most of these quantities may be determined for both pure mesogens and mixtures. The main instruments which may be used for the measurement of transition temperatures are mettler oven, microscope hot stage and differential scanning calorimetry (DSC). The mettler oven can detect small changes in transmission as a function of temperature. For the liquid crystals one may expect a change of optical properties for every phase transition. Under these experimental conditions, the transmission of a liquid crystal increases with decreasing order parameter. This results, for solid and smectic states, in a low transparency, while the nematic and isotropic states have a higher one. In order to perform a measurement, a melting tube is filled with the respective substance and placed into the oven. Any change in transmittance is monitored by a built-in photodiode in the oven, giving a photocurrent which is proportional to the transparency of the sample. With increasing temperature a general increase in transmission is observed duc to the general decrease of the order parameter. The determination of transition temperatures and the characterization of liquid crystal phases can be done concurrently using a hot-stage under a polarising microscope. The transition temperatures are measured upon first heating of the sample. For the melting and clearing points the first change observed in the texture is of relevance. In case of smectic-smectic, smectic-nematic or smectic-isotropic liquid transitions the temperature at the end of the transition is quoted. Often transitions from the crystalline to a smectic texture cannot be distinguished, because no change in optical appearance results. So in these cases, when both the techniques oven and hot-stage fail, the transition temperature is deduced from the DSC analysis. When the hot-stage under the microscope is used, a temperature gradient over the glass slide may cause an additional error. The existence of such a gradient is revealed when the slide is slightly shifted by hand through the field of view. For example, when the nematic-isotropic liquid transition is observed both the nematic and isotropic liquid are coexisting, but are sited at different locations in the field of view. The black islands seen in the texture are the regions where the material is already isotropic, embedded in a nematic environment. The clearing point would be quoted as the temperature at which the first black region appears. The existence of a temperature gradient in hot-stage may account for a possible systematic error of ± 1-2K. When a high precision is required, the mettler oven may be employed. The two methods (oven and hot-stage) are the most common techniques which are applied in the determination of phase transitions of liquid crystals.

The most widely used technique in mesophase research is the calorimetric study of the phase transitions. From the molar heat of transition, q, the entropy change of transition is calculated from the familiar relationship $\Delta\sum = q/T$. A large number of measurements using classical adiabatic calorimetry are available. However, the bulk of thermodynamic data presently available has been obtained by dynamic calorimetry. By the very nature of instrumentation dynamic methods are much less accurate than adiabatic calorimetry. The expected accuracy is usually not much better than ± 1% and in some cases only ± 10%.

The application of the methods of differential thermal analysis (DTA) and differential scanning calorimetry (DSC) has provided extremely rich data on the temperatures, heats of transition and heat capacity of various phases. In DTA the sample and the reference material are heated at some linear rate. The absolute temperature and the differential temperature between the sample and the reference material are recorded. The area beneath the differential curve is related to calories via calibration with a material of known heat of fusion. Since the area is due to temperature difference, factors such as sample and instrument heat capacities are important. When adequately calibrated, the data may be determined from the curves within an accuracy of ± 1%. The DSC involves the comparison of the sample with an inert reference during a dynamic heating or cooling programme. It employs two furnaces, one to heat the sample under investigation and the other to heat an inert reference material (usually gold). The two furnaces are separately heated but are connected by two control loops to ensure that the temperature of both remains identical through a heating or cooling cycle. The heating or cooling rate for each is constantly identical. A balance between the sample and the reference material is maintained by adding heat via the filament. When a sample melts, for example, from a crystalline solid to a S_A phase, energy must be supplied to the sample to prevent an imbalance in temperature between the sample and the reference material. This energy is measured and recorded by the instrument as a peak on a baseline. The instrument is precalibrated with a sample of known enthalpy of transition, and this enables the enthalpy of transitin to be recorded for the material being examined. The sample is weighed into a small aluminium pan and then the pan is placed into a holder in a large aluminium block to ensure good temperature control. The sample can be cooled with the help of liquid nitrogen and a working temperature range of between -180°C and 600°C may be achieved.

Although, the DSC reveals the presence of phase transitions in a material by detecting the enthalpy change associated with each phase transition, the precise phase identification can not be done. However, the level of enthalpy change involved at the phase transition does provide some indication of the types of phases involved. Accordingly, DSC is used in conjunction with optical polarising microscopy to determine the mesophase types exhibited by a material. If a transition between mesophases has been missed by optical microscopy, then DSC may reveal the presence of a transition at a particular temperature or vice-versa. After DSC, the optical microscopy should be used to examine the material very carefully to furnish information on the phase structure, and to ensure that the transitions have not been missed by DSC. This procedure hopefully may lead to the likely identity of the mesophases. Accordingly, optical polarising microscopy and DSC are important complementary tools in the identification of the types of mesophases exhibited by a material.

The DSC often has been successful in revealing the presence of chiral liquid crystal phases. However, in certain chiral phases the story is not so successful. In case of blue phases and TGB_A^* phase, the range of stability of mesophases are often too small to provide a distinct enthalpy peak. The transition between the ferrielectric smectic C^* (S_{Cfe}^*) phase and antiferroelectric smectic C^* (S_{CA}^*) phase and their transitions with the S_C^* phase involve extremely small enthalpy values. This makes detection by DSC very difficult. However, the latest DSC equipment may enable the detection of even such remarkably small enthalpy transitions.

3.2 ANISOTROPIC PHYSICAL PROPERTIES

The liquid crystals exhibit anisotropy in many of their physical properties. Due to these anisotropies and their resulting interactions with the surrounding environments a number of phenomena are observed in a liquid crystalline phase which are absent in the isotropic liquid phase. In the following, we give a brief discussion on some of these properties.

It is well known that liquid crystals are very sensitive to an electric field. This property allows their applications to display and other optical device technology. Since the beginning of the century, the liquid crystalline electrooptical properties have been studied very actively which led to the discovery of many important phenomena such as the electrohydrodynamic instability, field-induced domain pattern, Frederiks transition, flexoelectric distortion and spatially periodic structures, ferroelectric switching, induced biaxiality, etc. Many physical parameters of a materials are based on the electrooptical measurements. Some of the results on the physical properties of nematic and smectic liquid crystals shall be summarized in the subsequent chapters. Here the basic concepts related with the important physical properties are discussed.

3.2.1 Optical Anisotropy : The Refractive Index

The optical anisotropy is an essential physical property for the optimization of liquid crystal mixtures for application in liquid crystal display devices.

The liquid crystals are optically anisotropic materials, i.e., the speed of propagation of light waves in the medium is no longer uniform but is dependent upon the direction and polarization of the light waves transversing the material; thus the material is found to possess different refractive indices in different directions. It is the ability of aligned liquid crystals to control the polarization of light which has resulted in the use of liquid crystals in displays. Materials in aligned uniaxial liquid crystals (e.g. N_u, S_A, ...) are frequently regarded as behaving in a manner similar to that of uniaxial crystals and are described as doubly refracting, or birefringent. Accordingly, uniaxial liquid crystals are found to exhibit two principal refractive indices, the ordinary refractive index n_0, and the extraordinary refractive index n_e. The ordinary refractive index n_0 is observed with a light wave where the electric vector oscillates perpendicular to the optic axis. The extraordinary refractive index n_e is observed for a linearly polarized light wave where the electric vector is vibrating parallel to the optic axis. The optic axis of the uniaxial mesophases is given by the director (see Fig. 3.1).

The optical anisotropy, or birefringence, is wavelength and temperature dependent and defined by the equation

$$\Delta n = n_e - n_0 = n_{||} - n_{\perp}, \tag{3.1}$$

where $n_{||}$ and $n_{\perp}$ are the components parallel and perpendicualr to the director, respectively. For rod-like molecules $n_{||} > n_{\perp}$; Δn is, therefore, positive and can be between 0.02 to 0.4. In the case of discotic molecules $n_{||} < n_{\perp}$ and thus negative birefringence is associated with discotic nematic or columnar phases. In case of chiral nematics the optic axis coincides with the helix axis which is perpendicular to the local director; $n_e = n_{\perp}$, and n_0 is a function of both $n_{||}$ and $n_{\perp}$ and depends on the relative magnitude of the wavelength with respect to the pitch. When the wavelength of the light is much larger than the pitch, one obtains for the chiral nematics,

$$n_0 = \left[\frac{1}{2}(n_{||}^2 + n_{\perp}^2)\right]^{1/2} \tag{3.2}$$

$$n_e = n_{\perp}$$

As one still usually has $n_{\parallel} > n_{\perp}$, $\Delta n < 0$; chiral nematics then have a negative birefringence.

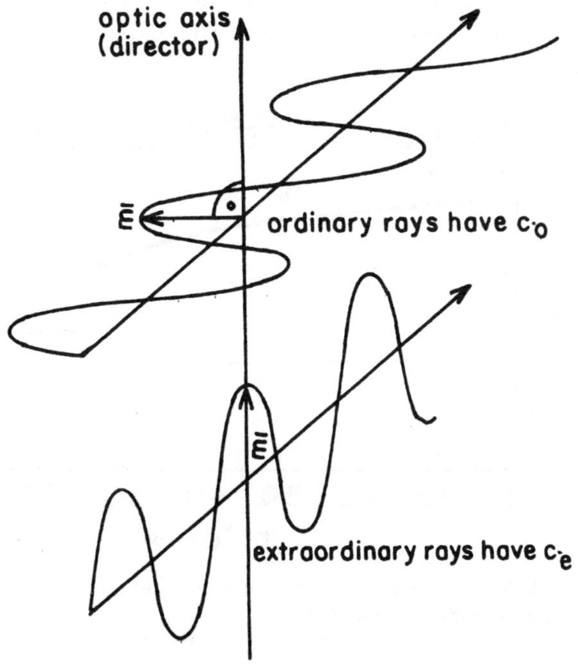

Fig. 3.1. Birefringence for a N_u phase possessing positive optical anisotropy $\Delta n > 0$.

For biaxial liquid crystals, all 3 principal refractive indices are different, but usually one (n_3) is significantly greater (or less) than the other two, and the uniaxial birefringence can be defined as

$$\Delta n = n_3 - \frac{1}{2}(n_2 + n_1) \tag{3.3}$$

and the biaxiality is

$$\delta\Delta n = n_2 - n_1 \tag{3.4}$$

As a consequence of the small degree of structural biaxiality of these phases, the biaxiality is small[1] (~ 0.01).

The optical anisotropy in liquid crystals is temperature dependent which can be described[2] as

$$\Delta n \sim \rho_0^{1/2}\,\overline{P}_2 \tag{3.5}$$

This expression shows a linear relation of Δn with the order parameter $\overline{P}_2$ and a square root dependence on the density ρ_0. However, it should be remembered that this is not necessarily true within the accuracy with which Δn can be obtained. The temperature dependent behaviour of refractive indices for PCH-5 is shown[3] in Fig. 3.2. It can be seen that it is mainly n_e which introduces the variation. Further, once the material has been heated above the clearing point the internal order of the liquid crystal is destroyed and the behaviour is as in an isotropic state.

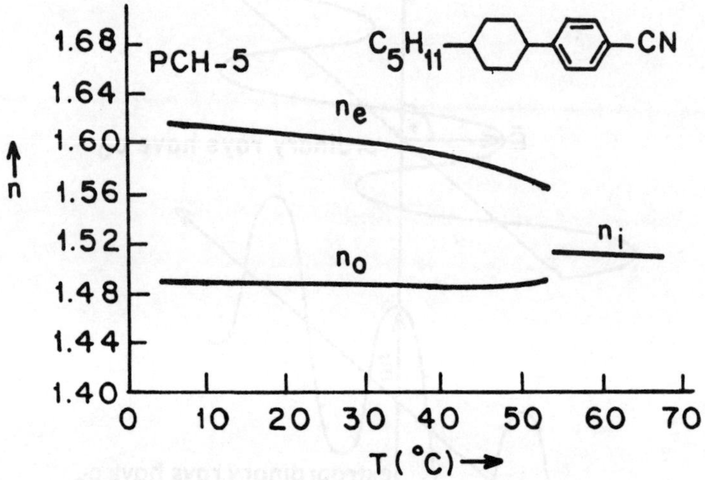

Fig. 3.2. Temperature dependence of the refractive indices n_e and n_0 for PCH-5 determined at 589.3 nm (Ref. 3).

The optical dispersion of a material is the dependence of the refractive index upon the wavelength. The determination of the refractive indices at different wavelengths in the visible spectrum enables a subsequent fit to the Cauchy equation[4],

$$n(\lambda) = n_\infty + \frac{\alpha}{\lambda^2} \tag{3.6}$$

where n_∞ is the refractive index extrapolated to infinite wavelength and α is a material specific coefficient. Using the fit, it is possible to calculate the respective refractive indices n_e and n_0 and consequently the optical anisotropy. Figure 3.3 shows the dispersion determined for the material PCH-5 at 293K. The dispersion is also temperature dependent.

The measurements of the refractive indices are conveniently carried out by the use of Abbe's double prism method. In this method, the liquid crystal is used as a thin film between the hypotenuse areas of two prisms. If the refractive index of the prism is greater than the indices of the liquid crystal, the boundary angles corresponding to total reflection of the ordinary as well as the extraordinary ray can be measured. The boundary associated with n_0 can be made more distinct by blocking the extraordinary light with a polarizer. Usually $n_0 < n_{prism} < n_e$, and n_e

cannot be determined unless special prisms are used. In practice, the Abbe refractometers employed for the measurement of optical anisotropy are arranged in two separate systems; one arrangement employs a measuring prism A_1 enabling the determination of the refractive indices within the range of 1.30 to 1.71. An additional setup is equipped with a special measuring prism A_2, which enables a measuring range from 1.45 to 1.85. In most cases the refractometer equipped with the measuring prism A_1 suffices. In case $n_{prism} < n_e$, it becomes necessary to determine n_0 and n_e using the measuring prism A_2. The moving scale of the measuring prism is calibrated in terms of the refractive index for the mean wavelength of the two sodium D lines at 589.3 nm. Values determined using the prism A_2 with radiation sources at other wavelengths and all values determined with the prism A_1 require correction.

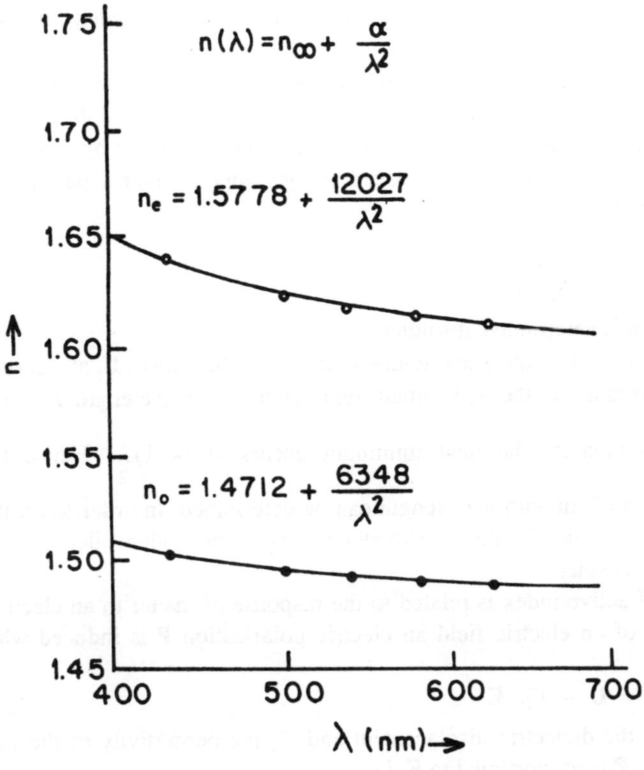

Fig. 3.3. Dispersion of the refractive indices n_e and n_0 for PCH-5 determined at 293 K (Ref. 3).

Alternatively, instead of measuring n_e, one can determine Δn directly by the use of interference technique[2]. One uses a uniform planar sample with its optic axis at an angle of 45° to the crossed polarizers. Normally incident light is split into two rays of equal intensity. After passage through the sample, the two rays have a phase difference and are made to interfere at the analyzer. The intensity of the transmitted light is monitored with a photodetector. If λ is the wavelength of monochromatic light, and t the thickness of the sample, the condition for the minimum intensity is

$$t\,\Delta n = k\lambda \tag{3.7}$$

where k is an integer. Δn can be determined by exploiting the eq. (3.7) in a variety of ways.

(a) For a fixed t and λ, the intensity of transmitted light can be recorded as a function of the temperature, which is varied continuously. This leads to a direct measurement of $\Delta n(T)$. However, at the nematic-isotropic liquid (NI) transition, Δn varies discontinuously. As a result, the value of k at the first minimum below nematic-isotropic liquid transition temperature (T_{NI}) is not known and hence to fix the absolute scale of Δn, one additional independent measurement is required.

(b) At fixed values of λ and T, a wedge-like sample can be used in which t varies along the X-axis from 0 to a value t_0. If t_0 is reached at a distance x′ from the apex, the thickness at a point x along the wedge is given by $t = x\, t_0/x'$. In this case a series of uniformly spaced fringes appear; their separation Δx can be measured,

$$\Delta n = \lambda\left(\frac{x'}{\Delta x\, t_0}\right)$$

Here Δn is determined absolutely.

(c) With a fixed thickness and temperature, λ can be varied. If, at a certain value of k, a minimum in the transmitted light occurs at a wavelength λ_1, on decreasing the wavelength, the next minimum occurs at $(k+1)\lambda_2$. Hence the relative variation of Δn with wavelength can be determined. In order to fix the absolute scale, the value of k must be known, and again one independent measurement of Δn is necessary.

The refractive index is related to the response of matter to an electric field. On application of an electric field an electric polarization **P** is induced which can be defined as

$$\mathbf{P} = \mathbf{D} - \epsilon_0\, \mathbf{E} \tag{3.8}$$

where **D** is the dielectric displacement and ϵ_0 the permittivity of the vacuum. For small fields, **P** is proportional to **E**, i.e.,

$$P_\alpha = \epsilon_0\, \chi^e_{\alpha\beta}\, E_\beta\,, \qquad \alpha,\beta = x,y,z \tag{3.9}$$

where $\chi^e_{\alpha\beta}$ is an element of the electric susceptibility tensor $\boldsymbol{\chi}^e$. In the electric case the permittivity tensor $\boldsymbol{\epsilon}$ is mostly used, which is related to the susceptibility by

$$\epsilon_{\alpha\beta} = \delta_{\alpha\beta} + \chi^{e}_{\alpha\beta} \tag{3.10}$$

In general, $\epsilon_{\alpha\beta}$ depends on the frequency and wave-vector of the applied field. Taking the director along the Z-axis, ϵ is diagonal, and the refractive indices have the following relations with the elements of the permittivity tensor in the optical frequency range :

$$n_{\parallel}^{2} = \epsilon_{zz} , \qquad n_{\perp}^{2} = \epsilon_{xx} = \epsilon_{yy} \tag{3.11}$$

3.2.2 Dielectric Anisotropy : The Dielectric Permittivity

Dielectric studies are concerned with the response of matter to the application of an electric field. This can be characterized by the dielectric permittivity ϵ. In a classical experiment a material is filled in a capacitor and the increase in the capacitance from a value C to a value ϵC is noted; ϵ is the (relative) permittivity of the material. The increase in capacitance is due to the polarization of the material by the field **E**. For an anisotropic material the resulting polarization (per unit volume) is given by eq. (3.9). Using eq. (3.10) for the permittivity, this gives

$$P_{\alpha} = \epsilon_0 (\epsilon_{\alpha\beta} - \delta_{\alpha\beta}) E_{\beta} \tag{3.12}$$

In a material consisting of non-polar molecules only an induced polarization occurs, which for static fields consists of two parts : the electronic polarization (which is also present at optical frequencies) and the ionic polarization. In materials with polar molecules, in addition to the induced polarization, an orientational polarization occurs, which arises due to the tendency of the permanent dipole moments to orient themselves parallel to the field. In isotropic fluids, the permittivity is isotropic, i.e., $\epsilon_{\alpha\beta} = \epsilon\, \delta_{\alpha\beta}$. In solids, on the other hand, the permittivity will usually be anisotropic, and the contribution of orientational polarization is less important due to rather fixed orientations of the molecules. In liquid crystals, there exist complicated situations of an anisotropic permittivity in combination with a fluid-like behaviour. Hence in the case of polar molecules, the orientational polarization may contribute significantly to the anisotropic permittivity.

We restrict our discussion mainly to the uniaxial liquid crystalline phases and choose the macroscopic Z-axis along the director. The principal elements of ϵ are then $\epsilon_{zz} = \epsilon_{\parallel}$ and $\epsilon_{xx} = \epsilon_{yy} = \epsilon_{\perp}$. The anisotropy of the dielectric permittivity is given by

$$\Delta\epsilon = \epsilon_{\parallel} - \epsilon_{\perp} \tag{3.13}$$

The mean dielectric permittivity $\bar{\epsilon}$ (for N_u phase) can be described by

$$\bar{\epsilon} = \frac{1}{3} (\epsilon_{\parallel} + 2\epsilon_{\perp}) \tag{3.14}$$

The dielectric permittivities are temperature and frequency dependent (see Fig. 3.4). The $\Delta\epsilon$ can be measured[2] by the use of a simple but effective, capacitative method which allows both the temperature dependency as well as the frequency dependency to be ascertained. In practice, a plane capacitor is used and the capacitance is measured with AC excitation. The frequency is chosen low enough (~ 1 kHz) to allow the measurement of essentially the static permittivities and is high enough to prevent electrochemical reactions or the formation of double layers at the electrodes. Often display-like cells are used which consist of two glass plates with well defined electrodes and kept a small distance apart by spacers.

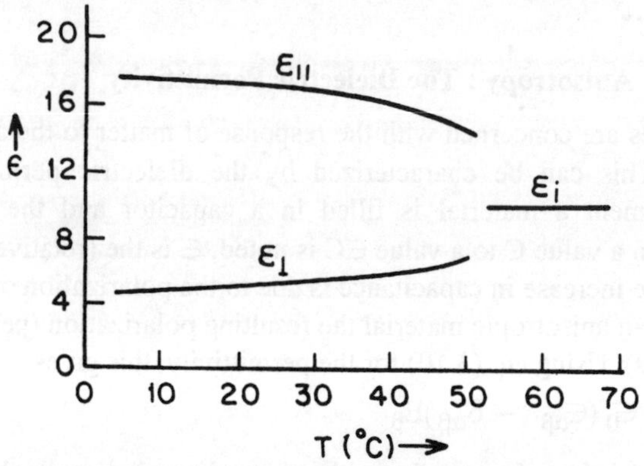

Fig. 3.4a. Temperature dependence of ∈ for PCH-5.

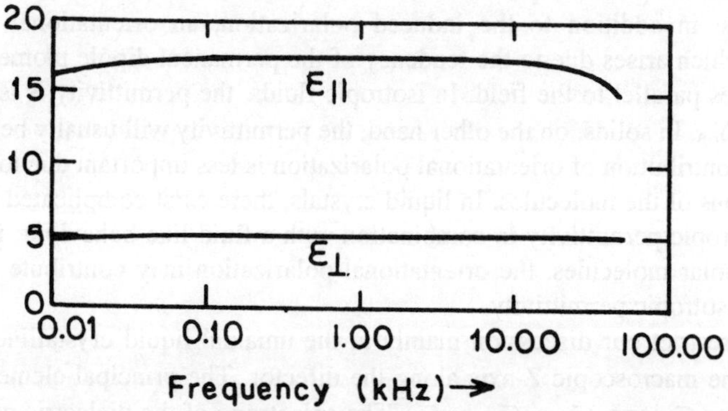

Fig. 3.4b. Frequency dependence of ∈ for PCH-5 at 298 K.

The parasitic capacitance C_p is determined by means of a calibration measurement using a liquid of known permittivity as dielectric. It is preferred to align the director with the magnetic field. This has the advantage that $\in_{\parallel}$ and $\in_{\perp}$ are measured with the same cell, thus increasing the accuracy of the measurement of $\Delta\in$. In that case nontransparent gold or copper electrodes can be used. The sample must not be too thin (say ≥ 100 μm for a magnetic induction of the order of 1T) so that the influence of the boundary layer of the liquid crystal, where the director is not perfectly aligned by the field, can be disregarded. In order to avoid the formation of electrohydrodynamic instabilities, which may disturb the uniform director pattern, the applied voltage must be small (< 1V).

A theoretical treatment of the anisotropic dielectric behaviour of liquid crystals, in general, must show how the macroscopically observed anisotropy arises from the anisotropies of the various molecular quantities. Such an approach turns out to be quite feasible. An alternative approach to derive molecular quantities from the observed macroscopic permittivities may be much more difficult. For the nonpolar compounds the situation is relatively simple. In the case of polar molecules dipolar interaction makes the situation rather complicated; the polarization is defined as

$$P_\beta = N\left(< \alpha_{\beta\gamma}\, E^i_\gamma > + < \bar{\mu}_\beta >\right), \tag{3.15}$$

where $\mathbf{E}^i$ is the internal field, i.e., the average field acting on a molecule and $<\boldsymbol{\mu}>$ indicates the average of the dipole moment $\boldsymbol{\mu}$ over the orientation of all the molecules in the presence of a directing electric field $\mathbf{E}^d$. The directing field is the effective field that tends to orient the permanent dipole moment. The fundamental equation for the permittivity of an anisotropic liquid follows directly from eqs. (3.12) and (3.15),

$$(\epsilon_{\beta\gamma} - \delta_{\beta\gamma})E_\gamma = (N/\epsilon_0)(< \alpha_{\beta\gamma} E^i_\gamma > + < \bar{\mu}_\beta >) \tag{3.16}$$

As the directing field energy is small, we can approximate,

$$\exp\left(-\frac{\mu_\gamma E^d_\gamma}{k_B T}\right) \sim 1 - \frac{\mu_\gamma E^d_\gamma}{k_B T} \tag{3.17}$$

Now $< \bar{\mu}_\beta >$ can be expressed as

$$< \bar{\mu}_\beta > = \frac{1}{k_B T} < \mu_\beta\, \mu_\gamma > E^d_\gamma, \tag{3.18}$$

where $< \mu_\beta\, \mu_\gamma > = 0$, if $\beta \neq \gamma$.

The central problem of the theory of permittivity is to establish relation between the externally applied field **E** and the internal and directing field, $\mathbf{E}^i$ and $\mathbf{E}^d$, respectively, and to calculate[5] these fields. According to Onsager $\mathbf{E}^d$ differs from $\mathbf{E}^i$ due to the presence of a reaction field **R**. The dipole polarizes its surrounding, which in turn leads to a reaction field, at the position of the dipole. The reaction field being parallel to the dipole can not direct the dipole, but it does contribute to $\mathbf{E}^i$. The internal field is now given by

$$\mathbf{E}^i = \mathbf{E}^d + \overline{\mathbf{R}} \tag{3.19}$$

These fields can only be calculated for certain assumed models of the dielectric. In the Onsager model a single molecule is considered in a spherical cavity, where its environment is taken as a continuum with the macroscopic properties of the dielectric. Maier and Meier[6] extended this theory to nematic liquid crystals in the frequency domain below 100 kHz. They ignored all anisotropies in the calculation of $\mathbf{E}^i$ and $\mathbf{E}^d$. Furthermore, this approximation is most useful for liquid crystals with values of $\overline{\epsilon}$ exceeding $\Delta\epsilon$, i.e., molecules where the dielectric anisotropy in units of $\overline{\epsilon}$ is small. Due to these reasons, on the molecular level, each liquid crystal can be envisaged as consisting of a multitude of electric dipole moments. The resulting moment of the molecule can be thought of as being the result of the permanent contribution $\boldsymbol{\mu}$ with components μ_ℓ and μ_t, and the induced contribution arising due to anisotropic polarizability $\boldsymbol{\alpha}$ with principal elements α_ℓ and α_t. The subscripts ℓ and t stands for the contributions longitudinal and perpendicular to the molecular axis with $\mu_\ell = \mu \cos\beta$ and $\mu_t = \mu \sin\beta$; β is defined as the angle between $\boldsymbol{\mu}$ and the direction of α_ℓ. Both the permanent and the induced dipole moments may be combined to give the following expression for the dielectric anisotropy[2]

$$\Delta\epsilon = \left(\frac{NhF}{\epsilon_0}\right)\left[\alpha_\ell - \alpha_t - \frac{F\mu^2}{2k_BT}(1 - 3\cos^2\beta)\right]\overline{P}_2 \tag{3.20}$$

The average dielectric constant is given by

$$\overline{\epsilon} = 1 + \left(\frac{NhF}{\epsilon_0}\right)\left(\overline{\alpha} + \frac{F\mu^2}{3k_BT}\right) \tag{3.21}$$

where $\overline{\alpha} = \frac{1}{3}(\alpha_\ell + 2\alpha_t)$, F and h are constants of proportionality called the reaction-field factor and the cavity factor. They are found to be

$$F = \left[1 - \frac{2N\overline{\alpha}(\overline{\epsilon}-1)}{3\epsilon_0(2\overline{\epsilon}+1)}\right]^{-1} \tag{3.22}$$

and

$$h = \frac{3\overline{\epsilon}}{2\overline{\epsilon}+1} \tag{3.23}$$

The values[7] of $\Delta\epsilon$ of technically useful materials range from between -6 to +50.

Obviously we cannot expect from the model to give quantitatively correct results because of all the approximations involved. Neverthless eq. (3.20) satisfactorily explains many essential features of the permittivity of nematic liquid crystals consisting of polar molecules. If there are liquid crystals with a polar substituent in a position such that $1 - 3\cos^2\beta = 0$ ($\beta = 54.7^\circ$) the permanent dipole moment contributes equally to both $\epsilon_\|$ and $\epsilon_\perp$; in this case $\Delta\epsilon$ is determined by the (positive) anisotropy of the polarizability. The dipole contribution to $\Delta\epsilon$ is positive for $\beta < 54.7^\circ$ and negative for $\beta > 54.7^\circ$. Depending on the magnitude of the dipole contribution, $\Delta\epsilon$ may become negative. The temperature dependence of $\Delta\epsilon$ is determined by the induced polarization, which is proportional to $\overline{P}_2$, and the orientation polarization, which varies as $\overline{P}_2/T$. The remaining factors are weakly dependent on temperature. When $\Delta\epsilon$ is negative or strongly positive, the $\overline{P}_2/T$ dependence of the dipole contribution to $\Delta\epsilon$ is predominant over the whole temperature range. As a result $|\Delta\epsilon|$ increases with decreasing temperature. If $\Delta\epsilon$ is positive and close to zero, the two contributions become approximately equal and just below T_{NI} the temperature variation of $\Delta\epsilon$ is governed by $\overline{P}_2$; $\Delta\epsilon$ increases with decreasing temperature. At lower temperatures, the variations of $\overline{P}_2$ are small and $\Delta\epsilon$ is determined predominantly by the $\overline{P}_2/T$ dependence of the orientation polarization. Consequently with decreasing temperature $\Delta\epsilon$ may be found first to increase and then after going through a maximum starts decreasing.

In smectic phases the interaction of a dipole moment with the dipoles of surrounding molecules works out differently from that in the nematic due to the non-isotropic distribution of the centers of mass. Consequently a different type of dipole-dipole correlation is observed in the smectic phases. For dipoles situated in the central part of the molecules, the distance between the dipoles of molecules in different smectic layers is much greater than the distance between neighbouring dipoles in the same layer. This means that for μ_ℓ antiparallel correlation with neighbours within the layers is dominant over parallel correlations with neighbours in adjacent layers. This contributes, on average, to antiparallel correlation and thus to a decrease of $\epsilon_\parallel$. Similarly the difference in distance between the two types of neighbours leads to a parallel correlation of μ_t and an increase of $\epsilon_\perp$. This can even lead to a change in sign of $\Delta\epsilon$ in certain cases as is shown[8] in Fig. 3.5.

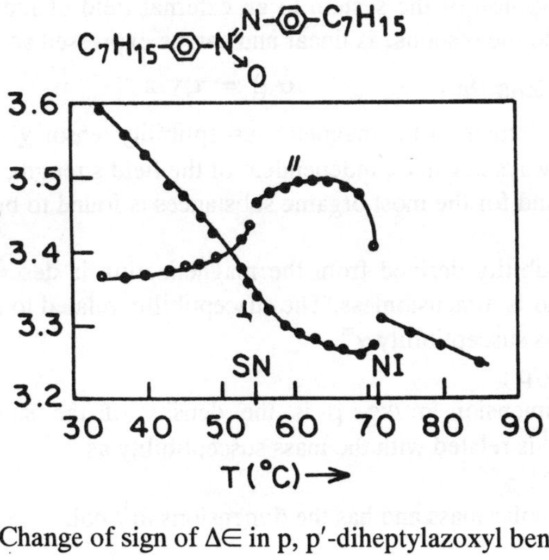

Fig. 3.5. Change of sign of $\Delta\epsilon$ in p, p′-diheptylazoxyl benzene.

3.2.3 Diamagnetic Anisotropy : The Magnetic Susceptibility

The liquid crystals are very sensitive to a magnetic field. Since the very beginning this property has been studied extensively and has helped in the understanding of the behaviour of the liquid crystalline materials.

There are three main types of magnetic materials diamagnetic, paramagnetic and ferromagnetic. In the absence of an applied field, in many materials, the magnetic effects associated with the orbital and spin motions of the atomic electrons cancel out exactly. When a changing magnetic field is applied to a diamagnetic material, changes are induced in the electron currents in the atom. As the field due to the induced current opposes the applied field, the net field is smaller than the applied field. In a paramagnetic materials, there exists a permanent residual magnetic dipole moment associated with the individual atoms. When a magnetic field is applied to a paramagnetic material, the dipoles tend to align themselves with their dipole moment along the field thereby increasing the total field inside the material. A ferromagnetic material is characterized by the strong interactions between the atomic dipoles of neighbouring atoms that are strong enough to influence a spontaneous alignment of the dipoles, even in the absence of a magnetic field.

The liquid crystals are diamagnetic like most organic materials. The diamagnetism exhibited by these systems is the result of additive contributions of the magnetic properties of the single molecular components. The application of an external magnetic field to a nematic results in a magnetization **M** which is defined in terms of magnetic induction **B** and the magnetic field strength **H** as

$$\mathbf{M} = \mu_0^{-1}\,\mathbf{B} - \mathbf{H} \qquad (3.24)$$

where μ_0 is the permeability of free-space. The magnetic susceptibility of a system describes the response of the system to an external field of induction **B**. For small values of the field the response is linear and can be expressed as

$$M_\alpha = \mu_0^{-1}\,\chi_{\alpha\beta}\,B_\beta\,; \qquad \alpha,\beta = x,y,z \qquad (3.25)$$

where $\chi_{\alpha\beta}$ is an element of the magnetic susceptibility tensor $\boldsymbol{\chi}$. For the diamagnetic materials $\boldsymbol{\chi}$ is always negative, independent of the field strength, largely independent of temperature and for the most organic substances is found to be of the order of 10^{-5} (SI units).

The susceptibility derived from the magnetization is described as the volume susceptibility and is dimensionless. The susceptibility related to the mass of material is called the mass susceptibility χ^m,

$$\chi^m = \chi/\rho\,; \qquad (3.26)$$

and has the dimension m^3/kg, ρ is the density of the substance. The molar susceptibility χ^M is related with the mass susceptibility as

$$\chi^M = M\,\chi^m \qquad (3.27)$$

where M is the molar mass and has the dimensions m^3/mol.

The diamagnetic properties of uniaxial mesophases can be described by the two susceptibilities $\chi_{\parallel}$ and $\chi_{\perp}$, which are the resulting susceptibilities when the magnetic field is applied parallel and perpendicular to the molecular long-axis or the director, respectively. The anisotropic diamagnetic susceptibility anisotropy is defined as

$$\Delta\chi = \chi_{\parallel} - \chi_{\perp}$$

For an arbitrary angle between $\mathbf{B}$ and $\hat{\mathbf{n}}$, the total magnetization can be written as

$$\mathbf{M} = \mu_0^{-1}\,[\chi_{\perp}\,\mathbf{B} + \Delta\chi(\mathbf{B}\cdot\hat{\mathbf{n}})\hat{\mathbf{n}}] \tag{3.28}$$

The diamagnetic anisotropy of all liquid crystals having an aromatic ring system has been found to be positive and of the order of magnitude 10^{-7} (SI units). The magnitude is found to decrease with every benzene ring which is substituted by a cyclohexane or cyclohexane derivative ring. A negative anisotropy is observed in purely cycloaliphatic liquid crystals. The diamagnetic anisotropy can be used as an order parameter to describe the degree of orientational order of a mesophase (see chapter 2).

3.2.4 Transport Properties

a. Viscosity coefficients

The resistance of a fluid system to flow when subjected to a shear stress is known as viscosity. It is a measure of the internal friction that arises due to velocity gradients if present in a system. With the help of a simple experiment it is possible to elucidate the viscosity of a fluid. We consider two flat plates separated by a thin fluid layer. If the lower plate is held anchored, it becomes necessary to apply a force in order to move the upper plate at a constant speed. This force is required to overcome the viscous forces due to the liquid. In the regime of the lamina or streamline flow the applied force F is found to be proportional to the area of the plates A and to the velocity of the upper plate Δv, and inversely proportional to the plate separation Δy,

$$F = \eta A\left(\frac{\Delta v}{\Delta y}\right) \tag{3.29}$$

The proportionality constant η is called the viscosity coefficient. When the layers lack the symmetry shown in Fig. 3.6, the velocity gradient $\Delta v/\Delta y$ must be replaced by $\delta v/\delta y$ and eq. (3.29) assumes the form

$$F = \eta A\left(\frac{\delta v}{\delta y}\right) \tag{3.30}$$

and is called Newton's equation.

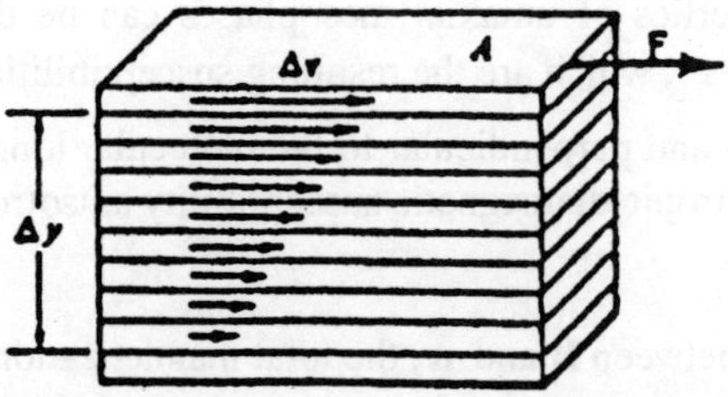

Fig. 3.6. Viscous shear in a fluid.

Many simple isotropic materials can be accurately described by the eq. (3.30). The viscosities may be expressed as either dynamic viscosities η or kinematic viscosities ν; they are related as

$$\nu = \frac{\eta}{\rho} \tag{3.31}$$

As many common nematic liquid crystals have densities in the range 0.98-1.02 g cm^{-3} at 20°C, usually a distinction between ν and η has not to be made.

The viscosity of any material is a collective property resulting from the interaction of the molecules with one another. A strong dependence of the viscosity on the molecular environment is therefore present. The nematic liquid crystal material often appears to flow as easily as a simple Newtonian liquid of similar molecular structure. However, the situation becomes very complicated when the state of molecular alignment in the nematic phase is taken into consideration. In a liquid crystal the viscosity exhibits an anisotropy. For a complete characterization of a nematic liquid crystal, five different viscosity coefficients are necessary. Three of these represent conventional shear flow with different arrangements of the director to the shear direction, while the other two take director rotation or coupling between the director and the flow pattern into account (see Fig. 3.7).

a.1 Translational viscosity

In liquid crystals five anisotropic viscosity coefficients may result, depending on the relative orientation of the director with respect to the flow of the material. When an oriented nematic liquid crystal is placed between two plates which are then sheared, following relevant cases are to be studied (Fig. 3.7),

(i) η_1 : The director is perpendicular to the flow pattern and parallel to the velocity gradient.

(ii) η_2 : The director is parallel to the flow pattern and perpendicular to the velocity gradient.

(iii) η_3 : The director is perpendicular to the flow pattern and perpendicular to the velocity gradient.

These three viscosity coefficients are known as the Miesowicz viscosities. They are difficult to determine. In addition to the coefficients η_1 and η_2 which are antisymmetric with regard to the flow direction and the velocity gradient, a symmetric viscosity coefficient η_{12} is also possible. This refers to the case when director is suspended at an angle of 45° with both the flow pattern and the velocity gradient. η_{12} $(= 2(2\eta_{45^0} - \eta_1 - \eta_2))$ is found to be very small.

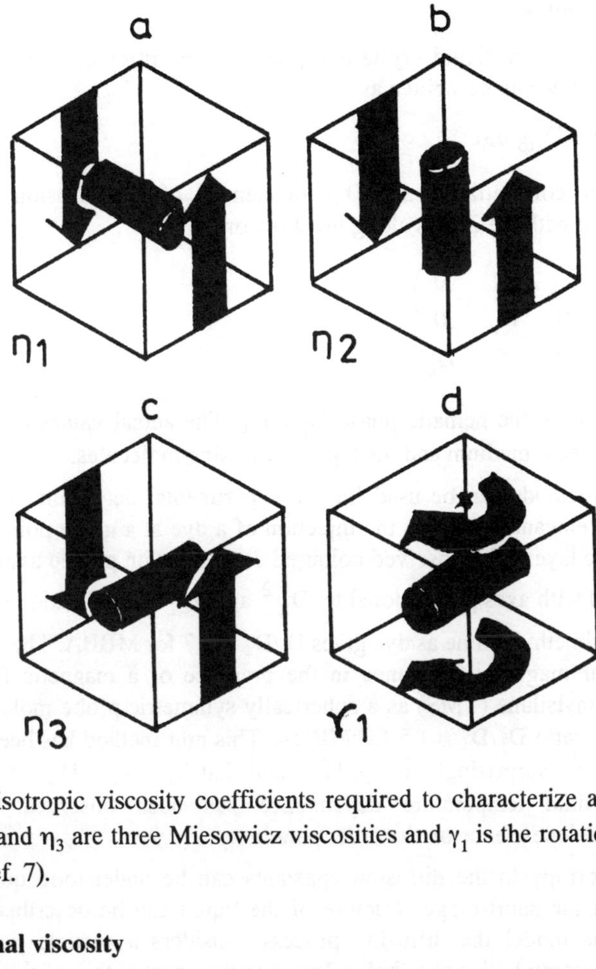

Fig. 3.7. Anisotropic viscosity coefficients required to characterize a nematic. η_1, η_2 and η_3 are three Miesowicz viscosities and γ_1 is the rotational viscosity (Ref. 7).

a.2 Rotational viscosity

When a rotation of the molecule around an axis perpendicular to the director takes place, an additional viscosity coefficient γ_1, known as rotational viscosity, results. This coefficient is very important when the reorientation of the director in an electric or magnetic field is discussed. In LCDs the switching time τ is approximately proportional to $\gamma_1 d^2$ with d representing the cell spacing. The value of γ_1 for technically important nematic materials is found to be in the range 0.02 Pa.s to about 0.5 pa.s. For many compound classes a correlation between the bulk and the rotational viscosity may exist.

Numerous methods can be used to measure the nematic viscosities. For a detail discussion of some of the most important ones the reader is referred to the Ref. 2.

b. Mass diffusion :

The mass diffusion in liquid crystalline phases is anisotropic. According to Fick's law, diffusion flow can be written as

$$J_\alpha = -D_{\alpha\beta}\, \partial_\beta c \tag{3.32}$$

where c is the concentration and $D_{\alpha\beta}$ an element of the diffusion tensor. For a uniaxial system with the Z-axis along the director

$$\mathbf{D} = \begin{pmatrix} D_\perp & 0 & 0 \\ 0 & D_\perp & 0 \\ 0 & 0 & D_\parallel \end{pmatrix} \tag{3.33}$$

It is found that for the nematic phase $D_\parallel > D_\perp$. The actual values of **D** depend on both the mesogenic medium and the type of diffusing molecules.

Various methods can be used for the experimental determination of diffusion constants. An elegant method is the injection of a dye at a given point in a uniform planar nematic layer. The observed coloured concentration profile assumes the form of an ellipsoid with axes proportional to $D_\parallel^{1/2}$ and $D_\perp^{1/2}$. Application of this method with nitroso-dimethylaniline as dye gives $D_\parallel/D_\perp = 1.7$ for MBBA. The application of pulsed nuclear magnetic resonance in the presence of a magnetic field gradient, using tetramethylsilane (TMS) as a spherically symmetric probe molecule, leads to an anisotropic ratio $D_\parallel/D_\perp = 1.5$ for MBBA. This nmr method has been extensively used by Noack[9]. Surprisingly at T_{NI} he found that $D_\parallel$, $D_\perp > D_{iso}$. Furthermore, a strict exponential temperature dependence has been observed, without any indication of a dependence on the order parameter.

The anisotropy in the diffusion constants can be understood qualitatively by assuming that the short-range structure of the liquid can be described by a quasi-lattice. In this model the diffusion process considers a particle jumping into a neighbouring empty lattice site (hole). Ignoring the exact nature of this process D_i is expressed as

$$D_i \sim f_i \exp(-\omega/k_B T) \tag{3.34}$$

where ω is the activation energy needed to create a hole and f_i is a structure factor related to the direct surroundings. In view of eq. (3.34) the temperature dependence of D_i is mainly due to this activation process. For ideal orientational order f_i will be inversely proportional to the area obtained as a result of the projection of a rod-like molecule onto a plane in the relevant direction. These areas can be expressed as

$$A_{\parallel} = \left(\frac{\pi}{4}\right) W^2, \quad \text{and} \quad A_{\perp} = LW.$$

So a maximum anisotropy can be expected of about

$$D_{\parallel} / D_{\perp} = L/W \tag{3.35}$$

For imperfect orientational order ($\overline{P}_2 < 1$), this equation can be expected to be modified analogously to the ratio $<\alpha_{\parallel}> / <\alpha_{\perp}>$ leading to

$$\frac{D_{\parallel}}{D_{\perp}} = \frac{(L/W)(1+2\overline{P}_2) + (2-2\overline{P}_2)}{(L/W)(1-\overline{P}_2) + (2+\overline{P}_2)} \tag{3.36}$$

where L and W are the length and width of a molecule, respectively.

For $L/W = 3$ and $\overline{P}_2 = 0.5$ this gives $D_{\parallel}/D_{\perp} = 1.75$.

In smectic liquid crystals, due to the layered structure, an additional potential barrier ΔU may be present which may hamper the diffusion process perpendicular to the layer. Consequently the ratio $D_{\parallel}/D_{\perp}$ may be expected to have the following form

$$\frac{D_{\parallel}}{D_{\perp}} = \left(\frac{L}{W}\right) \exp\left(-\frac{\Delta U}{k_B T}\right) \tag{3.37}$$

This may lead to a value $D_{\parallel}/D_{\perp} < 1$.

c. Electrical conductivity

The electrical conductivity of liquid crystal is mainly due to residual impurities, the nature of which is often unknown. This is directly related to the impurity diffusion via the Einstein relation,

$$\sigma = \frac{N q^2 D}{k_B T} \tag{3.38}$$

where N is the number of charge carriers with charge q. The electrical conductivity can be measured using the standard methods employed for the dielectric liquids in general. However, these measurements are not trivial due to the problems arising, for example, from electrode reactions, double layers leading to electrode polarization, separation of displacement and resistive currents, etc. The precise measurements of the electrical conductivity of liquid crystals with well controlled concentrations of impurity ions are relatively scarce. No large variations of the ratio

$\sigma_{||}/\sigma_{\perp}$ are observed for the nematics. Since the anisotropy of the conductivity is the driving force for the hydrodynamic instabilities of a nematic layer in an electric field, this ratio is very important.

3.3 ELASTIC CONSTANTS

The order existing in liquid crystals has interesting consequences on the mechanical properties of these materials. They exhibit elastic behaviour. Any attempt to deform the uniform alignments of the directors and the layers structure (in case of smectics only) results in an elastic restoring force. The constants of proportionality between deformation and restoring stresses are known as elastic constants[10,11].

The study of elastic constants in liquid crystals is of great practical significance and is interesting in its own right from the point of view of the understanding. In the first place, they appear in the description of virtually all phenomena where the variation of director is manipulated by external fields (display devices). Secondly, they provide unusually sensitive probes of the microscopic structure of the ordered states. Very useful information regarding the nature and importance of various anisotropies of the intermolecular potentials and of the spatial and angular correlation functions can be derived from the study of the order parameter fluctuations and defect stability in them. In a review[11], the author has provided a summarising overview of the developments in this field.

Let us consider a weakly deformed medium, i.e., the variations in the director orientations (e.g., Fig. 3.8(a), N_u phase) and in the layer dilation (Fig. 3.8(b), S_A phase) only become perceptible above a distance l_0 much greater than a_0 (the mean distance between molecules). These deformations are easily observed optically. If these variations are on the scale of the wavelength of light or even slower the deformed crystal appears locally as a perfect crystal, with fixed order parameters, but with variable preferred direction and layer dilation. This represents the case of long-wavelength deformation. Due to the spatial dependence of directors (and layer dilation) the mesophase gains excess free-energy (elastic or distortion energy) compared with the case of undeformed state. The free-energy of the ordered phase relative to that of undeformed phase of the same density may be written as

$$A = A_u + \Delta A_e \qquad (3.39)$$

where A_u is the free-energy of the uniform phase characterized by spatially invariant directors and undistorted layers and ΔA_e is the elastic (or distortion) free-energy. From the theoretical point of view, the curvature of directors $\hat{\mathbf{n}}(\mathbf{r})$ and $\hat{\mathbf{m}}(\mathbf{r})$ (in case of biaxial phases) and the dilation in layer thickness can be described in terms of continuum theory, the analogue of the classical elastic theory of a solid.

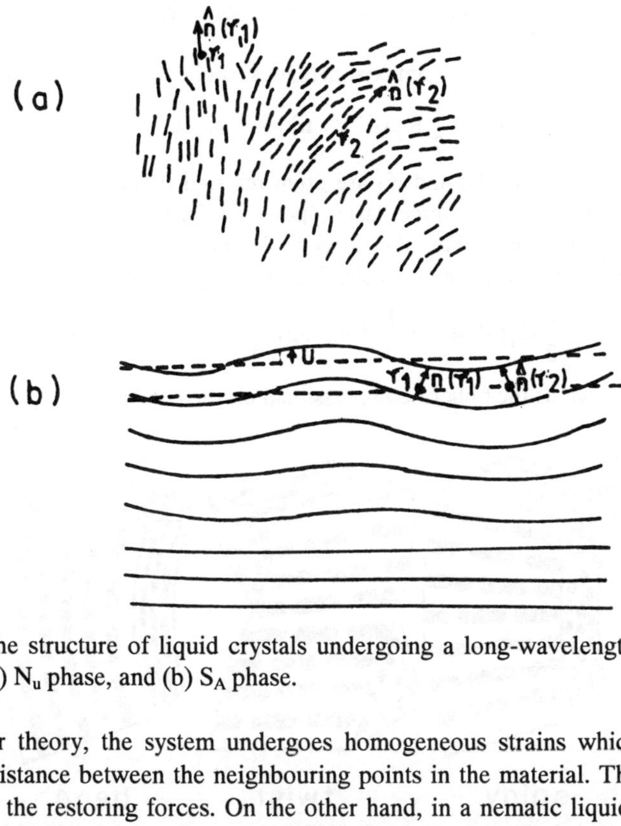

Fig. 3.8. The structure of liquid crystals undergoing a long-wavelength distortion, (a) N_u phase, and (b) S_A phase.

In the latter theory, the system undergoes homogeneous strains which cause the change in distance between the neighbouring points in the material. This change is opposed by the restoring forces. On the other hand, in a nematic liquid crystal, for example, the permanent forces opposing the change of distances between points do not exist. It is the restoring torques which are considered to oppose the curvature. Frank[10] called them as restoring stresses and assumed that these are proportional to the curvature strains. This is, in fact, equivalent to assuming that the free energy density is a quadratic function of curvature strains in which the analogues of elastic moduli appear as coefficients.

Since $a_0/l_0 << 1$ at any point (long-wavelength deformation), i.e., the components of the gradient of the directors obey the inequality

$$a_0 \, \nabla \hat{\mathbf{n}} << 1$$

$$a_0 \, \nabla \hat{\mathbf{m}} << 1 \qquad \text{(only in biaxial phase)}, \qquad (3.40)$$

the medium can be treated as a continuous one and the distortions can be described in terms of a continuum theory disregarding the details of structure on the molecular scale. The idea is that once a director is defined at some point, considering its slow variation with position it can be defined at other points by continuity and ΔA_e can be expressed in terms of elastic constants corresponding to the different modes of deformation.

The elastic free-energy for the uniaxial nematics can be written as (see sec. 5.1),

$$\Delta A_e = \int dr \left[\frac{1}{2}K_1(\nabla\cdot\hat{\mathbf{n}})^2 + \frac{1}{2}K_2(\hat{\mathbf{n}}\cdot\nabla\times\hat{\mathbf{n}})^2 + \frac{1}{2}K_3(\hat{\mathbf{n}}\times\nabla\times\hat{\mathbf{n}})^2\right] \tag{3.41}$$

where the constants K_1, K_2 and K_3 are often referred to as Frank elastic constants and are associated with the three basic types of deformation-splay, twist and bend (Fig. 3.9), respectively. For the smectic-A phase both the twist and bend deformations are prohibited and the elastic free-energy is expressed as (see sec. 5.1),

$$\Delta A_e = \int dr \left[\frac{1}{2}K_1(\nabla\cdot\hat{\mathbf{n}})^2 + \frac{1}{2}B(\partial u/\partial z)^2\right] \tag{3.42}$$

where B is known as the bulk or compressional elastic constant.

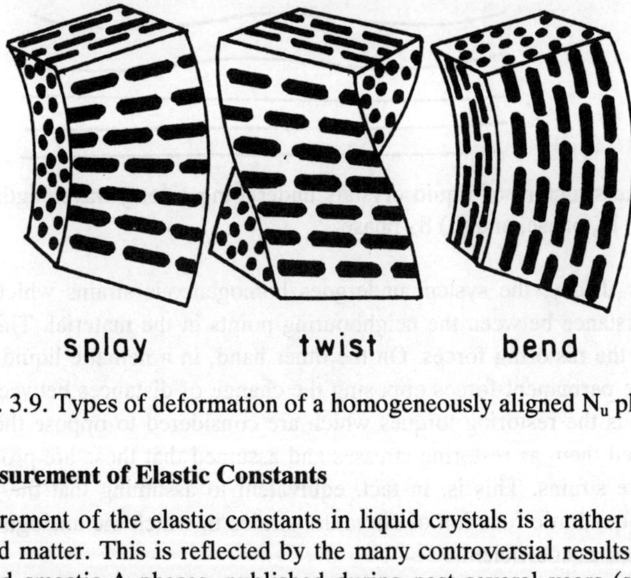

Fig. 3.9. Types of deformation of a homogeneously aligned N_u phase.

3.3.1 Measurement of Elastic Constants

The measurement of the elastic constants in liquid crystals is a rather delicate and complicated matter. This is reflected by the many controversial results for uniaxial nematic and smectic-A phases, published during past several years (see Ref. 11). Some of the important methods used for the measurement of elastic constants in liquid crystals are

(a) Frederiks transition

(b) light scattering

(c) alignment-inversion walls

(d) cholesteric-nematic transition.

The elastic constants K_1, K_2 and K_3 of N_u phase describe, respectively, the torques necessary to obtain the related splay, twist and bend deformations. Their order is about 2-20 pN. The simplest and the most straightforward method is via the observation of threshold voltages V_c (the voltage which must be applied to a cell to obtain a first change of the molecular order) or threshold magnetic fields H_c in different types of Frederiks transitions. The term Frederiks transition refers to the deformation of a layer with a uniform director pattern in an external field. In principle, three configurations are needed[12] (Fig. 3.10) :

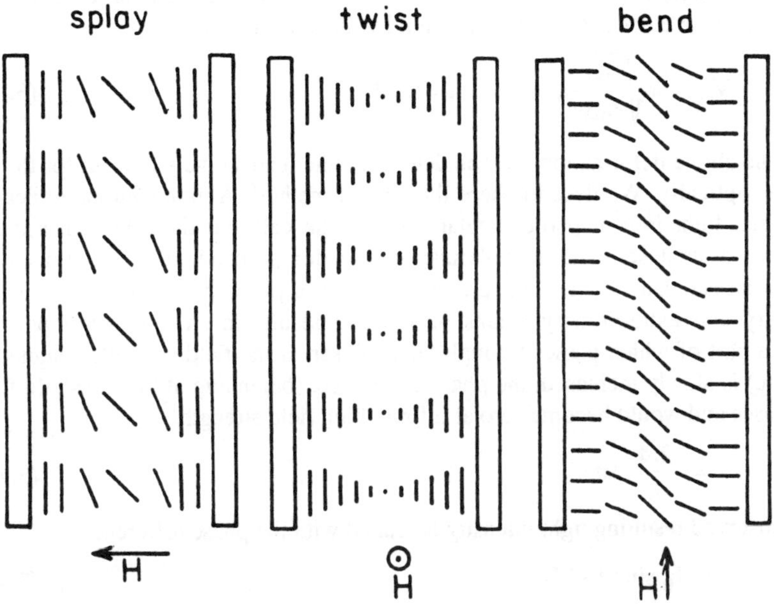

Fig. 3.10. Configurations required to obtain discrete deformations.

(i) A magnetic field, for example, is applied perpendicular to the surface of a homogeneous planar cell. Due to the diamagnetic anisotropy $\Delta\chi$ of the liquid crystal the molecules in the mid-plane of the cell turn parallel into the magnetic field, at sufficiently high field strength, whereas at the orientation layer the molecules stay planar. A splay deformation is induced between the surface and the mid-plane of the molecules.

(ii) If the magnetic field is applied parallel to the surface but perpendicular to the director of the liquid crystal, a twist between the surface and the mid-plane takes place.

(iii) In a homeotropic cell a magnetic field can also be parallel to the surface and perpendicular to the director. In this case a bend between the surface and the mid-plane molecules is induced.

The magnetic threshold field strengths required for the above three geometries of Fig. 3.10 (indicated as the splay, twist and bend modes) are given by

$$H_{c_i} = \frac{\pi}{t}\sqrt{\frac{K_i}{\mu_0^{-1}\Delta\chi}}, \qquad i=1,2,3 \tag{3.43}$$

where t is sample thickness.

For the case (i) with $\Delta\varepsilon > 0$ and case (iii) with $\Delta\varepsilon < 0$, the magnetic field can be replaced by an electric field. The threshold voltage is given by

$$V_{c_i} = \pi\sqrt{\frac{K_i}{\varepsilon_0\Delta\varepsilon}}, \qquad i=1,2,3 \tag{3.44}$$

The above deformations can be detected either capacitively (except for the case (ii)) or optically. We describe here the optical method. A monochromatic beam of polarized light is sent perpendicular through the cell and then through a second polariser as analyser, with the polarization direction perpendicular to that one of the first polariser. Due to the optical anisotropy Δn, a phase difference δ between the ordinary and extraordinary beams occurs and results in an elliptically polarized light, a part of which passes through the analyser. If the field strength (magnetic or electric) is slowly increased, the phase difference, (beginning at the threshold field), decreases and would become zero at infinite high field strength[13],

$$\delta = \frac{2\pi t}{\lambda}\Delta n \tag{3.45}$$

The observed resulting light intensity is related with the phase difference

$$I = I_0 \sin^2(\delta/2), \tag{3.46}$$

where I_0 is the maximum light intensity.

For the magnetic threshold H_{c_1} and the known cell thickness, eq. (3.43) gives the ratio $K_1/\Delta\chi$ as,

$$\frac{K_1}{\Delta\chi} = \frac{H_{c_1}^2\,\mu_0^{-1}\,t^2}{\pi^2}. \tag{3.47}$$

For the electric deformation, K_1 can be evaluated from eq. (3.44)

$$K_1 = \frac{V_{c_1}^2\,\varepsilon_0\,\Delta\varepsilon}{\pi^2}. \tag{3.48}$$

Thus, with a measured $\Delta\varepsilon$, eqs. (3.47) and (3.48) give $\Delta\chi$.

For the field strengths above the threshold, the deformation is not only a splay (Fig. 3.10(a)) but also a bend. Therefore, from a curve of I as a function of field strengh, the ratio K_3/K_1 can be evaluated numerically. With the known values of K_1, according to eq. (3.48), K_3 can be determined. In this method since only one cell is necessary the measurement is simplified.

If the polarization of the incoming light is parallel to, and the analyser is set perpendicular to the cell director, the resulting light intensity is zero in the undeformed state. In the twisted state the polarized light will be guided through the cell and will leave it more or less unchanged. Equation (3.43) gives

$$\frac{K_2}{\Delta\chi} = \frac{H_{c_2}^2\, \mu_0^{-1}\, t^2}{\pi^2} \qquad (3.49)$$

With $\Delta\chi$ determined from eqs. (3.47) and (3.48), K_2 can be evaluated.

From the above discussion, it is obvious that the determination of K_1 and K_2 by the threshold measurements requires in either case cells with parallel wall alignment, i.e., zero bias tilt angle between the nematic director and the cell boundaries.[14] Both the constants can be determined in the same cell of which the wall alignment can relatively easily be prepared. The direct threshold measurements for K_3 have to be performed in cells with homeotropic surface alignment. Keeping in view the high precision and the uniformity requirement, homeotropic boundaries are more difficult to prepare than homogeneous ones. So K_3 is normally determined in tangentially (parallel) aligned cells using either the initial slope of the field-dependent deformation-sensitive quantity used or the dependence of this quantity over the whole range of field-induced deformation. The quantity used to monitor the states of the deformation of the nematic layer is usually either the optical path difference between the ordinary and extraordinary light rays or the capacitance of the layer. The conductivity has also been used to monitor the deflection[15]. The determination of K_3 from the field-induced dependence of the deformation up to large fields requires numerical fitting of the measurements to the theoretical equations which are given in a parametric integral representation.[16,17] The K_3 estimated from the analytic expressions derived for small angle of deformation may lead to rather large errors if is not used properly[18,19]. The direct determination of K_2 from threshold measurements is quite elaborate because conventionally the monitoring of the deformation is done by conoscopic observation of the sample between the pole caps of a magnet.

The K_1 or K_3 can be determined by measuring the variation of birefringence for light incident normal to the film. With linearly polarized light incident and a suitable analyzer, a sudden change in the intensity of the transmitted light is observed at the threshold value of the field. Therefore, a measurement of the H_c gives K_1 or K_3 directly. As the field is gradually increased further, the intensity exhibits oscillations due to change in phase retardation. The threshold for a twist deformation cannot be detected optically when viewed along the twist axis. This is due to the large birefringence (δn) of the medium for this direction of propagation. Thus, for an experimental configuration in which the director is anchored parallel to the walls at either end and light is incident normal to the film, the state of polarization of the emergent beam is indistinguishable from that of the beam emerging from the untwisted nematic. For this reason a total internal reflection technique was used by Frederiks and Tsvetkov[20]; the light beam was allowed to fall at an appropriate angle on the specimen contained between a convex lens and a prism. A simple and more direct method has also been proposed by Madhusadana et al.[21]. When the ellipsoid of refractive index is viewed obliquely[22], the effective δn is reduced to a low value and the twist deformation produces a change in the state of polarization of the emergent beam which can be observed by the usual optical methods.

It is important to mention that if the magnetic field is not strictly perpendicular to the initial undisturbed orientation of the director, the distortion does not set in abruptly and the experimental deformation of elastic constants becomes somewhat unreliable. When the field is exactly applied at right angles, there is equal probability of the director turning through an angle θ or $-\theta$ with respect to the field. Consequently a number of disclination walls dividing the domains having different preferred orientations are formed in the specimen. This serves as a useful criterion for checking the alignment of the field.

The deformation at the threshold field (case (i) and (ii) of Fig. 3.10) may be detected conveniently by measuring the change in the capacitance. One of the essential requirement of this method which becomes a disadvantage is that relatively large areas have to be uniformly oriented. The cholesteric-nematic transition[23a] and alignment-inversion walls[23b] methods can also be used to estimate the values of elastic constant. The limited amount of systematic experimental results available for the elastic constants of nematics suggest that geometrical factors such as length and width of the molecules have a great influence on these properties.

An important question is whether electric or rather magnetic fields should be used for the deflection of the nematic director. The deflection by the electric fields is usually more straightforward than the deflection by the magnetic fields. In the latter case even the accurate measurement of the field at the location of the sample may be a tricky procedure due to field inhomogeneities, temperature dependence of the Hall-probe and so on. However, the advantages of the electric-field-induced alignment may be diminished by effects due to the conductivity of the materials. An exponential increase in the conductivity with temperature takes place which depends very critically on even slight ionic impurities that may contaminate the cell surfaces. Therefore, the conductivity effects often play an important role and have to be accounted properly in the measurements as well as in the evaluation of results.

The bulk elastic constant of S_A phase can be measured directly[24] by imposing a stress σ_{zz} on one of the boundaries of a S_A homeotropic sample and by measuring how it is transmitted to the other boundary. However, in an experiment it is more interesting to impose a displacement δt[25]. The strain $\partial u/\partial z$ is simply given by $(\delta t/t)$ and the transmitted stress by $B(\delta t/t)$. In fact, this should be a static experiment, but the defect motions relax stress and only the short-time behaviour gives B.

The possibility of a Frederiks transition in S_A phase with the application of magnetic field was considered by Rapini[26]. The sample is confined between parallel glass plates with the boundary condition that the molecules at the surface remain parallel to the surfaces. The layers are thus normal to the surface at the glass. The magnetic field **H** is applied perpendicular to the glass plates. It is observed that a transition does occur at a critical field H_c. For $H > H_c$ the distortion increases very slowly with increasing field and it is extremely hard to detect it. This transition can be discussed by assuming a distortion of the form

$$u(x,z) = u_0 \sin k_z z \cos k_x x\,, \tag{3.50}$$

where $k_z = \pi/t$. The elastic free energy averaged over the sample thickness, can be obtained as[22,24]

$$< \Delta A_e > = \frac{u_0^2 k_x^2}{8}\left\{B\left[(k_z/k_x)^2 + k_x^2 \lambda_A^2\right] - \Delta\chi H^2\right\} \tag{3.51}$$

where $\lambda_A = (K_1/B)^{1/2}$.

Instability will occur first for that value of distortion wave vector at which the square bracket in eq. (3.51) is minimum. This optimum value is

$$k_x^2 = k_z/\lambda_A = \pi/\lambda_A t\,. \tag{3.52}$$

It is obvious that the optimal wavelength of the distortion $(2\pi/k_x)$ is equal to the geometric mean of the sample thickness t and of the microscopic length λ_A. In case of cholesterics this has been verified experimentally[27]. But no experiments have yet been performed on smectics. The problem is that in the case of smectics we need to have very large samples, or very large fields. This can be understood by writing the critical field H_c obtained when the elastic and magnetic terms cancel exactly in eq. (3.51). We obtain.

$$\Delta\chi H_c^2 = 2B\left(\frac{k_z}{k_x}\right)^2 = 2\pi\left(\frac{B\lambda_A}{t}\right) \tag{3.53}$$

This equation derived by Hurault[28] shows that H_c is proportional to $t^{-1/2}$. This behaviour is quite different from that of observed conventional Frederiks transition ($H_c \sim t^{-1}$). Here the H_c decreases more slowly with the t. Writing eq. (3.53) as

$$\Delta\chi H_c^2 = 2\pi K_1/\lambda_A t \tag{3.54}$$

and using the estimates $K_1 = 10^{-6}$ dyn, $\lambda_A = 20$ A, t = 1 mm and $\Delta\chi = 10^{-7}$, one gets $H_C \sim 60$ kG which is rather a large value. In order to bring the value of H_C into a more convenient range (20 kG), smectic samples of thickness 1 cm is required which have been prepared. However, one serious problem is that above the H_C the distortion amplitude remains very small.

A similar type of distortion can be created more easily by increasing the separation between the glass plates[29,30] (mechanical means). Taking u to be independent of y the elastic free-energy density, eq. (3.42), reduces to

$$\Delta A_e = \frac{1}{2} B(\partial u/\partial z)^2 + \frac{1}{2} K_1(\partial^2 u/\partial x^2)^2 \,. \tag{3.55}$$

However, a second-order contribution to the layer dilation along Z-axis arises because of the bending of the layers which alters the effective layer spacing in that direction. Hence, eq. (3.55) reads,

$$\Delta A_e = \frac{1}{2} B\left[\frac{\partial u}{\partial z} - \frac{1}{2}\left(\frac{\partial u}{\partial x}\right)^2\right]^2 + K_1\left(\frac{\partial^2 u}{\partial x^2}\right)^2 . \tag{3.56}$$

The displacement u now is given by

$$u = sz + u_0 \sin k_z z \cos k_x x \,, \tag{3.57}$$

where $s = \Delta t/t$ with Δt being the plate displacement. The critical value of strain is given by

$$S_c = 2\pi\lambda_A/t \,. \tag{3.58}$$

The plate displacement $\Delta t = 2\pi\lambda_A \sim 150$ Å, and thus can be easily obtained in an experiment. The above conclusions have been verified experimentally[29,31]. With the help of piezoelectric ceramics the plate separation was increased in a controlled manner. At a critical value of dilation two transient bright spots were observed in the laser diffracted beam which confirmed the onset of a spatially periodic distortion above a threshold strain. A transient periodic pattern was also visible under a microscope. It was found that $k_x^2 \propto t^{-1}$ and in case of CBOOA at 78°C, $\lambda_A = 22\pm3$ Å which gives $B = 2\times10^7$ cgs, if $K_1 \sim 10^{-6}$ dyn.

Some of the elastic constants of S_C phase have been measured by constructing suitable sample geometry. It has been shown by Young et al.[32] that the free-standing film geometry is quite suitable for singling out the hydrodynamic variable Ω_z. In case the wavelengths are large as compared to film thickness (which is always the case in light scattering experiments) free-standing smectics C behave essentially like a two-dimensional nematics[33]. Many interesting experiments have been performed in the distorted phase of cholesterics, chiral smectics, columnar and polymeric mesophases.

3.4. EFFECTS OF CHEMICAL STRUCTURE ON THE PHYSICAL PROPERTIES

The physical properties of mesogenic materials depend on the chemical structure of the molecules. Some of these properties are listed[7] in Tables 3.1-3.3. These values have been obtained by employing an extrapolation technique using host mixture ZLI-3086. From these tables following obvious conclusions can be derived :

(i) An increase in $\in_{\parallel}$ is also accompanied by a related relative increase in $\in_{\perp}$. The multiple polar functional groups in the 3- and 4- position contribute most to the magnitude of $\Delta\in$. For fluorinated compounds intermediate values of $\Delta\in$ are observed. An alkyl or alkoxy group does not influence $\Delta\in$ much. The magnitude of $\Delta\in$ increases due to the additional dipole moments built into the rings. The molecules with increased polarizibility, for example the cyanobiphenyl, have higher values of $\Delta\in$ as compared to the compounds with low polarizabilities (e.g. bicyclohexanes). The influence of the linking groups on the $\Delta\in$ values is noticeable. The variation of $\Delta\in$ in case of three rings system is most complex. A change in the middle ring affects $\Delta\in$, but the concise predictions are not straightforward.

(ii) The variation of the position of the polar substituent within the molecule does not affect Δn much (Table 3.1). The mesogenic compounds composed of benzene rings have higher values of Δn than the respective cyclohexane counterparts (Tables 3.2 and 3.3). The introduction of an alkyl group in between two benzene rings may reduce Δn because the polarizability in the direction of the director is thereby reduced. There are many ways[7] to influence Δn. A very effective means to increase $n_{\parallel}$ and thereby Δn is to employ triple (tolane) bonds. Other available way is the inclusion of an ester group and/or a methoxy group.

(iii) The kinematic viscosity ν listed in Tables 3.1-3.3 has been measured at 20^0C. It is seen that ν is significantly affected by the chemical structure of the molecules. The most pronounced influence is of the core structure and of the linking group. The values of ν are higher for the three rings systems than the two rings compounds.

Table 3.1. Effect of the variation of the position of polar substituent within the molecule on the dielectric, optical and viscous properties of liquid crystals.

Structure	$\varepsilon_{\parallel}$	$\varepsilon_{\perp}$	$\Delta\varepsilon$	n_0	n_e	Δn	ν [$mm^2.g^{-1}$]
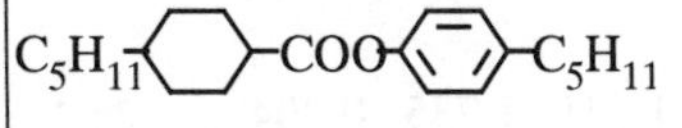	2.6	3.4	-0.8	1.479	1.547	0.068	13
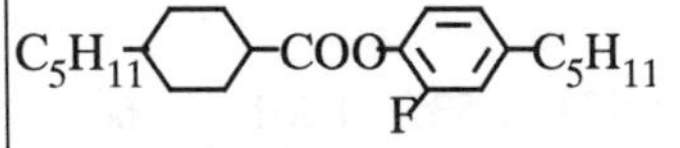	2.9	3.8	-0.9	1.470	1.531	0.061	14
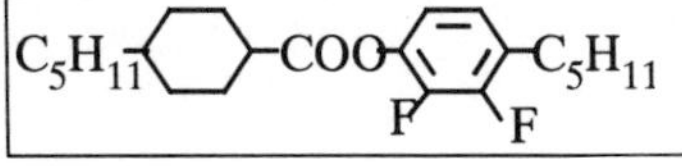	3.5	5.6	-2.1	1.472	1.529	0.057	24

Table 3.2. Effect of the terminal polar substituent and core structure on the dielectric, optical and viscous properties of liquid crystals.

Structure	$\varepsilon_\parallel$	$\varepsilon_\perp$	$\Delta\varepsilon$	n_0	n_e	Δn	ν [mm^2.g^{-1}]
C_5H_{11}-◯-◯(F)-CN	26.5	8.8	17.7	1.491	1.583	0.092	28
C_5H_{11}-◯-◯-CN	17.7	4.8	12.9	1.493	1.605	0.122	22
C_5H_{11}-◯-◯-CF$_3$	16.9	6.0	10.9	1.467	1.507	0.040	8
C_5H_{11}-◯-◯-NCS	15.8	5.0	10.8	1.516	1.691	0.175	12
C_5H_{11}-◯-◯-O-CHF$_2$	13.4	5.8	7.6	1.483	1.541	0.058	7
C_5H_{11}-◯-◯-O-CF$_3$	12.4	5.3	7.1	1.469	1.515	0.046	4
C_5H_{11}-◯-◯-F	8.5	5.3	3.2	1.499	1.523	0.024	3
C_5H_{11}-◯-◯-C$_3$H$_7$	2.6	4.2	-1.6	1.491	1.534	0.043	6
C_5H_{11}-◯-◯-O-CH$_3$	7.0	4.7	2.3	1.494	1.565	0.071	8
C_5H_{11}-◯(N=N)-◯-CN	42.0	8.0	34.0	1.511	1.735	0.224	55
C_5H_{11}-◯(O-O)-◯-CN	25.4	8.0	17.4	1.474	1.615	0.141	47
C_5H_{11}-◯-◯-CN	22.0	5.9	16.1	1.531	1.735	0.204	24
C_5H_{11}-◯-◯-CN	8.2	3.7	4.5	1.474	1.535	0.061	60

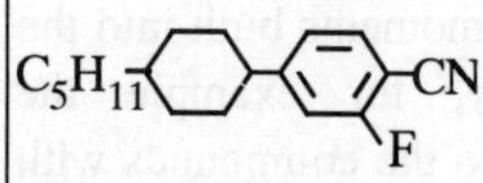

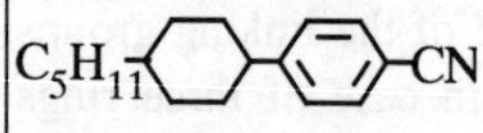

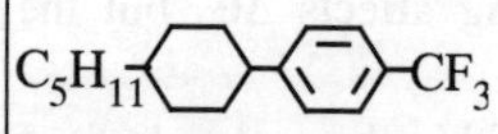

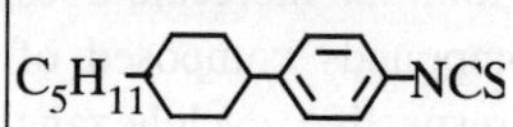

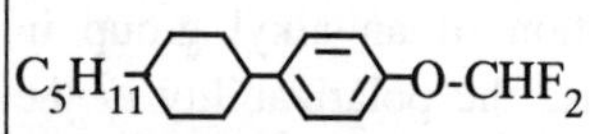

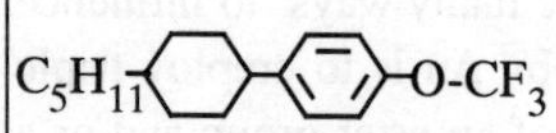

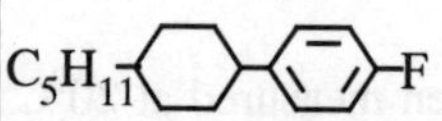

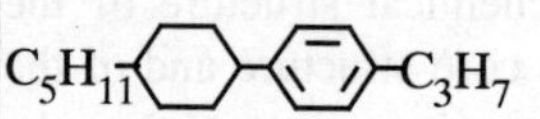

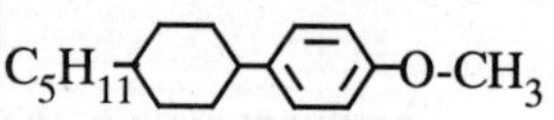

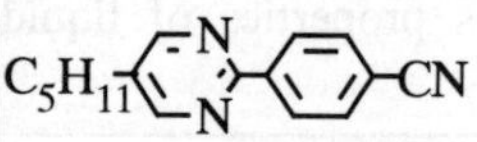

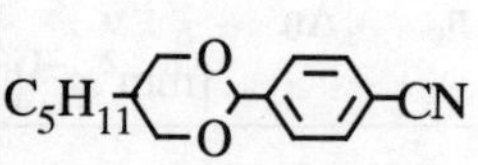

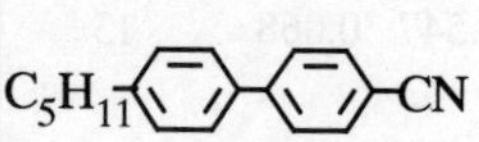

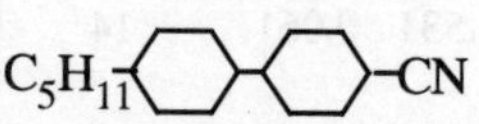

Table 3.3. Effect of the linking group and core structure of three rings system on the dielectric, optical and viscous properties of liquid crystals.

Structure	$\varepsilon_{\parallel}$	$\varepsilon_{\perp}$	$\Delta\varepsilon$	n_0	n_e	Δn	ν [$mm^2.g^{-1}$]
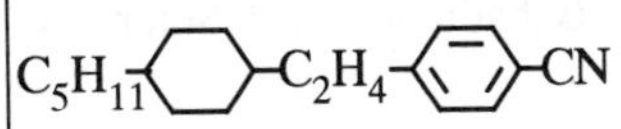	16.7	4.4	12.3	1.498	1.615	0.117	20
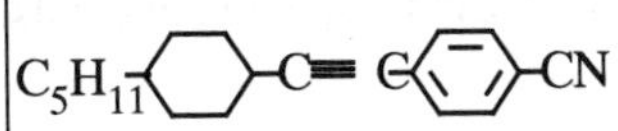	15.9	4.3	11.6	1.499	1.696	0.197	34
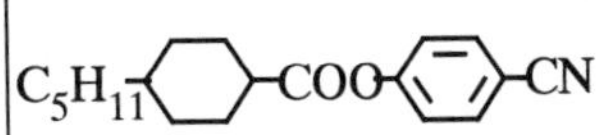	15.0	5.7	9.3	1.480	1.609	0.129	44
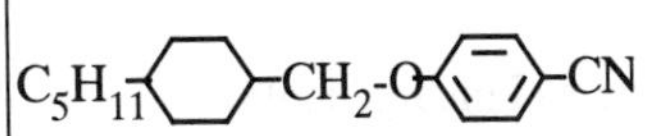	15.2	7.0	8.2	1.489	1.605	0.116	85
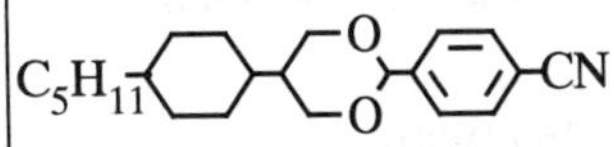	17.1	3.8	13.3	1.463	1.677	0.214	130
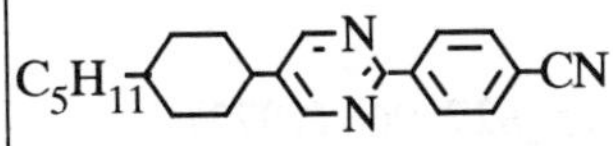	26.1	3.3	23.0	1.500	1.740	0.240	200
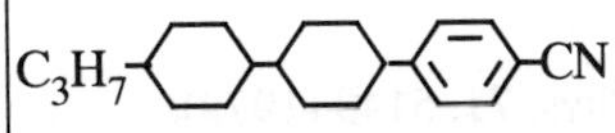	15.4	2.2	13.2	1.483	1.695	0.212	94
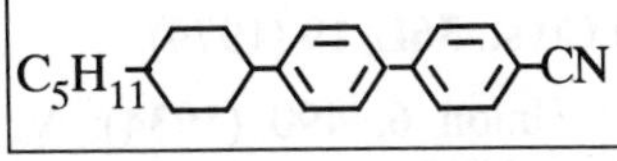	13.3	2.8	10.5	1.512	1.766	0.254	78

REFERENCES

1. F. Yang, and J.R. Sambles, Liquid Crystal **13**, 1 (1993).
2. W.H. de Jeu, Physical Properties of Liquid Crystalline Materials (Gordon and Breach Science Publishers, New York, London, 1980).
3. U. Finkenzeller, T. Geelhaar, G. Weber, and L. Pohl, Liquid Crystalline Reference compounds, Proc. of the 12th Int. Liquid Crystal Conf., Freiburg, West Germany (1988) p. 84.
4. F.G. Smith, and J.H. Thomson, Optics, 2nd ed. (John Wiley and Sons Ltd., Chichester, 1988) p. 59.

5. C.J.F. Böttcher, Theory of Electric Polarization, Vol. I (Elsevier, Amsterdam, 1973); and C.J.F. Böttcher, and P. Bordewijk, Theory of Electric Polarization, Vol. II (Elsevier, Amsterdam, 1978), p. 289.
6. W. Maier, and G. Meier, Z. Naturforsch **A16**, 262 (1961).
7. L. Pohl, and U. Finkenzeller, Physical Properties of Liquid Crystals, in, Liquid Crystals : Applications and Uses, Vol. 1, ed., B. Bahadur (World Scientific, 1990) Chapt. 4, p. 139.
8. G. Vertogen, and W.H. de Jeu, Thermotropic Liquid Crystals, Fundamentals (Springer-Verlag, 1988). Chapt. 10, p. 190.
9. F. Noack, Mol. Cryst. Liquid Cryst. **113**, 247 (1984).
10. F.C. Frank, Disc. Faraday Soc. **25**, 19 (1958).
11. S. Singh, Phys. Rep. **277**, 283 (1996).
12. A. Beyer, and U. Finkenzeller, Merck Liquid Crystals (January, 1995).
13. A. Scharkowski, H. Schmiedel, R. Stannarius, and E. Weißhuhn, Z. Naturforsch. **a45**, 37 (1990).
14. P.R. Gerber, and M. Schadt, Z. Naturforsch **a35**, 1036 (1980).
15. F. Schneider, Z. Naturforsch. **a28**, 1660 (1973).
16. H. Gruler, T.J. Scheffer, and G. Meier, Z. Naturforsch_**a27**, 966 (1972).
17. H. Deuling, Mol. Cryst. Liquid Cryst. **19**, 123 (1972).
18. H. Schadt, B. Scheuble, and J. Nehring, J. Chem. Phys. **71**, 5140 (1979).
19. R.J.A. Tough, and E.P. Raynes, Mol. Cryst. Liquid Cryst. **56L**, 19 (1979).
20. V. Frederiks, and V. Tsvetkov, Phys. Z. Soviet Union **6**, 490 (1934); V. Frederiks, and V. Zolina, Trans. Faraday Soc. **29**, 919 (1933).
21. N.V. Madhusudana , P.P. Karat, and S. Chandrasekhar, in Proc. Int. Liquid Crystal Conf., Bangalore, December (1973); ed., S. Chandrasekhar, Pramana Supplement **1**, 225 (1973).
22. S. Chandrasekhar, Liquid Crystals (Cambridge Univ. Press, Cambridge, 1992).
23. (a) G. Durand, L. Leger, F. Rondelez, and M. Veyssie, Phys. Rev. Lett. **22**, 227 (1969);
    (b) F. Brochard, J. Phys. (Paris) **33**, 607 (1972); L. Leger, Solid State Commun. **11**, 1499 (1972).
24. P.G. de Gennes, and J. Prost, The Physics of Liquid Crystals (Clarendon Press, Oxford, 1993).
25. R. Bartolino, and G. Durand, Phys. Rev. Lett. **39**, 1346 (1977).
26. A. Rapini, J. Phys. (Paris) **33,** 237 (1972); **34,** 629 (1973).

27. F. Rondelez, and J.P. Hulin, Solid State Commun. **10**, 1009 (1972).
28. J.P. Hurault, J. Chem. Phys. **59**, 2068 (1973).
29. M. Delaye, R. Ribotta, and G. Durand, Phys. Lett. **44A**, 139 (1973).
30. N.A. Clark, and R.B. Meyer, App. Phys. Lett. **22**, 493 (1973).
31. G. Durand, in, Proc. Int. Liquid Crystal Conf., Bangalore, December (1973), ed. S. Chandrasekhar, Pramana Supplement **1,** 23 (1973).
32. C. Young, R. Pindak, N.A. Clark, and R.B. Meyer, Phys. Rev. Lett. **40**, 773 (1978); R. Pindak, C. Young, R. Meyer and N.A. Clark, Phys. Rev. Lett. **45**, 1193 (1980).
33. P.G. de Gennes, Symp. Faraday Soc. **15**, 16 (1971).

4

Nematic Liquid Crystals

The nematic liquid crystals are technologically the most important of the many different types of liquid crystalline phases. Their anisotropic ordering and properties have been extensively utilized for a variety of applications such as in display devices, gas-liquid chromatography, separation of geometric isomers, thermal mapping, the determination of the polarization of optical transitions of solutes, optical waveguides, etc. The primary aim of this chapter and the subsequent chapter 5 is to describe some of the most important aspects of the chemistry and physics of the calamitic nematic liquid crystals.

4.1 ESSENTIAL FEATURES OF UNIAXIAL NEMATICS

A schematic representation of the uniaxial nematic phase is shown in Fig. 1.2. The essential features can be summarized as follows :

(a) The molecules are, on the average, aligned with their long-axes parallel to each other. Macroscopically there appears to exist complete rotational symmetry around the axis $\hat{\mathbf{n}}$. This is reflected in all the macroscopic tensor properties.

(b) There is no long-range correlation between the centers of mass of the molecules, i.e., they possess three-dimensional translational symmetry. This is evidenced from the X-ray diffraction pattern in which there is only diffuse scattering. This property determines the fluid character of the phase.

(c) The axis of uniaxial symmetry has no polarity. Although the constituent molecules may be polar, this does not lead to a macroscopic effect. This means that the states $\hat{\mathbf{n}}$ and $-\hat{\mathbf{n}}$ are equivalent.

(d) In an actual sample the orientation of $\hat{\mathbf{n}}$ is imposed by the boundary conditions and possibly also by the external fields. Without special measures the boundary conditions will vary over a substrate. This leads to variations in the director pattern, and thus to differences in birefringence. With a thin layer of a nematic between crossed polarizers under a microscope this is seen as a characteristic pattern or texture (see chapter 11). Threads are often observed in these textures. The threads are discontinuities in a director pattern that, in general, varies continuously over distances much greater than the molecular dimensions.

The duel nature of a nematic phase (liquid-like but uniaxial) are clearly seen in the nmr spectrum, the uniaxial symmetry causes certain line splittings (see sec. 2.3.1) and the fluid like behaviour is reflected in the fact that the lines are relatively narrow as a consequence of rapid molecular motions.

4.2 NEMATICS OF DIFFERENT SYMMETRY

As indicated in sec. 1.4.1 a breaking of rotational symmetry around the director $\hat{\mathbf{n}}$ may lead to the formation of a biaxial nematic (N_b) phase. The different symmetry groups which are subgroups of O(3) are, in principle, admissible; orthorhombic, triclinic, hexagonal or cubic. Contrary to the N_u phase, the rotation around the long-axis in N_b phase is strongly hindered. The N_b phase shown in Fig. 1.2b has been depicted as an orthorhombic fluid whose preferred molecular orientation is described by an orthonormal triad of director field. It is, however, important to mention that an uniaxial phase can exist with the biaxial building blocks and a biaxial phase with the uniaxial molecules.

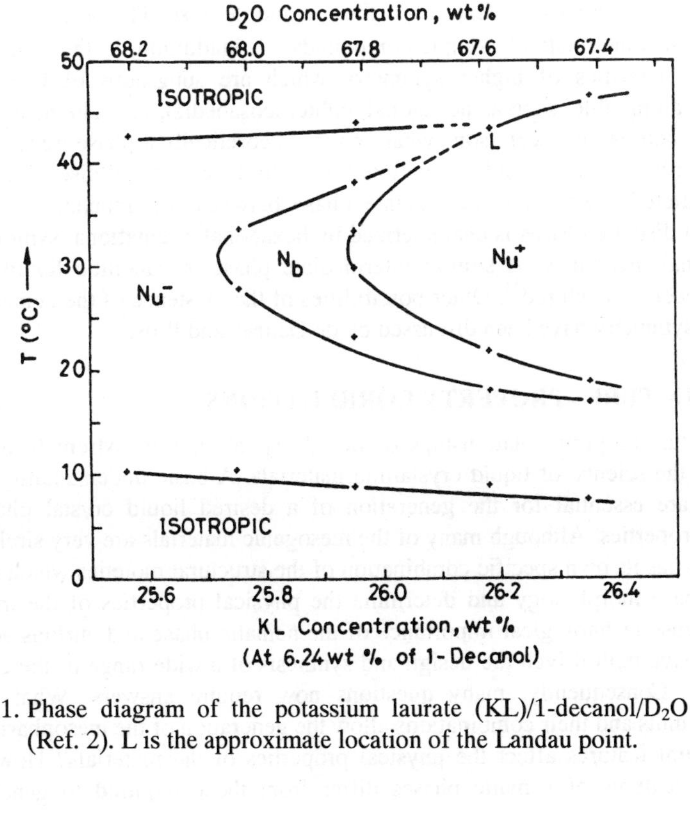

Fig. 4.1. Phase diagram of the potassium laurate (KL)/1-decanol/D_2O system (Ref. 2). L is the approximate location of the Landau point.

The existence of N_b phase was predicted on a theoretical basis by Freiser[1]. He showed that the simplest generation of interaction employed in the Maier-Saupe (MS) theory leads to a first-order uniaxial nematic-isotropic liquid (NI) transition followed, at lower temperature, by a second-order transition to biaxial state. The experimental discovery of the predicted N_b phase has been by Saupe and co-workers[2-4] and others[5,6]. These authors studied the phase diagram and critical properties of the ternary system potassium laurate-1-decanol D_2O over concentration range where nematic phases are likely to occur (Fig. 4.1). It was shown that in the limited concentration range the observed phase sequences on heating/cooling may be isotropic, positive uniaxial nematic (N_u^+), biaxial nematic (N_b) and negative uniaxial nematic (N_u^-). Thus an intermediate N_b phase is formed for a certain concentration range, while in other ranges a direct first-order $N_u^+ N_u^-$ transition occurs. The $N_u^+ N_b$ or $N_u^- N_b$ transition appears to be second-order. A suggestion was made by Chandrasekhar[7] that a thermotropic N_b phase can be prepared by bridging the gap between calamitic and discotic mesogens, i.e., by synthesizing a mesogen that combines the features of the rod and the disc. The N_b phase has been observed in some relatively simple compounds[8–11]. Inaddition, to the orthorhombic N_b phase, nematics of higher symmetry which are subgroups of O(3) are, in principle, admissible, such as hexagonal, cubic, icosahedral, etc. The nematics with these symmetries may seem somewhat exotic. Theoretically a phase transition from a nematic phase with six fold orientational order to the hexagonal discotic phase has been predicted[12]. Such an intermediate phase between the uniaxial nematic and hexagonal discotic phases is characterized by hexagonal orientational symmetry and translational invariance. A similar intermediate phase in calamitic liquid crystals also has been conjectured[13]. Other possibilities of the existence of the nematics with different symmetry have been discussed by de Gennes and Prost[14].

4.3 STRUCTURE - PROPERTY CORRELATIONS

The structure-property relationships of liquid crystals are an extremely important aspect of the science of liquid crystalline materials. A basic understanding of such relations are essential for the generation of a desired liquid crystal phase with specific properties. Although many of the mesogenic materials are very similar (Fig. 1.7), each has its own specific combination of the structural moieties which confer a certain phase morphology and determine the physical properties of the materials. The immense technological importance of the nematic phase and curious academic interests have both driven the design and synthesis of a wide range of the calamitic mesogens. Consequently, many questions now require answers. What type of structural units and their combinations allow the generation of the mesophases? How do structural features affect the physical properties of the materials? How do the structural features of nematic phases differ from those required to generate the smectic phases? How do the structural features vary from one smectic phase to another? What structural features are responsible for the generation of chiral mesophases? What are the roles of core structures, terminal moieties, linking groups and lateral substituents in determining the liquid crystallinity? In this section, we shall examine some of these questions and discuss in detail how the molecular structure and combination of structural units affect the generation and physical properties of uniaxial nematic phase (N_u or N); specific examples of terminal cyano-substituted materials shall be considered. The structure-property correlations of other mesophase types will be discussed in subsequent chapters.

4.3.1 Core Structures

It is difficult to give an absolute definition of a core. However, a core is usually defined as the rigid unit which is constructed from the linearly linked ring units; it also includes any linking groups and any lateral substituents connected to the rings.

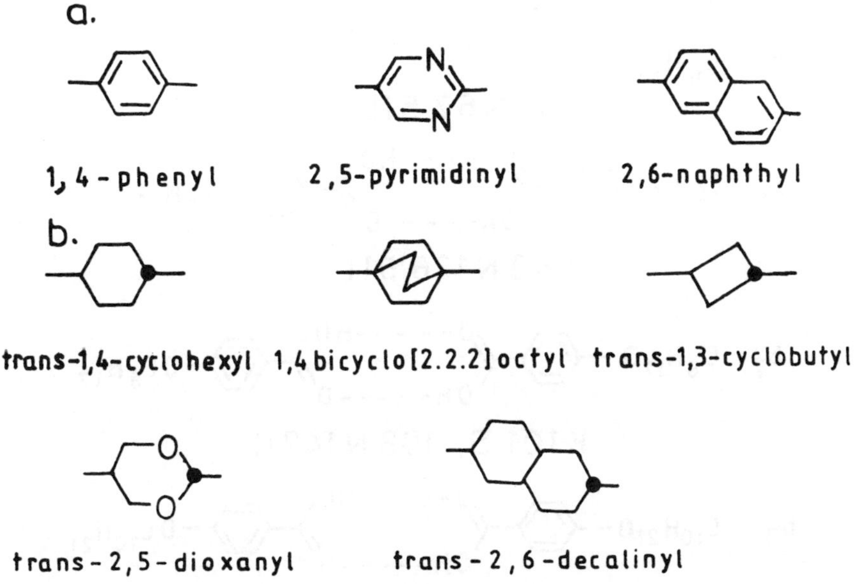

Fig. 4.2. Selected (a) aromatic core units, (b) alicyclic core units.

Most calamitic liquid crystals possess aromatic rings, commonly 1,4-phenyl; 2,5-pyrimidine and 2,6-naphthalene (Fig. 4.2a). The mesophases are also exhibited by many alicyclic materials where the cores are constructed solely of alicyclic rings (Fig. 4.2b). Thus a question arises what kinds of core combinations generate a nematic phase? It has been found that a large number of core combinations give the nematic phase either on their own or in combination with one or more smectic phases. Consequently, it is difficult to make a general rule about the types of cores which can generate the nematic phase or a particular smectic phase. From a carefully chosen comparative examples, following trends become obvious :

If the molecular structure is conducive towards the mesophase formation but some such structural features are present which prevent the lamellar packing (i.e. smectogenic packing), then the nematic phase is likely to be generated. The nematic phase tends to be favoured when the terminal chains are short. For a given core structure, it is very often the terminal chains that dictate the types of mesophase exhibited; in some cases both the nematic and smectic phases are exhibited.

Table 4.1. Effects of dimerization on the mesophase formation.

a.

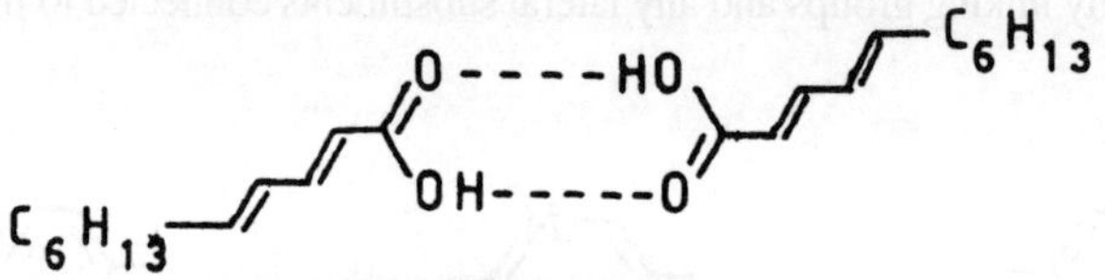

K32 N62.5IL

b_1

K88 N126.5IL

b_2

K101 S_C 108 N147IL

b_3

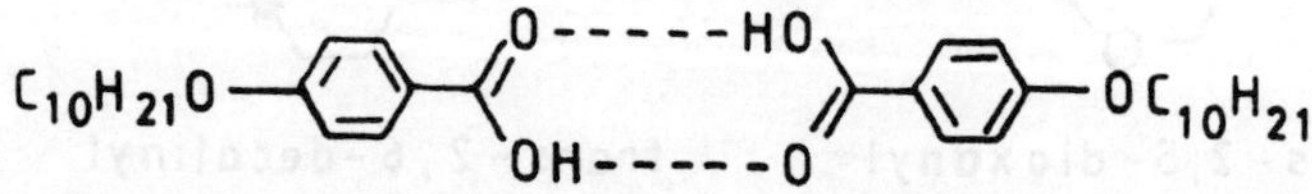

K97 S_C 122 N142IL

Usually the core of a nematogen is made up of two or more 1,4 disubstituted six-membered rings. However, exceptions to this rule has been observed; acyclic and one-ring compounds have been synthesized which exhibit nematic phase upon dimerization (Table 4.1). In these compounds hydrogen bonding is responsible for giving an elongated unit with a rigid central core and two flexible terminal chains. A central linear core is generated due to intermolecular hydrogen bonding and the structure is significantly extended by the two trans-substituted alkenic units at both the sides of the nematogen. Here the actual molecular structure is certainly not long and lath-like and hence would not be expected to exhibit mesomorphism, but a long lath-like structure is created due to dimerization through the hydrogen bonding. Why do the compounds T4.1a and T4.1b1 not exhibit a smectic phase also or solely a smectic phase? Here the compound given in Table 4.1 serial number a is referred as compound T4.1a. There are many reasons which prevent the molecular packing in a layer-like arrangement; accordingly only the nematic phase is exhibited. The terminal chain may be short enough relative to the length of the rigid core. The molecular shape often leads to steric problems for lateral attractions, the presence of polar groups must cause repulsion but polarity is often necessary to form smectogens because the polar interactions stabilise the lamellar packing arrangement of the molecules. The lamellar packing is stabilized by longer terminal chains. Compounds T4.1b2 and T4.1b3 have two long terminal alkoxy chains and generate nematic as well as smectic C phases.

The effects of the core changes upon the nematic phase stability (T_{NI}) and the nematic range of some two-ring and three-ring systems have been shown in Tables 4.2 and 4.3, respectively. From a careful comparative study of the listed materials with the 'benchmark' cyanobiphenyl (compound T4.2a) some trends become reasonably clear. The use of a heterocyclic pyrimidine ring (T4.2b) gives a moderate increase in T_{NI} and nematic range. It has been conjectured that this increase may be due to the removal of the steric hindrance in the inter-ring-bay region from the protons of the compound T4.2a and hence the two-rings can adopt a planar arrangement without interannular twisting of the parent system. This gives an enhanced longitudinal polarisability and hence higher values of T_{NI} and nematic range.

The influence of the alicyclic core changes on the stability of the nematic phase can be seen in compounds T4.2c-T4.2h. In these materials one phenyl ring has been replaced by a non-aromatic ring (for example, cyclohexane, bicyclooctane, dioxane). It can be seen that the use of a trans-cyclohexane ring (compound T4.2c) gives an increase of 20°C in T_{NI} and 24°C in nematic range as compared to the parent compound T4.2a. Further the values of the T_{NI} and nematic range are increased by 65°C and 35°C, respectively, when a phenyl ring is replaced with a bicyclooctane ring (compound T4.2e). These results do not support the long held early belief that the increase in the values of the T_{NI} and the nematic range are related with the increase in the value of anisotropic polarisability.

Table 4.2. Effects of core changes on the transition temperatures of some two-ring nematogens[15-22]. The dot • at one of the bonding sites of the rings denotes the trans-isomer.

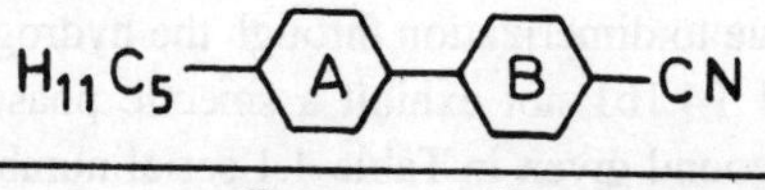

S.No.	A B	Transition temperatures (°C)
a.	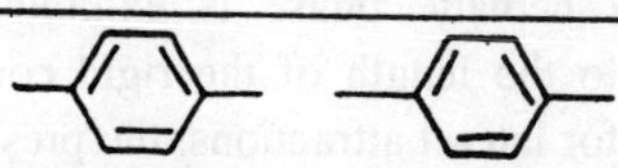	K 24 N 35 IL
b.	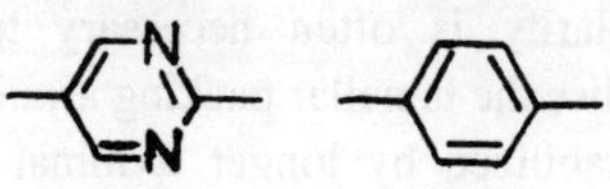	K 71 (N 52) IL
c.		K 31 N 55 IL
d.	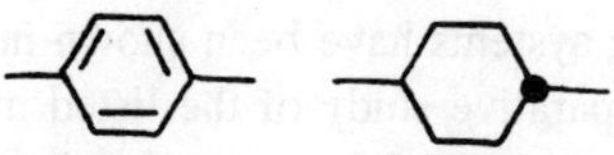	K< 20 (N-25) IL
e.		K 62 N 100 IL
f.	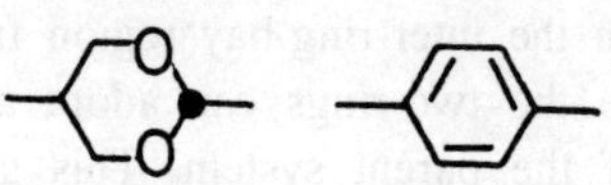	K 56 (N 52) IL
g.	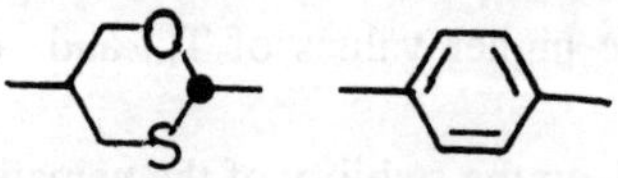	K 74 (N 19) IL
h.	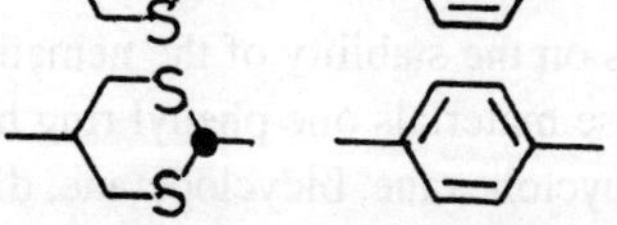	K 98 IL
i.	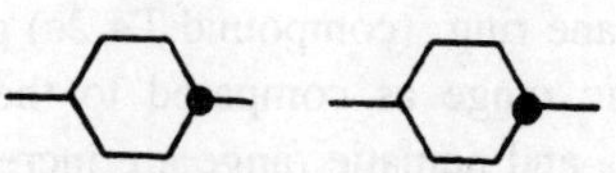	K 62 N 85 IL
j.	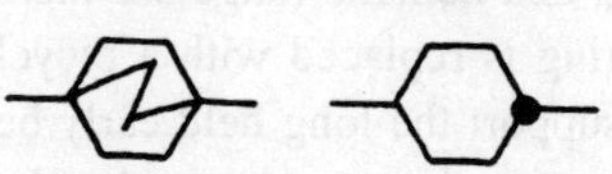	K 104 N 129 IL

Table 4.3. Effects of core changes on the transition temperatures of some three-ring nematogens[20-24].

a.	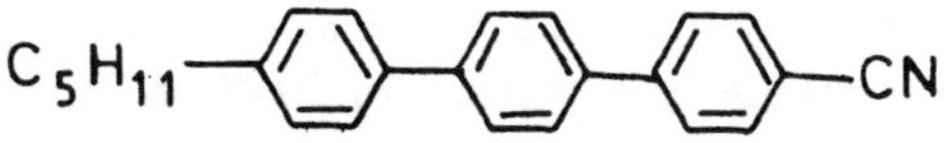	K130N239IL
b.	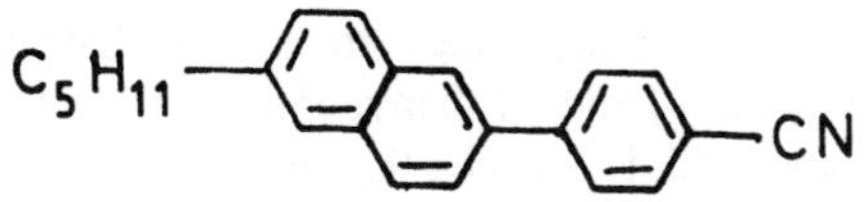	K84N126.5IL
c.	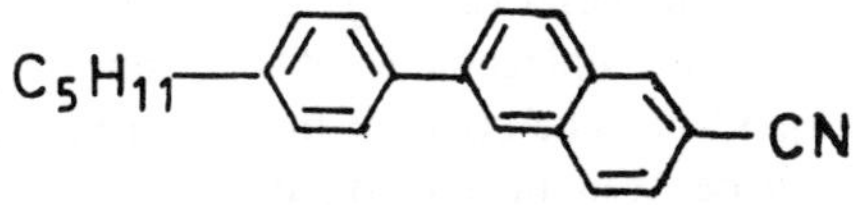	K68N130IL
d.	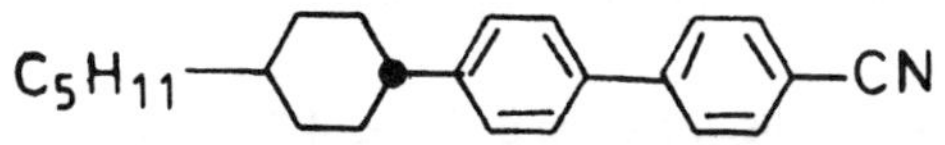	K96N222IL
e.	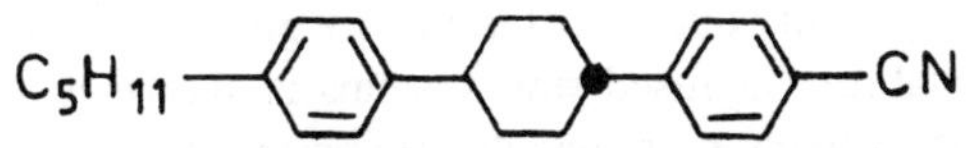	K80N160IL
f.	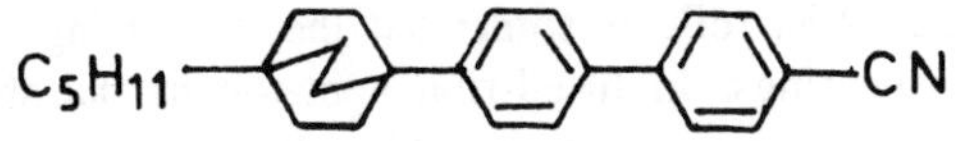	K159N269IL

However, the anisotropic polarisability still continues to play an important role. Why does the replacement of a phenyl ring by an alicyclic ring enhance the stabilization of the nematic phase? Perhaps it has to do with the ability of alicyclic rings to pack in an efficient, space filling manner while maintaining the liquid crystalline order. The use of two oxygen (dioxane, compound T4.2f) gives a little difference in T_{NI} but reduces the nematic range significantly as compared to the parent cyclohexane (T4.2c). The longitudinal molecular packing is disrupted in case of compound T4.2g because of increased molecular breadth. The T_{NI} value is reduced despite the increase in polarisability but the nematic range is incresed appreciably. The two-sulfurs system (compound T4.2h) does not exhibit a nematic phase, despite substantial supercooling, because the molecular packing is disrupted due to large size of the two sulfurs. It is important to mention here that only the trans-isomer of the 1,4-disubstituted cyclohexanes possess linear structures and hence exhibit mesophases. The isomeric-cis materials have a bent molecular shape which does not favour a mesogenic molecular packing. The replacement of both the phenyl rings with trans-cyclohexane rings (compound T4.2i) causes significant reduction in polarisability but further stabilizes the nematic phase. Now if one of these cyclohexane rings is replaced with a bicyclooctane ring (compound T4.2j) the T_{NI} and the nematic range are further increased. The effects of the end-group conjugation[17,21] on the nematic phase stability can be seen in compounds T4.2c and T4.2d. The T_{NI} is drastically lowered when a phenylene ring is separated from linking or end groups.

Addition of one more aromatic ring (compound T4.3a) to the compound T4.2a extends the linear core and hence increases the length to breadth ratio and anisotropic plarisability. As a result a large increase of 204°C in T_{NI} and 109°C in the nematic range is observed. In case of fused-ring systems (compounds T4.3b and T4.3c) the values of the T_{NI} and the nematic range are intermediate between the respective values of biphenyl unit (T4.2a) and terphenyl compound (T4.3a). The use of the cyclohexane ring (compound T4.3d) reduces the nematic phase stability and its range as compared with the parent material (T4.3a). A reverse trend is seen in compound T4.3f; in this material the replacement of one phenyl ring by a bicyclooctane ring increases the T_{NI} value. The effect of the inter-ring conjugation[20] on the nematic phase stability can be seen in compound T4.3d and T4.3e. The values of the T_{NI} and the nematic range are drastically lowered when the phenyl rings are not in conjugation.

It can be seen from the above discussion and Tables 4.2 and 4.3 that the general order of the relative nematic stability due to different core groups is as shown in Fig. 4.3a. The order of phase stability in fused ring systems is shown in Fig. 4.3b. A comparative study shows the effects of molecular shape and polarisability on the phase stability. The sequence 4.3a reflects somewhat the decreasing rigidity and increasing polarisability of the rings. In fused-ring nematogens the shape of the partially saturated systems dominates, that is, the polarisability increases as compared to the fully unsaturated ring due to partial hydrogenation of one of the fused rings. The resulting structure is neither linear nor fully staggered and hence the molecules cannot pack efficiently in the nematic phase.

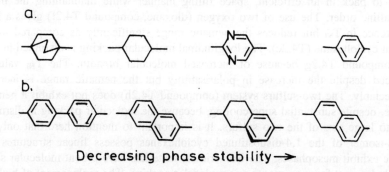

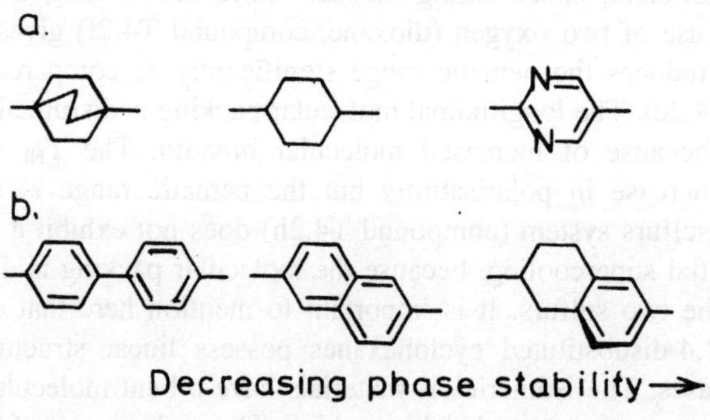

Fig. 4.3. The relative phase stability due to (a) different core groups, (b) fused-ring core groups.

The effects of the core changes on the physical properties (viscosity, dielectric anisotropy and birefringence) are shown in Table 4.4. The physical properties play very important role in selecting a material for applications. The viscosity should be low in order to enable fast switching at low voltage. In a twisted nematic display the dielectric anisotropy is required to be reasonably high for the fast switching at low voltage. It is the terminal cyano substituent which is responsible for the value of $\Delta\in$ but the core plays a very important role. The ratio $\Delta\in/\in_\perp$ is required to be low in case of some complex displays using multiplexable nematic mixtures. For the electrically controlled birefringent display a nematic material is required to have a negative value of $\Delta\in$ and low viscosity. A moderately high birefringence of 0.1 to 0.2 is necessary for the twisted nematic displays whereas a low birefringence (< 0.1) is beneficial for the cholesteric-nematic phase-change device and some supertwist devices. In the switching process the ratio of elastic moduli (K_3/K_1) plays a very important role and this ratio should be low.

Table 4.4. Effects of core changes on the viscosities, dielectric anisotropies and birefringence of some terminal cyano nematogens.

	η(cp) at 20°C	$\Delta\epsilon$	Δn
a. 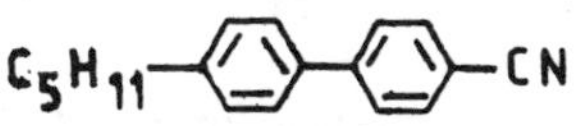	35	+8.5 at $\Delta T = 0.98$	0.18
b. 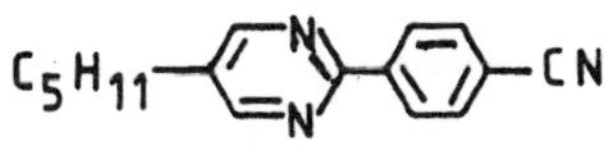	50	+18	0.18
c. 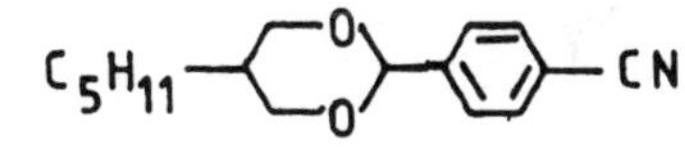	46	+11 at 42°C	0.08
d. 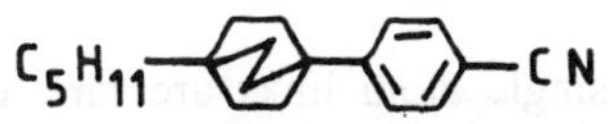	90	+10 at 25°C	0.14
e. 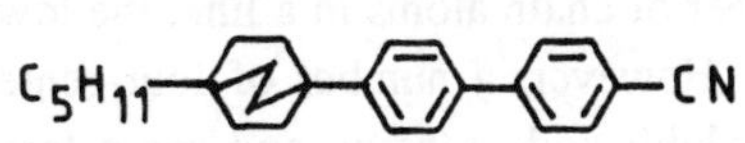	90	+13.3	0.14

4.3.2 Linking Groups

The structural units, other than a direct bond, which connect one part of a core to another are known as the linking or bridging groups. These groups can also be used to link the terminal chains to the core structure. The use of a bridging group introduces flexibility in the material and hence is a useful method for reducing the transition temperatures and destabilizing the mesophase. However, in many cases it may influence the thermal, chemical or photochemical stability or coloration[22a]. As the linking group provides a point of link up in the synthesis of a material, it is often easier to synthesize a material with linking groups than materials with direct bonds[25]. A number of bridging groups have been used in the synthesis of mesogenic compounds (see Table 4.5).

Table 4.5. Various types of linking groups used in mesogenic materials.

X–N=N–Y	X–C(=O)–O–Y	X–CH₂–O–Y
azo	ester	methyleneoxy
X–CH₂–CH₂–Y	X≡Y	X–CH=N–Y
dimethylene	acetylene	imine (Schiff's base)
X–CH=CH–Y (trans)	X–CH=CH–C(=O)–O–Y	
ethylene	cinnamate	

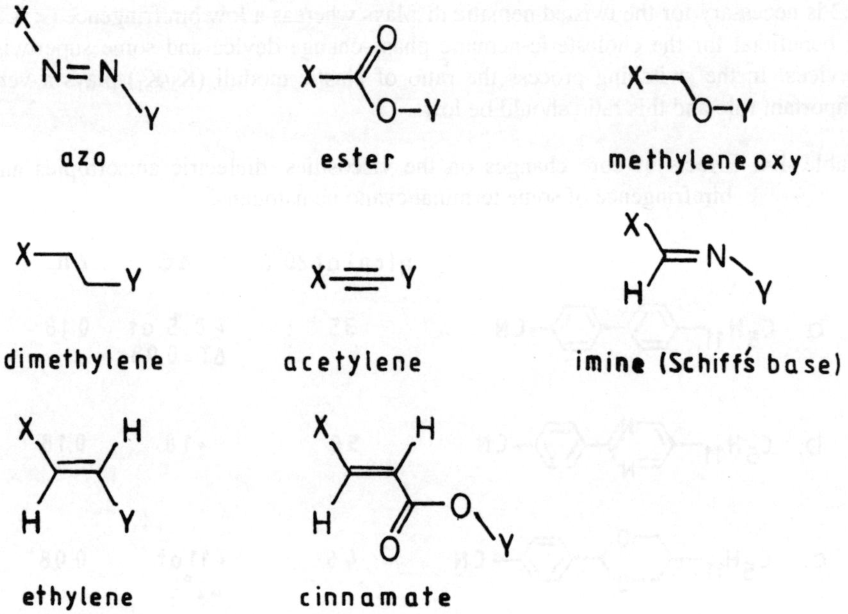

Examples of useful liquid crystals with single atom links are rare and their phase ranges and stabilities are limited[26]. The number of atoms in the link also affects the flexibility; the more the number of chain atoms in a link, the lower is the population of the linear conformations. However, a number of four-center linked mesogens have been prepared which exhibit wide ranging and room temperature nematic phases[27].

The introduction of a linking group into a mesogen increases the over all molecular length and the polarizability anisotropy and hence may improve the stability and range of mesogen. The main factors that govern the effects of linking groups on a mesogens are the geometry and the flexibility of the link, and the compatibility of the link with the core in terms of polarisability and π-conjugation. The links are required to have a favourable geometry in which both the ends must be at least parallel and preferably colinear.

The effects of some selected linking groups on the transition temperatures of nematogens are shown in Table 4.6. When compared with the cyano-substituted parent system (compound T4.6a), compounds T4.6b-T4.6d, with azo(–N=N–) and imine (Schiff's base –CH=N–) linking groups, exhibit much higher phase stability and melting points. These linking groups confer a stepped core structure in which the linearity is maintained, the broadness tends to disrupt lamellar packing. The ester(–CO_2–) linking group is a versatile and most commonly used linking unit in liquid crystals because it is relatively stable, easily synthesized and can generate liquid crystals with low melting points. It is a planar linking group with a degree of polarisability due to π electron associated with the carbonyl group. It is not a completely conjugative unit, and hence can be used to connect two aromatic units or an alicyclic unit to an aromatic unit. In comparison with the parent analogue (compound T4.6a) the presence of ester unit in compound T4.6e has clearly increased the nematic phase stability but the melting point has risen by 40°C; consequently only a monotropic nematogen is generated. The ester link, while maintaining the linearity, confers a stepped structure. The ester group increases the polarisability which may be responsible for the high melting point. The saturated tetrahedral linking groups (compounds T4.6f and T4.6g) are incompatible with the other structural combinations and hence very low nematic phase stability results in both cases. When these groups (ester, methyleneoxy, dimethylene) are used to link a trans-cyclohexane ring to a phenyl ring somewhat a different picture emerges. Since the ester group in compound T4.6h extends the molecular length and increases polarisability, a high value of the T_{NI} is observed. In compounds T4.6i and T4.6j the tetrahedral dimethylene and methyleneoxy groups do not separate the polarisable regions and accordingly the T_{NI} values are not low. The material T4.6j is an enantiotropic nematogen. The ethylene linkage (stilbene) in compound T4.6k is fully conjugative group that increases the longitudinal polarisability and extends the molecular length while preserving the linearity; accordingly a very high T_{NI} value results. However, such linkage may lead to an instability. The tolane (acetylene) linkage (compound T4.6l) maintains the rigidity, linearity and polarisability of the core and extends the molecular length. The T_{NI} value is lower than the stilbene and a small nematic range is exhibited because the conjugativity of the linkage is less effective and the polarisability is lower.

Table 4.6. Effects of some selected linking groups on the transition temperatures of nematogens.

a. C_5H_{11}—⟨☐⟩—⟨☐⟩—CN   K 24 N 35 IL

b. C_5H_{11}—⟨☐⟩—CH=N—⟨☐⟩—CN   K 46.4 N 75 IL

c. C_5H_{11}—⟨☐⟩—N=N—⟨☐⟩—CN   K 89 (N 86.5) IL

d. C_5H_{11}—⟨☐⟩—N=N(O)—⟨☐⟩—CN   K 91 N 126 IL

e. C_5H_{11}—⟨☐⟩—COO—⟨☐⟩—CN   K 64.5 (N 55.5) IL

f. C_5H_{11}—⟨☐⟩—O—⟨☐⟩—CN   K 49 [N −20] IL

g. C_5H_{11}—⟨☐⟩—CH$_2$—⟨☐⟩—CN   K 62 [N −24] IL

h. C_5H_{11}—⟨◯⟩—COO—⟨☐⟩—CN   K 48 N 79 IL

i. C_5H_{11}—⟨◯⟩—O—⟨☐⟩—CN   K 74 (N 49) IL

j. C_5H_{11}—⟨◯⟩—CH$_2$—⟨☐⟩—CN   K 30 N 51 IL

k. C_5H_{11}—⟨☐⟩—HC=CH—⟨☐⟩—CN   K 55 N 101 IL

l. C_5H_{11}—⟨☐⟩—C≡C—⟨☐⟩—CN   K 79.5 (N 70.5) IL

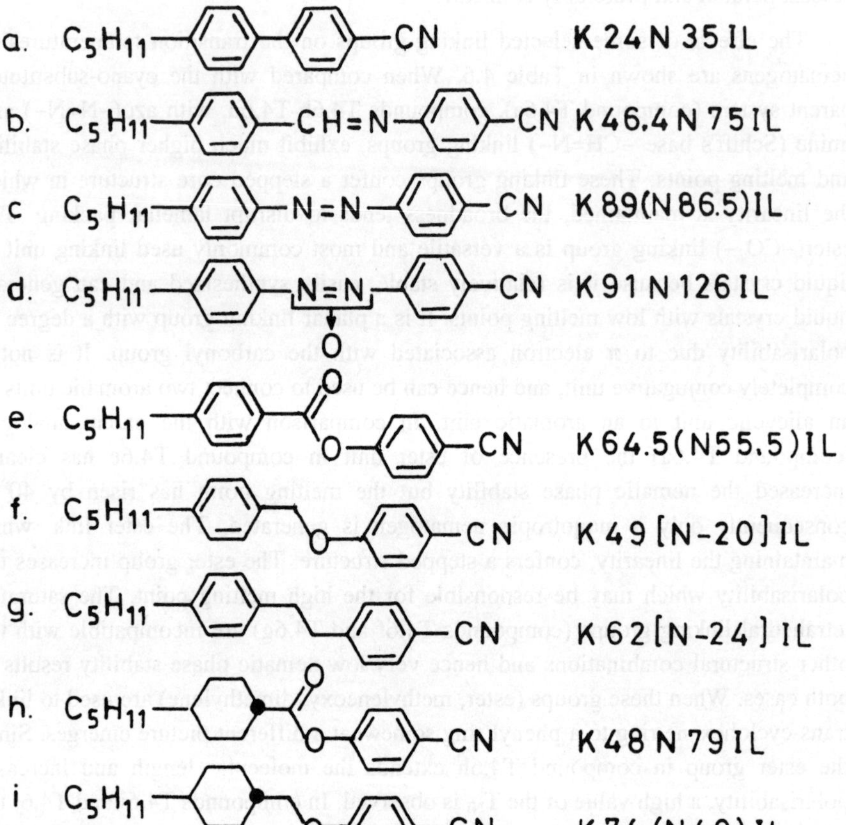

4.3.3 End Groups

The end groups, other than hydrogen, are virtually always used in the generation of a liquid crystal material. Since the calamitic systems have two ends, the end groups may be different or similar. Many terminal units have been employed in the generation of liquid crystals but the most successful route[25] is to use either a small polar substituent (most important one is the cyano group) or a fairly long, straight hydrocarbon chain (usually alkyl or alkoxy). The role of these groups is to act as either a flexible extension to the core or as a dipolar moiety to introduce anisotropy in physical properties. Further, the terminal moieties are believed to be responsible for stabilizing the molecular orientations essential for the generation of a mesophase. The physical properties are also strongly dependent on the terminal units.

The effects of the end groups on the nematic phase stability can be seen in Tables 4.7 and 4.8. Table 4.7 shows the transition temperatures of seven alkylcyanobiphenyl homologues. The core here is the 4,4′-disubstituted biphenyl unit which normally does not confer high mesophase stability. It can be seen that although all the compounds have the same core yet considerable variations in the transition temperatures are observed with the length of the end chains. The longer alkyl chains extend the molecular length and interact with one another which enhances the nematic phase stability and promotes smectic phase formation. Contrary to it the flexibility of the end chains tends to disrupt the molecular packing required for a mesophase generation. However, such flexibility tends to reduce the melting point. The rigidity of the compound T4.7a is very high due to its short end chains. As the length of the end chains increases the melting point reduces considerably, but due to increased van der Waals attraction the melting points increase. As a result for a low melting point a compromise of the chain length is required. With increasing chain length the T_{NI} decreases but again starts increasing when the terminal chains become very long. However, in many materials an odd-even effect is observed; the odd membered chains generate higher T_{NI} values than the even membered chains.

The alkoxycyanobiphenyls exhibit a similar trend[25] as that of alkyl-substituted analogues. However, throughout the series higher melting points and T_{NI} values are exhibited by the alkoxy compounds. A significant effect on the mesomorphic phase behaviour is observed due to chain branching which helps in introducing chirality into a molecule. However, it causes a disruption in the molecular packing and hence often reduces melting points and phase stability. The position of the end chains also play very crucial role in the generation of liquid crystal phases. Compound T4.8a has a high T_{NI} value whereas the compound T4.8b does not exhibit liquid crystal phases. A high smectic phase stability is shown by the compound T4.8c but the analogous oxygen containing compound T4.8d exhibits a nematic phase of low stability.

Table 4.7. Transition temperatures for alkylcyanobiphenyl homologues.

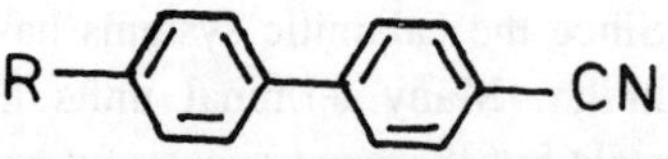

No.	R	Transition temperatures (°C)
a.	CH_3	K 109 N 45 IL
b.	C_2H_5	K 75 N 22 IL
c.	C_3H_7	K 66 N 25.5 IL
d.	C_4H_9	K 48 N 16.5 IL
e.	C_5H_{11}	K 24 N 35 IL
f.	C_6H_{13}	K 14.5 N 29 IL
g.	C_7H_{15}	K 30 N 43 IL

Table 4.8. Effect of the position of the end chains on the transition temperatures.

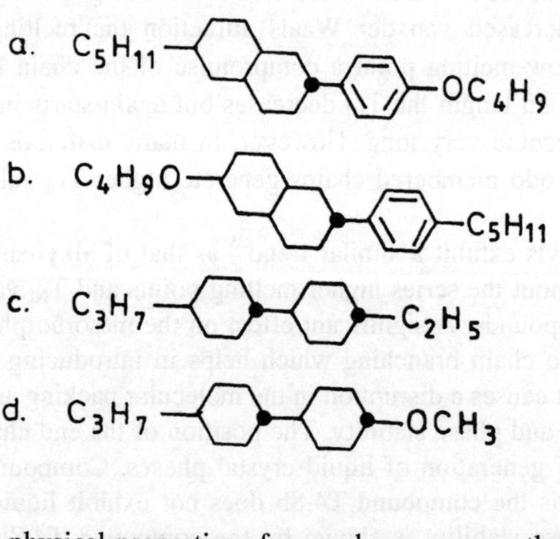

a. K 70.2 N 96 IL

b. K 37.5 IL

c. K -5 S_B 67 IL

d. K 10 N 17 IL

The physical properties of mesophases are greatly influenced by the choice of end groups. A polar terminal unit, for example the cyano group, provides a positive dielectric anisotropy whereas in case of a nonpolar alkyl chain it is the other structural features which determines the sign and value of the dielectric anisotropy. The viscosity of a material is influenced by the steric and polarity factors of the end groups. The effects of the size of the end group on the physical properties of four terminal halo-mesogens[28] are shown in Table 4.9. It is seen that as the size of the end group is increased the mesophase becomes more viscous and the values of the melting point and the T_{NI} also are increased. It has also been found that the use of long end chains and branched terminal chains gives an increase in viscosity because of their steric influence with each other. The polar oxygen within the alkoxy chains and carbonyl group further increase the viscosity.

Table 4.9. The influence of the size of the end group on the physical properties of some terminal halo-mesogens[28].

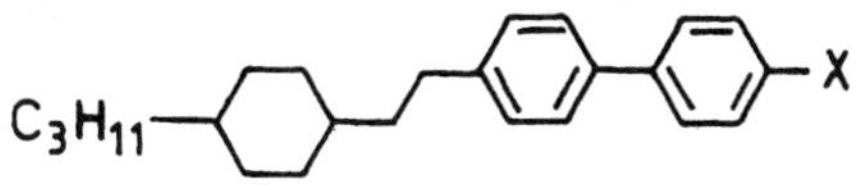

X	Transition temperatures (°C)	Viscosity (cp) 20°C
H	K67 N82 IL	22.7
F	K76 N125 IL	25.1
Cl	K100 N158 IL	46.6
Br	K125 N163 IL	63.0

4.3.4 Lateral Groups

A lateral group is one that is attached off the linear axis of the molecule, usually on the side of the aromatic core (see Fig. 1.7). A number of different lateral groups (e.g., F, Cl, Br, I, CN, NO_2, CH_3, CF_3, C_2H_5, C_6H_{13}, C_7H_{15}, etc.) have been used in the synthesis of mesogenic materials. In recent years, many useful materials have been synthesized by the suitable choice of the lateral substituents.

The physical properties of nematic mesogens are strongly influenced when a lateral group is appended to a nematic core. Considering the steric properties, a lateral substituent effectively widens the core and increases the intermolecular separation. This leads to a reduction in the lateral interactions[29] and hence the nematic phase stability is reduced. The electronic effects, the parallel and perpendicular components of anisotropic properties, depend upon the angle between the core and the substituent. The lateral groups may increase either the parallel component, or the perpendicular component, or both, to a greater or lesser extent. As an example, the effects of lateral cyano group on the physical properties of nematogens have been shown in Table 4.10. The most significant influence is seen when the lateral and terminal nitriles are mutually ortho. A large increase in the value of dielectric anisotropy is observed in this case, presumably because the lateral substituent sterically hinders some of the antiparallel correlation and adds to some degree to the effect of end group. The meta nitrile reduces the dielectric anisotropy but increases the melting point and the T_{NI} because perhaps the two dipolar groups are at least partially in opposition to one another.

Table 4.10. The effects of lateral cyano group on the physical properties of nematic mesogens[30].

a.
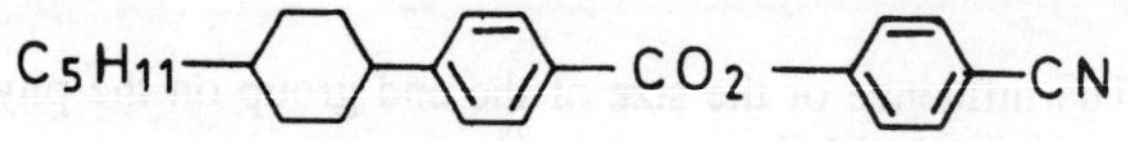

K111 N225IL ΔЄ=14.3

b.
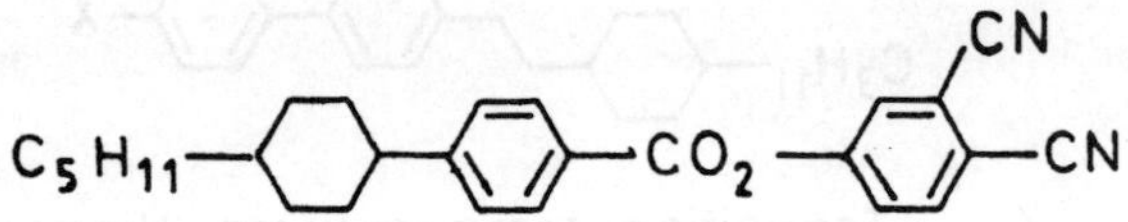

K85.2N143.9IL ΔЄ=28.2

c.
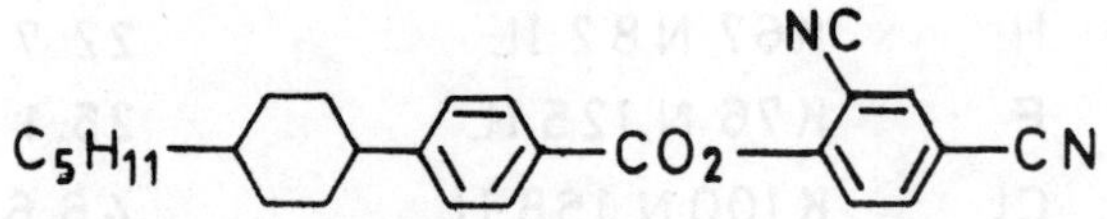

K132.2N178.7IL ΔЄ=3.2

The most commonly used lateral group is the fluoro substituent. The fluoro substituent is small (1.47 A) in size and is of high polarity (electronegativity ~4.0). This unique combination of steric and polarity effects has helped in synthesizing materials of specific physical properties without disrupting the phase stability too much. The chloro substituent generates a greater dipole than the fluro one but due to large size (1.75 A) it is of little use as a lateral substituent in the generation of mesogens with tailored properties. The role of the lateral fluoro substitution on the physical properties of nematogens can be seen in Table 4.11. A single lateral fluoro substitution (compound T4.11b) causes a significant reduction in the values of the melting point and the T_{NI}. The substitution of a second fluoro unit (compound T4.11c) has a little effect on the melting point but a dramatic depression in the T_{NI} is observed. The second lateral fluoro substituent affects the dielectric anisotroy significantly; $\in_{\perp}$ is reduced but $\in_{\parallel}$ is increased considerably giving a high value of $\Delta\in$ as compared to other two compounds. The dipole moment increases with the lateral substitution.

Table 4.11. The effects of lateral fluoro substitution on the physical properties of nematic mesogens.

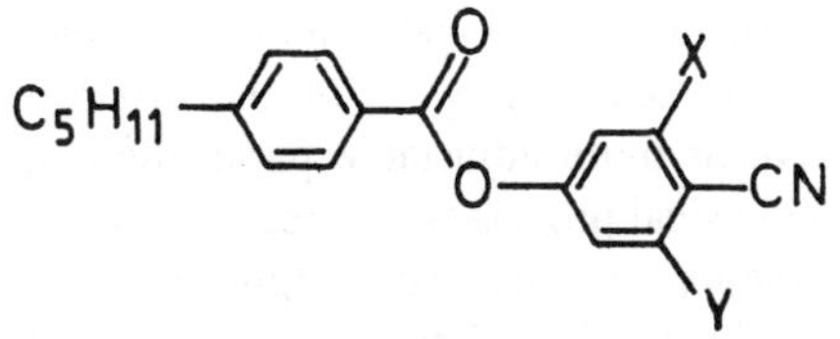

	X	Y	Viscosity (cp)	Dipole (D)	Δε
a.	H	H	47	5.9	40.1
b.	F	H	33	6.6	41.0
c.	F	F	-	7.2	61.0

A number of mesogenic materials have been generated[25] with the use of large lateral groups (e.g., CH_3, C_2H_5, C_6H_{13}, C_7H_{15}, $C_{16}H_{33}$ etc.). It has been found that a lateral methyl substituent reduces the melting point and the T_{NI} considerably. The slightly larger ethyl lateral group causes depression in T_{NI} to a similar extent but the fall in melting point is not so pronounced. As the alkyl chain becomes longer it adopts the conformation that runs along the mesogenic core; this causes no reductions in the melting points and minimal depressions in T_{NI} are observed.

4.4 STATISTICAL THEORIES OF THE NEMATIC ORDER

The nematic liquid crystal is fluid and at the same time anisotropic because while preserving their parallelism the molecules slide over one another freely. The experimental observations using various techniques show that the order parameters decrease monotonically as the temperature is raised in the mesophase range and drop abruptly to zero at the transition temperature. In case of uniaxial nematic the order parameter $\overline{P}_2$ drops abruptly to zero from a value in the range of 0.25 to 0.5 depending on the mesogenic material at the T_{NI}. Thus the NI transition is first-order in nature, though it is relatively weak thermodynamically because only orientational order is lost at the T_{NI} and the heat of transtion is only 1 kJ/mol. This in turn leads to large pretransitional abrupt increases in certain other thermodynamic properties, such as the specific heat, thermal expansion and isothermal compressibility of the medium near the T_{NI}. The change of entropy and volume associated with this transition are typically only a few percent of the corresponding values for the solid-nematic transition.

The theory for the nematic phase at and in the vicinity of their phase transitions has been developing in several directions[31]. In one of the most applied approaches one uses the phenomenological theory of Landau and de Gennes[32–34] in which the Helmholtz free energy is expressed in powers of the order parameters and its gradients. In the process, five or more adjustable parameters, associated with the symmetry of the system and physical processes, are required to be determined by the experiments. While this theory is physically appealing and mathematically convenient, it has many drawbacks, including the lack of quantitative predicting power about the phase diagram. In another approach, mainly developed by Faber[57], the nematic phase is treated as a continuum, in which a set of modes involving periodic distortion of an initially uniform director field is thermally excited. All orientational order is assumed to be due to mode excitations. For a system of N molecules, 2N modes are counted corresponding to all rotational degrees of freedom. This theory works well near the solid-nematic transition but fails close to nematic-isotropic transition. In the molecular field theories[58-121], one begins with a model in the form of interparticle potentials and proceeds to calculate the solvent mediated anisotropic potential acting on each individual molecule. Such calculations require full knowledge of pair correlation functions. For a potential which mimics all the important features of the molecular structure this approach includes lengthy and complicated mathematical derivations and numerical computations. As a consequence, too many simplifying approximations are made in the choice of the models and in the evaluation of correlation functions and transition properties.

The initial version of mean-field theory is due to Onsager[58] which ascribes the origin of nematic ordering to the anisotropic shape of molecules, i.e., to the repulsive interactions. The Maier-Saupe (MS) theory[78] and its modifications and extensions[79–93] attribute the formation of the ordered phase to the anisotropic attractive interactions. In reality, of course, both of these mechanism is operative. Thus, in the van der Waals type theories[99–112] both anisotropic hard core repulsions and angle dependent attractions are explicitly included. Another type of molecular theory has been developed by Singh[113] and others[114–121]. These works are based on the density functional approach which allows writing formally exact expressions for thermodynamic functions and one-particle distribution functions in terms of direct correlation functions. The molecular interactions do not appear explicitly in the theory. These type of theories are physically more reasonable than those of Maier-Saupe or repulsion dominant hard-particle theories. However, the major difficulty with the theories based on the density functional approach is associated with the evaluation of direct correlation functions which requires the precise knowledge of intermolecular interactions. Only approximate methods are known for this purpose.

We shall discuss[31] briefly here the basic ideas involved in the aforesaid theories and its application to the uniaxial nematic-isotropic (NI) phase transitions. The computer simulation studies of phase transitions in liquid crystals will be given in chapter 6 (section 6.9).

4.4.1 Landau-de-Gennes Theory of the Uniaxial Nematic-Isotropic (NI) Phase Transitions

Based on the Landau's general description of phase transitions[35,36], de Gennes[32] developed the phenomenological model for the NI phase transitions, the so-called Landau-de-Gennes (LDG) theory[32–34]. The strengths of the LDG theory are its simplicity and its ability to capture the most important elements of the phase transitions. The model is constructed independent of the detailed nature of the interactions and of the molecular structure. It has been applied to many other transitions[37] and embellished in many ways. First we discuss the essential ingredients of the Landau theory and then its application to the NI transitions will be summarised.

A. The basic ideas of Landau theory

The Landau theory is concerned with a phenomenological description of the phase transitions. These transitions involve a change of symmetry. Generally, the more symmetric (less ordered) phase corresponds to higher temperature and the less symmetric (more highly ordered) one to lower temperatures. The difference in symmetry between the two phases can be represented by the order parameters which are constructed in such a way that they are zero in the more symmetric phase. The mathematical description of the theory is based on the idea that the thermodynamic quantities of the less symmetric phase can be obtained by expanding the thermodynamic potential in powers of the order parameters and its spatial variations in the neighbourhood of order-disorder transition point and that sufficiently close to the transition only the leading terms of the series are important so that the said expansion becomes a single low order polynomial. The motivation for this simple and most elegant speculation is derived from the continuity of the change of state at a phase transition of the second order, i.e., the order parameters show values near the transition point. Hence the Landau's original procedure is, in principle, restricted to second order phase transitions. The thermodynamic behaviour of the order parameters in the less symmetric phase is then determined from the condition that their values must minimize the postulated expansion of the thermodynamic potential.

In order to clarify the Landau's original arguments[35,36] and to discuss the main ingredients of this model we begin by considering a macroscopic system whose equilibrium state is characterized by a spatially invariant, dimensionless, scalar order parameter Q. Though this situation is not directly related to the NI phase transition, it permits to understand the basic ingredients of the Landau approach. The general form of the thermodynamic potential G(p,T,Q) is postulated, near the transition point to be written as

$$G(p,T,Q) = G(p,T,0) + h_1 Q + \frac{1}{2} AQ^2 + \frac{1}{3} BQ^3 + \frac{1}{4} CQ^4 + \cdots \qquad (4.1)$$

where G(p,T,0) is the thermodynamic potential for a given temperature and pressure of the state with $Q = 0$. The numerical coefficients are introduced for the reason of convenience. The coefficients h_1, A,B,C,........ are functions of p and T. The equilibrium state can be obtained by minimizing G with respect to Q for the fixed p and T. In other words, the thermodynamic behaviour of the order parameter follows from the stability conditions,

$$\frac{dG}{dQ} = 0, \qquad \frac{d^2G}{dQ^2} > 0 \tag{4.2}$$

At the transition temperature T_C these stability conditions are

$$\frac{dG}{dQ} = h_1 = 0, \qquad \frac{d^2G}{dQ^2} = A = 0 \tag{4.3}$$

because of the coexistence of both the phases. This implies that (i) $h_1 = 0$, (ii) $A = a\,(T\text{-}T_C)$ with $a = \left(\frac{dA}{dT}\right)_{T_C} > 0$. The first condition is derived from the requirement that the high temperature phase with Q = 0 must give rise to an extremem value of G(p,T,Q). The second condition follows from the behaviour of G at Q = 0 for T above and below the transition temperature T_C. The function G(p,T,0) must have a minimum for $T > T_C$, i.e., $A > 0$, and a relative maximum for $T < T_C$, i.e., $A < 0$. The Landau theory postulates that the phase transition can be described by the following expression for the difference in the thermodynamic potential of the two phases,

$$\begin{aligned} \Delta G &= G(p,T,Q) - G(p,T,0) \\ &= \frac{1}{2}\,a(T-T_C)Q^2 - \frac{1}{3}\,BQ^3 + \frac{1}{4}\,CQ^4 + \cdots \end{aligned} \tag{4.4}$$

Here the negative sign with B has been chosen for the reason of convenience. It is further assumed that the coefficients B,C,..., must be weakly temperature dependent so that they can be treated as temperature independent coefficients.

The thermodynamic behaviour of the system follows directly from the stability condition,

$$\frac{dG}{dQ} = 0 = a(T\text{-}T_C)Q - BQ^2 + CQ^3 + \cdots \tag{4.5}$$

Equation (4.5) has the following solutions near the transition point

(i) Q = 0, the high temperature phase , (4.6a)

(ii) $$Q = \frac{B \pm [B^2 - 4aC(T-T_C)]^{1/2}}{2C} \tag{4.6b}$$

Now the discontinuity in Q at the transition, requires B = 0, i.e., for the low temperature Q reads

$$Q = \pm\left[\frac{a(T_C - T)}{C}\right]^{\frac{1}{2}} \tag{4.7}$$

Thus for the reason of stability the coefficient C must be positive. For a phase which remains invariant by replacing Q by − Q, as for example, the binary alloys, only coefficients belonging to even powers in Q survive. Thus in case of a second-order phase transition, eq. (4.1) takes the form

$$G(p,T,Q) = G(p,T) + \frac{1}{2} A(T)Q^2 + \frac{1}{4} CQ^4 + \cdots \qquad (4.8)$$

It is important to mention that the Landau theory can be extended to the first-order phase transition by two possible mechanism. The first one corresponds to the condition B = 0 and C < 0, the other one considers the presence of a third-order term BQ^3 in the expansion for G(p,T,Q). In the former case a stabilizing sixth-order term with coefficient E > 0 in eq. (4.1) is required.

So far only spatially uniform systems have been considered above. The eqs. (4.1) and (4.8) can be generalized to include spatial variations of the order parameters by replacing Q by Q(r) and including the contribution of the interaction term $\gamma[\nabla Q(r),T]$ in a power series in $\nabla Q(r)$ and retaining only the leading term

$$\gamma[\nabla Q(r),T] \cong \frac{1}{2} h_2(T) [\nabla Q(r)]^2 \qquad (4.9)$$

In order that the spatially uniform state be the state of lowest free-energy, $h_2(T) > 0$ and that near the critical temperature, $h_2(T)$ must be approximated as a temperature-independent constant.

The above development is based on the hypothesis that the expansion (4.1) makes sense and G is an analytic function of p, T and Q. However, there exists, a priori, no reason to believe that these requirements are true because in the neighbourhood of a critical point correlations between fluctuations of the order parameter are of great importance. The problem with this theory is that the coefficients appearing in the expansion are phenomenological and their dependence on the molecular properties are not determined. It is to be expected that these coefficients have singularities as a function of p and T. These singularities present great difficulties. As a result, it is assumed that the presence of singularities does not affect the terms of the expansion. In addition, this theory does not contain any information about the molecular interactions.

Despite above mentioned difficulties, Landau theory has been successfully applied[31–56] to a great variety of physical phenomenon. It reveals the role of symmetry in physics. Mathematically it is simpler than the mean-field (MF) theory. The inclusion of spatial variation of order parameters gives it a new dimension not found in the MF theory.

The first indication about the extension of Landau theory to the liquid crystals can be witnessed in the original work of Landau[35] itself where a short paragraph is devoted to the form of the probability density that defines a nematic state. Subsequently, the approach has been used extensively[14,31] to provide a phenomenological justification to explain almost all the observed facts in the area of liquid crystals, for example, multilayer ordering in smectic phases, various incommensurate modulations, ferroelectricity, elastostatics, interfacial phenomena, etc. Indenbom and others[38–41] applied the group theoretical and thermodynamic concepts forming the Landau theory to study the mesophase system.

B. The uniaxial nematic-isotropic (NI) phase transition : Landau-de-Gennes theory

The nematic state is described by the symmetric tensor order parameter **Q** with zero trace, i.e., $Q_{\alpha\alpha} = 0$. Since the thermodynamic potential is a scalar, the expansion can only contain terms that are invariant combinations of the elements $Q_{\alpha\beta}$ of the order parameter. In general, the expansion reads

$$g = g_0 + \frac{1}{2} A\, \mathrm{Tr}(Q^2) + \frac{1}{3} B\, \mathrm{Tr}(Q^3) + \frac{1}{4} C(\mathrm{Tr}\, Q^4) + \frac{1}{5} D[\mathrm{Tr}(Q^3)] \times [\mathrm{Tr}(Q^2)] + \frac{1}{6} E[\mathrm{Tr}(Q^2)]^3 + E'[\mathrm{Tr}\, Q^3]^2 + \cdots \tag{4.10}$$

where g and g_0 represent the Gibbs free-energy density of the nematic and isotropic phases, respectively. The term linear in $Q_{\alpha\beta}$ does not appear in the expansion due to the different symmetry of the two phases. In principle, the gradient terms have to be added.

From the general considerations we note the following possibilities about the expansion (4.10) :

(i) The absence of linear term in $Q_{\alpha\beta}$ allows for the existence of an isotropic phase. In case an external field is present a linear term has to be included which makes the isotropic phase impossible.

(ii) Since the NI transition is first-order, odd terms of order three and higher are allowed.

(iii) There are two independent sixth-order terms. The presence of the E′ term introduces the possibility of a biaxial nematic phase. Since for the moment we are interested only in a uniaxial phase E′ term will be omitted.

(iv) The NI phase transition takes place in the neighbourhood of A = 0. Therefore, it is assumed that the temperature dependence of free-energy is contained in the coefficient A alone and that other coefficients can be regarded as temperature independent. To describe the phase transition we write

$$A = a(T - T^*_{NI}) \tag{4.11}$$

with a as a positive constant and T^*_{NI} a temperature close to the NI transition temperature T_{NI}.

In order to make comparison with molecular statistical calculations which are often performed at constant density, we consider the Helmholtz free-energy density f instead of g. For the uniaxial nematic phase, we write the expansion

$$f = f_0 + \frac{1}{2} A\, \mathrm{Tr}(Q^2) + \frac{1}{3} B\, \mathrm{Tr}(Q^3) + \frac{1}{4} C\, \mathrm{Tr}(Q^4) + \frac{1}{5} D[\mathrm{Tr}(Q^2)\, \mathrm{Tr}(Q^3)] + \frac{1}{6} E[\mathrm{Tr}(Q^2)]^3 \tag{4.12}$$

For calculating the minimum of the free-energy, we use the order parameter in diagonal form

$$Q = \begin{vmatrix} -\frac{1}{2}(x+y) & 0 & 0 \\ 0 & -\frac{1}{2}(x-y) & 0 \\ 0 & 0 & x \end{vmatrix} \qquad (4.13a)$$

Here the condition of zero trace is automatically fulfilled. In addition, it allows the possibility that all the three eigenvalues are different, $x \neq 0$, $y \neq 0$ (biaxial nematics), $x \neq 0$, $y = 0$ (uniaxial nematics) and $x = 0$, $y = 0$ (isotropic liquids).
The invariants of **Q** are

$$\mathrm{Tr}(Q^2) = \tfrac{1}{2}(3x^2 + y^2)$$

$$\mathrm{Tr}(Q^3) = \tfrac{3}{4}x(x^2 - y^2)$$

Choosing the uniaxial ordering along the Z-axis (the director) and $y = 0$, the free energy density is written as

$$f = \tfrac{3}{4}Ax^2 + \tfrac{1}{4}Bx^3 + \tfrac{9}{16}Cx^4 + \tfrac{9}{40}Dx^5 + \tfrac{9}{16}Ex^6 \qquad (4.13b)$$

Here the free-energy is normalized such that $f_0 = 0$.

We consider two cases. In the simplest model $D = E = 0$. For the minimum to be at finite $x = x_{min}$, $C > 0$ and the sign of B is opposite to the sign of $x_{o\,min}$. So for a calamitic nematic $B < 0$ and for a discotic nematic $B > 0$. It is convenient to express the minimum value of x_{min} in terms of $\overline{P}_2\left(\equiv \frac{3}{2}x_{min}\right)$. With these provisos, eq. (4.13b) reads

$$f = \tfrac{1}{3}a(T - T^*_{NI})\overline{P}_2^{\,2} - \tfrac{2}{27}B\overline{P}_2^{\,3} + \tfrac{1}{9}C\overline{P}_2^{\,4} \qquad (4.14)$$

The equilibrium value of $\overline{P}_2$ is obtained by minimizing the free-energy (4.14) with respect to $\overline{P}_2$; the solutions are

$$\overline{P}_2 = 0, \qquad \text{the isotropic phase} \qquad (4.15)$$

$$\overline{P}_{2\pm} = \frac{B}{4C}\left\{1 \pm \left[1 - \frac{24aC(T - T^*_{NI})}{B^2}\right]^{\frac{1}{2}}\right\} \qquad (4.16)$$

The correct solution which can describe the temperature dependence of the order parameter in the nematic phase is the $\overline{P}_{2+}$ solution. The transition temperature T_{NI} can be calculated using the condition $f = f_0$, i.e.,

$$a(T_{NI} - T^*_{NI})\overline{P}_{2NI}^{\,2} - \tfrac{2}{9}B\overline{P}_{2NI}^{\,3} + \tfrac{1}{3}C\overline{P}_{2NI}^{\,4} = 0 \qquad (4.17)$$

and the second relation between $\overline{P}_{2NI}$ and T_{NI} is given by

$$a(T_{NI} - T^*_{NI})\overline{P}_{2NI} - \tfrac{1}{3}B\overline{P}^2_{2NI} + \tfrac{2}{3}C\overline{P}^3_{2NI} = 0 \qquad (4.18)$$

Equations (4.17) and (4.18) yields

$$\tfrac{1}{9}B\overline{P}^3_{2NI} = \tfrac{1}{3}C\overline{P}^4_{2NI} \qquad (4.19)$$

Thus the following two solutions are possible :

(i) $\overline{P}_{2NI} = 0, \qquad T_{NI} = T^*_{NI}$ (4.20a)

(ii) $\overline{P}_{2NI} = \frac{B}{3C}, \qquad T_{NI} = T^*_{NI} + \frac{B^2}{27a\,C}$ (4.20b)

Obviously, the result $\overline{P}_{2NI} = 0$ at $T = T^*_{NI}$ corresponds to the $\overline{P}_{2-}$ solution, whereas the $\overline{P}_{2+}$ solution gives $\overline{P}_{2NI} = (B/3C)$ at the higher temperature $T_{NI} = T^*_{NI} + \frac{B^2}{27aC}$. It turns out that $\overline{P}_{2NI+}$ solution represents the thermodynamically stable solution. Equation (4.16) determines a third temperature T^+_{NI} given by

$$T^+_{NI} = T^*_{NI} + \frac{B^2}{24a\,C} \qquad (4.21)$$

In case $T > T^+_{NI}$ the solutions $\overline{P}_{2+}$ and $\overline{P}_{2-}$ no longer apply because of their imaginary behaviour. In conclusion, the LDG theory distinguishes four different temperature regions :

(i) $T > T^+_{NI}$: the minimum corresponds to an isotropic phase, $\overline{P}_2 = 0$

(ii) $T_{NI} < T < T^+_{NI}$: the minimum of the free-energy is still given by $\overline{P}_2 = 0$, i.e., the isotropic phase is the thermodynamically stable state. There exists a relative minimum at $\overline{P}_2 = \overline{P}_{2+}$ and a relative maximum at $\overline{P}_2 = \overline{P}_{2-}$. As a result an energy barrier of height $f(\overline{P}_{2-}) - f(\overline{P}_{2+})$ exists between the two minima $\overline{P}_2 = 0$ and $\overline{P}_2 = \overline{P}_{2+}$. It follows that a metastable nematic phase can be obtained in this temperature region due to overheating. At T^+_{NI} the height of barrier becomes zero. For $\overline{P}_{2+}(T^+_{NI}) = \overline{P}_{2-}(T^+_{NI}) = (B/4C)$ the associated point in the free-energy curve represents a point of inflection.

(iii) $T^*_{NI} < T < T_{NI}$, the minimum corresponds to a nematic phase. There exists a local minimum corresponding to a possible supercooled isotropic state.

The transition entropy at T_{NI} is given by

$$\Delta\Sigma = \left.\frac{\partial(f - f_0)}{\partial T}\right|_{T=T_{NI}} = \frac{1}{3}a\overline{P}_{2NI}^{\,2} = \frac{aB^2}{27C^2} \tag{4.22}$$

The latent heat per unit volume, ΔH, is given by

$$\Delta H = \frac{aB^2T_{NI}}{27C^2} \tag{4.23}$$

Thus the coefficients a, B and C can be determined from the experimental values of $\overline{P}_{2NI}$, T_{NI} and ΔH. Table 4.12 lists the values of these parameters for MBBA.

Table 4.12. Values of parameters in the Landau expansion for MBBA.

Parameter	Value
a	42×10^3 J/m^3/k
B	64×10^4 J/m^3
C	35×10^4 J/m^3

(iv) $T < T^*_{NI}$: the minimum corresponds to a nematic phase. The $\overline{P}_{2+}$ solution corresponds to the lowest free-energy density, whereas the $\overline{P}_{2-}$ solution to a relative minimum and $\overline{P}_2 = 0$ gives rise to a relative maximum.

The height h′ of the energy barrier at $T = T_{NI}$ between the isotropic $\overline{P}_2 = 0$ state and the nematic $\overline{P}_2 = \overline{P}_{2NI}$ state is given by

$$h' = \frac{B^4}{11664C^3} \tag{4.24}$$

In general, all the real nematics contain impurities. It is known that the solute impurities which are not sufficiently rod like and rigid in molecular structure depress the NI transition temperature. Due to the presence of non-mesogenic impurities in the pure nematic the NI transition is broadened and a two phase region appears. The width of the two phase region is proportional to the entropy of the solvent. With the increase of the solute concentration the T_{NI} decreases. This problem has been studied theoretically using a variety of models. Using the LDG theory the existence of this two phase region has been explained[53] and shown that it indicates the tricritical behaviour of the NI transition. Further, the role of density variations on the NI transition has been investigated[51,52]. It has been shown that the density fluctuation alters the character of NI transition and makes it very weakly first-order.

C. The influence of external fields on the NI transition

In the absence of external magnetic or electric fields, the nematic and the isotropic liquid do not have the same symmetry. The application of fields may change the character of the NI phase transition. The influence of an applied field is to induce orientational order in the isotropic phase that grows with increasing field intensity. For the case of positive dielectric anisotropy, the first-order phase boundary in the temperature- applied field plane terminates in a field-induced critical point. With increasing external fields, the jump at the transition vanishes at the field-induced critical point. The increase in the orientational order in the isotropic liquid results in an enhancement of the NI transition temperature, δT_{NI}. As a result, it is now possible to pass through the field-induced critical point and observe the state beyond which the nematic and the paranematic states are indistinguishable. The application of magnetic field to the nematics with a negative magnetic anisotropy induces, in general, biaxial ordering and so a biaxial solution of **Q** is required.

In this section, the role of external fields on the physical properties of NI phase transition will be examined within the framework of the LDG theory. Suppose that a static magnetic field **H** is applied to the system. Many applications of liquid crystals are strongly dependent on their response to such external perturbations. The application of field leads to an extra orientation dependent term in the free- energy density of eq. (4.12)

$$f_m = -\frac{1}{2}H_\alpha H_\beta \chi_{\alpha\beta} \tag{4.25}$$

Expressing $\chi_{\alpha\beta}$ in the order parameter elements this can be written as

$$f_m = -\frac{1}{2}\bar{\chi}H^2 - \frac{1}{2}\Delta\chi_{max}H_\alpha H_\beta Q_{\alpha\beta} \tag{4.26}$$

where $\bar{\chi} = \frac{1}{3}\chi_{\alpha\alpha}$. The first term may be omitted because it is independent of the molcular ordering. The sign of $\Delta\chi_{max}$ must be positive to stabilize the uniaxial symmetry. A negative sign of $\Delta\chi_{max}$ refers to the nematic ordering in which the director is perpendicular to the field. Consequently, the field direction becomes a second axis and the phase is biaxial. Table 4.13 shows the field effect on possible phases. When an electric field is applied to the system the contributing term to the free-energy density has the same form as eq. (4.26) with $\bar{\chi}$ and $\Delta\chi_{max}$ replaced by the average permittivity and the maximum permittivity anisotropy, respectively.

Table 4.13. The breaking of symmetry of a phase due to a magnetic field.

Phase in zero field h = 0	Phase in nonzero field $h \neq 0$	
	$\Delta\chi_{max} > 0$	$\Delta\chi_{max} < 0$
IL	N_u^+	N_u^-
N_u^+	N_u^+	N_b
N_u^-	N_b	N_u^-

When $\Delta\chi_{max} > 0$ the phase remains uniaxial. With the field **H** and the director $\hat{\mathbf{n}}$ along the Z-axis and the order parameter defined by eq. (4.13a) with y = 0, the LDG free-energy density, containing terms upto fourth order in order parameter, reads

$$f = -\frac{1}{2}\Delta\chi_{max} H^2 x + \frac{3}{4}Ax^2 + \frac{1}{4}Bx^3 + \frac{9}{16}Cx^4 \tag{4.27}$$

Minimization of free-energy gives

$$a\left(T - T_{NI}^*\right) = h/\overline{P}_2 - \frac{1}{3}B\overline{P}_2 - \frac{2}{3}C\overline{P}_2^2 \tag{4.28}$$

where $h = (1/2)\ \Delta\chi_{max} H^2$. Since the $\overline{P}_2(T)$ curve (Fig. 4.4) has a negative slope, the order of the N_u^+ increases. It is clear that if $h \neq 0$, the $\overline{P}_2 = 0$ is never a solution of eq. (4.28), instead of an isotropic phase a small induced N_u^+ ordering is obtained. In order to differentiate between usual N_u^+ phase and the induced N_u^+ phase, the later one is often called the paranematic phase. The value of $\overline{P}_2$ in the paranematic phase is small and can be obtained from eq. (4.28) by ignoring the coefficients B and C,

$$\overline{P}_2(h) = \frac{h}{a\left(T - T_{NI}^*\right)} \tag{4.29}$$

Figure 4.4 shows[34a] the variation of order parameter with $T - T_{NI}^*$ for different values of field variable h. It can be seen that the jump of the order parameter at the NI phase transition is directly related to the value of the field. For smaller values of the field there exists a first-order phase transition between the paranematic and the nematic phases. The order parameter jump decreases with increasing field until the critical value h_c of the field is reached where there is no jump anymore. At this point transition becomes second-order. For fields higher than h_c there is no phase transition and the nematic and the paranematic phases are indistinguishable.

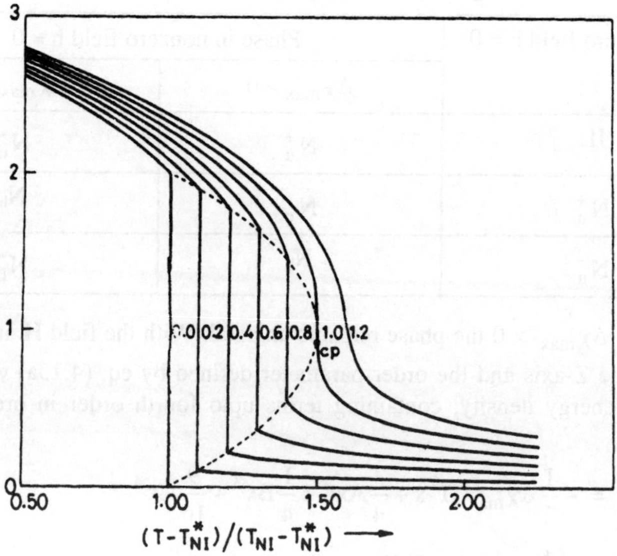

Fig. 4.4. The variation of order parameter with temperature for different values of the field variable. The dashed line represents the NI coexistence curve, cp is critical point and the number on the curves are the values of the field variables.

The location of critical point is given by

$$f'(x) = f''(x) = f'''(x) = 0 \tag{4.30a}$$

or equivalently

$$h_{cp} = -\frac{B^3}{324C^2} = \frac{1}{12}C\left(\bar{P}^0_{2NI}\right)^3 \tag{4.30b}$$

$$T_{cp} - T^0_{NI} = \frac{1}{54aC} = \frac{1}{2}\left(T^0_{NI} - T^*_{NI}\right) \tag{4.30c}$$

$$\bar{P}_{2c} = -\frac{B}{6C} = \frac{1}{2}\bar{P}^0_{2NI} \tag{4.30d}$$

Here the superscript 0 refers to the zero field. At the phase transition there are two minima x_1, x_2 of equal energy with the condition,

$$f(x_1) = f(x_2)$$

$$f'(x_1) = f'(x_2) = 0$$

The solution gives

$$x_{1,2} = x_{cp}\left[1 \pm \sqrt{3-2\gamma}\,\right] \tag{4.31a}$$

where

$$\gamma = \left(\frac{27aC}{B^2}\right)(T - T_{NI}^*) = 1 + \frac{h}{2h_{cp}} \tag{4.31b}$$

The shape of the coexistence curve of the nematic and the paranematic phase is parabolic. The shift in the transition temperature is proportional to h

$$T_{NI}(h) - T_{NI}(o) = \frac{2h}{a\,\overline{P}_{2NI}^{0}} \tag{4.31c}$$

As is obvious from the above discussion, three general predictions can be made from the theory

(i) The existence of a paranematic phase with field-induced orientational order.
(ii) The increase of T_{NI} with the increasing field and
(iii) The existence of a magnetic (electric) critical point.

However, it is important to mention that experimentally the above effects are very small. Taking maximum field and typical anisotropy of the order of $|\Delta\chi_{max}| \approx 10^{-7}$ (CGS) one gets, for $T_{NI} - T_{NI}^* = 1K$, the induced nematic order at T_{NI} of the order of 10^{-5}–10^{-4} which is very small as compared to the typical values 0.3-0.5 at the other side of the NI phase transition. Helfrich[42] was first to observe an increase in T_{NI} by applying the electric field. Rosenblatt[43] reported the magnetic experiment. The observed effect is so weak that the possibility of observing the magnetic critical point in thermotropic nematics is quite remote. Contrary to it, the first evidence for the existence of electrically induced critical point was shown by Nicastro and Keyes[44].

When $\Delta\chi < 0$, the application of magnetic field, in general, induces biaxial ordering. So in this case a biaxial solution of Q is required. Retaining the terms upto fourth order in $Q_{\alpha\beta}$ and keeping the field along the Z-axis, the free-energy, with the order parameters x and y, can be expressed as

$$f(x,y) = -hx + \frac{3}{4}Ax^2 + \frac{1}{4}Bx^3 + \frac{9}{16}Cx^4 + \left(\frac{1}{4}A - \frac{1}{4}Bx + \frac{3}{8}Cx^2\right)y^2 + \frac{1}{8}Cy^4 \tag{4.32}$$

Applying the same rule as discussed above, an N_bN_u transition is found; the transition is first-order for $h > h_{tcp}$ and second-order for $h < h_{tcp}$. The point $h = h_{tcp}$ is a tricritical point. For the same value of $|\Delta\chi_{max}|$, $H_{tcp} = (9/4)\,H_{cp}$.

The possibility of the existence of a critical region at the first-order transition line of the NI transition has been investigated[45] in the context of epsilon (ε) expansion. It has been observed that the LDG expansion with tensorial order parameter, which has a BQ^3 interaction in addition to CQ^4 (and h = 0), has a critical value $B = B_C(C,A)$ below which there is no transition. At the critical value the system undergoes a second-order transition with no symmetry breaking. Above the critical value of B the transition is of first-order. This is in contradiction with the prediction of Landau's theory. The results holds also for d(dimension) > 4, since it depends only on the fact that $A_C = 0$. Thus if at T_{NI} the behaviour is critical, assuming that in the scaling law the non-analytic behaviour appears at the critical point (fluid-like critical point) and also on the spinodal curve, the critical indices of the absolute stability limit of the nematic phase (metastable) are $\beta_1 = \beta$ and $\alpha_1 = \gamma_1 = 1 - \beta_1$. Thus it is obvious that there is a possibility of critical behaviour with d = 3 for the NI phase transition. This argument has been verified experimentally[46].

The renormalization group (RG) technique has been used to calculate[47,48] the $T_{NI} - T^*_{NI}$ of the NI transition. The model free-energy of LDG type can be written as

$$f = \int \left[\frac{1}{4} (AQ_{ij}^2 + \nabla_k Q_{ij} \nabla_k Q_{ij}) - BQ_{ij}Q_{jk}Q_{k\ell} + C(Q_{ij}Q_{ij})^2 - H_{ij}Q_{ij} \right] d^d x \tag{4.33}$$

Here $d^d x$ indicates a functional integral in d dimension over the tensor field $\mathbf{Q} = \mathbf{Q}(x)$. The model (4.33) was studied extensively by ε (= 4-d) expansion technique[49]. This model relies on the fact that the MF approximation is exact for d > 4. It is a perturbation expansion about the solution for d = 4. The fixed point of the RG corresponds to a second order phase transition with B = 0. The cubic coupling was found to be a relevant term, so B was treated as perturbation. The scaling form of the equation of state in the second order of ε expansion is obtained[47],

$$\frac{H}{Q^\delta} + \frac{\beta}{Q^\omega} = f'(x') \tag{4.34}$$

where $x' = \dfrac{t}{Q^{1/\beta}}$. The result for

$$f'(x) = 1 + x + \epsilon f_1(x) + \epsilon^2 f_2(x) \tag{4.35a}$$

is

$$\delta = 3 + \varepsilon, \quad \omega = 1 + \frac{7\varepsilon}{13}, \quad \beta = \frac{1}{2} - \frac{3\varepsilon}{26} \tag{4.35b}$$

Here t be the reduced temperature, $t = (T - T^*_{NI})/T^*_{NI}$. It is the temperature at which the second-order phase transition would take place if B = 0.

This method needs only one experimental data input namely the jump in order parameter at T_{NI}. The results obtained are still for away from experimental findings. The calculation has been extended near the coexistence curve[48] which is defined as the region of small external field and below the critical temperature. The critical exponents have been evaluated[50] numerically for the d = 3 LDG model near the isolated critical point on the NI transition line from the renormalization group theory. It has been found that the critical exponents $\gamma = 1.277$ and $\nu = 0.638$ are in fair agreement with the best ε expansion result, $\gamma = 1.277$ and $\nu = 0.64$.

D. Landau theory of the NI transition : inclusion of fluctuations

Though the singular like behaviour of various quantities at the NI transition is yet a puzzle, the solution could be obtained within the framework of the LDG theory. Since the nature of NI transition is first-order, the part of critical region, closest to the critical point is not accessible to the experiments. It was suggested by Keyes and Shane[54] that the critical exponents for quantities diverging towards T_{NI}^{*}, before being cut off by a first-order transition at the T_{NI}, should be the characteristic of tricritical point (tcp). The difference between critical and tricritical behaviour of the NI transition is difficult to verify. The value of the exponent β (= 0.247 ± 0.01) obtained for 8CB strongly supports the suggestion of tricritical character of the NI phase transition. On the other hand, the experimental value of specific heat near the NI phase transition of MBBA indicates that the behaviour of this transition is near tricritical and does not appear to agree with the LDG model (4.12).

In the nematic phase the director fluctuations are critical. The strongly developed director fluctuations could alter the character of the NI transition and make it very weakly first-order. The RG approach has been applied[55] to the description of fluctuation behaviour near the isolated critical point on the NI transition line. Wang and Keyes[56] have calculated the fluctuation of five components of the orientational order parameter of a nematic liquid crystal involving several types of critical and multicritical points. The study of pretransitional (fluctuation) phenomena near the first-order phase transition may provide an answer for their closeness to the second-order transition. Light scattering and magnetic birefringence measurements in the isotropic phase of nematogens present strong pretransitional effects and can be understood within the framework of the LDG theory.

The two important ingredients that influence the thermodynamic functions near a phase transition are the order parameter fluctuation amplitude and the spatial correlations between the fluctuations of the order parameter. The average size of the range of correlations between the fluctuations defines the so called correlation length ξ. Far away from the critical point ξ is of the order of the intermolecular distance. On the other hand, near a second-order phase transition the correlations decrease very slowly with distance, indicating a divergence of ξ at the critical temperature. The Landau description of the fluctuation based phenomena can be expected to be valid if the Ginzburg criterian is satisfied,

$$b\xi^3 >> k_B T \tag{4.36}$$

where b is the height of the barrier separating the ordered and disordered states. Close to the T_{NI} the important fluctuations involve regions of volume ξ^3. The inequality (4.36) then states that thermally activated fluctuations leading from ordered to disordered regions or vice versa do not occur with a significant probability. As a result, the LDG expansion is expected to be valid in case the fluctuations are small.

The fluctuations can be usually included into the Landau theory by calculating the free-energy density at each temperature as a function of both the order parameter and its spatial derivatives. For the low-energy (long-wavelength) fluctuations only the lower-order spatial derivatives of the order parameter are considered. The free-energy density can be expanded with respect to both **Q(r)** and its derivatives. As there is no way of forming a scalar quantity linear in $\partial_r Q_{\alpha\beta}$, where $\partial_r = \frac{\partial}{\partial x_r}$. Also there is no scalar combination of $Q_{\alpha\beta}$ and its spatial derivatives. So the lowest order spatial derivative invariants of $Q_{\alpha\beta}$ have the form

$$(\partial_\alpha Q_{\beta\gamma})^2,\ (\partial_\beta Q_{\beta\gamma})^2$$

As a result the LDG expansion that can describe the effect of the local fluctuations in the isotropic phase reads

$$f[T,p,\mathbf{Q}(\mathbf{r}),\partial_i\,\mathbf{Q}(\mathbf{r})] = f_i + \frac{1}{2}AT\mathrm{r}\,\mathbf{Q}^2 + \frac{1}{3}B\,\mathrm{Tr}\,\mathbf{Q}^3 + \frac{1}{4}C(\mathrm{Tr}\mathbf{Q}^2)^2$$

$$+\ \frac{1}{5}D[\mathrm{Tr}(\mathbf{Q}^2)(\mathrm{Tr}\,\mathbf{Q}^3)] + \frac{1}{6}E(\mathrm{Tr}\mathbf{Q}^2)^3 + E'(\mathrm{Tr}\mathbf{Q}^3)^2 - \frac{1}{2}\Delta\chi_{max}\,\mathbf{HQH}$$

$$-\ \frac{1}{12\pi}\Delta\varepsilon_{max}\,\mathbf{EQE} + \frac{1}{2}L_1(\partial_\alpha\,Q_{\beta\gamma})^2 + \frac{1}{2}L_2(\partial_\beta\,Q_{\beta\gamma})^2 \qquad (4.37)$$

where f_i represents the free-energy density of the isotropic phase without local fluctuations. Two new expansion parameters L_1 and L_2 which are closely related to the elastic constants, have appeared. Equation (4.37) is the basic equation of the generalized Landau-de-Gennes (GLDG) theory of the NI transition that includes long-wavelength fluctuations of the order parameter. The GLDG theory has various applications, of which the most important ones are the pretransitional effects, for example, the magnetically induced birefringence (Cotton-Mouton effect) and light scattering. The important quantities of interest are the correlation functions relevant to the pretransitional phenomena and if these correlation functions are not too large the Gaussian approximation is expected to be a good approximation. Concentrating only on the pretransitional light scattering and assuming B = C ... = 0, it follows directly from eq. (4.37) that the local fluctuations in the isotropic phase give rise to a local free-energy density of the type

$$f(\mathbf{r}) = f_i + \frac{1}{2}a(T - T^*_{NI})\,Q_{\alpha\beta}\,(\mathbf{r})\,Q_{\beta\alpha}(\mathbf{r}) + \frac{1}{2}L_1[\partial_\alpha\,Q_{\beta\gamma}(\mathbf{r})]$$

$$[\partial_\alpha\,Q_{\beta\gamma}\,(\mathbf{r})] + \frac{1}{2}L_2[\partial_\alpha\,Q_{\alpha\gamma}\,(\mathbf{r})]\,[\partial_\beta\,Q_{\beta\gamma}\,(\mathbf{r})] \qquad (4.38)$$

where $Q_{\alpha\beta}(\mathbf{r})$ can be expanded in Fourier series,

$$Q_{\alpha\beta}(\mathbf{r}) = \sum_{k} Q_{\alpha\beta}(\mathbf{k})\, e^{i\mathbf{k}\cdot\mathbf{r}} \tag{4.39a}$$

with

$$Q_{\alpha\beta}(\mathbf{k}) = \frac{1}{V} \int d\mathbf{r}\, Q_{\alpha\beta}(\mathbf{r})\, e^{-i\mathbf{k}\cdot\mathbf{r}} \tag{4.39b}$$

Substituting eq. (4.39a) into eq. (4.38) and integrating over the volume of the system, one obtains the following contribution of the fluctuation to the total free-energy

$$\begin{aligned} f_f = {} & \frac{1}{2} Va(T - T_{NI}^*) \sum_{k} Q_{\alpha\beta}^*(\mathbf{k})\, Q_{\alpha\beta}(\mathbf{k}) \\ & + \frac{1}{2} VL_1 \sum_{k} k^2\, Q_{\alpha\beta}^*(\mathbf{k})\, Q_{\alpha\beta}(\mathbf{k}) \\ & + \frac{1}{2} VL_2 \sum_{k} k_\alpha\, k_\beta\, Q_{\alpha\gamma}^*(\mathbf{k})\, Q_{\beta\gamma}(\mathbf{k}) \end{aligned} \tag{4.40}$$

where $k^2 = k_\alpha k_\alpha$. It is obvious that the value of f_f strongly depends on the values of the amplitude $Q_{\alpha\beta}(\mathbf{k})$. The validity of the LDG theory here is based on the postulate that the appearance of a given set of amplitudes $\{Q_{\alpha\beta}(k)\}$ is described by a probability distribution of the form

$$P\{Q_{\alpha\beta}(\mathbf{k})\} = \frac{1}{Z} \exp(-\beta f_f) \tag{4.41}$$

For $\mathbf{k} \neq 0$, the amplitudes are complex quantities,

$$Q_{\alpha\beta}(\mathbf{k}) = r_{\alpha\beta}(\mathbf{k}) + i\, s_{\alpha\beta}(\mathbf{k}) \tag{4.42}$$

where $r_{\alpha\beta}(\mathbf{k})$ and $s_{\alpha\beta}(\mathbf{k})$ are real variables. These variables satisfy the relations

$$r_{\alpha\beta}(\mathbf{k}) = r_{\alpha\beta}(-\mathbf{k})\,; \qquad s_{\alpha\beta}(\mathbf{k}) = -s_{\alpha\beta}(-\mathbf{k})$$

and

$$r_{\alpha\beta}(\mathbf{k}) = r_{\beta\alpha}(\mathbf{k})\,; \qquad s_{\alpha\beta}(\mathbf{k}) = s_{\beta\alpha}(\mathbf{k})$$

Since the trace of the fluctuation tensor $\mathbf{Q}(\mathbf{k})$ is zero,

$$r_{\alpha\alpha}(\mathbf{k}) = s_{\alpha\alpha}(\mathbf{k}) = 0\,.$$

For obvious reasons the relevant thermal averages based upon the tensor elements $Q_{\alpha\beta}(0)$ are required to be calculated. These averages are obtained to be of the form

$$\left\langle Q_{xy}^2(0)\right\rangle = \left\langle Q_{xz}^2(0)\right\rangle = \left\langle Q_{yz}^2(0)\right\rangle = \frac{k_B T}{2Va(T - T_{NI}^*)} \tag{4.43a}$$

$$\left\langle Q_{xx}^2(0)\right\rangle = \left\langle Q_{yy}^2(0)\right\rangle = \left\langle Q_{zz}^2(0)\right\rangle = \frac{4}{3}\left\langle Q_{xy}^2(0)\right\rangle = \frac{4}{3}\left\langle Q_{xz}^2(0)\right\rangle$$

$$= \frac{4}{3}\left\langle Q_{yz}^2(0)\right\rangle = \frac{2k_BT}{3Va(T-T_{NI}^*)} \tag{4.43b}$$

The relevant thermal averages $\left\langle r_{\alpha\beta}^2(\mathbf{k})\right\rangle$ and $\left\langle s_{\alpha\beta}^2(\mathbf{k})\right\rangle$ are obtained as

$$\left\langle r_{\alpha\beta}^2(\mathbf{k})\right\rangle = \left\langle s_{\alpha\beta}^2(\mathbf{k})\right\rangle \tag{4.43c}$$

$$\left\langle r_{xx}^2(\mathbf{k})\right\rangle = \left\langle r_{yy}^2(\mathbf{k})\right\rangle = \left\langle r_{zz}^2(\mathbf{k})\right\rangle = \frac{k_BT}{3V[a(T-T_{NI}^*) + L_1k^2]} \tag{4.43d}$$

and

$$\left\langle r_{xy}^2(\mathbf{k})\right\rangle = \left\langle r_{xz}^2(\mathbf{k})\right\rangle = \left\langle r_{yz}^2(\mathbf{k})\right\rangle = \frac{k_BT}{4V[a(T-T_{NI}^*) + L_1k^2]} \tag{4.43e}$$

In these equations the elastic constant L_2 has been taken zero. The correlation functions of the kind $\langle Q_{xx}(0)\, Q_{xx}(\mathbf{R})\rangle$ and $\langle Q_{xy}(0)\, Q_{xy}(\mathbf{R})\rangle$, which are of prime interest, can now be calculated in a straight forward way. The distance dependence of these functions is described in terms of correlation length ξ, for example,

$$\left\langle Q_{xx}(0)\, Q_{xx}(\mathbf{R})\right\rangle = \frac{k_BT}{6\pi L_1 R}\exp(-R/\xi) \tag{4.44}$$

with ξ given by

$$\xi = \left[\frac{L_1}{a(T-T_{NI}^*)}\right]^{1/2} \tag{4.45}$$

The correlation length ξ is a measure of the distance over which the local fluctuations are correlated. $\xi \to 0$ at infinitely high temperature and diverges as $T \to T_{NI}^*$. Near $T = T_{NI}^*$ this divergent behaviour of ξ is responsible for the so-called pretransitional phenomena, for example, the strong increase in the light scattering cross-section. With known values of the thermal averages, the depolarization ratio, i.e., the ratio between the intensities of the scattered light with polarizations parallel and perpendicular to the incident light at a scattering wave vector q = 0, is given by

$$\frac{I_{\parallel}(0)}{I_{\perp}(0)} = \frac{\left\langle Q_{xx}^2(0)\right\rangle}{\left\langle Q_{xy}^2(0)\right\rangle} = \frac{4}{3} \tag{4.46}$$

The calculation becomes more complicated when the L_2 term is also included in evaluating thermal averages like $\langle r_{\alpha\beta}(\mathbf{k})\; r_{\gamma\delta}(\mathbf{k})\rangle$. In this case three correlation lengths appear

$$\xi_1 = \left[\frac{L_1}{a(T-T_{NI}^*)}\right]^{1/2} \tag{4.47a}$$

$$\xi_2 = \left[\frac{L_1 + \frac{1}{2} L_2}{a(T - T^*_{NI})}\right]^{1/2} \tag{4.47b}$$

$$\xi_3 = \left[\frac{L_1 + \frac{2}{3} L_2}{a(T - T^*_{NI})}\right]^{1/2} \tag{4.47c}$$

Owing to the positive sign of the correlation lengths, L_1 and L_2 must satisfy the relation

$$L_1 + \frac{2}{3} L_2 > 0 \tag{4.48}$$

Thus only the largest distance makes sense. This means if $L_2 < 0$, ξ_1 is the correlation length, whereas ξ_3 is the correlation length when $L_2 > 0$. It is important to mention that the discussion presented here is not expected to present a quantitatively correct description of the pretransitional phenomena. However, the comparison with the experiment and detail discussion on the results within and beyond the Gaussian approximation are given elsewhere[34].

4.4.2 The Hard Particle or Onsager-Type Theories

In the year 1949, Onsager[58] showed that a system of long rigid rods exhibits a transition from an isotropic phase to a denser anisotropic phase. The calculation of Onsager was based on a cluster expansion for the free-energy given as a functional of the distribution in orientation of the rigid particles. The model was applied later by Zwanzig[59] to a certain idealized situation and the rigid rod transition was verified to the order of seventh virial. A Pade' analysis of Zwanzig approach by Runnels and Colvin[61] has shown that the transition observed by Zwanzig is very stable, and in three-dimensions at least certainly of first-order. A slightly different mean-field calculation for the hard rod problem based on the well known lattice model was made by Flory and coworkers[62,63]. The above treatments[58–63] of hard rod systems are valid only for very long rods with length to breadth ratio $x_o \geq 100$ which is typical of polymeric systems. For the shorter rods ($x_o \approx 3$–5 or almost $x_o < 10$) at high densities an application of scaled particle theory (SPT) due to Cotter[64,65] has been very convenient for evaluating the excess free-energy. Other treatments of hard rod systems include y-variable expansion[66,67], application of density functional theory[113] (DFT), the functional scaling approach[71], computer simulations (see chapter 6), role of molecular flexibility on the density change at the transition[72–74] etc.

The Onsager's line of approach boils down to a discussion of thermodynamic properties of a system of hard rods. Let us consider a fluid of N rods of length L and diameter d_0 in volume V at temperature T. Assuming that a rod can take only ν discrete orientations, the configurational partition function can be written[64,65] as

$$Q_N = \frac{1}{N!}\frac{1}{\nu^N}\sum_{\Omega}\int d\mathbf{r}\exp[-\beta U_N] \tag{4.49}$$

where the summation is over the orientations of the rod. U_N, the potential energy of interaction, is approximated as the sum of pair potentials,

$$U_N = \sum_{1<i<j<N} u_{ij}$$

Here N! allows for the indistinguishability of the particles and ν^N for the number of orientational states.

For hard particles,

$$u_{ij} = u(\mathbf{r}_{ij},\Omega_i,\Omega_j) = \infty \quad \text{, if i and j would overlap}$$
$$= 0 \quad \text{, otherwise} \tag{4.50}$$

Since the configurational space available for each molecule is V

$$\int d\mathbf{r}\exp(-\beta U_N) = V^N \exp[-\beta\phi_N(\mathbf{\Omega})] \tag{4.51}$$

where the function ϕ_N depends only on the orientations. For a given configuration of the system, if there are N_1 molecules along Ω_1, N_2 molecules along Ω_2, ..., N_ν molecules along Ω_ν, ϕ_N becomes a function of occupation number and

$$Q_N = \left(\frac{V^N}{N!\nu^N}\right)\sum_{N_1=0}^{N}\sum_{N_2=0}^{N}\cdots\sum_{N_\nu=0}^{N}\frac{N!}{\prod_{i=1}^{\nu} N_i!}\exp[-\beta\phi_N(N_1,N_2,...,N_\nu)] \tag{4.52}$$

with $\sum_{i=1}^{\nu} N_i = N$. Introducing the mole fraction $s_i = N_i/N$ of various components and using the maximum term approximation, the configurational Helmholtz free-energy of a system of hard rods can be expressed as

$$\frac{\beta A_{hr}}{N} = -\frac{1}{N}\ell n\, Q_N$$

$$= \ell n\,\rho_0 - 1 + \sum_{i=1}^{\nu}\tilde{s}_i\,\ell n(\tilde{s}_i) + \frac{\beta}{N}\phi_N(\tilde{N}_1,\tilde{N}_2,...,\tilde{N}_\nu) \tag{4.53}$$

where tildes denote that the distribution corresponds to the maximum term of Q_N and $\phi_N(\tilde{N}_1,\tilde{N}_2,...,\tilde{N}_\nu)$, as first noted by Zwanzig[59], is the excess Helmholtz free-energy relative to an ideal gas of a system of molecules having fixed orientations.

Considering the continuous distribution of angles, replacing s_i by $f(\Omega_i)$ and finally converting from sums back to integrals, one obtains

$$\frac{\beta A_{hr}}{N} = \ell n\, \rho_0 - 1 + \int f(\Omega) \ln [4\pi f(\Omega)]\, d\Omega + \frac{\beta}{N} \phi_N \{f(\Omega)\} \tag{4.54}$$

Thus the major work is to derive the excess free-energy function ϕ_N. The two most common approaches which have been used to evaluate ϕ_N are the cluster or virial expansion technique of Onsager[58] and the scaled particle theory[64,65].

Onsager made a virial expansion of ϕ_N and retained terms only up to the second virial coefficient, i.e., ϕ_N was approximated as

$$\phi_N \approx \frac{1}{2} \rho_0 \sum_i \sum_j s_i\, s_j\, V_{exc}(\Omega_i, \Omega_j) \tag{4.55}$$

where $V_{exc}(\Omega_i, \Omega_j)$ ($\equiv V_{exc}(\Omega_{ij})$) is the mutual exclusion volume (or covolume) of the two rods with orientations Ω_i and Ω_j, respectively. Converting from sums to integrals, eq. (4.54) reads

$$\frac{\beta A_{hr}}{N} = \ell n\, \rho_0 - 1 + \int f(\Omega) \ln [4\pi f(\Omega)]\, d\Omega + \frac{1}{2} \rho_0 \int f(\Omega_1) f(\Omega_2) V_{exc}(\Omega_{12}) d\Omega_1 d\Omega_2 \tag{4.56}$$

In order to calculate $V_{exc}(\Omega_{12})$ the shape of the rods must be specified. For the long rods ($L >>> d_o$), where the end effect is ignored, one obtains[14]

$$V_{exc}(\Omega_{12}) = 2L^2 d_0 \left|\sin \Omega_{12}\right| \tag{4.57}$$

where (Ω_{12}) is the angle between the axes of two molecules with orientations Ω_1 and Ω_2. The most convenient shape of rods, used by Onsager[58] and many others[64,65,101], is a spherocylinder, which is a right circular cylinder capped on each end by a hemisphere of the same radius. In this case[58,101(b)]

$$V_{exc}(\Omega_1, \Omega_2) = 8v_o + 4al^2 \left|\sin \Omega_{12}\right| \tag{4.58}$$

where a, l and v_0 are, respectively, the radius, cylindrical length and volume of a spherocylinder.

The singlet orientational distribution function can be determined by minimizing eq. (4.56) subject to the condition,

$$\int f(\Omega)\, d\Omega = 1 \tag{4.59}$$

Thus $f(\Omega)$ can be obtained as a solution of the variational equation

$$\ell n[4\pi f(\Omega_1)] = \alpha_\ell - 1 - 8v_0\rho_0 - 4al^2 \rho_0 \int f(\Omega_2)\, d\Omega_2 \left|\sin \Omega_{12}\right| \tag{4.60}$$

where α_ℓ is a Lagrange multiplier to be determined from the normalization condition eq. (4.59). Equation (4.60) is a non-linear integral equation which ought to be solved numerically for $f(\Omega)$. However, it is difficult to solve the eq. (4.60) exactly. Onsager obtained an approximate variational solution which is based on the trial function of the form

$$f_1 = (\text{const}) \cosh(\alpha \cos\theta) \tag{4.61}$$

where α is a variational parameter and θ is the angle between the molecular axis and the nematic axis. It was found[58] that in the region of interest α turns out to be large (~ 20) and the system exhibits an abrupt first-order phase transition from isotropic ($\alpha = 0$) to nematic ($\alpha \geq 18.6$) phase characterized by

$$\phi_{nem} = 4.5\,\frac{d_0}{L}\,, \qquad \phi_{iso} = 3.3\,\frac{d_0}{L}\,, \qquad \overline{P}_{2NI} \approx 0.84 \tag{4.62}$$

where $\phi\left(= \frac{1}{4}\pi\rho_0 L d_0^2\right)$ represents the volume fraction of the rods. The relative density change $\frac{\Delta\rho}{\rho_{nem}}$, with $\Delta\rho = \rho_{nem} - \rho_{iso}$, at the transition in the full Onsager calculation is about 25%. These values are much too high as compared to the experimental results. The transition predicted by Onsager's approach was confirmed later by Zwanzig[59] who calculated higher virial coefficients up to the seventh restricting the molecules to take only three mutually perpendicular orientations.

Clearly the Onsager's approach only provides a qualitative insight into the effect of repulsive interactions as far as nematics are concerned. A detail discussion on the validity of this approach is given elsewhere[75]. It is claimed that it gives a single and qualitatively correct picture of the order-disorder transition due to the shape of the molecules. Straley[76] has argued that the truncation of the cluster expansion series after the second virial coefficient can be justified quantitatively only for very long rods ($x_0 > 100$). For the shorter rods ($x_0 \leq 40$) qualitatively reliable results are expected. The numerical solutions obtained by Lasher[60] in the limit of very long rods show that the NI transition is characterized by $\Delta\rho/\rho_{nem} \approx 0.21$ and $\overline{P}_{2NI} \approx 0.784$.

Flory and Ronca[63] considered the application of a lattice model to treat the hard rod problem and made the calculations, in the limit of long rods, at the relatively high densities.

In Flory's lattice approach[62] each rod with parameter L/d_0 is construed to consist of ℓ_0 segments, one segment being accommodated by a cell of the lattice. The preferred axis of a given domain is taken along one of the principal axes of the lattice. Disorder in the orientation of the rods with respect to the preferred axis of the system is expressed by a parameter y determined by the projection of the rod of parameter L/d_0 in a plane perpendicular to the preferred axis. The calculation carried out for the phase equilibrium between anisotropic (nematic) and isotropic phases gives higher values for the volume fractions at the transition as compared to Onsager's results.

$$\phi_{nem} \simeq 12.5\frac{d_0}{L}, \qquad \phi_{iso} \simeq 8\frac{d_0}{L}; \qquad \left(\frac{L}{d_0} \geq 10\right) \tag{4.63}$$

The value of $\overline{P}_{2NI}$ is not quite meaningful in view of the crude approximations in treating f(Ω). It turns out to be even larger than the Onsager solution. The combined covolume of solute species in the isotropic phase at the coexistence was found to be 7.89, whereas the critical value of the axial ratio for the coexistence of the two phases in the neat liquid $x_{ocrit} \simeq 6.417$.

From the above discussion, it is clear that the use of the Onsager[58–60] and Flory[62,63] approaches to study the NI phase transition in real nematics characterised by $x_0 \sim 3$–5 (or almost $x_0 < 10$) and high density is completely unjustified. However, these approaches are useful to describe the phase transitions in polymer liquid crystals and their solutions in suitable solvents. For real nematics composed of spherocylindrical molecules the scaled particle theory (SPT)[64,65] has been used to calculate the excess free-energy function $\phi_N(\tilde{N}_1, \tilde{N}_2, ..., \tilde{N}_v)$. The principal quantity in SPT is the work function $W_i(\sigma_1, \sigma_2, \rho_0)$ defined as the reversible work of adding a scaled spherocylinder of radius $\sigma_1 a$, cylindrical length $\sigma_2 \ell$ and fixed orientation Ω_i. W_i is related to the configurational Gibbs free-energy through the exact relation

$$\frac{\beta G}{N} = \sum_i s_i[\ell n(s_i \rho_0) + \beta W_i(1,1,\rho_0)] \tag{4.64}$$

When the scaling parameters $\sigma_1 \to 0$ and $\sigma_2 \to 0$, that is, when the scaled particle shrinks to a point, it has been shown that[64]

$$\exp[-\beta W_i] \sim 1 - \rho_0 \sum_j s_j V_{ij}(\sigma_1, \sigma_2) \tag{4.65}$$

where $V_{ij}(\sigma_1, \sigma_2)$ is the volume excluded to an unscaled molecule j with orientation Ω_j by the scaled particle i with orientation Ω_i at some fixed point. At the opposite extreme when σ_1 and σ_2 both are very large, $W_i(\sigma_1, \sigma_2)$ approaches the reversible PV work required to creat a macroscopic spherocylindrical cavity in the fluid,

$$\lim_{\substack{\sigma_1 \to \infty \\ \sigma_2 \to \infty}} W_i = \left[\pi(\sigma_1 a)^2 \sigma_2 \ell + \frac{4\pi}{3}(\sigma_1 a)^2\right] P \tag{4.66}$$

Also

$$V_{ij}(\sigma_1, \sigma_2) = \frac{4\pi}{3} a^3 (1+\sigma_1)^3 + \pi a^2 \ell (1+\sigma_1)^2 (1+\sigma_2) + 2al^2(1+\sigma_1)\sigma_2 \left|\sin \Omega_{ij}\right| \tag{4.67}$$

The $W_i(1,1,\rho_0)$ was calculated by interpolating between these two limits, i.e., between very large and very small values of σ_1 and σ_2, and the phase transition was located by equating the Gibbs free energy and pressure of the ordered and disordered phases. It was predicted that as x_0 increases, the transition densities decrease, the relative density change increases substantially and the order parameter $\overline{P}_{2NI}$ increases slightly. In the case of very short spherocylinders ($x_0 < 2$) the value of relative density change is less than 1%.

A comparison of the SPT equation of state with the results of MC and MD simulations[122–125] in the isotropic phase shows that for x_0 = 2 and 3, SPT overestimates the pressure at the high densities. Savithramma and Madhusudana[104] obtained much better results by extending a method originally proposed by Andrews[77] for hard spheres to spherocylinders. It is based on the idea that the reciprocal of the thermodynamic "activity" is simply the probability of being able to insert a particle into a system without overlapping with other particles. Another method which improves the agreement with the simulation results is due to Barboy and Gelbart[66]. It is based on the expansion of free energy in terms of the "y-variables", instead of the standard one in densities using virial coefficients, defined as

$$y = \rho_0 / (1 - v_0 \rho_0)$$

The expansion coefficients of y^n are related to the virial coefficients. Mulder and Frenkel[67] applied the y-expansion technique to study the NI transition in a system of ellipsoidal particles by restricting the expansion to y^2.

Parson[68] derived an expression for the free-energy of a hard rod system using "decoupling approximation". Assuming that the system interacts through a pair potential $u(\mathbf{r},\Omega_{12})$, the free-energy was derived as

$$\frac{\beta A_{hr}}{N} = \langle \ell n\, f(\Omega) \rangle - \frac{1}{6} \beta \int d\mathbf{r}\, d\Omega_1\, d\Omega_2\, f(\Omega_1) f(\Omega_2)$$

$$\times \int_0^n dn'\, r \frac{\partial u}{\partial r} g(\mathbf{r},\Omega_{12};n) \tag{4.68a}$$

where $\langle \ell n\, f(\Omega) \rangle = \int d\Omega\, f(\Omega)\, \ell n\, f(\Omega)$.

For a system composed of molecules whose only anisotropy is in their shape the pair potential can be taken of the special form $u(\mathbf{r},\Omega_{12}) = u(r/\sigma(\hat{r},\Omega_{12})) = u(r/\sigma)$, where σ is an angle-dependent range parameter such that $\sigma = \sigma(\hat{r},\Omega_{12})$ and $\mathbf{r}$ is the interparticle separation. In the decoupling approximation $g(\mathbf{r},\Omega_{12})$ scales as $g(r/\sigma)$. This amounts a complete decoupling between the translational and orientational degrees of freedom. This is exact at low density, since $g \approx e^{-\beta u}$, but cannot be exact at higher density. Using the decoupling approximation, eq. (4.68a) can be written as

$$\frac{\beta A_{hr}}{N} = <\ell n\, f(\Omega)> + \frac{1}{2} \int d\Omega_1\, d\Omega_2\, f(\Omega_1)\, f(\Omega_2)\, V_{exc}(\Omega_{12}) \int_0^n \alpha(n')\, dn' \tag{6.68b}$$

where

$$\alpha(n) = - \int_0^\infty dy\, y^3 \frac{\partial u}{\partial y} g(y) \tag{4.68c}$$

and

$$V_{exc}(\Omega_{12}) = \frac{1}{3} \int d\hat{r}\, \sigma^3(\hat{r}, \Omega_{12}) \tag{4.68d}$$

Here the translational degrees of freedom appear entirely in the coefficient $\alpha(n)$. When $n \to \infty$, obviously $\alpha(n) = 1$ for the hard particle fluid and one is back to the Onsager result. Using Berne and Pechukas[70] expression for $\sigma(\mathbf{r}, \Omega_{12})$

$$\sigma^2(\hat{r}, \Omega_{12}) = d_0^2 \left[1 - \chi \frac{(\hat{r}\cdot\hat{e}_1)^2 + (\hat{r}\cdot\hat{e}_2)^2 - 2\chi(\hat{r}\cdot\hat{e}_1)(\hat{r}\cdot\hat{e}_2)(\hat{e}_1\cdot\hat{e}_2)}{1 - \chi^2(\hat{e}_1\cdot\hat{e}_2)^2} \right]^{-1} \tag{4.69}$$

Equation (4.68d) for fixed relative orientation $\hat{e}_1 \cdot \hat{e}_2 = \cos\theta_{12}$ gives

$$V_{exc} = 8\,(1-\chi^2)^{-1/2}\,(1-\chi^2 \cos^2\theta_{12})^{1/2} \tag{4.70}$$

Here $\hat{e}_1$ and $\hat{e}_2$ are the unit vectors along the symmetry axes of the two interacting spheroids, $d_0 = 2a$ and the shape anisotropy parameter,

$$\chi = \frac{x_0^2 - 1}{x_0^2 + 1} \tag{4.71}$$

The decoupling approximation transforms the system into a fluid of spheres interacting with a central potential modulated by an angle-dependent excluded volume term. The calculations were done[68] for the hard rod fluid using the equation of state for the transformed hard sphere fluid. It was found that a transition to an orientationally ordered state occurs at a critical packing fraction η_c which decreases with x_0. For $x_0 \leq 3.5$, η_c is so high that the system tends to crystallize before the ordered liquid state is reached. Using perturbation theory, the calculations were done for a soft-rod fluid, where the transformed system of spheres interact with a potential $u(y) = \varepsilon/y^m$. When $m \to \infty$, this reduces to the hard sphere case. The transition temperature was obtained as a function of density and molecular shape and the order parameter has been found to obey the scaling law $Q = Q[\eta(\varepsilon/k_\theta T)^{3/m}]$. For m = 12 this gives good agreement with the experiment and show that most of the features of the transition can be explained with purely repulsive interactions.

Lee[71] introduced a functional scaling concept to study the stability of nematic ordering as a function of molecular shape anisotropy. The free-energy functional for a system of hard spherocylinder was constructed from a direct generalization of an analytic equation of state for a hard spheres under a simple functional scaling via the excluded volume of two hard nonspherical particles

$$J(\eta) \rightarrow J(\eta)\,\frac{1}{8}\,<V_{exc}(\hat{e}_1,\hat{e}_2)/v_0> \tag{4.72}$$

This is equivalent to the decoupling approximation[68]. The generalized free-energy of a system of hard spherocylinders in a closed form was obtained as

$$\frac{\beta A_{hsc}}{N} = \beta\mu_0(T) + \ell n\,\rho_0 - 1 + <\ell n\,[4\pi f(\Omega)]>$$
$$+\frac{\eta(4-3\eta)}{(1-\eta)^2}\left[1+\frac{3}{2\pi}\left(\frac{(L/d_0)^2}{1+3L/2d_0}\right)\langle|\sin\Omega_{12}|\rangle\right] \tag{4.73}$$

where

$$\langle|\sin\Omega_{12}|\rangle = \int d\Omega_1\,d\Omega_2\,f(d\Omega_1)\,f(d\Omega_2)\,|\sin\Omega_{12}|\ ,$$

$\eta = \rho_0 v_0$ is the packing fraction. For very long rods eq. (4.73) reduces exactly to Onsager's relation in the low density limit. Numerical calculations were performed for a variety of length to diameter ratio x_0 of hard spherocylinders. It was found that as the ratio x_0 increases from one to infinity, $\eta_{max}(x_0)$ increases from 0.740 to 0.907. Further, the discontinuity in the packing fraction at the NI transition becomes smaller as the molecular shape anisotropy decreases. The theory was extended[71b] to calculate the transition properties of hard ellipsoids of revolution. The free-energy expression of a system of N hard ellipsoids reads

$$\frac{\beta A_{her}}{N} = \beta\mu_0(T) + \ell n\,\rho_0 - 1 + <\ell n\,[4\pi f(\Omega)]>$$
$$+\frac{\eta(4-3\eta)}{(1-\eta)^2}\left\langle(1-\chi^2)^{-\frac{1}{2}}(1-\chi^2\cos^2\theta_{12})^{\frac{1}{2}}\right\rangle \tag{4.74}$$

Calculations were done for the thermodynamic parameters at the NI transition for hard ellipsoids with x_0 = 2.75 and 3.0 and the results are compared in Table 4.14 with those of several other approaches. It can be seen that the functional scaling results agree well with the simulation values[126,127] than other approaches.

Table 4.14. Comparison of the NI transition parameters for the hard ellipsoids with $x_0 = 2.75$ and 3.0.

x_0	Quantities	MC[126,127]	Baus, et al.[129]	Singh and Singh[117b]	Mulder Frenkel[67]	Marko[130]	Lee[71b]
2.75	η_{nem}	0.570	0.512	0.347	0.462	0.518	0.552
	η_{iso}	0.561	0.501	0.329	0.449	0.517	0.544
	$\Delta\eta/\eta_{nem}$	0.016	0.021	0.052	0.028	0.002	0.014
	$\overline{P}_2$	–	0.548	0.532	0.552	–	0.517
3.0	η_{nem}	0.517	0.484	0.330	0.437	0.494	0.517
	η_{iso}	0.507	0.472	0.309	0.420	0.493	0.508
	$\Delta\eta/\eta_{nem}$	0.019	0.022	0.063	0.039	0.002	0.017
	$\overline{P}_2$	–	0.561	0.547	0.568	–	0.533

From the above discussion, it is clear that all rigid rod calculations predict too low a value for the mean density and too large a value for the density discontinuity at the transition. The experimental value for $\Delta\rho/\rho_n$ for the nematic phase transitions is of the order of 1%. In fact, there is a basic difficulty with all the hard particle theories. Their properties are "athermal" and entirely density dependent. For a system of hard particles it follows that $\Gamma = \infty$, whereas experimentally $\Gamma = 4$ for PAA. Clearly, therefore, a proper description of the nematic phase requires inclusion of attractive interactions in the theoretical treatments.

4.4.3. The Maier - Saupe (MS) Type Theories

Maier and Saupe[78] assumed that the nematic ordering is caused by the anisotropic part of the dispersion interaction between the molecules. The shape anisotropy of the molecules was entirely ignored. In accordance with the symmetry of the structure, viz., the cylindrical distribution about the preferred axis and the absence of polarity, the orientational energy of a molecule i can be approximated as

$$u_i = -\overline{u}_2 V^{-2} \overline{P}_2 P_2(\cos\theta_i) \qquad (4.75)$$

where $\overline{u}_2$ is taken to be a constant independent of pressure, volume and temperature. The V^{-2} dependence is due to the dispersion interaction. However, it is well accepted that the exact nature of the interactions need not be specified for the development of the theory. All that is required to obtain the results of MS theory is an anisotropic potential with a particular dependence on the molecular orientations[89]. Using the V^{-2} dependence, the Helmholtz free-energy can be expressed as

$$\frac{\beta A}{N} = \frac{1}{2}\beta\overline{u}_2 V^{-2}\overline{P}_2(\overline{P}_2+1) - \ell n \int_0^1 \exp\left[\frac{3}{2}\beta\overline{u}_2 V^{-2}\overline{P}_2 \cos^2\theta_i\right] d(\cos\theta_i) \ldots \quad (4.76)$$

The consistency relation is obtained by the minimization of the free-energy,

$$\left(\frac{\partial A}{\partial \overline{P}_2}\right)_{V,T} = 0 \quad (4.77a)$$

or

$$3\overline{P}_2\left(\frac{\partial\langle\cos^2\theta_i\rangle}{\partial\overline{P}_2}\right) - 3\langle\cos^2\theta_i\rangle + 1 = 0 \quad (4.77b)$$

and

$$\overline{P}_2 = \langle P_2(\cos\theta_i)\rangle \quad (4.77c)$$

The minimisation of the free-energy occurs at that value of $\overline{P}_2$ which satisfies the consistency relation. The calculations lead to a first -order NI transition at

$$\frac{u_2}{k_B T_{NI} V_c^2} = 4.541 \quad (4.78a)$$

and

$$\overline{P}_{2NI} = 0.4292 \quad (4.78b)$$

where V_c is the molar volume of nematic phase at T_{NI}. The following expression for the volume change at T_{NI} is obtained[78,82]

$$\Delta V = -2A\left(\frac{\partial A}{\partial V}\right)_{T=T_{NI}} \quad (4.79)$$

or

$$\frac{\Delta V}{V} = \left(\frac{V^2}{\beta\overline{u}_2\overline{P}_{2NI}^2}\right)\left[2\ell n\int_0^1 \exp\left\{\frac{3}{2}\beta\overline{u}_2 V^{-2}\overline{P}_{2NI}\cos^2\theta\right\} d(\cos\theta) - \beta\overline{u}_2 V^{-2}\overline{P}_{2NI}(\overline{P}_{2NI}+1)\right] \quad (4.80)$$

If ΔV is known from the experiment, $\overline{P}_{2NI}$ can be determined. However, this method gives only 1-2 percent of change in the values of $\overline{P}_{2NI}$. Because the theory contains only one unknown parameters, $\overline{u}_2$, the order parameter $\overline{P}_{2NI}$ and the entropy change $\Delta\Sigma_{NI}$ are predicted to be the universal properties of all the nematogens. Experimentally, $\overline{P}_{2NI}$ varies in the range of ~ 0.25 to 0.5 for different compounds[95]. The agreement between theory and experiment can be improved by extending the MS theory in a variety of ways. These include the addition of terms to the MS potential function to allow for the higher rank interactions[83a] and deviation from molecular cylindrical symmetry[83b,93] as well as by taking the MF approximation to higher order[87]. It has been found that molecular biaxility decreases the value of $\overline{P}_2$ at the transition, and leads to better agreement with experiment. The assumption that the molecule is a rigid rod is an oversimplification. Marcelja[84a], Photinos et al.[84b], and Luckhurst[128] have analysed the influence of the flexible end-chain on the ordering process. In a Flory type calculation Marcelja has included the configurational statistics of end chains and could explain the "odd-even" effect of both T_{NI} and $\overline{P}_{2NI}$ as a homologous series is ascended. Luckhurst[128] refined these calculations for the compounds with two rigid cyanobiphenyl moieties linked by flexible spacers in which the odd-even alternation in T_{NI} is about 100^0C.

Humphries et al.[83a] employed a more general form, than eq. (4.75), for the orientational pseudo potential, by adding the fourth legendre polynomial, as follows

$$u_i = -\overline{u}_2 V^{-m} \left[\overline{P}_2 P_2(\cos\theta_i) + m'\overline{P}_4 P_4(\cos\theta_i)\right] \tag{4.81}$$

where m and m′ are adjustable parameters to be determined from a comparison with the experiment. m′ is a measure of the relative importance of the fourth Legendre polynomial term. They were able to explain the variation of $\overline{P}_{2NI}$ and $\Delta\Sigma/Nk$ by a single parameter m′ and obtained agreement with the experimental temperature dependence of the order parameter when m′ = 4. Further, taking account of the deviation from spherical symmetry of the pair spatial correlation function of the molecules, they modified the orientational pseudo potential as

$$u_i = -\overline{u}_2 V^{-m} [1 + \delta\overline{P}_2]\overline{P}_2 P_2(\cos\theta) \tag{4.82}$$

Here δ denotes the deviation from the spherical symmetry of the spatial correlation function. If δ = 0 and m = 2 this form immediately reduces to eq. (4.75). It is important to note that unless δ is equal to 0, this potential is theoretically inconsistent because the solution of the self-consistent equation does not give minima in the free energy of the system.

The volume dependence of u_i has been analysed by Cotter[80] in the light of Widom's[79] idea. The empirical data show that the volume dependence of u_i may be different from that considered in eq. (4.75). It has been suggested that for the thermodynamic consistency of the MF model $u_i \propto V^{-m}$, where m is a number. Now eq. (4.78a) becomes

$$\frac{\overline{u}_2}{k_B T_{NI} V_c^m} = 4.541 \tag{4.83a}$$

but $\overline{P}_{2NI}$ remains unchanged. Thus the value of $\overline{P}_{2NI}$ does not depend too critically on the exponent m, because the isobaric change of volume with temperature over the entire nematic range is usually only of the order of 1-2 percent. However, the value of exponent m becomes very important while analysing the influence of pressure on the transition parameters. McColl and Shih[96] determined successfully a parameter Γ defined as

$$\Gamma = -\rho_0(\partial\overline{P}_2/\partial\rho_0)_T / T(\partial\overline{P}_2/\partial T)_{\rho_0} = \left(\frac{\partial \ln T}{\partial \ln \rho_0}\right)_{\overline{P}_2} \quad (4.83b)$$

This parameter is taken as a particularly sensitive probe of the relative importance of attractive and repulsive interactions. A plot of ln T against ln ρ_0 at constant $\overline{P}_2$ is virtually linear. The pressure studies[96–98] in case of PAA have shown that the value of Γ is 4 and the thermal range of nematic phase at constant volume is about 2.5 times that at constant pressure. The $\overline{P}_{2NI}$ is almost independent of pressure. The studies of pressure induced mesomorphism in PAA suggest that empirically m = 4 which has no theoretical justification. In view of Cotter's argument[80], m = 1, irrespective of the nature of the pair potential.

There are, however, several unsatisfactory features of the MS theory. In general, the MF models overestimate the strength of the transition. The predicted heat of transition from the nematic to isotropic phase is given by

$$H = T_{NI}[(\alpha/K)\Delta V - \overline{P}_2(V_c, T_{NI})] \quad (4.84)$$

where α and K are, respectively, the coefficients of thermal expansion and isothermal compressibility of the isotropic phase at T_{NI}. The theoretical values of it are usually found to be about 2 or 3 times higher than the experimental values. Large discrepancies are also found in the values of the specific heat C_v and the isothermal compressibility K in the nematic phase. Further, the calculations based on the MF models give $(T_{NI} - T^*_{NI})/T_{NI} \approx 0.9$ which leads to $T_{NI} - T^*_{NI} \approx 30^\circ$–$40^\circ$. The reason for these discrepancies is that the MF method neglects completely the effect of short-range order. Making use of the Bethe approximation and cluster variation methods the influence of short-range order on the NI transition in the MS model has been studied by several workers[81,85–88]. These approximations are identical in the isotropic phase, i.e., they give the same value of T^*_{NI}, related with the pretransitional phenomena, magnetically induced birefringence and the scattering of light by orientational fluctuations. These phenomena are qualitatively well described by these approximations. In these works, each molecule is assumed to have γ_n nearest neighbour ($\gamma_n \geq 3$); no two nearest neighbours are taken to be the nearest neighbours of each other. While each such central molecule i is subjected to a potential only due to the γ_n outer molecules of the cluster an outer molecule j is also subjected to a mean-field due to the rest of the medium. In the two-site cluster (TSC) approximation the thermal averages are evaluated with the two particle distribution function[86]

$$P_2(\hat{e}_i, \hat{e}_j) = \frac{1}{Z_{12}} \exp\left[\beta J P_2(\hat{e}_1 \cdot \hat{e}_2) + (\gamma_n - 1)\beta J \overline{s}(P_2(e_{1z}) + P_2(e_{2z}))\right] \quad (4.85a)$$

with

$$Z_{12} = \int d\hat{e}_1 \int d\hat{e}_2 \exp[\beta J P_2(\hat{e}_1 \cdot \hat{e}_2) + (\gamma_n - 1)\beta J \bar{s}(P_2(e_{1z}) + P_2(e_{2z}))] \quad (4.85b)$$

Here $\bar{s}$ is a variational parameter and J the interaction coupling constant.
The internal energy per particle is given by

$$U = -\frac{1}{2}\gamma_n J \bar{\sigma}_s \quad (4.86)$$

where $\bar{\sigma}_s$, the short-range order parameter, measures the correlation between the orientations of the neighbouring molecules

$$\bar{\sigma}_s = \langle P_2(\hat{e}_1 \cdot \hat{e}_2) \rangle \quad (4.87)$$

The calculations yield $\bar{P}_{2NI} \approx 0.40$ and the ratio $(T_{NI} - T^*_{NI})/T_{NI} \approx 0.05$; the exact values depend on the lattice type, the number of nearest neighbours. Haegen et al.[92] have extended the TSC approximation to a 4-particle cluster (FSC) approximation and obtained the result $(T_{NI} - T^*_{NI})/T_{NI} \approx 0.04$; again the exact value depends on the lattice type.

Another important origin of the discrepancy between the experimental and MS theoretical values is the assumption that the molecule is cylindrically symmetric so that it is sufficient to define one order parameter $\bar{P}_2$. However, in real nematogens most molecules are lath-shaped and have a biaxial character. Hence two order parameters are required to describe the uniaxial nematic phase composed of biaxial molecules. The second order parameter is defined as

$$D = \frac{3}{2}\left\langle \sin^2\theta \cos 2\Psi \right\rangle \quad (4.88)$$

The mean-field theories have been developed[83a,93] by including a term in the potential function proportional to D, in addition to the usual term. It is shown that the molecular biaxiality decreases the values of $\bar{P}_{2NI}$ and gives a better agreement with the experiment.

Luckhurst and Zannoni[89b] have discussed the question "why is the Maier-Saupe theory of nematic liquid crystals so successful"? Their arguments can be summarised as follows. The MS theory is founded on a MF treatment of long-range contributions to the potential function and ignores the important short-range forces. The theory has been particularly successful in accounting, not only for the order-disorder transition, but also for the orientational properties of the nematic liquid crystals. It is founded on the MF approximation applied to a weak anisotropic pair interaction arising due to dispersion force and predicts the universal properties of the nematogens. Despite its qualitative and even semi-quantitative successes, the MS theory has been widely criticised[80,90,91]. One of the major criticism is due to its foundation on the London dispersion forces. A second attack on the theory is concerned with its complete neglect of anisotropic short-range repulsive forces. This neglect is completely unjustified as it is accepted that the scalar repulsive forces are mainly responsible for determining the organisation in simple fluids[94]. However, the results of hard rod theories are in marked contrast with the MS theories. On the one hand, theories based on the short-range repulsive forces, which are expected to make the dominant contribution to the potential, give results in poor accord with experiment. While, on the other hand, the MS theory founded on long-range dispersion forces, provides a satisfactory description of the orientational properties of real nematogens. A formal solution to this dilemma is provided by the molecular field theories based on a complete general form of the total potential. This is separated into a scalar component and an anisotropic part which is then written in terms of the Pople expansion. The orientational part of the single particle potential is found to be a series whose first term has the same form as the MS result with $\overline{u}_2$ containing contributions from all the anisotropic forces. In this respect, the MS theory is equivalent in form to one which includes both long- and short-range contributions to the anisotropic intermolecular potential. There are several unsatisfactory features of this analysis which are discussed elsewhere[89b]. As a solution to these difficulties, it has been suggested that both the short- and long-range forces are important in determining the molecular organisation in a nematic phase but that they operate at quite different levels. The short-range forces are responsible for the formation of highly ordered groups or clusters of molecules. The possibility of cluster formation has already been mentioned in the original MS theory[78] although with a different purpose. In the view of Luckhurst and Zannoni[89b] the net effect of the existence of clusters would be to reduce the anisotropy associated with the short-range forces while increasing the influence of long-range forces. Further, the clusters are not destroyed at the NI phase transition. The anisotropic forces between clusters are, therefore, responsible for the orientational properties of the nematic mesophase. The parameter $\overline{u}_2$ in the effective potential is not then determined by the interaction of two molecules, but by two clusters. From this idea the failure of theories based on the repulsive forces can also be understood. The fundamental unit in these calculations is a single particle and at the transition the orientational order is completely destroyed producing a large increase in entropy. Contrary to it for the cluster model the transition unlocks the ordering of the clusters but not that of the molecules within the cluster; consequently the transition entropy is small. In the language of statistical mechanics the cluster can be described in terms of the spatial and orientational pair distribution function. Then the Luckhurst and Zannoni[89b] idea is equivalent to the assumption that the distribution function exhibits short-range order, over several molecular distances, and this order is not destroyed at the NI transition. As a result, the variation in the orientational properties on passing from one phase to the other will be determined by the long-range intermolecular forces corresponding to distances over which the distribution function does change. Consequently, the MS theory, founded on the long-range forces, provides a good description of nematic liquid crystals, whereas theories based on short-range forces are invariably less successful.

As evident from the above discussion, both short and long range forces are important in determining the molecular organisation in a liquid crystal. Therefore, a realistic theoretical model should be constructed using both the repulsive and attractive interactions between molecules.

4.4.4 The van der Waals (vdW) Type Theories

For the calamitic nematics several attempts have been made to develop theories[99–120] in which both the intermolecular repulsions and attractions are explicitly included. Most of these, known as van der Waals (vdW) type theories, differ only in evaluating the properties of a reference system interacting via repulsive force, and are almost identical as far as treating the attractive interaction is concerned. Basic to these works is the recognition that the predominant factor in determining the liquid crystalline stability is geometric and that the role of the attractive interactions is, to a first approximation, merely to provide a negative, spatially uniform mean-field in which the molecules move. The theories[115-120] based on the density functional approach shall be discussed in sec. 4.4.5. Here first we present a perturbation method[105a], within the mean-field approximation, to describe the equilibrium properties of nematics. All the vdW type theories can be derived from this scheme by considering only the first-order perturbation term. We shall then discuss the salient features of all these works.

In developing a perturbation theory, one begins by writing the pair potential energy of interaction, $u(\mathbf{x}_i,\mathbf{x}_j)$, as a sum of two parts - one part is known as reference potential, $u^{(0)}$, and the other perturbation potential, $u^{(p)}$, i.e.,

$$u(\mathbf{x}_i,\mathbf{x}_j) = u^{(0)}(\mathbf{x}_i,\mathbf{x}_j) + m_p\, u^{(p)}(\mathbf{x}_i,\mathbf{x}_j) \qquad (4.89)$$

Here $u^{(0)}$ is chosen to include the rapidly varying short-range repulsive interactions, while $u^{(p)}$ represents the more smoothly varying long-range attractions, m_p is a perturbation parameter.

The configurational integral of the system is written as

$$Q_N = \frac{1}{N!\,(4\pi)^N} \int d\mathbf{r}^N \int d\Omega^N \exp\left[-\beta U_N(\mathbf{x}_1,\mathbf{x}_2,\ldots,\mathbf{x}_N)\right] \qquad (4.90)$$

where U_N is approximated as the sum of pair potentials. The angular integration in eq. (4.90) can be approximated to arbitrary accuracy by dividing the unit sphere into arbitrary small sections of solid angle $\Delta\Omega$($n = \frac{4\pi}{\Delta\Omega}$ is the number of discrete orientations) and summing over all possible orientational distributions $\{N_1,\ldots,N_p, \ldots$

$N_n\}$ where N_p is the number of molecules having orientations falling in pth solid angle and $\sum_{p=1}^{n} N_p = N$. Thus

$$Q_N = \frac{1}{N!(4\pi)^N} \sum_{N_1} \ldots \sum_{N_n} \frac{N!(\Delta\Omega)^N}{N_1! \ldots N_n!} \int d\mathbf{r}^N \exp[-\beta U_N(\mathbf{r}^N, \tilde{N}_1, \ldots, \tilde{N}_n)] \qquad (4.91a)$$

$$\approx \left(\frac{\Delta\Omega}{4\pi}\right)^N \left[\prod_{p=1}^{n} \tilde{N}_p!\right]^{-1} \int d\mathbf{r}^N \exp[-\beta U_N(\mathbf{r}^N, \tilde{N}_1, \ldots, \tilde{N}_n)] \qquad (4.91b)$$

where the set $\{\tilde{N}_1, \ldots, \tilde{N}_n\}$ corresponds to the maximum term in (4.91a). Assuming the pairwise additivity of interaction potential and dividing the potential into two parts (eq. (4.89)), one obtains

$$Q_N = \left(\frac{\Delta\Omega}{4\pi}\right)^N \left[\prod_{p=1}^{n} \tilde{N}_p!\right]^{-1} \int d\mathbf{r}^N \exp[-\beta u_N^{(0)}(\mathbf{r}^N, \tilde{N}_1, \ldots, \tilde{N}_n)]$$

$$\times \exp[-\beta m_p u_N^{(p)}(\mathbf{r}^N, \tilde{N}_1, \ldots, \tilde{N}_n)] \qquad (4.92)$$

and

$$\frac{\partial \ell n\, Q_N}{\partial m_p} = -\frac{1}{2V}\beta \sum_{p=1}^{n} \sum_{p'=1}^{n} \tilde{N}_p \tilde{N}_{p'} \int d\mathbf{r}\, u^{(p)}(\mathbf{r}, \Omega_p, \Omega_{p'})\, g(\mathbf{r}, \Omega_p, \Omega_{p'}) \qquad (4.93)$$

where

$$g(\mathbf{r}, \Omega_p, \Omega_{p'}) = \frac{V^2}{Q_N} \left(\frac{\Delta\Omega}{4\pi}\right)^N \left[\prod_{p=1}^{n} \tilde{N}_p!\right]^{-1}$$

$$\times \int d\mathbf{r}^N \exp[-\beta U_N(\mathbf{r}^N, \tilde{N}_1, \ldots, \tilde{N}_n, m_p] \qquad (4.94)$$

Integration of eq. (4.93) leads to

$$\ell n\, Q_N = \ell n\, Q_N^0 - \frac{\beta}{2V} \int_0^1 dm_p \sum_{p=1}^{n} \tilde{N}_p \sum_{p'=1}^{n} \tilde{N}_{p'} \int d\mathbf{r}\, u^{(p)}(\mathbf{r}, \Omega_p, \Omega_{p'})\, g(\mathbf{r}, \Omega_p, \Omega_{p'}, m_p) \qquad (4.95)$$

Choosing a continuous function $f(\Omega)$ to describe the orientational distribution such that

$$\tilde{N}_p = N\, f(\Omega_p)\, d\Omega_p \qquad (4.96)$$

the Helmholtz free-energy can be expressed as

$$\frac{\beta A}{N} = \frac{\beta A^{(0)}}{N} + \frac{1}{2}\rho_0 \beta \int_0^1 dm_p \int f(\Omega_p)\, d\Omega_p \int f(\Omega_{p'})\, d\Omega_{p'} \int d\mathbf{r}$$

$$u^{(p)}(\mathbf{r},\Omega_p,\Omega_{p'})\, g(\mathbf{r},\Omega_p,\Omega_{p'},m_p) \tag{4.97}$$

Writing the expansions

$$g(\mathbf{r},\Omega_p,\Omega_{p'},m_p) = g^{(0)}(\mathbf{r},\Omega_p,\Omega_{p'}) + m_p\, g^{(1)}(\mathbf{r},\Omega_p,\Omega_{p'}) + \cdots \tag{4.98a}$$

$$A = A^{(0)} + m_p\, A^{(1)} + m_p^2\, A^{(2)} + \ldots \tag{4.98b}$$

and inserting these expressions into eq. (4.97), one obtains on equating the coefficients of m_p^r from both sides

$$\frac{\beta A^{(r)}}{N} = \frac{1}{2}\beta\rho_0 \int f(\Omega_p)\, d\Omega_p \int f(\Omega_{p'})\, d\Omega_{p'} \int d\mathbf{r} \times u^{(p)}(\mathbf{r},\Omega_p,\Omega_{p'})\, g^{(r-1)}(\mathbf{r},\Omega_p,\Omega_{p'}) \tag{4.99}$$

where r denotes the order of perturbation. All the zeroth order terms refer to quantities corresponding to the reference system. Now, if we define the effective one-body potential as

$$\Psi^{(r)}(\Omega) = \rho_0 \int f(\Omega_{p'})\, d\Omega_{p'} \int d\mathbf{r}\, u^{(p)}(\mathbf{r},\Omega_p,\Omega_{p'})\, g^{(r-1)}(\mathbf{r},\Omega_p,\Omega_{p'}) \tag{4.100}$$

eq. (4.97) reads

$$\frac{\beta A}{N} = \frac{\beta A^{(0)}}{N} + \frac{1}{2}\beta \int f(\Omega_p)\, d\Omega_p \left[\sum_{r=1}^{\infty} \Psi^{(r)}(\Omega)\right] \tag{4.101}$$

All the vdW types mean-field theories can be derived from eq. (4.101) by considering only the first-order perturbation term which is written as

$$\Psi^{(1)}(\Omega_1) = \rho_0 \int f(\Omega_2)\, d\Omega_2 \int d\mathbf{r}\, u^{(p)}(\mathbf{r},\Omega_1,\Omega_2)\, g^{(0)}(\mathbf{r},\Omega_1,\Omega_2) \tag{4.102}$$

where $g^{(0)}(\mathbf{r},\Omega_1,\Omega_2)$ is the pair correlation function (PCF) for the reference system. The contribution of the reference system $A^{(0)}$ is evaluated in different ways in these works[99–112].

In the derivation of free-energy as a function of $f(\Omega)$ by Gelbart and Baron[101a] the molecular shape repulsions are treated within the approximation of scaled particle theory and the long-range attractions enter directly through a mean-field average.

Basic to the Gelbart and Baron approach is the formulation of the effective one-body potential as an average over the pair attraction which excludes all relative positions denied to a pair of molecules because of their anisotropic hard-cores. The Helmholtz free-energy reads

$$\frac{\beta A}{N} = \frac{\beta A_{hr}}{N} + \frac{1}{2}\beta\int d\Omega_1\, f(\Omega_1)\,\overline{\Psi}^{(1)}_{GB}(\Omega_1) \tag{4.103}$$

with

$$\overline{\Psi}^{(1)}_{GB}(\Omega_1) = \rho_0 \int f(\Omega_2)\, d\Omega_2 \int d\mathbf{r} \exp[-\beta u_{hr}(\mathbf{r},\Omega_1,\Omega_2)]\, u_{att.}(\mathbf{r},\Omega_1,\Omega_2) \tag{4.104}$$

For A_{hr} the relation derived, for a system of hard spherocylindrical molecules of length L and diameter 2a, by Cotter[64b] using SPT was used by Gelbart and Baron

$$\frac{\beta A_{hr}}{N} \cong \frac{\beta A_{hsc}}{N} = \int d\Omega\, f(\Omega)\ln[4\pi f(\Omega)] + \frac{2q'v_0\rho_0}{(1-v_0\rho_0)^2}\left(1+\frac{1}{2}qv_0\rho_0\right)$$

$$\int d\Omega_1\, f(\Omega_1)\int d\Omega_2\, f(\Omega_2)\sin\theta_{12} + \{\text{terms independent of } f(\Omega)\} \tag{4.105}$$

where $v_0 = \pi a^2 L + \frac{4}{3}\pi a^3$ is the rod volume, $q = \frac{4}{3}\pi a^3 / v_0$ and $q' = aL^2 / v_0$. It is important to note here that the eq. (4.105) was rederived by Cotter[100b] (see eq. (4.111)).

Using both the isotropic and orientation dependent parts of the attractive potential, i.e.,

$$u_{att}(\mathbf{r},\Omega_1,\Omega_2) = u^{iso}_{att}(\mathbf{r}) + u^{aniso}_{att}(\mathbf{r},\Omega_1,\Omega_2) \tag{4.106}$$

it has been shown[101b] that the main contribution to the angular dependence of the attractive energy arises from the coupling between u^{iso}_{att} and the hard rod exclusion. Since $\exp(-\beta u_{hr})$ can assume only two values 0 and 1, eq. (4.104) can be written as

$$\overline{\Psi}^{(1)}_{GB}(\Omega_1) = \rho_0 \int f(\Omega_2)\, d\Omega_2 \int_{\xi_0(\Omega_1,\Omega_2)} d\mathbf{r}\,[u^{iso}_{att}(\mathbf{r}) + u^{aniso}_{att}(\mathbf{r},\Omega_1,\Omega_2)] \tag{4.107}$$

It is obvious that both parts, isotropic as well as anisotropic, of the attractive pair potential contribute to the orientation dependence of the mean-potential $\overline{\Psi}^{(1)}$. Thus, even in case of a spherically symmetric attractive interactions the effective potential felt by a single molecule would be orientation dependent. Contrary to it, this would not be true in case the hard cores are replaced by the hard spheres because then the domain of integration ξ_0 becomes independent of Ω_1 and Ω_2 , and only u^{aniso}_{att} contributes to the orientation dependence of $\overline{\Psi}^{(1)}$.

Taking the form of attractive pair potential

$$u_{att}(\mathbf{r},\Omega_1,\Omega_2) = -\frac{1}{r^6}[C_{iso} + C_{aniso}\cos^2\Omega_{12}] \tag{4.108}$$

where C_{iso} and C_{aniso} are the interaction constants and expanding $\overline{\Psi}^{(1)}$ as

$$\overline{\Psi}^{(1)}_{GB}(\rho_0,\Omega) = A_0\,\rho_0 + A_2\rho_0\overline{P}_2P_2(\cos\theta) + A_4\rho_0\overline{P}_4P_4(\cos\theta) \tag{4.109}$$

Gelbart and Gelbart[101b] characterized $\xi_0(r,a,l)$ analytically, then evaluated the integrals of eq. (4.107) numerically and obtained the coefficients A_L's in terms of C_{iso} for $x_0 = 1$, 2, 3 and 4.2, and taking for each case $C_{iso}/C_{aniso} = 8$, 50 and 250. Using Onsager's[58] one parameter representation of $f(\Omega)$

$$f(\Omega) = \xi_0\cosh(\xi_0\cos\theta)/4\pi\sinh\xi_0$$

the order parameters were calculated; ξ_0 was chosen to give $\overline{P}_2 = 0.52$, $\overline{P}_4 = 0.13$, $\overline{P}_6 = 0.017$ for $x_0 = 3$ and $C_{iso}/C_{aniso} = 8$.

Cotter[100] considered the application of van der Waals approach to a model system of hard spherocylinders subjected to a spatially uniform mean-field potential,

$$\overline{\Psi}^{(1)}_{cot} = -\varepsilon_0\,\rho_0 - \varepsilon_2\,\rho_0\overline{P}_2P_2(\cos\theta) \tag{4.110}$$

where ε_0 and ε_2 are the positive energy parameters. Using the rederived[100b] scaled particle expression for A_{hr}

$$\frac{\beta A_{hr}}{N} \equiv \frac{\beta A_{hsc}}{N} = \langle \ell n[4\pi f(\Omega)\rangle + \ell n\left(\frac{\rho_0}{1-v_0\rho_0}\right) + \frac{3v_0\rho_0}{(1-v_0\rho_0)}$$
$$+\frac{[(4+q-\frac{1}{2}q^2)v_0^2\rho_0^2 + 2rv_0\rho_0[3-(1-q)v_0\rho_0]\langle|\sin\Omega_{12}|\rangle}{3(1-v_0\rho_0)^2} \tag{4.111}$$

with $$\langle|\mathrm{Sin}\,\theta_{12}|\rangle = \frac{\pi}{4} - \frac{5\pi}{32}\overline{P}_2^2 \tag{4.112}$$

Extensive numerical calculations were carried out[100b] for the NI transition parameters $x_0 = 3$, $v_0 = 230$ A^3, $\varepsilon_0/v_0k_B = 25000$K and $\varepsilon_2/v_0k_B = 2000$K. The results were compared with the experimental data of PAA (see Table 4.15). A satisfactory qualitative agreement between theory and experiment was found. However, no satisfactory quantitative agreement with experiment was found; the predicted values of the relative density discontinuity, entropy of transition, slope of the P-T coexistence curve and the order parameter at the transiton $\overline{P}_{2NI}$ are too large, while the mean reduced density at the transition is too small.

Table 4.15. Comparison of the NI transition in PAA and in the model systems. The symbols have their meanings as defined in the text.

		Ypma and Vertogen[102]			Savithramma and Mudhusudana[104]		Singh and Singh[105]		Convex peg model[108]	
Quantity	Cotter[100b]	Spherical molecules $\alpha_2=0$	Non-spherical molecules $\alpha_2=0.0168$	TSC approx. $\alpha_2=0.0167$ $\gamma_n=4$	SPT results $x_0=1.75$	Andrews method results $x_0=2.05$	Ref.[105a] $x_0=1.80$	TSC approx.[105d] $x_0=2.0$ $\gamma_n=4$	Aspect ratio 3:1.45:1 (biaxial ellipsoid) $\varepsilon=$ 737.8K	PAA
ε_0/v_0k	25000 k	39000	39000	45000	54181	49650	46058	2830	–	–
$\varepsilon_2/v_0 k$	2000 k	3130	1850	2300	1764	1655	1602	283	–	–
T_{NI}	410.4	408	409	409	409	409	409	409	411	409
$\overline{P}_{2NI}$	0.542	0.437	0.473	0.393	0.455	0.454	0.457	0.411	0.36	0.36–0.4
η_{nem}	0.454	0.593	0.596	0.621	0.62	0.62	0.62	0.484	0.523	0.62
$\Delta\rho/\rho_{nem}$	0.040	0.0038	0.010	0.0039	0.0057	0.0061	0.0067	0.061	0.012	0.0035
$\Delta\Sigma/Nk$	0.887	0.542	0.590	0.295	–	–	0.488	0.651	0.37	0.218
$\Gamma(T_{NI})$	3.90	1.00	3.96	3.96	3.94	3.85	4.0	2.30	4.82	4.0
$\left(\frac{dT_{NI}}{dp}\right)_{p=1bar}$	175K/k bar	19.7	48.6	35.8	–	–	37.2	–	112	43.0

Baron and Gelbart[101c] applied their GvdW theory to systems with spherocylindrical hard cores and attractive forces described by eq. (4.108), and made the calculations using eq. (4.111) for A_{hr} and approximating $f(\Omega)$ by the Onsager representation. The use of Onsager's relation for $f(\Omega)$ introduces substantial errors and so the quantitative agreement between theory and experiment can not be expected. However, it is important to note that the GvdW approach[101], in which $\overline{\Psi}^{(1)}$ is determined from the model pair potential, is clearly superior in comparison to the use of an essentially phenomenological pseudo potential (eq. 4.110) by Cotter[100b]. Further, despite its quantitative inadequacies, the van der Waals approach indicates that the anisotropy of the short-range intermolecular repulsions plays a major role in determining nematic order and stability, and can not be neglected, even to a first approximation. For a model system with hard cores plus attractions, anisotropic hard cores are clearly necessary for explaining the behaviour of nematogens in even a qualitative manner.

A number of computer simulation studies[122–125] on system of hard spherocylinders with $x_0 = 2$ and 3 have been reported. A comparison of these results in the isotropic phase with the results of SPT shows that while SPT gives reasonably good values at the low densities, it overestimates the pressure as the density is increased. Savithramma and Madhusudana[104] extended the Andrews method[77] to derive the thermodynamic properties of an ensemble of spherocylinders. They used this extended model to study the NI transition properties and made calculations in the MF approximation for a system of hard spherocylinders, by using the virial coefficients of the isotropic phase as well as for hard spherocylinders superimposed by the attractive potential of the form (4.110). In the calculation they adjusted the potential parameters so as to get T_{NI} = 409K and $\eta_{nem} \approx 0.62$. From these calculations it was found that (i) the calculation based on the extended Andrews model can be carried out up to $x_0 = 2.9$, while those based on SPT could only be made up to $x_0 = 2.45$, (ii) the trend of different transition quantities as a function of x_0 are similar to those as given by SPT, (iii) in case of Andrews model, $\Gamma = 4$ for $x_0 = 2.075$, which is a substantial improvement over $x_0 = 1.75$ of SPT, and (iv) the calculation[104] leads to reasonably good agreement with the experimental data.

Ypma and Vertogen[102] suggested that a nematic can be looked upon as a normal liquid and derived an equation of state based on the following point of view. The spatial ordering in the liquid was considered to be determined by the hard-core repulsion which was, to a first approximation, replaced by a hard-sphere repulsion. The orientational ordering, resulting due to the effect of the eccentricity of the hard core and the superimposed anisotropic attractive interaction, were treated in two ways. First, a molecular-field approximation was applied. Second, using a two-particle orientational distribution function the short-range orientational order was investigated. The molecules were assumed to interact through the van der Waals dispersion forces. Assuming two particular choices of the attractive potential

$$u_{att}(\mathbf{r},\Omega_1,\Omega_2) = -r^{-6}(\varepsilon_0 + \varepsilon_2 P_2(\cos\theta_{12})) \quad (4.113a)$$

and

$$u_{att}(\mathbf{r},\Omega_1,\Omega_2) = -(\varepsilon_0 + \varepsilon_2 P_2(\cos\theta_{12}))/V \quad (4.113b)$$

the total Helmholtz free-energy was expressed as

$$\frac{\beta A}{N} = \frac{\beta A_{hs}}{N} + \langle \ell n[4\pi f(\Omega)] \rangle - \frac{12\eta\left(1+\frac{1}{2}\eta\right)}{(1-\eta)^2}\alpha_2 \overline{P}_2^2 + \frac{1}{2}\beta\rho_0 \int d\mathbf{r}\, g_{hs}(\mathbf{r})\, (\varepsilon_0(r) + \varepsilon_2(r)\overline{P}_2^2) \quad (4.114)$$

where

$$\frac{\beta A_{hs}}{N} = \ell n\left(\frac{\rho_0}{1-\eta}\right) + \frac{3}{2(1-\eta)^2} - \frac{5}{2} \quad (4.115)$$

and α_2 is a model parameter.

Thus in the approach of Ypma and Vertogen the mean-potential was expressed in terms of the hard-sphere radial distribution function $g_{hs}(r)$

$$\overline{\psi}^{(1)}_{YV} = \rho_0 \int f(\Omega_1)\, d\Omega_1 \int d\mathbf{r}\, g_{hs}(r)\, [\varepsilon_0(r) + \varepsilon_2(r)\overline{P}_2 P_2(\cos\theta_1)] \quad (4.116)$$

Extensive numerical calculations were done for the model potential (4.113) for α_2 = 0 (spherical molecules) as well as for $\alpha_2 \neq 0$ nonspherical molecules. The use of eq. (4.113b) allows an interesting comparison with the results of Cotter[100b]. Similar trends of the variation of NI transition properties were observed as that of Cotter's work. It can be seen from the Table 4.15 that except for Γ the overall agreement of the hard sphere model ($\alpha_2 = 0$) is better than that of Cotter's model. Although, Cotter's basic assumption of a spherocylindrical molecular shape seems more realistic than a spherical shape, it is expected that her model may predict much a worse result if the excluded volume of two sphereocylinders is accounted for more accurately. Similar results were obtained by the hard sphere model with distance - dependent interaction potential (4.113a) which corresponds to the van der Waals dispersion forces.

The influence of a slightly nonspherical molecular shape ($\alpha_2 \neq 0$) on the NI transition properties was studied by taking the model potential (4.113b) and treating the orientational coordinates in MF approximation. For a given set of interaction strengths the variation of packing fraction, order parameter, relative density change, etc., were studied as a function of the eccentricity parameter α_2. It was found that even small values of α_2 have a strong influence on the thermodynamic properties at the transition. With increasing α_2 the steric hindering becomes more effective, shifting the phase transition to lower densities with increasing density change and jump of the order parameter. Further, except for Γ and (dT_{NI}/dp), results for the transition properties have worsened somewhat as compared to hard-sphere model, but still are in better agreement with experiment than the Cotter's results for spherocylinders. The extensive calculations were also performed to analyse the effects of short-range orientational correlations on the NI transition. It was found that the short range orientational correlations have a strong influence on the transition properties and the overall agreement with experiment is quite satisfactory with a smaller number of nearest neighbours γ_n = 3 or 4.

The first-order perturbation theory eq. (4.101) was applied by Singh and co-workers[105] to study in detail the NI transition in a model system of hard ellipsoidal molecules with a superimposed attractive potential described by the dispersion and quadrupolar interactions. The following form for the attractive potential was adopted,

$$u_{att}(\mathbf{r},\Omega_1,\Omega_2) = -C_{id}\, r^{-6} - (C_{ad}\, r^{-6} + C_{aq}\, r^{-5})\, P_2(\cos\theta_{12}) \tag{4.117}$$

The first term represents the isotropic component of the dispersion interaction and the second and third terms describe, respectively, the anisotropic dispersion and quadrupolar interactions. C_{id}, C_{ad} and C_{aq} are their respective constants.

The following expression for the Helmholtz free-energy was obtained[105]

$$\frac{\beta A}{N} = \frac{\beta A_{hr}}{N} + \frac{1}{2}\beta \int f(\Omega)\, d\Omega\, \psi_{ss}^{(1)}(\Omega) \tag{4.118}$$

where

$$\frac{\beta A_{hr}}{N} = \ell n\, \rho - 1 + \langle \ell n[4\pi f(\Omega)] \rangle + \frac{\eta(4-3\eta)}{(1-\eta)^2} \left[F_0(\chi) - F_2(\chi)\overline{P}_2^{\,2} - F_4(\chi)\overline{P}_4^{\,2}\right] \tag{4.119}$$

and

$$\psi_{ss}^{(1)}(\Omega) = -\in_0 - \in_2 \overline{P}_2 P_2(\cos\theta) - \in_4 \overline{P}_4 P_4(\cos\theta) \tag{4.120}$$

Here $F(\chi)$'s are χ-dependent polynomials and $\in$'s are molecular parameters. Extensive numerical calculations were done for two model potentials; the first one assumes $C_{aq} = 0$, while the second $C_{ad} = 0$. These choices enable us to investigate separately the effects of anisotropic dispersion and quadrupole interactions on the transition properties. From these calculations it was found that the x_0 and the interaction parameters have a strong influence on the thermodynamic properties, and the effects are more pronounced in case of quadrupolar interaction. It can be seen from the Table 4.15 that the values reported by Savithramma and Madhusudana[104] and ours[105a] are almost identical and are in quite good agreement with the experimental data of PAA. We also investigated[105d] the influence of short-range orientational correlation on the transition properties by using a two-site cluster (TSC) variation method. Though the results demonstrate that the short-range orientational order has strong influence on the NI transition properties, the theoretical values are overall in poor agreement with the experiment as compared to our earlier work.

Tjipto-Margo and Evans[108] have incorporated the convex peg potential into a vdW theory of NI phase transition. In the convex peg model, the molecules are envisioned to have a hard (biaxial) core embedded in a spherically symmetric square well,

$$u(\mathbf{r},\Omega_1,\Omega_2) = \begin{cases} \infty, & \text{for } r \in V_{exc}(\Omega_1,\Omega_2) \\ -\varepsilon, & \text{for } r \le \sigma \end{cases} \tag{4.121}$$

Equation (4.121) represents repulsion if the cores overlap (reside within the excluded volume V_{exc}) and attraction if the centre-to-centre distance is less than or equal to the sum of the largest semi-major axes (σ). Thus the anisotropies in the convex peg potential are derived from both its repulsive and attractive regions. For the formal calculations the hard core was represented as a general convex body, whereas in the numerical work by a biaxial ellipsoid[108b]. In accordance with the GvdW theories the repulsive interactions were treated to all orders using a resummation technique and the attractive interactions were incorporated to the lowest order. The thermodynamic system was considered to be a fluid of hard biaxial ellipsoids with semi-axes lengths a, b and c. The hard ellipsoid fluid can sustain three stable fluid phases, the uniaxial and biaxial nematic phases and the isotropic fluid. For the formation of biaxial phase, the a, b and c axes lengths of the biaxial ellipsoid must closely approximate the geometric mean condition $a^2 = bc$ and the fluid density must reside in a small density domain. Tjipto-Margo and Evans[108a] did not put these restrictions and consequently, set the two macroscopic biaxial order parameters to zero in the calculations. Global phase diagrams were determined and the NI transition properties were calculated (see Table 4.15). It was found that (i) the model predicts one uniaxial nematic and two isotropic phases (vapour and liquid), (ii) the nematic-vapour, the nematic-isotropic liquid, the vapour-isotropic liquid phases as well as a nematic-vapour-isotropic liquid triple point coexist, (iii) due to the differing strengths of the attractive forces for the prolate and oblate bodies, the polate-oblate symmetry is broken in this model. However, in the limit of very high temperature and very high densities this symmetry is exhibited, (iv) the 'first-orderness' of the NI transition increases as temperature decreases and (v) the NI phase transition properties agree satisfactorily with the experimental data of PAA.

Flory and Ronca[110] extended the lattice model treatment[63] to a system of hard rod molecules subjected to orientation-dependent mutual attractions. They derived the energy relations for a system at constant volume, by considering interactions between pair of segments in contact, rather than in terms of interactions between entire molecules. A characteristic temperature T^* was defined to measure the intensity of these interactions. The orientational energy of the system as a whole was basically of the MS type. Numerical calculations were reported with the findings : Steric effects of molecular shape asymmetry, embodied in the axial ratio x_0, are of foremost importance. The reduced temperature $\tilde{T}_{NI} = T_{NI} / x_0 T^*$ at which the NI transition takes place in the neat liquid decreases with decrease in x_0 below its athermal limit $x_{0crit} = 6.417$ for $\tilde{T}_{NI}^{-1} = 0$. Both the transition entropy and the orientational heat capacity C_p are monotonic through the transition; C_p diverges at a temperature appreciably above T_{NI}, where the metastable anisotropic state becomes unstable. The application of lattice model has also been considered by Dowell[106a] to examine the question - what happens to the relative stabilities of isotropic liquid, nematic, smectic A_1 and smectic S_{A_d} phases formed by rigid cores and partially flexible tails? She assumed site-site hard repulsions, then added soft repulsions and London dispersion attractions using segmental Lennard-Jones (12-6) potentials, and finally added dipolar interactions which include dipole-dipole forces and segmental dipole-induced dipole forces. The results obtained show that it is not necessary to invoke dipolar forces to have stable liquid crystal phases and that the dipolar forces increase the stability of these phases in systems with Lennard-Jones interactions. The dipolar forces also shift the temperature ranges of these phases closer to ambient. In an another work[106b], the NI transition properties were calculated for molecules composed of rigid cores having semi-flexible tails and interacting through

Lennard-Jones potential. The calculated values of transition properties are in agreement with the experimental data of PAA. These works[106] elucidate the importance of realistic intermolecular potentials, particularly the role of soft repulsions in describing an order-disorder transition between two phases.

An orientation-averaged pair correlation (OAPC) theory was proposed by Woo and coworkers[112] which is based on the Bogoliubov-Born-Green-Kirkwood-Yvon (BBGKY) theory and takes account of the pair spatial correlations arising from the intermolecular attractions and repulsions. In this theory Lennard-Jones (12-6) potential was employed but the anisotropy of the pair correlation function was not considered. A model calculation was carried out which treats spatial correlations between molecules as for a classical isotropic liquid and orientational correlations in the mean-field approximation. Generalized BBGKY equations were solved for a simple pairwise potential chosen to fit certain experimental data on PAA. The theory was reformulated[112b] on the basis of a general probability distribution function $P_N(1,..., N)$. It was shown that both the MF theory and the OAPC theory arise from applying low-order approximations to P_N. In the MF theory spatial correlations were introduced and the decoupling of spatial and orientational dependences were assumed. Numerical calculations were carried out for a more general pairwise potential, than that used by Rajan and Woo[112a], by choosing potential parametrs to mimic PAA and MBBA. Nakagawa and Akahane[109] introduced the pair spatial correlation function approximately into the MF theory in terms of an orientation-averaged pair potential and studied its effects on the NI phase transition in the decoupling approximation. The approximated pair spatial correlation function[131] is much simpler than Woo's and coworkers but involves the deviation from the spherical symmetry. The numerical calculations were done for a model potential of Lennard Jones (12-6) type and retaining the leading coefficients of the orientation - averaged pair potential. It was found that the anisotropy of a pairwise intermolecular potential increases both the long-range orientational order and the NI transition temperature. The order parameter change near the transition point is in good agreement with the data for PAA but the transition entropy and C_v do not agree satisfactorily.

The molecular-field theories have been reformulated by Vertogen and de Jeu[132]. Based on the so-called spherical version of the models, the equation of state for nematics was derived as a generalization of the theory of simple liquids[133] and the calculations were performed in the MF approximation. Starting from this equation of state for nematics the two well known approaches, Onsager[58] and Maier-Saupe[78], can be recovered by a vector of variable length, but such that its thermally averaged length is always equal to one. In view of the mathematical difficulties it is extremely hard to arrive at a reliable equation of state for a simple model of nematics from first principles. Therefore, only a qualitatively correct equation of state has been discussed by Vertogen and de Jeu. In complete analogy to the theory of simple liquids[133], the partition function of the model system is given by

$$Z = Z_{hr}\left\langle \exp\left[\frac{1}{2}\beta \sum_{\kappa\neq\ell} u_{att}(r_{\kappa\ell}, \Omega_\kappa \Omega_\ell)\right]\right\rangle_0 \tag{4.122}$$

where Z_{hr} denotes the partition function of the hard-rod model and the thermal average is taken with respect to the hard-rod system. Thus the calculation of the equation of state contains two parts : the calculation of the partition function of the hard-rod system and the calculation of the thermal average for an assumed attractive interaction. The problem thus boils down to the calculation of

$$Z_{hr} = \frac{1}{N!(\lambda\tau)^{3N}} \int \prod_{i=1}^{N} dr_i \, d\Omega_i \exp\left[-\frac{1}{2}\beta \sum_{\kappa,\ell=1}^{N} u_{hr}\right] \tag{4.123}$$

where the constant τ is given by

$$\tau = \frac{h}{[(2\pi)^{5/3} k_B T I_1^{2/3} I_3^{1/3}]^{1/2}} \tag{4.124}$$

and

$$\lambda = h/(2\pi m k_B T)^{1/2}$$

Here I_1 and I_3 are the principal moments of inertia of the cylinder with I_3 about the cylinder axis. The calculation of the partition function is still an unsolved problem. The integration over the position coordinates $\mathbf{r}_i$ as well as over the Eulerian angles Ω_i cannot be carried out analytically. The usual way is to first deal with the integration over the position coordinates. According to the vdW approximation the integral over the position coordinates eq. (4.123) can be approximated in a qualitative sense by

$$Z(\Omega_1, \Omega_2, ..., \Omega_N) = \frac{1}{N!} \prod_{i=1}^{N} \left[V - \frac{1}{2} \sum_{\substack{j\neq i \\ j=1}}^{N} V_{exc}(\Omega_{12}) \right] \tag{4.125}$$

An approximate expression for the excluded volume of two cylinders has been given by Vertogen and de Jeu[132]. A comparison between this expression and the exact Onsager result shows that the approximation has the correct behaviour for all values of x_0.

The final step now concerns the calculation of eq. (4.123) expressed as

$$Z_{hr} = \frac{1}{(\lambda\tau)^{3N}} \int Z(\Omega_1, \Omega_2, ..., \Omega_N) \, d\Omega_1 d\Omega_2 ... d\Omega_N \tag{4.126}$$

with

$$Z(\Omega_1,\Omega_2,...,\Omega_N) = \frac{V^N}{N!} \exp\left\{\sum_{i=1}^{N} \ell n\left[1 - \frac{1}{2V}\sum_{\substack{j\neq i\\ j=1}}^{N} V_{exc}(\Omega_{12})\right]\right\} \quad (4.127)$$

Now applying the MF approximation the Helmholtz free-energy of the hard-rod system is approximated as

$$\frac{\beta A_{hr}}{N} = 3\ell n\,\lambda + 3\ell n\,\tau + \ell n\,\rho_0 - 1 - \ell\, n\left[1 - \frac{1}{2}\rho_0 \langle V_{exc}(\Omega_{12})\rangle\right] + \langle \ell n\, f(\Omega)\rangle \quad (4.128)$$

It is extremely difficult to evaluate the thermal average over the hard rod system and so has to be approximated

$$\left\langle \exp\left[\frac{1}{2}\beta \sum_{\kappa\neq\ell} u_{att}(r_{\kappa\ell},\Omega_\kappa,\Omega_\ell)\right]\right\rangle_0 = \exp\left[\frac{1}{2}\beta \sum_{\kappa\neq\ell} \langle u_{att}(r_{\kappa\ell},\Omega_\kappa,\Omega_\ell)\rangle_0\right] \quad (4.129)$$

where

$$\langle u_{att}(\mathbf{r}_{\kappa\ell},\Omega_\kappa,\Omega_\ell)\rangle_0 = \frac{1}{V^2}\int d\mathbf{r}_\kappa\, d\mathbf{r}_\ell d\Omega_\kappa d\Omega_\ell\, g_{hr}(\mathbf{r}_{\kappa\ell},\Omega_\kappa,\Omega_\ell) u_{att}(\mathbf{r}_{\kappa\ell},\Omega_\kappa,\Omega_\ell) \quad (4.130)$$

In the spirit of vdW theory, eq. (4.129) can be approximated as

$$\left\langle \exp\left[\frac{1}{2}\beta \sum_{\kappa\neq\ell} u_{att}(r_{\kappa\ell},\Omega_\kappa,\Omega_\ell)\right]\right\rangle_0 = \exp\left[\frac{1}{2}N\beta\rho_0\bar{u}_0 + \frac{1}{2}N\beta\rho_0\bar{u}_2\langle P_2(\cos\theta_{12})\rangle\right] \quad (4.131)$$

Starting from the approximate expression[132]

$$V_{exc}(\Omega_{12}) = A_0 - A_1 P_2(\cos\theta_{12}) \quad (4.132)$$

for the excluded volume of two rods, the Helmholtz free energy in the spherical version of the model is obtained[132] as,

$$\begin{aligned}\frac{\beta A}{N} = {} & 3\ell n\,\lambda + 3\ell n\,\tau + \ell n\,\rho_0 - \frac{7}{4} - \ell n\left[1 - \frac{1}{2}\rho_0(A_0 - A_1\bar{P}_2^{\,2})\right] \\ & - \frac{1}{2}\beta\rho_0\bar{u}_0 - \frac{1}{2}\beta\rho_0\bar{u}_2\bar{P}_2^{\,2} - \frac{1}{4}B\bar{P}_2 - \frac{3}{4}\left(B^2\bar{P}_2^{\,2} - \frac{2}{3}B\bar{P}_2 + 1\right)^{1/2} \\ & + B\bar{P}_2^{\,2} - \frac{2}{3}\ell n\frac{2\pi}{3} - \ell n(1-\bar{P}_2) - \frac{1}{2}\ell n(1+2\bar{P}_2)\end{aligned} \quad (4.133)$$

Using the thermodynamic relation the equation of state is given by

$$\beta p = \frac{\rho_0}{1-\frac{1}{2}\rho_0(A_0 - A_1\overline{P}_2^2)} - \frac{1}{2}\beta\rho_0^2\overline{u}_0 - \frac{1}{2}\beta\rho_0^2\overline{u}_2\overline{P}_2^2 \qquad (4.134)$$

In the above equations A's, B and $\overline{u}$'s are potential parameters.

As expected the equation of state (4.134) is found to describe the NI transition rather well from the qualitative point of view. The experimental curve of order parameter and the values of transition density can be reproduced fairly well by an appropriate choice of the parameters. However, the required parameter values do not correlate with the molecular structure. Thus the derived equation of state is unable to describe the structure-property relationship in a physically satisfactory way. The discrepancies are not at all surprising in view of the simplicity of the model. Some of these can be attributed to rather poor approximation methods, which overestimate a number of quantities, e.g., the influence of the excluded volume. Another most important source of the difficulties in interpretation is the poor representation of the intermolecular interactions and the neglect of molecular flexibility.

4.4.5 Application of Density Functional Theory (DFT) to the NI Phase Transition

The density functional theory[113–121,134–151] is developing as a cost effective procedure for studying the physical properties of nonuniform systems. The theory was pioneered by Kohn and others[135,136], and Mermin[137] for a quantum mechanical many-body system. Its initial classical version and development are by Lebowitz and Percus[138], and the present form by Saam and Ebner[139], and Yang et al[140]. The general formalism of the classical version of DFT has played useful role in the development of approximate integral equations for the pair distribution functions of a uniform fluid[141]. The theory has subsequently been used (see Ref. 113) profitably in the study of freezing of a variety of fluids, solid-melt interfaces, nucleation, liquid crystals, two dimensional systems, molten salts, binary mixtures, aperiodic crystals or glasses, quasi-crystals, the elastic properties of solids and liquid crystals, spectroscopic properties, dislocations and other topological defects, flexoelectricity in liquid crystals, interfacial phenomena and wetting transitions, etc. In a review, Singh[113] has given an excellent account of DFT, its application to some of the above development and prospects. The application of the theory to study the elastostatics in liquid crystals has been given in a self-contained manner by the author[152]. A few other articles and monographs[142–150] are also available. Thus the theory has been clearly discussed several times in literature and the essentials of the approach are well documented. Therefore, we present[31] here a brief account of the theory relevant to the phase transition studies in liquid crystals.

In the absence of an external field, the singlet distribution function is written as

$$\rho(\mathbf{x}) = \frac{1}{\Lambda}\exp\left[\beta\mu + C(\mathbf{x})\right] \qquad (4.135)$$

where Λ is the cube of the thermal wavelength associated with a molecule, $-k_BTC(\mathbf{x})$ is the solvent mediated potential field acting at $\mathbf{x}$. The single-particle direct correlation function $C(\mathbf{x})$ is a functional of $\rho(\mathbf{x})$ and is related to the Ornstein-Zernike (OZ) direct correlation function by the relation,

$$\frac{\delta C(\mathbf{x}_1)}{\delta \rho(\mathbf{x}_2)} = C(\mathbf{x}_1, \mathbf{x}_2) \tag{4.136}$$

The excess Helmholtz free-energy,

$$\beta \Delta A = \beta (A - A_{id}) \tag{4.137a}$$

is also related to C(x) through the relation,

$$\frac{\delta(\beta \Delta A)}{\delta \rho(\mathbf{x})} = - C(\mathbf{x}) \tag{4.137b}$$

Here βA_{id} is the ideal gas part of the reduced Helmholtz free-energy. These are the starting equations of the density functional theory and have been used to develop a variety of approximate forms for the free-energy functionals. The functional integration of eq. (4.136) from some initial density ρ_f (of isotropic liquid) to final density $\rho(\mathbf{x})$ (of the ordered phase) gives

$$C(\mathbf{x}_1;[\rho(\mathbf{x})]) - C(\mathbf{x}_1;\rho_f) = \int \tilde{C}(\mathbf{x}_1, \mathbf{x}_2; \rho(\mathbf{x}))\, \Delta\rho(\mathbf{x}_2)\, d\mathbf{x}_2 \tag{4.138}$$

where

$$\tilde{C}(\mathbf{x}_1, \mathbf{x}_2; \rho(\mathbf{x})) = \int_0^1 d\alpha\, C(\mathbf{x}_1, \mathbf{x}_2; (\rho_f + \alpha[\Delta\rho(\mathbf{x})])) \tag{4.139}$$

and

$$\Delta\rho(\mathbf{x}) = \rho(\mathbf{x}) - \rho_f \tag{4.140a}$$

$$\Delta\rho(\mathbf{x}_i) = \rho(\mathbf{x}_i) - \rho_f \tag{4.140b}$$

In the above equations [ ] bracket indicates the functional dependence of the quantities on the single particle distribution function. Parameter α characterizes a path in density space along which the chemical potential remains constant. From the functional Taylor expansion, it follows that

$$C(\mathbf{x}_1, \mathbf{x}_2; (\rho_f + \alpha[\Delta\rho(\mathbf{x})])) = C(\mathbf{x}_1, \mathbf{x}_2; \rho_f) + \alpha \int d\mathbf{x}_3\, C(\mathbf{x}_1, \mathbf{x}_2, \mathbf{x}_3; \rho_f)\, \Delta\rho(\mathbf{x}_2)\, \Delta\rho(\mathbf{x}_3) + \ldots \tag{4.141}$$

Combining eqs. (4.138) and (4.141), we obtain

$$\ln(\rho(\mathbf{x}_1)/\rho_f) = \int d\mathbf{x}_2\, C(\mathbf{x}_1, \mathbf{x}_2; \rho_f)\, \Delta\rho(\mathbf{x}_2) + \frac{1}{2} \int d\mathbf{x}_2\, d\mathbf{x}_3\, \Delta\rho(\mathbf{x}_2)\, C(\mathbf{x}_1, \mathbf{x}_2, \mathbf{x}_3; \rho_f)\, \Delta\rho(\mathbf{x}_3) \tag{4.142}$$

This is a nonlinear equation and relates the single particle density distribution of the ordered phase to the direct correlation functions of the coexisting liquid.
The functional integration of (4.137b) from ρ_f to $\rho(\mathbf{x})$ and use of the eqs. (4.138)-(4.141), leads to,

$$\beta(\Delta A - \Delta A_f) = -\int d\mathbf{x}\, C(\mathbf{x};\rho_f)\,\Delta\rho(\mathbf{x}) - \frac{1}{2}\int d\mathbf{x}_1\, d\mathbf{x}_2\, \Delta\rho(\mathbf{x}_1)\, C(\mathbf{x}_1,\mathbf{x}_2;\rho_f)\,\Delta\rho(\mathbf{x}_2) - \frac{1}{6}\int d\mathbf{x}_1\, d\mathbf{x}_2\, d\mathbf{x}_3\, \Delta\rho(\mathbf{x}_1)\,\Delta\rho(\mathbf{x}_2)\, C(\mathbf{x}_1,\mathbf{x}_2,\mathbf{x}_3;\rho_f)\,\Delta\rho(\mathbf{x}_3) + \ldots.. \tag{4.143}$$

The Grand thermodynamic potential, which is generally used to locate the transition, is defined as

$$-W = \beta A - \beta\mu \int d\mathbf{x}\, \rho(\mathbf{x}) \tag{4.144}$$

Substituting the value of βA, we obtain correct to second order in $\Delta\rho(\mathbf{x})$

$$\Delta W = W - W_f = \int d\mathbf{x}\, (\rho(\mathbf{x})\, \ell n\, [\rho(\mathbf{x})/\rho_f] - \Delta\rho(\mathbf{x})) - \frac{1}{2}\int d\mathbf{x}_1\, d\mathbf{x}_2\, \Delta\rho(\mathbf{x}_1)\, C(\mathbf{x}_1,\mathbf{x}_2;\rho_f)\,\Delta\rho(\mathbf{x}_2) - \frac{1}{6}\int d\mathbf{x}_1\, d\mathbf{x}_2\, d\mathbf{x}_3\, \Delta\rho(\mathbf{x}_1)\,\Delta\rho(\mathbf{x}_2)\, C(\mathbf{x}_1,\mathbf{x}_2,\mathbf{x}_3;\rho_f)\,\Delta\rho(\mathbf{x}_3) \tag{4.145a}$$

where W_f is the grand canonical thermodynamic potential of the isotropic liquid. Combining eqs. (4.142) and (4.145a), we get for ΔW an expression which is found to be convenient in many applications,

$$\Delta W = -\int d\mathbf{x}\, \Delta\rho(\mathbf{x}) + \frac{1}{2}\int d\mathbf{x}_1\, d\mathbf{x}_2\, (\rho(\mathbf{x}_1) + \rho_f)\, C(\mathbf{x}_1,\mathbf{x}_2;\rho_f)\,\Delta\rho(\mathbf{x}_2) + \frac{1}{6}\int d\mathbf{x}_1\, d\mathbf{x}_2\, d\mathbf{x}_3\, (2\rho(\mathbf{x}_1) + \rho_f)\,\Delta\rho(\mathbf{x}_2)\, C(\mathbf{x}_1,\mathbf{x}_2,\mathbf{x}_3;\rho_f)\,\Delta\rho(\mathbf{x}_3) \tag{4.145b}$$

Equations (4.142) and (4.145) are the basic equations of the theory of freezing and of interfaces of ordered phase and its melt. Of interest is the solutions of $\rho(\mathbf{x})$ of eq. (4.142) having the symmetry of ordered phase. These solutions, inserted in eq. (4.145), give the grand potential difference between the ordered and liquid phases. The phase with the lowest grand potential is taken as the stable phase. Phase coexistence occurs at the value of ρ_f which makes $\Delta W = 0$ for the ordered and liquid phases.

The grand thermodynamic potential is considered to be a functional of the singlet distribution $\rho(\mathbf{r},\Omega)$. Equation (4.145) can be written as

$$\Delta W = W - W_f = \Delta W_1 + \Delta W_2 \tag{4.146}$$

with

$$\frac{\Delta W_1}{N} = \frac{1}{\rho_f V} \int \mathbf{dr}\, d\Omega(\rho(\mathbf{r},\Omega)\ \ell n(\rho(\mathbf{r},\Omega)/\rho_f) - \Delta\rho(\mathbf{r},\Omega)) \tag{4.147a}$$

and

$$\frac{\Delta W_2}{N} = -\frac{1}{2\rho_f V} \int \mathbf{dr}\, d\Omega_1\, d\Omega_2\, \Delta\rho(\mathbf{r}_1,\Omega_1)\, C(\mathbf{r},\Omega_1,\Omega_2)\, \Delta\rho(\mathbf{r}_2,\Omega_2) \tag{4.147b}$$

The density of the ordered phase can be obtained by minimizing ΔW with respect to arbitrary variation in the ordered phase density subject to the constraint that there is one molecule per lattice site (for perfect crystal) and/or orientational distribution is normalized to unity. Thus

$$\ell n\,(\rho(\mathbf{r}_1,\Omega_1)/\rho_f) = \alpha_\ell + \int \mathbf{dr}_2\, d\Omega_2\, C(\mathbf{r},\Omega_1,\Omega_2;\rho_f)\, \Delta\rho(\mathbf{r}_2,\Omega_2) \tag{4.148}$$

where α_ℓ is a Lagrange multiplier which appears in the equation because of constraint imposed on the minimization. For locating the transition one attempts to find the solution of $\rho(\mathbf{r},\Omega)$ of eq. (4.148) which has the symmetry of the ordered phase. Below a certain liquid density, say ρ', the only solution is $\rho(\mathbf{r},\Omega) = \rho_f$. Above ρ' a new solution of $\rho(\mathbf{r},\Omega)$ can be obtained which corresponds to the ordered phase. The phase with lowest grand potential is taken as the stable phase. The transition point is determined by the condition ΔW = 0. Implicit in this approach is an assumption according to which the system is either entirely liquid or entirely ordered phase, no phase coexistence is permitted. This signifies the mean-field character of the theory.

For axially symmetric molecules, the singlet density of the nematic phase can be expressed as

$$\rho(\mathbf{r},\Omega) = \rho_n\, f(\Omega) \tag{4.149}$$

with

$$\rho_n = \rho_f(1 + \Delta\rho^*) \tag{4.150}$$

and

$$f(\Omega) = 1 + \sum_{\ell \ge 2}{}' (2\ell+1)\, \overline{P}_\ell\, P_\ell(\cos\theta) \tag{4.151}$$

where $\Delta\rho^* = (\rho_n - \rho_f)/\rho_f$ is the relative change in the density at the transition and ρ_n be the number density of the nematic phase.

For the nematic phase, it is convenient to use the following ansatz for f(Ω) :

$$f(\Omega) = A_0 \exp\left[\sum_\ell \alpha_\ell\, P_\ell(\cos\theta)\right] \tag{4.152}$$

A_0 is determined from the normalization condition (4.59). When $\alpha_\ell \to 0$, $f(\Omega) \to 1$ corresponding to the isotropic phase.

The entropy term in eq. (4.147a) can be reduced to the form

$$\frac{\Delta W_1}{N} = -\Delta\rho^* + (\rho_n/\rho_f)\left\{\ell n\left[\frac{\rho_n A_0}{\rho_f}\right] + \sum_{\ell\geq 2}{}'\alpha_\ell \overline{P}_\ell\right\} \quad (4.153)$$

The interaction term $\Delta W_2/N$ is evaluated using eq. (4.152) for $f(\Omega)$,

$$\frac{\Delta W_2}{N} = -\frac{1}{2}\Delta\rho^{*2}\,\hat{C}^0_{00} - \frac{(1+\Delta\rho^*)^2}{2}\sum_{\ell_1,\ell_2\geq 2}{}' \overline{P}_{\ell_1}\overline{P}_{\ell_2}\hat{C}^{(0)}_{\ell_1\ell_2} \quad (4.154)$$

where $\hat{C}^{(0)}_{\ell_1\ell_2}$ is the structural parameter. Expanding $C(\mathbf{r},\Omega_1,\Omega_2)$ in spherical harmonics in space-fixed frame

$$C(\mathbf{r},\Omega_1,\Omega_2) = \sum_{\ell_1\ell_2\ell}\sum_{m_1m_2m} C_{\ell_1\ell_2\ell}(r)\,C_g(\ell_1\ell_2\ell;\, m_1m_2m)$$
$$Y_{\ell_1m_1}(\Omega_1)\,Y_{\ell_2m_2}(\Omega_2)\,Y^*_{\ell m}(\hat{\mathbf{r}}) \quad (4.155)$$

from eq. (4.148) one finds the expression for the order parameter as

$$(1+\Delta\rho^*)\,\delta_{\ell 0} + \overline{P}_\ell = \int d\Omega_1\, P_\ell(\cos\theta_1)\exp[\alpha_\ell + \rho_f \sum_{\ell_1\ell_2}{}'((2\ell_1+1)(2\ell_2+1))^{1/2}$$
$$\times\, C_g(\ell_1\ell_2 0;000)\; P_\ell(\cos\theta_1)\,\overline{P}_\ell \int dr\, r^2\, C_{\ell_1\ell_2 0}(r)\,] \quad (4.156)$$

In order to find the solutions of above equations, the values of $C_{\ell_1\ell_2\ell}(r)$ are required which can be obtained by solving the integral equations[117c–f].

4.4.5.1 Modified weighted-density approximation (MWDA)

The theory of weighted density approximation (WDA) was originally proposed by Nordholm et. al.[153], and Tarazona[154]; later refined by Curtin and Ashcroft[155], and finally modified (MWDA) by Denton and Ashcroft[156].

The Helmholtz free energy of an inhomogeneous system can be expressed as the sum of two contributions : an ideal gas part A_{id} (known exactly), and an excess contribution A_{ex} due to interaction between the particles,

$$A[\rho(\mathbf{r},\Omega)] = A_{id}[\rho(\mathbf{r},\Omega)] + A_{ex}[\rho(\mathbf{r},\Omega)] \quad (4.157)$$

Here both terms are a unique functional[142] of the one-particle density $\rho(\mathbf{r},\Omega)$. The first term is given by

$$A_{id}[\rho(\mathbf{r},\Omega)] = \beta^{-1}\int_V d\mathbf{r}\, d\Omega\, \rho(\mathbf{r},\Omega)\,\{\ell n\,[\rho(\mathbf{r},\Omega)\Lambda] - 1\} \quad (4.158)$$

The second term of eq. (4.157) is the excess Helmholtz free-energy of the nonuniform system. Unlike the ideal gas contribution, this term is not known exactly, and is the focus of attention in the approximations presented by density functional theories. In the MWDA the excess free energy of the nonuniform system is approximated by the excess free energy of a uniform system, but evaluated at a weighted density $\hat{\rho}$;

$$A_{ex}[\rho(\mathbf{r},\Omega)] = A_{ex}^{MWDA}[\rho(\mathbf{r},\Omega)] = Nf_0(\hat{\rho}) \qquad (4.159)$$

where $f_0(\hat{\rho})$ is the excess free-energy per particle of a uniform system at density $\hat{\rho}$. The weighted density is defined by[156]

$$\hat{\rho} = \frac{1}{N}\int d\mathbf{r}_1\, d\Omega_1\, \rho(\mathbf{r}_1,\Omega_1) \int d\mathbf{r}_2\, d\Omega_2\, \rho(\mathbf{r}_2,\Omega_2)\, w(\mathbf{r}_1,\mathbf{r}_2,\Omega_1,\Omega_2;\hat{\rho}) \qquad (4.160)$$

with the constraint that the weight function $w(\mathbf{r}_1,\mathbf{r}_2,\Omega_1,\Omega_2,\hat{\rho})$ must satisfy the normalization condition

$$\int d\mathbf{r}_1\, d\mathbf{r}_2\, d\Omega_1\, d\Omega_2\, w(\mathbf{r}_1,\mathbf{r}_2,\Omega_1,\Omega_2;\hat{\rho}) = 1 \qquad (4.161)$$

For the molecular fluid, the explicit form for w can be written as

$$w(\mathbf{r}_1,\mathbf{r}_2,\Omega_1,\Omega_2;\hat{\rho}) = -\frac{1}{2f_0'(\hat{\rho})}\left[\beta^{-1} C(\mathbf{r}_1,\mathbf{r}_2,\Omega_1,\Omega_2;\hat{\rho}) + \frac{1}{V}\hat{\rho}\, f_0''(\hat{\rho})\right] \qquad (4.162)$$

The primes on $f_0(\hat{\rho})$ indicate derivatives with respect to density. For the nematic phase, one obtains

$$\hat{\rho} = \rho_n\left[1 - \frac{\pi x_0 d_0^3}{12\hat{\eta}\,\beta\, f_0'(\hat{\rho})}\left(\overline{P}_2^{\,2}\, \hat{C}_{22}^{(0)} + \overline{P}_4^{\,2}\, \hat{C}_{44}^{(0)}\right)\right] \qquad (4.163)$$

Having computed $\hat{\rho}$, the next step is to substitute $\hat{\rho}$ into eq. (4.159) to calculate A_{ex}^{MWDA}. Using eqs. (4.149) and (4.152), the nonuniform ideal gas contribution $A_{id}[\rho(\mathbf{r},\Omega)]$ becomes

$$\beta A_{id}[\rho(\mathbf{r},\Omega)] = V\rho_n\left[\ell n(A_0\, \rho_n\, \Lambda) - 1 + \sum_{\ell}{}' \alpha_\ell\, \overline{P}_\ell\right] \qquad (4.164)$$

The total free-energy of the nematic phase is now written as

$$\beta A = \beta N f_0(\hat{\rho}) + V\rho_n\left[\ell n(A_0\, \rho_n\, \Lambda) - 1 + \sum_{\ell}{}' \alpha_\ell\, \overline{P}_\ell\right] \qquad (4.165)$$

4.4.5.2 Calculations and results

The usual approach to the above theory is to first obtain some approximation to evaluate the DPCF, $C(\mathbf{r},\Omega_1,\Omega_2)$, appearing in eq. (4.146). Next to choose a parametrization of the one body density that is appropriate for the kind of expected phase transitions. With these two requirements specified, the functional (4.146) is completely defined and the calculation proceeds by the minimization of ΔW with respect to the parameters in $\rho(\mathbf{r},\Omega)$. When a configuration other than the isotropic liquid satisfies $\Delta W = 0$, the ordered and isotropic phases coexist and all the transition properties can be identified.

The ordered phase (translationally and/or orientationally) coexists with the isotropic liquid when

$$\left(\frac{\partial}{\partial \xi_i}\right)(\Delta W / N) = 0 \qquad (4.166)$$

and

$$\Delta W / N = 0 \qquad (4.167)$$

where ξ_i are variational parameters appropriate for the phase under investigation. Equations (4.166) and (4.167) show a stability condition and a phase coexistence condition. The DFT calculations are done by minimizing eq. (4.146) with respect to the variational parameters ρ_n, α_2 and α_4 (eq. (4.148)). The coexistence point is then located by varying ρ_f with $\Delta W/N = 0$. In MWDA the effective density $\hat{\rho}$ is computed first and then the free energy (eq. (4.165)) is minimized with respect to ρ_n, α_2 and α_4. The transition densities of the coexisting phases are determined by equating the pressures and chemicals potentials of the two phases.

The structural parameters for the nematic phase can be defined as

$$\hat{C}^{(0)}_{\ell_1\ell_2} = (2\ell_1+1)(2\ell_2+1)\rho_f \int d\mathbf{r}\, d\Omega_1\, d\Omega_2\, C(\mathbf{r},\Omega_1,\Omega_2)\, P_{\ell_1}(\cos\theta_1)\, P_{\ell_2}(\cos\theta_2) \qquad (4.168)$$

Using eq. (4.156) after simplification eq. (4.168) can be written as

$$\hat{C}^{(0)}_{\ell_1\ell_2} = \left[\frac{(2\ell_1+1)(2\ell_2+1)}{4\pi}\right]^{1/2} \rho_f\, C_g(\ell_1\ell_2 0, 000) \int d\mathbf{r}\, r^2\, C_{\ell_1\ell_2 0}(\mathbf{r}) \qquad (4.169)$$

The theory was successfully applied to the study of crystallization of particles interacting via spherical pair potentials including the case of hard spheres[157–160]. Singh and Singh[117b] have pioneered the study of transitions to the plastic and nematic states, within the decoupling approximation, in the hard ellipsoidal system using this theory. For a plastic crystal which has a crystalline lattice for the center of mass of molecules but orientational distribution like liquid

$$\rho(\mathbf{r},\Omega) \equiv \rho(\mathbf{r}) = [1 + \Delta\rho^* + \sum_q \mu_q\, e^{i\mathbf{G}_q\cdot\mathbf{r}}]$$

where $\Delta\rho^*$ is the fractional change in density due to transition. The Fourier coefficients μ_q are the order parameters of the isotropic-plastic transiton. The calculation for isotropic-plastic transition was performed by assuming the crystalline lattice of the plastic phase to be fcc with lattice parameter a determined self-consistently by the relation

$$a = (4/\rho_0)^{1/3}; \qquad \rho_p = \rho_\ell(1 + \Delta\rho^*)$$

where ρ_p is the mean number density of the plastic phase and ρ_ℓ is the density of coexisting isotropic liquid.
It was predicted that the equilibrium positional freezing (plastic) on fcc lattice takes place for the value of $\hat{C}^1_{0,0}$ [$=1-1/S(|G_m|)$, where $S(|G_m|)$ is the first peak in the structure factor of the centre of mass] ≈ 0.67 or $S(|G_m|) \approx 3.07$. The equilibrium orientational freezing (nematic) takes place when the orientational correlation $\hat{C}^0_{2,2} \approx 4.45$. Further, the plastic phase stabilizes first for $0.57 \leq x_0 \leq 1.75$ and the nematic phase for $x_0 < 0.57$ and $x_0 > 1.75$. Though these values are in reasonable agreement with simulation results[126b], this work suffers with certain uncertainties, the most important of which is the lack of information about $C(\mathbf{r},\Omega_1,\Omega_2)$, which even in the case of the isotropic state is a function of four independent variables. This forces the use of what is essentially a guess for $C(\mathbf{r},\Omega_1,\Omega_2)$. Other problems with the initial application of the density functional approach are that the oriented crystal phase was not considered and that the parametrization of the plastic crystal phase was done so that the rather narrow real-space lattice peaks associated with hard core crystallization could not be generated. In an attempt to remedy some of these problems Marko[130] applied DFT to study the properties of plastic phase at the isotropic-plastic transition, of oriented solid phase at the isotropic-solid transition, and of nematic phase at the isotropic-nematic transition in a fluid of prolate hard ellipsoids. In this study the direct correlations were computed with a variational technique based on the Percus-Yevick (PY) equation for the correlation functions. Trial correlation functions are chosen and then optimal solution is obtained using the PY equation. The calculation procedure concerns with the direct calculation of excess grand potential for the ordered phase, followed by the resulting nonlinear functional. The first step is to specify the one-body distribution in terms of a few parameters which can describe the ordered phase reasonably. For the plastic phase, the trial one body density is taken as a sum of Gaussian distributions centered on real-space lattice sites and is described by three parameters[130a]

$$\rho(\mathbf{x}) = \rho_\ell\, \Delta(\pi\sigma_\ell^2)^{-3/2} \sum_{t\in T_f} \exp[-\sigma_\ell^{-2}(r-at)^2] \tag{4.170}$$

where ρ_ℓ is the ordered phase density, σ_ℓ is the lattice site distribution width and Δ is the volume per lattice site. In case of fcc lattice $\Delta = a^3/\sqrt{2}$. T_f is the set of fcc lattice vectors with unit nearest-neighbour spacing.
The trial density for the nematic was taken as

$$\rho(\mathbf{x}) = \rho_\ell \exp\left[a_1(\hat{e}\cdot\hat{z})^2 + a_2(\hat{e}\cdot\hat{z})^4\right]\left[\int_0^1 dx' \exp(a_1 x'^2 + a_2 x'^4)\right]^{-1} \tag{4.171}$$

where a_1 and a_2 parametrize the orientational distribution function. The trial DCF and pair distribution were assumed to be of the form[130b]

$$C(\mathbf{r},\Omega_1,\Omega_2) = C_0(r^*)(1 + \alpha P_2(\cos\theta_{12})) \tag{4.172a}$$

and

$$g(\mathbf{r},\Omega_1,\Omega_2) = g_0(r^*)(1 + \alpha P_2(\cos\theta_{12})) \tag{4.172b}$$

where C_0 and g_0 are, respectively, the hard sphere DCF and PY correlation function[161], and $r^* = r/D(\Omega_{12})$. The integration of the PY equation squared over the hard-core region yields a functional that quantifies the accuracy of the trial solution, and thus should be minimized with respect to α to obtain the best solution. The integration over only the core region emphasizes the DCF rather than the pair distribution, and limits the integration to a finite region, which is important numerically. The error functional was defined as

$$I(\alpha) = \int d\mathbf{x}_1\{1 + C(\mathbf{x}_1,0;\alpha) + \rho\int d\mathbf{x}_2\, C(\mathbf{x}_2,0;\alpha)\,[g(\mathbf{x}_1,\mathbf{x}_2;\alpha) - 1]\}^2 \tag{4.173}$$

$I(\alpha)$ was computed numerically for prolate ellipsoids with anisotropies x_0 = 1.0, 1.22, 1.53, 2.0 and 3.0 and for the packing fractions η = 0.1, 0.2, ..., 0.7 over appropriate ranges of α (η = 0.7405 is the close-packing density). This correction makes the DCF more negative, corresponding to a larger free-energy cost, for parallel ellipsoidal configurations. For perpendicular configurations, the correction makes the DCF less negative, corresponding to a smaller free-energy cost of these configurations.

The DCF thus evaluated were used in free-energy functional (4.146) with the plastic crystal and nematic distributions (eqs. (4.170) and (4.171)) and global minima were determined over ranges of liquid density. The isotropic-plastic, isotropic-solid and isotropic-nematic transition properties were calculated. It was found that for small anisotropies (x_0 < 1.5) the isotropic fluid to plastic-crystal densities are unaffected by the small correction. However, the results show an increase in transition density as the anisotropy increases which is in accordance with the MC simulations but indicates the opposite trend as observed by Singh and Singh[117b]. For 2 ≤ x_0 ≤ 3, although the isotropic-nematic transition density agrees well with the simulation results[126b], the density discontinuity is too small, and the results of density functional calculations depend crucially on the approximations made for the pair DCF. Thus it is feasible to obtain information about the DCF by the numerical means, even in case of orientation dependent interactions.

A general method was developed by Perera et al[162] to solve numerically the hypernetted chain (HNC) and Percus-Yevick (PY) integral equation theories for fluids of hard nonspherical particles forming nematic phase. The explicit numerical results were given for fluids of hard ellipsoids of revolution[162a] with x_0 varying from

1.25 to 5.0 as well as for fluids of hard spherocylinders[162b] with x_0 varying from 2 to 6. Theoretical results were compared with the available MC data[122–125,126b] for the equation of state and the pair correlation function. The DCF for the isotropic phase were obtained. It was shown that the DFT results are strongly dependent upon the approximations used for the isotropic direct pair correlation function. Using these results the isotropic-nematic transition properties were calculated[118] for the hard ellipsoids and spherocylinders. For the hard ellipsoids all results were obtained using HNC DCF because no phase changes were found with PY values. It was found that for $x_0 = 3$ the MC transition density is about 20% higher whereas the MC fractional density change is smaller than the density functional results. These values are compared in Table 4.16 with other values. For spherocylinders with $x_0 = 6$ it was observed that the second order DFT theory combined with the HNC DCF yields a transition density which is about 14% lower than the MC values. In addition to hard particles fluid the calculations were also carried out[118] for fluids characterized by the pair potentials of a generalized MS type

$$u(1,2) = u_0(r) + u_2(r)\, P_2(\cos\theta_{12}) \tag{4.174}$$

with

$$u_2(r) = -4\bar{s}\,\varepsilon(\sigma/r)^6\,, \qquad \text{(model I)} \tag{4.175a}$$

and

$$u_2(r) = -4\bar{s}\,\varepsilon\left[\left(\frac{\sigma}{r}\right)^{12} + \left(\frac{\sigma}{r}\right)^{6}\right], \qquad \text{(model II)} \tag{4.175b}$$

Here $u_0(r)$ was taken to be the usual Lennard-Jones interaction and $\bar{s}$ is a parameter determining the strength of the anisotropic interaction. The transition properties for both the models were calculated and it was found that in both cases the transition temperatures are considerably higher than the absolute stability limits given by the HNC theory. Further, the fractional changes in density found for both models (eqs. 4.175)) are about an order of magnitude larger than those obtained for hard ellipsoids and spherocylinders.

Singh and coworkers[117d] solved the PY integral equation for two model fluids : HER fluid represented by a Gaussian overlap model and a fluid the molecules of which interact via a Gay-Berne model potential. These model systems are expected to capture some of the basic features of real ordered phases. For example, a system of HER has been found[126b] to exhibit four distinct phases, isotropic fluid, nematic fluid, plastic solid and ordered solid. The simulation results[163,164] show that the Gay-Berne (GB) pair potential is capable of forming nematic, smectic A, smectic B and an ordered solid in addition to the isotropic liquid.

Table 4.16. Comparison of the NI phase transition properties for hard ellipsoids of length-to-breadth ratio $x_0 = 3$, $P^* = Pv_0/k_BT$, $\mu^* = \mu/k_BT$.

Quantity	MC simulation[126a]	Singh and coworkers		Singh and Singh[117b]	Marko[130b]	Perera et al[118]	Holyst and Poniewierski[167]	Singh et al[120]
		Ref. 117e	Ref. 117f					
η_{iso}	0.507	0.466	0.471	0.309	0.493	0.418	0.454	0.475
η_{nem}	0.517	0.479	0.490	0.322	0.494	0.436	0.474	0.494
$\Delta\rho/\rho_{nem}$	0.019	0.028	0.040	0.042	0.003	0.041	0.042	0.038
$\bar{P}_{2NI}$	0.50	0.745	0.552	0.547	0.017	0.657	0.485	0.546
$\bar{P}_{4NI}$	–	0.474	0.266	0.197	–	0.358	–	0.196
P^*	9.786	6.429	6.662	–	–	–	–	–
μ^*	25.150	17.982	18.480	–	–	–	–	–

The GB pair potential model is written as

$$u(\mathbf{r},\Omega_1,\Omega_2) = 4\varepsilon(\hat{r},\Omega_1,\Omega_2)\left[\left(\frac{\sigma_0}{r-\sigma(\hat{r},\Omega_1,\Omega_2)+\sigma_0}\right)^{12} - \left(\frac{\sigma_0}{r-\sigma(\hat{r},\Omega_1,\Omega_2)+\sigma_0}\right)^{6}\right] \tag{1.176}$$

where $\varepsilon(\hat{r},\Omega_1,\Omega_2)$ and $\sigma(\hat{r},\Omega_1,\Omega_2)$ are angle dependent strength and range parameters and are defined as

$$\varepsilon(\hat{r},\Omega_1,\Omega_2) = \varepsilon_0(1-\chi^2[\hat{e}_1\cdot\hat{e}_2]^2)^{-\frac{1}{2}} \times\left\{1-\chi'\frac{(\hat{r}\cdot\hat{e}_1)^2+(\hat{r}\cdot\hat{e}_2)^2-2\chi'(\hat{r}\cdot\hat{e}_1)(\hat{r}\cdot\hat{e}_2)(\hat{e}_1\cdot\hat{e}_2)}{1-\chi'^2(\hat{e}_1\cdot\hat{e}_2)^2}\right]^2 \tag{4.177a}$$

and

$$\sigma(\hat{r},\Omega_1,\Omega_2) = \sigma_0\left[1-\chi\frac{(\hat{r}\cdot\hat{e}_1)^2+(\hat{r}\cdot\hat{e}_2)^2-2\chi(\hat{r}\cdot\hat{e}_1)(\hat{r}\cdot\hat{e}_2)(\hat{e}_1\cdot\hat{e}_2)}{1-\chi^2(\hat{e}_1\cdot\hat{e}_2)^2}\right]^{-\frac{1}{2}} \tag{4.177b}$$

where χ is defined by eq. (4.71) and $\chi' = (\sqrt{k'}-1)/(\sqrt{k'}+1)$; k' is the ratio of the potential well depths for the side-by-side and end-to-end configurations. It is important to note that in this calculation[117d] the OZ equation using PY closure was solved numerically considering 30 harmonic coefficients whereas only 14 coefficients were included by Ram and Singh[117c] in their calculation. Though the results found for these models, HER and GB, are in qualitative agreement with the simulation results[163,165], the quantitative agreement is not satisfactory. Secondly, the HNC and PY approximations do not give thermodynamic consistency for the virial and compressibility routes to calculate pressures. So in an another work Singh and coworkers[117e] developed a thermodynamically consistent (TC) integral equation for the pair correlation functions of molecular fluids which interpolates continuously between the HNC and PY approximations. The TC integral equation is a generalization of Rogers and Young[166] method devised for atomic fluids to an angle dependent pair potential. More importantly the thermodynamic consistency between the virial and compressibility equation of state has been achieved through a suitably chosen adjustable parameter. The solutions obtained by using TC equation have been found to be in accurate agreement with the simulation results.

Taking the values of the spherical harmonic coefficients of the DPCF as obtained from the TC closure relation NI phase transition properties have been calculated[117f] for the HER fluid represented by a Gaussian overlap model and GB fluids using the DFT and MWDA methods. The results for hard ellipsoids ($x_0 = 3.0$) are compared in Table 4.16 with MC simulation and various other calculations. It can be seen that the work of Singh and Singh[117b], using functional Taylor expansion and decoupling approximation, yields too small densities at the transition. Marko's[130] version gives an almost correct density but predicts an extremely small density jump and order parameter. Further, the transition densities obtained by using the DFT and MWDA approximation agree very well with the MC simulation[126a] results. The fractional density change is overestimated as compared to simulation values. The transition points at two temperatures $T^* = 0.95$ and 1.25 were determined for the GB fluid. It has been found that the NI transition is predicted at higher density whereas the fractional density changes are small as compared to MD results[163a]. Further, the coexistence densities increase with increasing temperature. At a lower temperture the fractional density change is in better agreement with the MD values as compared to its value at a higher temperature. Further, the DFT and MWDA overestimate the pressure and chemical potential as the temperature is increased.

REFERENCES

1. M.J. Freiser, Phys. Rev. Lett. **24**, 1041 (1970); Mol. Cryst. Liquid Cryst. **14**, 165 (1971).

2. L..J. Yu, and A. Saupe, Phys. Rev. Lett. **45**, 1000 (1980).

3. A. Saupe, P. Boonbrahm, and L.J. Yu, J. Chim. Phys. **80**, 7 (1983).

4. P. Boonbrahm, and A. Saupe, J. Chem. Phys. **81**, 2076 (1984).

5. F. Biscarini, C. Chiccoli, P. Pasini, F. Semeria, and C. Zannoni, Phys. Rev. Lett. **75**, 1803 (1995).

6. P.O. Quist, Liquid Crystals **18,** 623 (1995).

7. S. Chandrasekhar, Mol. Cryst. liquid Cryst. **124**, 1 (1985).

8. S. Chandrasekhar, B.K. Sadashiva, S. Ramesha, and B.S. Srikanta, Pramana-J. Phys. **27**, L713 (1986).

9. S. Chandrasekhar, B.K. Sadashiva, B.R. Ratna, and V.N. Raja, pramana-J. Phys. **30**, L49 (1988).

10. J. Malthete, L. Liebert, and A.M. Levelut, C.R. Acad. Sci. (Paris) **303**, 1073 (986).

11. K. Praefcke, B. Kohne, D. Singer, D. Demus, G. Pelzl, and S. Diele, Liquid Crystals **7**, 589 (1990).

12. C. Giannessi, Phys. Rev. **A28**, 350 (1983); ibid **A34**, 705 (1986).

13. J. Toner, Phys. Rev. **A27**, 1157 (1983).

14. P.G. de Gennes, and J. Prost, The Physics of Liquid Crystal (Clarendon Press, Oxford, 1993).

15. G.W. Gray, K.J. Harrison, and J. A. Nash, Electron. Lett. **9**, 130 (1973).

16. L. Pohl, R. Eidenschink, J. Krause, and D. Erdmann, Phys. Lett. **A60**, 421(1977).

17. R. Eidenschink, D. Erdmann, J. Krause, and L. Pohl, Angew. Chem. Int. Ed. Engl. **16**, 100 (1977).

18. T. Inukai, H. Inoue, and H. Sato, US Patent **4**, 211, 666 (1978).

19. G.W. Gray, Abstract 12 Freiburger Arbeitstagung 1983 (1982).

20. R. Eidenschink, J. Krause, and L. Pohl, German Patent **2**, 701, 591 (1978).

21. A Villiger, A. Boller, and M. Schadt, Z. Naturforsch. **34**, 1535 (1979).

22. G.W. Gray, and S.M. Kelly, (a) J. Chem. Soc. Perkin Trans II **1981**, 26(1981); (b) Angew. Chem. Int. Ed. Engl. **20**, 383 (1981).

23. G.W. Gray, K.J. Harrison, and J.A. Nash, J. Chem. Soc. Chem. Commun. **1974**, 431 (1974).

24. G.W. Gray, J. Physique Coll. **36**, C1-337 (1975).

25. P.J. Collings, and M. Hird, Introduction to Liquid Crystals, Chemistry and Physic (Taylor and Francis Ltd., London, 1997) Chapt. 8, p. 147.

26. K. Mitsuhiro, S. Hayashi, K. Nishimura, S, Kusabayashi, and S. Takenaka, Mol. Cryst. Liquid Cryst. **106**, 31 (1984).

27. S.M. Kelly, Liquid Crystals **10**, 261, 273 (1991).

28. H. Takatsu, K. Takeuchi, and H. Sato, Mol. Cryst. Liquid Cryst. **100**, 345 (1983).

29. W. Weissflog, and D. Demus, Cryst. Res. Technol. **18**, K 21 (1983); ibid **19**,55 (1984); Mol. Cryst. Liquid Cryst. **129**, 235 (1985).

30. M.A. Osman, and T. Huynh- Ba, Mol. Cryst. Liquid Cryst. **82**, 331(1983).

31. S. Singh, Phys. Rep. **324**, 107 (2000).

32. P.G. de Gennes, Mol. Cryst. Liquid Cryst. **12**, 193 (1971).

33. P. Sheng, and E.B. Priestley, The Landau-de-Gennes Theory of Liquid Crystal Phase Transitions, in, Introduction to Liquid Crystals, eds., E.B. Priestley, P.J. Wojtowicz, and P. Sheng (Plenum, New York, 1974) Chapt. 10, p. 143.

34(a). E.F. Gramsbergen, L. Longa and W.H. de Jeu, Phys. Rep. **135**, 195 (1986).

(b) G. Vertogen, and W.H. de Jeu, Thermotropic Liquid Crystals, Fundamentals (Springer–Verlag, 1988) Chapt. 10, p. 221.

35. L.D. Landau, Phys. Z Sowjefunion **11**, 26 (1937) (Collected papers of L.D. Landau, ed., D. Ter Haar (Gordon and Breach Science Publishers, New York, 2nd ed., 1967) p. 193).

36. L.D. Landau, and E.M. Lifshitz, Statistical Physics, Vol. 1, 3rd ed. (Pergamon, Oxford, 1980).

37. J.C. Toledano, and P. Toledano, The Landau Theory of Phase Transitions (World Scientific, Singapore, 1987).

38. V.L. Indenbom, S.A. Pikin, and E.B. Liginov. Sov. Phys. Crystallogr. **21**, 632 (1976).

39. S.A. Pikin, and V.L. Indenbom, Sov. Phys. Usp. **21**, 487 (1978).

40. V.L. Indenbom, E.B. Liginov, and M.A. Osipov, Sov. Phys. Crystallogr. **26**, 656 (1981).

41. V.L. Indenbom, and E.B. Liginov, Sov. Phys. Crystallogr. **26**, 526 (1981).

42. W. Helfrich. Phys. Rev. Lett. **24**, 201 (1970).

43. C. Rosenblatt, Phys. Rev. **A24**, 2236 (1981).

44. A.J. Nicastro, and P.H. Keyes, Phys. Rev. **A30**, 3156 (1984).

45. P.K. Mukherjee, and T.B. Mukherjee, Phys. Rev. **B52**, 9964 (1995).

46. A.D. Rzoska, S.J. Rzoska, and J. Ziolo, Phys. Rev. **E54**, 6452 (1996).

47. P.K. Mukherjee, J. Saha, B. Nandi, and M. Saha, Phys, Rev. **B50**, 9778 (1994).

48. P.K. Mukherjee, and M. Saha, Phys. Rev. **E51**, 5745 (1995).

49. R.G. Priest, and T.C. Lubensky Phys. Rev. **B13** , 4159 (1976).

50. P.K. Mukherjee, and M. Saha Mol. Cryst. Liquid Cryst. **307**, 103 (1997).

51. P.K. Mukherjee, T.R. Bose, D. Ghosh, and M. Saha, Phys. Rev. **E51**, 4570 (1995).
52. B. Nandi, P.K. Mukherjee, and M. Saha, Mod Phys. Lett. **B10**, 777 (1996).
53. P.K. Mukherjee, Mod. Phys. Lett. **B11**, 107 (1997).
54. P.H. Keyes, and J.R. Shane, Phys. Rev. Lett. 42, 722 (1979).
55. P.B. Vigman, A.I. Larkin, and V.M. Filev, Sov. Phys. JETP **41**, 944 (1976).
56. Z.H. Wang, and P.H. Keyes, Phys. Rev. **E54**, 5249 (1996).
57. T.E. Faber, Proc. Roy Soc. London Sr. **A375**, 579 (1982).
58. L. Onsager, Ann. N.Y. Acad. Sci. **51**, 627 (1949).
59. R. Zwanzig, J. Chem. Phys. **39**, 1714 (1963).
60. G. Lasher, J. Chem. Phys. **53**, 4141 (1970).
61. L.K. Runnels, and C. Colvin, J. Chem. Phys. **53**, 4219 (1970).
62. P.J. Flory, Proc. Roy Soc. **A234**, 73 (1956).
63. P.J. Flory, and G. Ronca, Mol. Cryst. Liquid Cryst. **54**, 289 (1979).
64. (a) M.A. Cotter, and D.E. Martire, J. Chem. Phys. **52**, 1902, 1909, 4500 (1970). (b) M.A. Cotter, Phys. Rev. **A10**, 625 (1974).
65. M.A. Cotter, Hard Particle Theories of Nematics, in, The Molecular Physics of Liquid Crystals, eds., G.R. Luckhurst, and G.W. Gray (Acad. Press Inc., London, 1979) Chapt. 7, p. 169.
66. B. Barboy, and W.M. Gelbart, J. Chem. Phys. **71**, 3053 (1979).
67. B.M. Mulder, and D. Frenkel, Mol. Phys. **55**, 1193 (1985).
68. J.D. Parson, Phys. Rev. **A19**, 1225 (1979).
69. M. Warner, Mol. Cryst. Liquid Cryst. **80**, 79 (1982).
70. B.J. Berne, and P. Pechukas, J. Chem. Phys. **56**, 4213 (1972).
71. S.D. Lee, (a) J. Chem. Phys. **87**, 4972 (1987); (b) ibid **89**, 7036 (1988).
72. A. Wulf, and A.G. de Racco, J. Chem. Phys. **55**, 12 (1971).
73. (a) F. Dowell, and D.E. Martire, J. Chem. Phys. **68**, 1088, 1094 (1978); (b) F. Dowell, Phys, Rev, **A28**, 3520 (1983).
74. S. Tang, and G.T. Evans, J. Chem. Phys. **99**, 5336 (1993).
75. A. Saupe, J. de Phys. Colloq. **40**, C3 (1979).
76. J.P. Straley, Mol. Cryst. Liquid Cryst. **24**, 7 (1973).
77. F.C. Andrews, J. Chem. Phys. **62**, 272 (1975).

78. W. Maier, and A. Saupe, Z. Naturforsch **a13**, 564 (1958); **a14**, 882 (1959); **a15**, 287 (1960).

79. B. Widom, J. Chem. Phys. **39**, 2808 (1963).

80. M.A. Cotter, Mol. Cryst. Liquid Cryst. **39**, 173 (1977).

81. T.J. Krieger, and H.M. James, J. Chem. Phys. **22**, 796 (1954).

82. S. Chandrasekhar, and N.V. Madhusudana, Acta Cryst. **A27**, 303 (1971).

83. (a) R.L. Humphries, P.G. James, and G.R. Luckhurst, J. Chem. Soc. Faraday Trans II **68**, 1031 (1972); (b) G.R. Luckhurst, C. Zannoni, P.L. Nordio, and U. Segre, Mol. Phys. **30**, 1345 (1975).

84. (a) S. Marcelja, J. Chem. Phys. **60**, 3599 (1974).

(b) D.J. Photinos, E.T. Samulski, and H. Toriumi, J. Chem. Phys. **94**, 4688, 4694 (1990); **94**, 2758 (1991).

85. N.V. Madhusudana, and S. Chandrasekhar, Solid state comm. **13**, 377 (1973).

86. J.G.J. Ypma, G. Vertogen, and H.T. Koster, Mol. Cryst. Liquid Cryst. **37**, 57 (1976).

87. P. Sheng, and P.J. Wojtowicz, Phys. Rev. **A14**, 1883 (1976).

88. N.V. Madhusudana, K.L. Savithramma, and S. Chandrasekhar, Pramana **8,** 22 (1977).

89. (a) G.R. Luckhurst, Molecular Field Theories of Nematics, in, The Molecular Physics of Liquid Crystals, eds., G.R. Luckhurst, and G.W. Gray (Acad. Press Inc., London, 1979) Chapt. 4, p. 85.

(b) G.R. Luckhurst, and C. Zannoni, Nature **267**, 412 (1977).

90. A. Wulf, J. Chem. Phys. **64**, 104 (1976).

91. J.I. Kaplan, and E. Drauglis, Chem. Phys. Lett. **9**, 645 (1971).

92. R. Van der Haegen, J. Debruyne, R, Luyckx, and H.N.N. Lekkerkerker, J. Chem. Phys. **73**, 2469 (1980).

93. (a) J.D. Bunning, D.A. Crellin, and T.E. Faber, Liquid Crystals **1**, 37 (1986); (b) B. Bergersen, P. Palffy-Muhoray, and D.A. Dunmur, Liquid Crystals **3**, 347 (1988).

94. H.C. Anderson, D. Chandler, and J.D. Weeks, Adv. Chem. Phys., **34**, 105 (1976).

95. A. Beguin, J.C. Dubois, P. Le Barny, J. Billard, F. Bonamy, J.M. Buisine, and P. Cuvelier, "Sources of thermodynamic data on mesogens", Mol. Cryst. Liquid Cryst. **115**, 1 (1984).

96. J.R. McColl, and C.S. Shih, Phys. Rev. Lett. **29**, 85 (1972).

97. B. Deloche, B. Cabane, and D. Jerome, Mol. Cryst. Liquid Cryst. **15**, 197 (1971).

98. R.G. Horn, J.de Physique **39**, 199 (1978); R.G. Horn, and T.E. Faber, Proc. Roy. Soc. (London) **A368**, 199 (1979).

99. N.V. Madhusudana, Theories of Liquid Crystals, in, Liquid Crystals : Applications and Uses, Vol. 1, ed., B. Bahadur (World Scientific, 1990) Chapt. 2, p. 37.

100. (a) M.A. Cotter, The van der Waals Approach to Nematic Liquid Crystals, in, The Molecular Physics of Liquid Crystals, eds., G.R. Luckhurst, and G.W. Gray (Acad. Press Inc., London) Chapt. 8, p. 181;

(b) M.A. Cotter, J. Chem, Phys. **66**, 1098, 4710 (1977).

101. (a) W.M. Gelbart, and B.A. Baron, J. Chem. Phys. **66**, 207 (1977);

(b) W.M. Gelbart, and A. Gelbart, Mol. Phys. **33**, 1387 (1977);

(c) B.A. Baron, and W.M. Gelbart, J. Chem. Phys. **67**, 5795 (1977);

(d) W.M. Gelbart, and A. Ben-Shaul, J. Chem. Phys. **77**, 916 (1982).

102. J.G.J. Ypma, and G. Vertogen, Phys. Rev. **A17**, 1490 (1978).

103. W. Warner, J. Chem. Phys. **73**, 6327 (1980).

104. K.L. Savithramma, and N.V. Madhusudana (a) Mol. Cryst. Liquid Cryst. **62**, 63 (1980); (b) ibid **97**, 407 (1983).

105. (a) S. Singh, and Y. Singh , Mol. Cryst. Liquid Cryst. **87**, 211 (1982);

(b) S. Singh, and K. Singh , Mol. Cryst. Liquid Cryst.**101**, 77 (1983);

(c) K. Singh, and S. Singh , Mol. Cryst. Liquid Cryst.**108**, 133 (1984);

(d) S. Singh, T.K. Lahiri, and K. Singh, Mol. Cryst. Liquid Cryst. **225**, 361 (1993).

106. (a) F. Dowell, Phys. Rev. **A28**, 1003 (1983);

(b) F. Dowell, Phys. Rev. **A31**, 2464, 3214 (1985).

107. P. Palffy-Muhoray, and B. Bergersen, Phys. Rev. **A6**, 2704 (1987).

108. B. Tjipto-Margo, and G.T. Evans (a) Mol. Phys. **74**, 85 (1991). (b) J. Chem. Phys. **94**, 4546 (1990).

109. M. Nakagawa, and T. Akahane, Mol. Cryst. Liquid Cryst. **90**, 53 (1982).

110. P.J. Flory, and R. Ronca, Mol. Cryst. Liquid Cryst. **54**, 311 (1979).

111. A. Bellemans, Phys. Rev. Lett. **21**, 527 (1968).

112(a). V.T. Rajan, and C.W. Woo, Phys. Rev. **A17**, 382 (1978);

(b) L. Feijoo, V.T. Rajan, and C. W. Woo, Phys. Rev. **A19**, 1263 (1979).

113. Y. Singh, Phys. Rep. **207**, 351 (1991).

114. R. Evans, Density Functionals in the Theory of Nonuniform Fluids, in, Fundamentals of Inhomogeneous Fluids, ed., D. Henderson (New York, Dekker, 1992) Chap. 3, p. 85.

115. M.D. Lipkin, and D.W. Oxtoby, J. Chem. Phys. **79**, 1939 (1983).

116. T.J. Sluckin, and P. Shukla, J. Phys. **A16**, 1539 (1983).

117. (a) Y. Singh, Phys. Rev. **A30**, 583 (1984); (b) U.P. Singh, and Y. Singh, Phys. Rev. **A33**, 2725 (1986); (c) J. Ram, and Y. Singh, Phys. Rev. **A44**, 3718 (1991); (d) J. Ram, R.C. Singh, and Y. Singh, Phys. Rev. **E49**, 5117 (1994); (e) R.C. Singh, J. Ram, and Y. Singh, Phys, Rev. **E54**, 977 (1996); (f) R.C. Singh, Equilibrium Properties of Molecular Liquids and Liquid Crystals, Ph.D Thesis (Banaras Hindu University, 1998).

118. A. Perera, G.N. Patey, and J.J. Weis, J. Chem. Phys. **89**, 6941 (1988).

119. A.M. Somoza, and P. Tarazona, J.Chem. Phys. **91**, 517 (1989); Phys. Rev. **A41**, 965 (1990).

120. U.P. Singh, U. Mohanty, and Y. Singh, Physica **A158**, 817 (1989).

121. H. Graf, and H. Lowen, J. Phys. Condens. Matter **11**, 1435 (1999).

122. J. Vieillard-Baron, Mol. Phys. **28**, 809 (1974).

123. D.W. Rebertus, and K.M. Sando, J. Chem. Phys. **67**, 2585 (1977).

124. T. Boublik, I. Nezbeda, and O. Tynko, Czech. J. Phys. **B26**, 1081 (1976).

125. P.A. Monson, and M. Rigby, Mol. Phys. **35**, 1337 (1978).

126. (a) D. Frenkel, and B.M. Mulder, Mol. Phys. **55**, 1171 (1985); (b) D. Frenkel, B.M. Mulder, and J.P. McTague, Phys. Rev. Lett. **52**, 287 (1984); (c) D. Frenkel, H.N.W. Lekkerkerker, and A. Stroobants, Nature **322**, 822 (1988).

127. (a) R. Eppenga, and D. Frenkel, Mol. Phys. **52**, 1303 (1984); (b) D. Frenkel, and R. Eppenga, Phys. Rev. **A31**, 1776 (1985).

128. G.R. Luckhurst, in, Recent Advances in Liquid Crystalline polymers, ed., L.L. Chapoy (Elsevier, London, 1986) p. 105.

129. (a) M. Baus, J.L.Colot, X.G. Wu, and H. Xu, Phys. Rev, Lett. **59**, 2184 (1987); (b) J.L. Colot, X.G. Wu, H. Xu, and M. Baus, Phys. Rev. **A38**, 2022 (1988).

130. J.F. Marko, (a) Phys. Rev. Lett. **60**, 325 (1988); (b) Phys. Rev. **A39**, 2050 (1989).

131. (a) D.R. Evans, G.T. Evans, and D.K. Hoffman, J. Chem. Phys. **93**, 8816 (1990); (b) G.T. Evans, and E.B. Smith, Mol. Phys. **74**, 79 (1991).

132. G. Vertogen, and W.H. de Jeu, Thermotropic Liquid Crystals, Fundamentals (Springer–Verlag, 1988) Chapt. 13, p. 245.

133. J.A. Barker, and D. Henderson, Rev. Mod. Phys. **48**, 587 (1976).
134. C.N. Likos, and N.W. Ashcroft, J. Chem. Phys. **99**, 9090 (1993).
135. P. Honenberg, and W. Kohn, Phys. Rev. **B136**, 864 (1964).
136. W. Kohn, and L.J. Sham, Phys. Rev. **A140**, 1133 (1965).
137. D. Mermin, Phys. Rev. **A137**, 1441 (1964).
138. J.L. Lebowitz, and J.K. Percus, J. Math. Phys. **4**, 116 (1963).
139. W.F. Saam, and C. Ebner, Phys. Rev. **A15**, 2566 (1977).
140. A.J.M. Yang, P.D. Fleming, and J.H. Gibbs, J. Chem. Phys. **64**, 3732 (1976).
141. J.K. Percus, in, The Equilibrium Theory of Classical Fluids, ed., H.L. Frisch, and J.L. Lebowitz (Benjamin, New York, 1964).
142. R. Evans, Adv. Phys. **28**, 143 (1979).
143. F.F. Abraham, Phys. Rep. **53**, 93 (1979).
144. J.S. Rowlinson, and B. Widom, Molecular Theory of Capillarity (Clarendon Press, Oxford, 1982).
145. A.D.J. Haymet, Ann. Rev. Phys. Chem. **38**, 89 (1987).
146. M. Baus, J. Phys. Condens Matter **2**, 2111 (1990).
147. R.G. Parr, and W. Yang, Density Functional Theory of Atoms and Molecules (Oxford Univ. Press, New York, 1989).
148. R.O. Jones, and O. Gunnarsson, Rev. Mod. Phys. **61**, 689 (1989).
149. T. Ziegler, Chem. Rev. **91**, 651 (1991).
150. J.K. Labanowski, and J. Andzelm, eds., Density Functional Methods in Chemistry (Springer, New York, 1991).
151. B.G. Johnson, P.M.W. Gill, and J.A. Pople, J. Chem. Phys. **98**, 5612 (1993).
152. S. Singh, Phys. Rep. **277**, 283 (1996).
153. S. Nordholm, M. Johnson, and B.C. Freasier, Aust. J. Chem. **33**, 2139 (1980).
154. P. Tarazona, (a) Mol. Phys. **52**, 81 (1989); (b) Phys. Rev. **A31**, 2672 (1985); erratum **32**, 3148 (1985).
155. W.A. Curtin, and N.W. Ashcroft, Phys. Rev. **A32**, 2909 (1985).
156. A.R. Denton, and N.W. Ashcroft, Phys. Rev. **A39**, 426, 4701 (1989).
157. T.V. Ramakrishnan, and M.Yussouff, Phys. Rev. **B19**, 2775 (1979).
158. A.D.J. Haymet, J. Chem. Phys. **78**, 4641 (1983).
159. G.L. Jones, and U. Mohanty, Mol. Phys. **54**, 1241 (1985).
160. M. Baus, and J.L. Colot, Mol. Phys. **55**, 653, (1985).

161. W.R. Smith, and D. Henderson, Mol. Phys. **3**, 411 (1970).

162. (a) A. Perera, P.G. Kusalik, and G.N. Patey, J. Chem. Phys. **87**, 1295 (1987); ibid **89**, 5969 (1988); (b) A. Perera, and G.N. Patey, J. Chem. Phys. **89**, 5861 (1988).

163. (a) E. De Miguel, L.F. Rull, M.K. Chalam, K.E. Gubbins, and F.V. Swol, Mol. Phys. **72**, 593 (1991); (b) E. De Miguel, L.F. Rull, M.K. Chalam, and K.E. Gubbins, Mol. Phys. **74**, 405 (1991).

164. G.R. Luckhurst, R.A. Stephens, and R.W. Phippen, Liquid Crystals **8**, 451 (1990).

165. J. Talbot, A. Perera, and G.N. Patey, Mol. Phys. **70**, 285 (1990).

166. F.J. Rogers, and D.A. Young, Phys. Rev. **A30**, 999 (1984).

167. R. Holyst, and A. Poniewierski, Mol. Phys. **68**, 381 (1989).

5

Nematic Liquid Crystals : Elastostatics and Nematodynamics

The purpose of this chapter is to describe the elastic and dynamical properties of nematic liquid crystals. Many of the most important physical properties exhibited by the nematics, such as its unusual flow properties or its response to external perturbations, can be studied by considering the liquid crystal as a continuous medium. The foundations of the continuum model were laid by Oseen[1] and Zöcher[2] in the year 1933. These authors developed a static theory which proved to be quite successful. Later Frank[3] reexamined the Oseen's treatment and described it as a theory of curvature elasticity. The dynamical theories were put forward first by Anzelius[4] and Oseen[1], but the general conservation laws and constitutive equations were developed by Ericksen[5,6] and Leslie[7] for describing the mechanical behaviour of the nematic state. In the following sections, we shall describe the elastic properties and the fluid dynamics of nematic liquid crystals.

5.1 ELASTOSTATICS IN NEMATICS

A detail account of the curvature elasticity in liquid crystals has been given by the author in a review[8]. The fundamental equation of the continuum theory for liquid crystalline phases write the elastic free-energy density in terms of elastic constants associated with the possible modes of deformation.

5.1.1 Elastic Continuum Theory for Nematics

The basic idea of the elastic continuum theory has been discussed in chapter 3 (see sec. 3.3). In this section the elastic free-energy ΔA_e will be expressed in terms of elastic constants for the uniaxial nematic N_u , and the biaxial nematic N_b phases of nonchiral compounds.

5.1.1.1 Uniaxial nematic phase

On the basis of symmetry arguments, it can be shown that in a uniaxial nematic there are three independent elastic constants[2,9–12] associated with different modes of distortion.

Phenomenologically, the free-energy density of the deformed medium $\Delta A_e/V$ can be constructed[8,12] from the following conditions :

(a) ΔA_e must vanish if $\nabla\hat{\mathbf{n}} = 0$. We assume that it can be expanded in powers of $\nabla\hat{\mathbf{n}}$.

(b) ΔA_e must be even in $\hat{\mathbf{n}}$; the states $\hat{\mathbf{n}}(\mathbf{r})$ and $-\hat{\mathbf{n}}(\mathbf{r})$ are indistinguishable.

(c) ΔA_e must be invariant in the continuous group of rotations about the local $\hat{\mathbf{n}}(\mathbf{r})$. There will exist no terms linear in $\nabla\hat{\mathbf{n}}$; the only terms of this kind that are invariant by rotations are

$$\begin{cases} \nabla\cdot\hat{\mathbf{n}}, & \text{due to requirement (b) this does not survive;} \\ \hat{\mathbf{n}}\cdot\nabla\times\hat{\mathbf{n}}, & \text{this changes sign by the transformation, } X \to -X,\ Y \to -Y,\ Z \to -Z. \end{cases}$$

In case of chiral nematics, which are not centro-symmetric, this term will survive.

(d) Terms of the type $\nabla\cdot\mathbf{u}$, where $\mathbf{u}$ is an arbitrary vector field, in ΔA_e may be ignored. This is a result of the identity

$$\int \nabla\cdot\hat{\mathbf{n}}\, d\mathbf{r} \equiv \int d\boldsymbol{\sigma}\cdot\mathbf{u}$$

where the integral on the right represents a surface integral and $d\boldsymbol{\sigma}$ is normal to the surface at each point. This identity shows that such terms contribute only to the surface energies and can be discarded for the present discussion of bulk properties. To construct[12] ΔA_e , we consider explicitly the spatial derivatives of $\hat{\mathbf{n}}(\mathbf{r})$; they form a tensor of rank two $\partial_\alpha n_\beta$ (where $\partial_\alpha \equiv \partial/\partial x_\alpha$) and may appropriately be called the curvature strain tensor[3]. It is convenient to separate this tensor into a symmetric part

$$e_{\alpha\beta} = \frac{1}{2}(\partial_\alpha n_\beta + \partial_\beta n_\alpha) \tag{5.1}$$

and an antisymmetric part, related to the $\nabla\times\hat{\mathbf{n}}$

$$\left(\nabla\times\hat{\mathbf{n}}\right)_z = \frac{\partial}{\partial x} n_y - \frac{\partial}{\partial y} n_x\,,\ \text{etc.} \tag{5.2}$$

In general, $e_{\alpha\beta}$ has six independent components but these are restricted due to the fact that $\hat{\mathbf{n}}$ is a unit vector. In order to define the components of this tensor, we introduce a local (orthogonal) frame of reference with the Z-axis parallel to the local direction of $\hat{\mathbf{n}}$. Now due to the identity

$$0 = \nabla(n_z^2 + n_x^2 + n_y^2) = 2n_z\,\nabla n_z + 0 = 2\nabla n_z \tag{5.3}$$

all gradients of n_z vanish. Hence

$$e_{zz} = 0$$

$$e_{zx} = \frac{1}{2} (\nabla \times \hat{\mathbf{n}})_y \tag{5.4}$$
$$e_{zy} = -\frac{1}{2} (\nabla \times \hat{\mathbf{n}})_z$$

Also

$$e_{xx} + e_{yy} + e_{zz} = e_{xx} + e_{yy} = \nabla \cdot \hat{\mathbf{n}} \tag{5.5}$$

In view of requirement (b), ΔA_e will be a quadratic function of the components $e_{\alpha\beta}$ and of $\nabla \times \hat{\mathbf{n}}$. It is convenient to express ΔA_e as

$$\Delta A_e = \Delta A_{e1} + \Delta A_{e2} + \Delta A_{e3} \tag{5.6}$$

where ΔA_{e1} contains terms quadratic in $e_{\alpha\beta}$, ΔA_{e2} terms quadratic in $\nabla \times \hat{\mathbf{n}}$ and ΔA_{e3} represents the cross-terms. For the present case the most general form[13] of ΔA_{e1} can be written as

$$\Delta A_{e1} = \lambda_1 e_{zz}^2 + \lambda_2 (e_{xx} + e_{yy})^2 + \lambda_3 e_{\alpha\beta} e_{\beta\alpha} + \lambda_4 e_{zz}(e_{xx} + e_{yy}) + \lambda_5 (e_{xz}^2 + e_{yz}^2) \tag{5.7}$$

In view of eq. (5.4), this reduces to

$$\Delta A_{e1} = \lambda_2 (\nabla \cdot \hat{\mathbf{n}})^2 + \lambda_3 e_{\alpha\beta} e_{\beta\alpha} + \frac{1}{4} \lambda_5 (\hat{\mathbf{n}} \times \nabla \times \hat{\mathbf{n}})^2 \tag{5.8}$$

We make use of the identity

$$e_{\alpha\beta} e_{\beta\alpha} = (\nabla \cdot \hat{\mathbf{n}})^2 + \partial_\alpha (n_\beta \partial_\beta n_\alpha) - \partial_\beta (n_\beta \partial_\alpha n_\alpha) + \frac{1}{2} (\nabla \times \hat{\mathbf{n}})^2 \tag{5.9}$$

The second and third terms must be discarded because of requirement (d). Since

$$(\nabla \times \hat{\mathbf{n}})^2 = (\hat{\mathbf{n}} \cdot \nabla \times \hat{\mathbf{n}})^2 + (\hat{\mathbf{n}} \times \nabla \times \hat{\mathbf{n}})^2 \tag{5.10}$$

ΔA_{e1} finally contains a sum of three contributions of the kind

$$(\nabla \cdot \hat{\mathbf{n}})^2, \quad (\hat{\mathbf{n}} \cdot \nabla \times \hat{\mathbf{n}})^2 \quad \text{and} \quad (\hat{\mathbf{n}} \times \nabla \times \hat{\mathbf{n}})^2 \tag{5.11}$$

The term ΔA_{e2}, which is quadratic in $\nabla \times \hat{\mathbf{n}}$, must have the form

$$\Delta A_{e2} = \mu_1 (\nabla \times \hat{\mathbf{n}})_z^2 + \mu_2 [(\nabla \times \hat{\mathbf{n}})_x^2 + (\nabla \times \hat{\mathbf{n}})_y^2]$$
$$= \mu_1 (\hat{\mathbf{n}} \cdot \nabla \times \hat{\mathbf{n}})^2 + \mu_2 (\hat{\mathbf{n}} \times \nabla \times \hat{\mathbf{n}})^2 \tag{5.12}$$

Finally the most general form for the cross term ΔA_{e3} is

$$\Delta A_{e3} = \nu (\hat{\mathbf{n}} \times \nabla \times \hat{\mathbf{n}})^2 \tag{5.13}$$

Regrouping eqs. (5.8), (5.12) and (5.13), we may express the distortion free-energy of the uniaxial nematics as

$$\Delta A_e = \int dr \left[\frac{1}{2} K_1 (\nabla \cdot \hat{\mathbf{n}})^2 + \frac{1}{2} K_2 (\hat{\mathbf{n}} \cdot \nabla \times \hat{\mathbf{n}})^2 + \frac{1}{2} K_3 (\hat{\mathbf{n}} \times \nabla \times \hat{\mathbf{n}})^2 \right] \tag{5.14}$$

This is the fundamental formula of the continuum theory for N_u phase; the coefficients K_1, K_2 and K_3 are often referred to as Frank elastic constants and are associated with the three basic types of deformation as shown in Fig. 3.9. The distortion free-energy density can be written as

$$f_d = \frac{1}{V}\Delta A_e = \frac{1}{2}K_1(\nabla\cdot\hat{\mathbf{n}})^2 + \frac{1}{2}K_2(\hat{\mathbf{n}}\cdot\nabla\times\hat{\mathbf{n}})^2 + \frac{1}{2}K_3(\hat{\mathbf{n}}\times\nabla\times\hat{\mathbf{n}})^2 \qquad (5.15)$$

The splay mode is characterized by a nonvanishing divergence of the vector field $\hat{\mathbf{n}}$ ($\nabla\cdot\hat{\mathbf{n}} \neq 0$). The twist and bend modes have $\nabla\times\hat{\mathbf{n}} \neq 0$. The degree of bending is given by the component of $\nabla\times\hat{\mathbf{n}}$ perpendicular to $\hat{\mathbf{n}}$, whereas the amount of twist is given by the component of $\nabla\times\hat{\mathbf{n}}$ parallel to $\hat{\mathbf{n}}$. Since it is possible to create deformations which are pure splay, pure twist and pure bend, each K_i must be positive[14].

5.1.1.2 Biaxial nematic phase

Several continuum theories[8,15–23] have been developed to describe the elastic and hydrodynamic properties of biaxial nematic phase. According to one given by Saupe[15], the hydrodynamics of a compressible biaxial nematic with local orthorhombic symmetry can be expressed in terms of 12 elastic constants (excluding three constants contributing only to the surface energy) and 12 viscosity coefficients. Saupe theory was rederived[21] by adopting the Ericksen-Leslie approach[5–7,11,24,25] developed for the uniaxial nematics. Kini and Chandrasekhar[26] have discussed the feasibility of determining some of the 12 elastic constants of an orthorhombic nematic by studying the elastic and viscous responses of the system under the action of external magnetic and electric fields. Using the formalism of tensor analysis, Govers and Vertogens[22] derived the expression for the distortion free-energy density involving 12 elastic constants. These authors have also showed that for a chiral phase an additional five independent twist terms are added to this energy. Hexagonal nematics are characterized by six elastic constants, while cubic nematics by the three elastic constants. Here we give[8,27] the elastic continuum theory for orthorhombic nematics.

Let us consider only small deformations and assume that the preferred direction of the orientation of molecules in an orthorhombic nematic phase is described by an orthonormal triad of director vector fields $\hat{\mathbf{n}}$, $\hat{\mathbf{m}}$ and $\hat{\ell}$. Let the orientation of the unstrained director triad at a point **r** be

$$\hat{\ell}_0 = (1,0,0), \quad \hat{\mathbf{m}}_0 = (0,1,0), \quad \hat{\mathbf{n}}_0 = (0,0,1). \qquad (5.16)$$

and the orientation of director triad at a neighbouring point **r** + **dr** be

$$\hat{\ell} = (1,\ell_y,\ell_z), \qquad \hat{\mathbf{m}} = (m_x,1,m_z), \qquad \hat{\mathbf{n}} = (n_x,n_y,1) \qquad (5.17)$$

It is important to mention that although the N_b phase is essentially described by two order parameters, for convenience, we use[15,26] three vectors and express[8,27] all the relevant quantities in terms of them. We refer to (X,Y) plane as the $(\ell_0, \mathbf{m}_0)$ plane, etc. As $\hat{\ell}$, $\hat{\mathbf{m}}$, $\hat{\mathbf{n}}$ are orthonormal, we get

$$m_x = -\ell_y, \qquad n_y = -m_z, \qquad \ell_z = -n_x \tag{5.18}$$

Here it should be noted that only three out of six perturbations are independent. This means that if we rotate the director triad about $\hat{\ell}$ by a small angle then $\hat{\mathbf{m}}$ and $\hat{\mathbf{n}}$ should both rotate about $\hat{\ell}$ by the same angle.

To second order in director gradients the elastic free-energy can be written as

$$\begin{aligned}\Delta Ae = \int d\mathbf{r} \Big[&\frac{1}{2} K_{LL} (m_{z,x})^2 + \frac{1}{2} K_{MM} (n_{x,y})^2 + \frac{1}{2} K_{NN} (\ell_{y,z})^2 \\ &+ \frac{1}{2} K_{LM} (\ell_{y,x})^2 + \frac{1}{2} K_{MN} (m_{z,y,})^2 + \frac{1}{2} K_{NL} (n_{x,z})^2 \\ &+ \frac{1}{2} K_{ML} (\ell_{y,y})^2 + \frac{1}{2} K_{NM} (m_{z,z,})^2 + \frac{1}{2} K_{LN} (n_{x,x})^2 \\ &- C_{LM}\, n_{x,x}\, m_{z,y} - C_{MN}\, \ell_{y,y}\, n_{x,z} - C_{NL}\, m_{z,z}\, \ell_{y,x} \\ &+ K_{0L} (\ell_{y,y}\, n_{x,z} - \ell_{y,z}\, n_{x,y}) + K_{0M} (m_{z,z}\, \ell_{y,x} - m_{z,x}\, \ell_{y,z}) \\ &+ K_{0N} (n_{x,x}\, m_{z,y} - n_{x,y}\, m_{z,x}) \Big] \end{aligned} \tag{5.19}$$

Here a subscript comma denotes partial differentiation with respect to the subscript (e.g., $n_{x,y} = \partial n_x/\partial y$). There is a complete symmetry in eq. (5.19) with respect to both $(\hat{\ell}, \hat{\mathbf{m}}, \hat{\mathbf{n}})$ and (x,y,z). In eq. (5.19) the twelve K's and the three C's are the curvature elastic constants. Three constants, K_{OL}, K_{OM} and K_{ON}. contribute only to the surface torque so that the elastic equilibrium is determined by the nine K's and three C's constants.

To first-order in director gradients these elastic constants have the following significance. K_{LL}, K_{MM} and K_{NN} represent, respectively, twists about $\hat{\ell}$, $\hat{\mathbf{m}}$ and $\hat{\mathbf{n}}$. K_{ML}, K_{NM} and K_{LN} represent splays of $\hat{\ell}$ in the $(\hat{\ell}_0, \hat{\mathbf{m}}_0)$ plane, of $\hat{\mathbf{m}}$ in the $(\hat{\mathbf{m}}_0, \hat{\mathbf{n}}_0)$ plane, and of $\hat{\mathbf{n}}$ in the $(\hat{\mathbf{n}}_0, \hat{\ell}_0)$ plane, respectively. K_{LM}, K_{MN} and K_{NL} represent, respectively, bends of $\hat{\ell}$ in the $(\hat{\ell}_0, \hat{\mathbf{m}}_0)$ plane, of $\hat{\mathbf{m}}$ in the $(\hat{\mathbf{m}}_0, \hat{\mathbf{n}}_0)$ plane, and of $\hat{\mathbf{n}}$ in the $(\hat{\mathbf{n}}_0, \hat{\ell}_0)$ plane. All the three C's coefficients have similar interpretation. C_{LM} represents a simultaneous splay of $\hat{\mathbf{n}}$ in the $(\hat{\ell}_0, \hat{\mathbf{n}}_0)$ plane and a bend of $\hat{\mathbf{m}}$ in the $(\hat{\mathbf{m}}_0, \hat{\mathbf{n}}_0)$ plane. It must be noted that when we refer to "twist about $\hat{\ell}$" we mean that $\hat{\ell}$ remains unaltered but the director triad is rotated by a small angle about $\hat{\ell}$ such that m_z and n_y appear ($n_y = -m_z$) and these are functions of **r**.

For the N_u phase with uniaxial molecular order elastic free-energy is expressed in terms of 3 elastic constants. Assuming that the director $\hat{\mathbf{n}}$ in the undistorted state is along the space-fixed (SF) Z-axis, one obtains

$$\Delta A_{e,u} = \frac{1}{2}\int d\mathbf{r}\Big[K_1(n_{x,x}+n_{y,y})^2 + K_2(n_{x,y}-n_{y,x})^2 + K_3(n_{x,z}^2+n_{y,z}^2) - 2(K_{22}+K_{24})(n_{x,x}\, n_{y,y} - n_{y,x}\, n_{x,y})\Big] \tag{5.20}$$

The last term of this equation reduces to a surface term. Comparing eqs. (5.19) and (5.20), we note that in the uniaxial phase $m_x = -\ \ell_y = 0$,

$$\begin{aligned} K_1 &= K_{MN} = K_{LN} \\ K_2 &= K_{LL} = K_{MM} \\ K_3 &= K_{NL} = K_{NM} \\ K_1 - K_2 &= C_{LM} \end{aligned} \tag{5.21}$$

and $\quad K_{NN} = K_{LM} = K_{ML} = C_{MN} = C_{LN} = 0$

Thus in going from uniaxial to biaxial phase the deformation modes of splay, twist and bend split each into two modes. In addition, six new modes are developed.

5.1.2 Theories for the Elasticity of Uniaxial Nematics

In this section, we shall describe, in brief, the status of the theoretical development for the elastic constants of uniaxial nematics. The details of these works are well documented elsewhere[8] and shall not be described here. However, we shall discuss the relatively recent works in some detail.

5.1.2.1 Application of Landau-de-Gennes theory

The Landau-de-Gennes (LDG) free-energy expansion (4.10) is only valid for fluctuations in a volume that is sufficiently so small that the variations of the order parameter over this volume could be ignored. The linear dimension of this volume must be smaller than the correlation for the order parameter fluctuations. If it is not true, the LDG Helmholtz free-energy expansion must contain gradient terms and we may write

$$\begin{aligned} A = A_0 &+ \frac{1}{2}a(T-T^*)Q_{ij}(\mathbf{r})\,Q_{ji}(\mathbf{r}) + \frac{1}{3}B\,Q_{ij}(\mathbf{r})\,Q_{jk}(\mathbf{r})\,Q_{ki}(\mathbf{r}) \\ &+ \frac{1}{4}C[\,Q_{ij}(\mathbf{r})\,Q_{ji}(\mathbf{r})]^2 + \frac{1}{4}C'\,Q_{ij}(\mathbf{r})\,Q_{jk}(\mathbf{r})\,Q_{k\ell}(\mathbf{r})\,Q_{\ell i}(\mathbf{r}) \\ &+ \cdots + \frac{1}{2}L_1\,\partial_i\,Q_{jk}(\mathbf{r})\,\partial_i\,Q_{jk}(\mathbf{r}) + \frac{1}{2}L_2\,\partial_i\,Q_{ij}(\mathbf{r})\,\partial_k\,Q_{kj}(\mathbf{r}) \end{aligned} \tag{5.22}$$

where i, j, k, ℓ = 1, 2, 3 denote the components along the three orthogonal axes of the coordinate system, $\partial_i = \partial/\partial x_i$ is the partial derivative with respect to spatial coordinate x_i, and summation over repeated indices is implied. Here the order parameter fluctuations are correlated over a distance of order $\xi' = \sqrt{L_1/[a(T-T^*)]}$.

Expressing the component of Q(**r**) as

$$Q_{ij}(\mathbf{r}) = \frac{1}{2}\bar{P}_2(\mathbf{r})\,[3n_i(\mathbf{r})\,n_j(\mathbf{r}) - \delta_{ij}] \tag{5.23}$$

We can rewrite eq. (5.22) in terms of the local order parameter $\bar{P}_2(\mathbf{r})$ and the local director $\hat{\mathbf{n}}(\mathbf{r})$,

$$\begin{aligned} A = A_0 &+ \frac{3}{4}a(T-T^*)\bar{P}_2^2(\mathbf{r}) + \frac{1}{4}B\bar{P}_2^3(\mathbf{r}) + \frac{9}{16}(C+\tfrac{1}{2}C')\bar{P}_2^4(\mathbf{r}) \\ &+ \frac{9}{4}L_1\bar{P}_2^2(\mathbf{r})[\partial_i n_j(\mathbf{r})][\partial_i n_j(\mathbf{r})] + \frac{9}{8}L_2\bar{P}_2^2(\mathbf{r})[\nabla\cdot\hat{\mathbf{n}}(\mathbf{r})]^2 \\ &+ \frac{9}{8}L_2\bar{P}_2^2(\mathbf{r})[\hat{\mathbf{n}}(\mathbf{r})\times(\nabla\times\hat{\mathbf{n}}(\mathbf{r}))]^2 + \frac{3}{4}L_1[\nabla\bar{P}_2(\mathbf{r})]^2 \\ &+ \frac{1}{8}L_2[\nabla\bar{P}_2(\mathbf{r})]^2 + \frac{3}{8}L_2[\hat{\mathbf{n}}(\mathbf{r})\cdot\nabla\bar{P}_2(\mathbf{r})]^2 \\ &+ \frac{3}{2}L_2\bar{P}_2(\mathbf{r})[\nabla\cdot\hat{\mathbf{n}}(\mathbf{r})][\hat{\mathbf{n}}(\mathbf{r})\cdot\nabla\bar{P}_2(\mathbf{r})] \\ &+ \frac{3}{4}L_2\bar{P}_2(\mathbf{r})\,[\hat{\mathbf{n}}(\mathbf{r})\times(\nabla\times\hat{\mathbf{n}}(\mathbf{r}))]\cdot\nabla\bar{P}_2(\mathbf{r}) \end{aligned} \tag{5.24}$$

In this equation there are four kinds of terms. The first four terms involve only the value of the orientational order parameter $\bar{P}_2(\mathbf{r})$. Next three terms are concerned with the spatial variation of $\hat{\mathbf{n}}(\mathbf{r})$. The next two terms take account of the spatial variation of $\bar{P}_2(\mathbf{r})$. The remaining terms represent the interaction between the spatial variations of $\bar{P}_2(\mathbf{r})$ and $\hat{\mathbf{n}}(\mathbf{r})$.

If we assume that eq. (5.24) is valid in the nematic phase and that the magnitude of the order parameter is everywhere equal to the average order parameter $\bar{P}_2$, the first four terms simply give some constant K′ and all the terms associated with the gradients of $\bar{P}_2(\mathbf{r})$ can be neglected. Further

$$\begin{aligned} \partial_i n_j\,\partial_i n_j = [\nabla\cdot\hat{\mathbf{n}}(\mathbf{r})]^2 + [\hat{\mathbf{n}}(\mathbf{r})\cdot(\nabla\times\hat{\mathbf{n}}(\mathbf{r}))]^2 + [\hat{\mathbf{n}}(\mathbf{r})\times\nabla\times\hat{\mathbf{n}}(\mathbf{r})]^2 \\ -\nabla\cdot[\hat{\mathbf{n}}(\mathbf{r})\,(\nabla\cdot\hat{\mathbf{n}}(\mathbf{r})) + \hat{\mathbf{n}}(\mathbf{r})\times\nabla\times\hat{\mathbf{n}}(\mathbf{r})] \end{aligned} \tag{5.25}$$

The last term of eq. (5.25) gives the surface contribution to the free-energy density upon integration and therefore can be neglected if we are only interested in bulk properties. Finally, eq. (5.24) reads

$$A = K' + \frac{9}{4}\overline{P}_2^2(\mathbf{r})[(L_1 + \frac{1}{2}L_2)[\nabla\cdot\hat{\mathbf{n}}(\mathbf{r})]^2 + L_1[\hat{\mathbf{n}}(\mathbf{r})\cdot\nabla\times\hat{\mathbf{n}}(\mathbf{r})]^2$$
$$+ (L_1 + \frac{1}{2}L_2)[\hat{\mathbf{n}}(\mathbf{r})\times\nabla\times\hat{\mathbf{n}}(\mathbf{r})]^2] \quad (5.26)$$

Equation (5.26) strongly resembles the continuum relation (3.41) for the elastic free-energy in a nematic; comparing the two equations one obtains

$$K_1 = \frac{9}{2}(L_1 + \frac{1}{2}L_2)\overline{P}_2^2 \quad (5.27a)$$

$$K_2 = \frac{9}{2}L_1\overline{P}_2^2 \quad (5.27b)$$

and

$$K_3 = K_1 \quad (5.27c)$$

It is obvious that $K_i \propto \overline{P}_2^2$, provided L_1 and L_2 do not vary strongly through the isotropic-nematic transition and that to order $\overline{P}_2^2$, $K_1 = K_3$ and all the three moduli vary with the temperature like $\overline{P}_2^2$. The prediction $K_1 = K_3$ is not consistent with the experimental observations[9,28]. However, this is an artifact of the present derivation in which only the gradient terms of the second-order tensor order parameter have been considered. If the higher rank order parameters are included in the free-energy expansion, all the three elastic constants will, in general, be different[29]. The necessary condition for the stability of the equilibrium nematic configuration is that for any deformation the free-energy (eq. 5.26)) must be positive. This requires that

$$L_1 > 0; \qquad L_1 + \frac{1}{2}L_2 > 0. \quad (5.28)$$

If now we fix $\hat{\mathbf{n}}(\mathbf{r})$ and allow $\overline{P}_2(\mathbf{r})$ to vary, following inequalities are obtained

$$L_1 + \frac{1}{6}L_2 > 0, \qquad \text{for } \hat{\mathbf{n}}(\mathbf{r}) \perp \overline{P}_2(\mathbf{r}) \quad (5.29a)$$

$$L_1 + \frac{2}{3}L_2 > 0, \qquad \text{for } \hat{\mathbf{n}}(\mathbf{r}) \parallel \nabla\overline{P}_2(\mathbf{r}) \quad (5.29b)$$

de Gennes[30] has shown that the orientation of the molecules at a nematic-isotropic liquid interface can depend on the sign of L_2. If $L_2 > 0$ the surface energy is less when molecules lie parallel to the interface. On the other hand, when $L_2 < 0$, the surface energy is less corresponding to the perpendicular molecular configuration.
Equations (5.22) and (5.24) can be generalized to study the effects of external magnetic and/or electric fields on the elastic behaviour of nematics. Let us assume

that a static magnetic field **H** is applied to the system. Since the nematics are normally diamagnetic, the magnetic field contribution[31] to the free-energy is given by

$$A_m = -\frac{1}{6}(\Delta\chi)_{max}\,\overline{P}_2(\mathbf{r})\,\{3[\mathbf{H}\cdot\hat{\mathbf{n}}(\mathbf{r})]^2 - H^2\} \tag{5.30}$$

Similarly, in the presence of an external electric field **E**, the free-energy expansion contains a term given by

$$A_E = -\frac{1}{24\pi}(\Delta\epsilon)_{max}\,\overline{P}_2(\mathbf{r})\,\{3[\mathbf{E}\cdot\hat{\mathbf{n}}(\mathbf{r})]^2 - E^2\} \tag{5.31}$$

Thus, in case of nematics subjected to external fields eq. (5.26) reads

$$\begin{aligned} A = K' &+ \frac{9}{4}\overline{P}_2^2\,\{(L_1+\frac{1}{2}L_2)[\nabla\cdot\hat{\mathbf{n}}(\mathbf{r})]^2 + L_1[\hat{\mathbf{n}}(\mathbf{r})\cdot\nabla\times\hat{\mathbf{n}}(\mathbf{r})]^2 \\ &+ (L_1+\frac{1}{2}L_2)\,[\hat{\mathbf{n}}(\mathbf{r})\times\nabla\times\hat{\mathbf{n}}(\mathbf{r})]^2\} \\ &- \frac{1}{6}(\Delta\chi)_{max}\,\overline{P}_2(\mathbf{r})\,\{3[\mathbf{H}\cdot\hat{\mathbf{n}}(\mathbf{r})]^2 - H^2\} \\ &- \frac{1}{24\pi}(\Delta\epsilon)_{max}\,\overline{P}_2(\mathbf{r})\,\{3[\mathbf{E}\cdot\hat{\mathbf{n}}(\mathbf{r})]^2 - E^2\} \end{aligned} \tag{5.32}$$

5.1.2.2 Theories based on Maier-Saupe (MS) or Onsager type molecular models

Several theoretical studies of elastic constants for the uniaxial nematics were reported[10,32–39] in which the pair potential was assumed to consist of either short-ranged repulsions, modelled by an anisotropic hard-core, or long-ranged dispersion attractions. The hard core repulsions enter into the free-energy only through the entropy terms, whereas the attractions are preassumed to contribute only to the mean-field averaged energy.

Assuming an attractive potential due to the dispersion force, Nehring and Saupe[10] calculated the ratios of the elastic constants of uniaxial nematic. They got the results : $K_i \propto \overline{P}_2^2$ and $K_1' : K_2 : K_3' : K_{24} : K_{13}^{(2)} = 5 : 11 : 5 : -9 : -6$. Here $K_1' = K_1 - 2K_{13}^{(2)}$ and $K_3' = K_3 + 2K_{13}^{(2)}$. The first three constants were found to be positive and were already calculated earlier[40]. The results for the K_{24} and $K_{13}^{(2)}$, which turned out to be negative, were new. In this calculation the second order splay-bend contributions to the deformation energy were included. In addition, it was found that for the pure splay, twist and bend moduli $K_1 : K_2 : K_3 = 5 : 11 : 5$ and that the neglected interaction terms which contributes only to the surface integrals do not influence these constants and that their ratios are temperature independent. Although K_i's are approximately proportional to $\overline{P}_2^2$ experimentally,

their order of magnitudes ($K_1 = K_3 < K_2$) contradicts with the inequalities $K_3 > K_1 > K_2$. It was argued that this discrepancy is due to the crudeness of the model.
On the other hand, Priest[32], Straley[33] and others[34-39] calculated the elastic constants by considering only the hard-core repulsion in the calamitic molecules and obtained the correct observed trend $K_3 > K_1 > K_2$. Priest's results

$$\frac{K_1}{\overline{\overline{K}}} = 1 + \Delta - 3\Delta'(\overline{P}_4 / \overline{P}_2) \quad (5.33a)$$

$$\frac{K_2}{\overline{\overline{K}}} = 1 - 2\Delta - \Delta'(\overline{P}_4 / \overline{P}_2) \quad (5.33b)$$

$$\frac{K_3}{\overline{\overline{K}}} = 1 + \Delta + 4\Delta'(\overline{P}_4 / \overline{P}_2) \quad (5.33c)$$

were found to be in good qualitative agreement with experiment. Here $\overline{\overline{K}} = (K_1 + K_2 + K_3)/3$, Δ and Δ' are positive constants which depend on the molecular properties. Thus in Priest's work the difference between K_1 and K_3 arises due to the contributions of terms containing $\overline{P}_4$ and in the limit $\overline{P}_4 \lll \overline{P}_2$, $K_1 = K_3$. For the case of hard spherocylinders Δ and Δ' were evaluated[32] by adopting a method which involves the evaluation of changes in translational entropy caused by three types of deformation

$$\Delta = 2[(\ell/w)^2 - 1] / \left[7\left(\frac{\ell}{w}\right)^2 + 20\right] \quad (5.34a)$$

$$\Delta' = 27\left[\frac{1}{16}\left(\frac{\ell}{w}\right)^2 - \frac{1}{6}\right] / \left[7\left(\frac{\ell}{w}\right)^2 + 20\right] \quad (5.34b)$$

where w and w + ℓ are the width and length of a spherocylinder, respectively. Most of the hard rod models[32-37] are strictly valid only for the very long and thin rods and they usually predict too large a value for the ratio K_3/K_2 and cannot reproduce the temperature dependences of the elastic constants.

Using the Onsager theory, Straley[33] obtained the result similar to those of Priest for the elastic constants of hard rod system. This work avoids the use of spherical harmonic expansions and is based on the assertions, (i) the response of a liquid crystal to a weak-orienting field which varies slowly in space is just that the preferred orientation aligns everywhere along that field and (ii) that the relative probability of a molecule at $\mathbf{r}$ having orientation $\hat{\mathbf{m}}$ is $f(\hat{\mathbf{m}} \cdot \hat{\mathbf{n}}(\mathbf{r}))$ where f represents the same distribution function which Onsager finds for the case that $\hat{\mathbf{n}}$ is independent of position. The extra free-energy due to this variation of $\hat{\mathbf{n}}(\mathbf{r})$ was calculated. However, the results obtained cannot be compared with the real systems and only the order of magnitude agreement could be obtained. Lee and Meyer[39] obtained orientational distribution functions by numerically solving Onsager's integral equation using a simple iterative method. These distribution functions were used to calculate the elastic constants by adopting Straley's method. The phase equilibrium and viscous properties were also calculated. The elastic constants K_i were calculated as a function of the concentration c. It was found that K_1 and K_2 are only slightly dependent on the order parameter but K_3 diverges to infinity at perfect order. The numerical values of these constants are different from those of Straley[33] because both works use different forms of the angular distribution function. Since the first derivatives of distribution function has to be used in the calculation of the elastic constants, this difference becomes numerically significant. Using a direct correlation function approach (see sec. 5.1.2.4), Poniewierski and Stecki[34] calculated the elastic constants of long hard spherocylinders in the Onsager approximation as a function of the reduced concentration (or dimensionless density) with the help of distribution given by Lasher[40]. The two results, of Lee and Meyer[39] and Poniewierski and Stecki[34], are almost the same where the concentrations overlap.

An alternative to the MS mean-field theory of orientational order in nematics was suggested by Faber[41]. It is a theory of disorder rather than order, i.e., the nematic is treated as a continuum with a perfectly aligned ground state in which a spectrum of distortion modes involving splay, twist and bend are thermally excited with amplitudes determined by the elastic constants. Expressions for $\overline{P}_2$ and $\overline{P}_4$ were derived in terms of K_1, K_2 and K_3. The angular distribution function predicted by this work is significantly different from that as predicted by MS theory. As a result the two theories lead to rather different results for the limiting value of $\overline{P}_2$ at T_{NI}. It is important to mention here that Faber's approach is based on the unjustified approximations, for example, the random phase approximation and q (wave number of the mode of excitation) independence of the elastic constants.

5.1.2.3 The van der Waals type and similar theories

Stecki and Poniewierski[42] calculated the elastic constants of N_u phase for a model of hard spherocylinders with superimposed r^{-6} attraction modulated by a Maier-Saupe $P_2(\cos\theta_{12})$ term. This work is based on the direct correlation function approach. Kimura et al.[43] derived the expression for the elastic constants in the MF approximation by assuming a similar model, i.e., the intermolecular potential is written as the sum of hard-rod repulsion, described by rigid spherocylinders (length L and diameter D) and MS type attraction. These authors could obtain the numerical results for D/L ≈ 3~5. The correct inequalities $K_3 > K_1 > K_2$ were predicted and the temperature dependences of K_i's and their ratios were found to be in accordance with the experiment. Further, it has been emphasized that the theory[43] can also account for the elastic properties of the lyotropic system of molecules having large ratio of L/D. For molecules with long flexible chains the theory was not adequate because it could not account for the effect of molecular deformation in the calculation.

Gelbart and Benshaul[44] used generalized van der Waals (GvdW) theory[45,46], which includes self-consistently both repulsions and attractions, to evaluate the elastic constants. In this approach the contributions of the short- and long-range pair potentials are coupled. Considering the rigid axial molecules it assumes that the short-ranged forces are dominated by hard (anisotropic) - core repulsion and the long-ranged attractions are comprised essentially of an angle dependent dispersion interaction. The hard-core entropy contributions to the free-energy were calculated by using the y-expansion[47] for the thermodynamic functions of hard particle liquids. The energetic contributions as determined by both the pair attractions and the hard-core excluded volume were calculated by the special mean-field averaging procedure[46]. The sum of these two kinds of contributions gives the molecular expressions for the K_i's. It is clear from this work that the entropic and energetic contributions to the elastic constants are comparable and each are dominated by $\overline{P}_2^2$ terms. the correct trend $K_3 > K_1 > K_2$ has been predicted. For PAA at T = 400K it was found that $K_1 : K_2 : K_3 = 2.6 : 1 : 3.2$ whereas the observed[9] ratios are 2.0 : 1.0 : 3.2. The relative magnitudes of K_i's were shown to depend sensitively on the size and shape of the molecular hard core and polarizability.

van der Meer et al.[48] have considered a model based on the distributed harmonic forces between the molecules. Both the attractive and repulsive forces have been considered as distributed along the molecules. This approach disregards the temperature dependence of the elastic constants and considers the special case of perfect orientational order $\overline{P}_2 = \overline{P}_4 = 1$. It is a variant on the ideas of Gelbart and coworkers (GvdW theory)[44-46], who studied the combined effect of the attractive and repulsive forces. The GvdW model assumes that the attractive forces act between the molecular centers and thus would be acceptable for single atoms or small molecules but not for large and elongated molecules. In the model of distributed harmonic forces it is assumed that the attractive and repulsive forces are distributed along the molecules and thus are only important for those parts of two molecules that are in close proximity. This fact is accounted for using the pair distribution functions. In fact, two models for the pair distribution function have been considered; nematic-like pair correlation, in which case there exists a random distribution of molecular centers of neighbours around the excluded volume of a central molecule, and smectic-like pair correlation characterized by the preferential distribution of molecular centers of neighbours in the same plane. It is important to mention here that the smectic-like pair correlations is different from the smectic fluctuations which occur as a pretransitional effect above a smectic-nematic phase transition. The smectic-like pair correlation is a short-range effect that is expected to occur in a mesogenic molecules having alkyl chains. Although this model has many obvious short-comings, it is successful in reproducing certain experimental trends.

The model independent of distribution function, always predicts $K_2/K_1 = 1/3$. In the case of nematic-like pair correlation K_3/K_1 is predicted to vary as x_0^2 which is in agreement with the experimental trends observed for molecules without alkyl chains. In the case of smectic-like pair correlation, simulated by taking a Gaussian distribution, K_3/K_1 decreases with the increasing degrees of this correlation.

Zakharov[49] evaluated the elastic constants and the order parameters by using a theory that is based on the method of conditional distribution (MCD)[50,51]. In this scheme the probability density of the distribution of a selected group of particles is considered with some restrictions on the distribution of the other particles. In the first approximation, all the states of the system, where each molecular cell is occupied by a particle, were taken into account. This introduces a concept of reduced distribution functions which obey infinite chains of integro-differential equations. For an arbitrary equation of the chain, based on the concept of the mean force potential (MFP)[51], a truncation procedure was adopted. This gives a closed system of integral equations for the MFPs. Their numerical solution exhibits certain qualitative features which are described below.

From the integration of the Gibbs canonical distribution a set of functions $P_i(i)$, $P_{jj}(ij)$, etc., were introduced. These functions determine the probability density of finding arbitrary particles at the points $\mathbf{r}_i$ associated with the volume v_i, $\mathbf{r}_j$ with v_j, etc., under the condition that all of the remaining cells contain one particle each. Using the concept of MFPs these functions can be expressed as

$$\mathbf{P}_i(\mathbf{r}_i, \boldsymbol{\Omega}_i) = \mathbf{Q}^{-1} \exp[-\beta\phi_i(\mathbf{r}_i, \boldsymbol{\Omega}_i)] \tag{5.35}$$

$$\mathbf{P}_{ij}(\mathbf{r}_i, \boldsymbol{\Omega}_i, \mathbf{r}_j, \boldsymbol{\Omega}_j) = \mathbf{Q}^{-2} \exp\{-\beta[u(\mathbf{r}_i, \boldsymbol{\Omega}_i, \mathbf{r}_j, \boldsymbol{\Omega}_j) + \phi_{ij}(\mathbf{r}_i, \boldsymbol{\Omega}_i, \mathbf{r}_j, \boldsymbol{\Omega}_j)]\} \tag{5.36}$$

where

$$\mathbf{Q} = \int_{v_i} d\mathbf{r}_i \int_{\alpha_i} d\boldsymbol{\Omega}_i \exp[-\beta\phi_i(\mathbf{r}_i, \boldsymbol{\Omega}_i)] \tag{5.37}$$

and ϕ_i and ϕ_{ij} represent the mean-force potentials. α_i is the volume associated with the orientation. The MFPs can be written as

$$\frac{\partial\phi_{ij}(\mathbf{r}_i, \boldsymbol{\Omega}_i)}{\partial\mathbf{r}_i} = \int_{v_j} d\mathbf{r}_j \int_{\alpha_j} d\boldsymbol{\Omega}_j \frac{\partial u(\mathbf{r}_i, \boldsymbol{\Omega}_i, \mathbf{r}_j, \boldsymbol{\Omega}_j)}{\partial\mathbf{r}_i} \frac{\mathbf{P}_{ij}(\mathbf{r}_i, \boldsymbol{\Omega}_i, \mathbf{r}_j, \boldsymbol{\Omega}_j)}{\mathbf{P}_i(\mathbf{r}_i, \boldsymbol{\Omega}_i)} \tag{5.38}$$

and

$$\frac{\partial\phi_{ij\ell}(\mathbf{r}_i, \boldsymbol{\Omega}_i, \mathbf{r}_j, \boldsymbol{\Omega}_j)}{\partial\mathbf{r}_i} = \int_{v_\ell} d\mathbf{r}_\ell \int_{\alpha_\ell} d\boldsymbol{\Omega}_\ell \frac{\partial u(\mathbf{r}_i, \boldsymbol{\Omega}_i, \mathbf{r}_j, \boldsymbol{\Omega}_j)}{\partial\mathbf{r}_i} \frac{\mathbf{P}_{ij\ell}(\mathbf{r}_i, \boldsymbol{\Omega}_i, \mathbf{r}_j, \boldsymbol{\Omega}_j, \mathbf{r}_\ell, \boldsymbol{\Omega}_\ell)}{\mathbf{P}_{ij}(\mathbf{r}_i, \boldsymbol{\Omega}_i, \mathbf{r}_j, \boldsymbol{\Omega}_j)} \tag{5.39}$$

Equation (5.39) connects the two-particle function with three particle functions. While the eq. (5.38) gives the mean-force acting on a molecule in cell i due to the molecule in cell j where the state of the latter is averaged, eq. (5.39) gives the mean-force acting on a molecule in cells i and j due to the molecule in cell ℓ, where the state of latter is averaged.

Decomposing the MFPs into irreducible parts[52], and neglecting the three-particle and higher-order correlations a closed system of nonlinear integral equations for the MFPs were obtained,

$$\exp[-\beta\phi_{ij}(\mathbf{r}_i,\mathbf{\Omega}_i)] = \int_{v_j} d\mathbf{r}_j \int_{\alpha_j} d\mathbf{\Omega}_j \exp\{\beta[\phi_{ji}(\mathbf{r}_j,\mathbf{\Omega}_j) - u(\mathbf{r}_i,\mathbf{\Omega}_i,\mathbf{r}_j,\mathbf{\Omega}_j)]\}\,\mathbf{P}_j(\mathbf{r}_j,\mathbf{\Omega}_j) \tag{5.40}$$

The method of solution of this five dimensional problem (5.40) is most complicated. However, its numerical solution can be obtained by an iterative method. The resulting solution determines the one-particle and two-particle functions, order parameters and the elastic constants

$$\frac{K_1}{\overline{\overline{K}}} = 1+\chi_1\left(5-9\left[\frac{<\cos^4\theta_i> - <\cos^6\theta_i>}{<\cos^2\theta_i> - <\cos^4\theta_i>}\right]\right) \tag{5.41a}$$

$$\frac{K_2}{\overline{\overline{K}}} = 1-\chi_1\left(1+3\left[\frac{<\cos^4\theta_i> - <\cos^6\theta_i>}{<\cos^2\theta_i> - <\cos^4\theta_i>}\right]\right) \tag{5.41b}$$

$$\frac{K_3}{\overline{\overline{K}}} = 1-4\chi_1\left(4-3\left[\frac{<\cos^4\theta_i> - <\cos^6\theta_i>}{<\cos^2\theta_i> - <\cos^4\theta_i>}\right]\right) \tag{5.41c}$$

with

$$\chi_1 = (\chi^2-1)/4(\chi^2+2); \qquad \chi = \frac{x_0^2-1}{x_0^2+1} \tag{5.42}$$

and

$$<\cos^2\theta_i> = \frac{1}{3}(2\overline{P}_2+1) \tag{5.43a}$$

$$<\cos^4\theta_i> = \frac{8}{35}\overline{P}_4 + \frac{4}{7}\overline{P}_2 + \frac{1}{5} \tag{5.43b}$$

$$<\cos^6\theta_i> = \frac{2}{29}\overline{P}_6 + \frac{2}{29}\overline{P}_4 + \frac{55}{116}\overline{P}_2 + \frac{33}{232} \tag{5.43c}$$

Knowing the solution of eq. (5.40) the elastic constants can be calculated. The volume dependences of order parameters and K_j's were studied for the Berne-Pechukas interaction potential,

$$u(\mathbf{r}_{12}, \mathbf{\Omega}_1, \mathbf{\Omega}_2) = 4\epsilon(\mathbf{r}_{12}, \mathbf{\Omega}_1, \mathbf{\Omega}_2)\left[\left(\frac{\sigma(\mathbf{r}_{12}, \mathbf{\Omega}_1, \mathbf{\Omega}_2)}{\mathbf{r}_{12}}\right)^{12} - \left(\frac{\sigma(\mathbf{r}_{12}, \mathbf{\Omega}_1, \mathbf{\Omega}_2)}{\mathbf{r}_{12}}\right)^{6}\right] \tag{5.44}$$

where the strength parameter

$$\epsilon(\mathbf{r}_{12}, \mathbf{\Omega}_1, \mathbf{\Omega}_2) = \epsilon_0 [1 - \chi^2 (\hat{e}_1 \cdot \hat{e}_2)^2]^{-1/2} \tag{5.44a}$$

and the range parameter $\sigma(\mathbf{r}_{12}, \mathbf{\Omega}_1, \mathbf{\Omega}_2)$ is given by the eq. (4.69).

It was observed that the order parameters decrease with increasing volume and temperature which is in accordance with the experimental[53] and simulation[54] results. While the observed values of $K_1/\overline{K}$ and $K_2/\overline{K}$ increase with increasing volume, the values of $K_3/\overline{K}$ decrease with it. Further $K_1/\overline{K}$ increases strongly with x_0 and $K_2/\overline{K}$ decreases with it, also $0.5 < K_3/K_1 < 3.0$ and $0.5 < K_2/K_1 < 0.8$.

5.1.2.4 Application of density functional theory

Adopting the procedure, based on the density functional approach[55], we express the elastic free-energy as[8]

$$\begin{aligned}\beta\, \Delta A_e[\rho] &= \beta(\Delta A[\rho] - \Delta A[\rho_0]) \\ &= -\frac{1}{2}\int d\mathbf{x}_1 \int d\mathbf{x}_2\, [\rho_e(\mathbf{x}_1)\rho_e(\mathbf{x}_2) - \rho_0(\mathbf{x}_1)\rho_0(\mathbf{x}_2)]\, C^{(2)}(\rho_f) \\ &\quad -\frac{1}{3}\int d\mathbf{x}_1 \int d\mathbf{x}_2 \int d\mathbf{x}_3\, \Delta\rho(\mathbf{x}_1)\Delta\rho(\mathbf{x}_2)\Delta\rho(\mathbf{x}_3)\, C^{(3)}(\rho_f)\end{aligned} \tag{5.45}$$

Here $\beta\Delta A[\rho_0]$ is the reduced Helmholtz free-energy of a system of undistorted phase of density ρ_0. $\rho_e(\mathbf{x}_i)$ and $\rho_0(\mathbf{x}_i)$ represent, respectively, the single-particle density distribution functions corresponding to the distorted and undistorted phases. From a knowledge of the density distribution functions and the direct correltion functions, explicit relations for the elastic constants of ordered phases can be derived by comparing eq. (5.45) with the elastic continuum relations (see Sec. 5.1.1).

Uniaxial nematic phase

Since the principal effect of an orientational stress is to cause the director to vary spatially in such a way that at each point in space the singlet orientational distribution has the same uniaxial form with only the axis varying, we write[44,57]

$$f(\mathbf{r},\mathbf{\Omega}) = f_0(\mathbf{r},\mathbf{\Omega})\,[1 + t_s(\mathbf{r},\mathbf{\Omega})] \tag{5.46}$$

where

$$f_0(\mathbf{r},\mathbf{\Omega}) = f(\hat{\mathbf{e}}\cdot\hat{\mathbf{n}}(\mathbf{r}))$$

The term t_s represents the stress-induced changes in the form of the orientational distribution function. In case of long-wavelength distortion this term makes no contribution to the Frank elastic constants.

Molecular expressions for K_i's were derived by Singh and Singh[57] by assuming the configurations as adopted by Priest[32], and Gelbart and Benshaul[44]. An arbitrary point at R = 0 in the deformed phase is chosen as the origin of the space-fixed coordinate system. The Z-axis is taken parallel to the director at the origin where the molecule 1 is assumed to be located. The molecule 2 is at the point $\mathbf{r}_{12}$ from the origin. The variations in $\hat{\mathbf{n}}(\mathbf{r}_{12})$, confined to a plane (say (X,Z) plane), can be described by

$$\hat{\mathbf{n}}(\mathbf{r}_{12}) = \hat{\mathbf{z}}\cos\theta_n + \hat{\mathbf{x}}\sin\theta_n \tag{5.47}$$

Since the deformation angle, $\theta_n(\mathbf{r}_{12})$, is the angle between the director at the $\mathbf{r}_{12}(\equiv \mathbf{r})$ and the director at the origin, we write

$$f(\mathbf{\Omega}, \mathbf{R}+\mathbf{r}) \equiv f(\mathbf{\Omega},\theta_n(\mathbf{r}))$$

In case the deformation angle is small (say $\Delta\theta$), $f(\Omega,\Delta\theta(\mathbf{r}))$ can be expanded as a power series in $\Delta\theta$.

Assuming the above structure, the expressions for the elastic constants were derived[57,60] in terms of the successive higher-order correlation functions of the isotropic liquid,

$$K_i = \sum_n K_i^{(n)} \tag{5.48}$$

where

$$K_i^{(0)} = -\rho_0^2 k_B T \int r^2\, d\mathbf{r} \int d\mathbf{\Omega}_1 \int d\mathbf{\Omega}_2\, \delta f(\mathbf{\Omega}_1,0)\, F_i(\hat{\mathbf{r}},\mathbf{\Omega}_1,\mathbf{\Omega}_2)\, C^{(2)}(\rho_0) \tag{5.49}$$

$$K_i^{(1)} = -\frac{2}{3}\rho_0^3 k_B T \int r^2\, d\mathbf{r} \int d\mathbf{r}_3 \int d\mathbf{\Omega}_2 \int d\mathbf{\Omega}_3\, \delta f(\mathbf{\Omega}_1,0)\, \delta f(\mathbf{\Omega}_3,0)\, F_i(\hat{\mathbf{r}},\mathbf{\Omega}_1,\mathbf{\Omega}_2)\, C^{(3)}(\rho_0) \tag{5.50}$$

etc., with

$$F_i(\hat{\mathbf{r}},\mathbf{\Omega}_1,\mathbf{\Omega}_2) = f'(\mathbf{\Omega}_2,0)\begin{pmatrix} -(\hat{\mathbf{r}}\cdot\hat{\mathbf{x}})(\hat{\mathbf{r}}\cdot\hat{\mathbf{z}}) \\ 0 \\ (\hat{\mathbf{r}}\cdot\hat{\mathbf{x}})(\hat{\mathbf{r}}\cdot\hat{\mathbf{z}}) \end{pmatrix} + \frac{1}{2} f''(\mathbf{\Omega}_2,0)\begin{pmatrix} (\hat{\mathbf{r}}\cdot\hat{\mathbf{x}})^2 \\ (\hat{\mathbf{r}}\cdot\hat{\mathbf{y}})^2 \\ (\hat{\mathbf{r}}\cdot\hat{\mathbf{z}})^2 \end{pmatrix} \tag{5.51}$$

Here $\hat{\mathbf{x}}$, $\hat{\mathbf{y}}$ and $\hat{\mathbf{z}}$ are the unit vectors along the SF, X, Y and Z axes and

$$\delta f(\mathbf{\Omega}_i,0) = f(\mathbf{\Omega}_i,0) - 1.$$

Since the three-body direct correlation function $C^{(3)}$ is not known even for the atomic fluids, the following relation was used to simplify the terms involving them

$$\frac{\delta C^{(2)}(\mathbf{r},\mathbf{\Omega}_1,\mathbf{\Omega}_2)}{\delta\rho_0} = \int C^{(3)}(\mathbf{r}_{12},\mathbf{r}_{13},\mathbf{r}_{23},\mathbf{\Omega}_1,\mathbf{\Omega}_2,\mathbf{\Omega}_3)\, d\mathbf{r}_3\, d\mathbf{\Omega}_3 \qquad (5.52)$$

For a system of hard ellipsoids of revolution (HER) the contributions of each term of the series (5.48) were evaluated[57] for the prolate ellipsoids ($x_0 = 3.0$ and 4.0) and oblate ellipsoids ($1/x_0 = 3.0$ and 4.0). It was found that the magnitude of $K^{(1)}_{i,HER}$ term is small for all i but the truncation of series at an early stage introduces error. Therefore, the contributions of the higher-order terms were evaluated using the Pade' [1,0] approximant. For prolate ellipsoids, which models the calamitic nematic phase, it was found that $K_{3,HER} > K_{1,HER} > K_{2,HER}$ and the ratio $K_{3,HER}/K_{1,HER}$ increases with x_0. In case of oblate ellipsoids, which simulate the discotic nematic phase, $K_{2,HER} > K_{1,HER} > K_{3,HER}$.

Numerical calculations for a nematic system of hard spherocylinders (HSC) using eqs. (5.49) and (5.50) have not been carried out because it was not possible to obtain the DPCF. Within the systematic approach of density functional formalism an alternative approach has been developed by Lee[36] which is based on a self-consistent functional scaling for the free-energy and the DPCF. However, it has been found that as the length-width ratio increases this approach reduces to Straley's[33] molecular theory in the low density limit. The distortion free-energy functional was constructed as

$$\begin{aligned}\beta\,\Delta A_e \approx\ & \frac{1}{2}\int d\mathbf{r}_1 \int d\mathbf{\Omega}_1 \int d\mathbf{r}_2 \int d\mathbf{\Omega}_2\, C^{(2)}(\mathbf{r}_{12},\mathbf{\Omega}_1,\mathbf{\Omega}_2) \\ & [\delta\rho_e(\mathbf{r}_1,\mathbf{\Omega}_1)\,\delta\rho_e(\mathbf{r}_2,\mathbf{\Omega}_2) - \delta\rho_0(\mathbf{r}_1,\mathbf{\Omega}_1)\,\delta\rho_0(\mathbf{r}_2,\mathbf{\Omega}_2)]\end{aligned} \qquad (5.53)$$

where $\delta\rho = [\rho(\mathbf{r},\mathbf{\Omega}) - \rho_0]$, is the density change between the ordered phase and the isotropic phase.

In the limit of long-wavelength distortion the director field varies slowly in space,

$$\begin{aligned}\delta\,\rho_e(\mathbf{r}+\Delta\mathbf{r},\mathbf{\Omega}) \approx\ & \delta\,\rho_e(\mathbf{r},\mathbf{\Omega}) + \Delta\mathbf{r}_\alpha\,\partial_\alpha\,\rho_e(\mathbf{r},\mathbf{\Omega}) \\ & + \frac{1}{2}\,\Delta\mathbf{r}_\alpha\,\Delta\mathbf{r}_\beta\,\partial_\alpha\,\partial_\beta\,\rho_e(\mathbf{r},\mathbf{\Omega}) + \cdots\end{aligned} \qquad (5.54)$$

Rewriting $\mathbf{r}_2 = \mathbf{r} + \mathbf{u}$ and applying the chain rule

$$\partial_\beta\,\rho(\mathbf{\Omega}_2\cdot\hat{\mathbf{n}}(\mathbf{r}_1)) \approx \rho'(\cos\theta_2)\,\mathbf{\Omega}_{2\gamma}\,\partial_\beta\,n_\gamma(\mathbf{r}_1)$$

and integrating over $\mathbf{r}_1$, one obtains

$$\beta\,\Delta A_e \approx \int A_{\alpha\mu}(\mathbf{r})\,\partial_\alpha\,n_\mu(\mathbf{r})\,d\mathbf{r} + \frac{1}{2}\int B_{\alpha\beta\mu}(\mathbf{r})\,\partial_\alpha\,n_\mu(\mathbf{r})\,d\sigma_\beta$$

$$+\frac{1}{2}M_{\alpha\beta\mu\nu}\int \partial_\alpha\,n_\mu(\mathbf{r})\,\partial_\beta\,n_\nu(\mathbf{r})\,d\mathbf{r} \tag{5.55}$$

where

$$A_{\alpha\mu}(\mathbf{r}) \approx -\frac{1}{2}\iiint \delta\rho(\boldsymbol{\Omega}_1\cdot\hat{\mathbf{n}}(\mathbf{r}_1))u^\alpha\,C^{(2)}(\mathbf{u},\boldsymbol{\Omega}_1,\boldsymbol{\Omega}_2)\rho'(\cos\theta_2)\boldsymbol{\Omega}_{2\mu}\,d\mathbf{u}\,d\boldsymbol{\Omega}_1\,d\boldsymbol{\Omega}_2 \tag{5.56}$$

$$B_{\alpha\beta\mu}(\mathbf{r}) \approx -\frac{1}{2}\iiint \delta\rho(\boldsymbol{\Omega}_1\cdot\hat{\mathbf{n}}(\mathbf{r}_1))u^\alpha\,u^\beta\,C^{(2)}(\mathbf{u},\boldsymbol{\Omega}_1,\boldsymbol{\Omega}_2)\rho'(\cos\theta_2)\boldsymbol{\Omega}_{2\mu}\,d\mathbf{u}\,d\boldsymbol{\Omega}_1\,d\boldsymbol{\Omega}_2 \tag{5.57}$$

$$M_{\alpha\beta\mu\nu} \approx \frac{1}{2}\iiint u^\alpha\,u^\beta\,C^{(2)}(\mathbf{u},\boldsymbol{\Omega}_1,\boldsymbol{\Omega}_2)\rho'(\cos\theta_1)\rho'(\cos\theta_2)\Omega_{1\mu}\,\Omega_{2\nu}\,d\mathbf{u}\,d\boldsymbol{\Omega}_1\,d\boldsymbol{\Omega}_2 \tag{5.58}$$

Neglecting the surface term B for the small surface effects and comparing eq. (5.55) with (5.14) for the nonpolar materials ($A_{\alpha\mu} = 0$), the following representations were obtained :

$$K_1 \equiv M_{1111} = M_{2222} = M_{3333} \tag{5.59a}$$

$$K_2 \equiv M_{1122} = M_{2211} = -\,M_{1221} = -\,M_{2112} \tag{5.59b}$$

$$K_3 \equiv M_{2233} = M_{3322} = M_{3311} = M_{1133} \tag{5.59c}$$

Defining the scaled modulus $K_\alpha D/k_BT$, and replacing $\rho(\cos\theta)$ by $cf(\cos\theta)$ the elastic constants were obtained as

$$K_\alpha \approx -\frac{1}{2}k_BT\,\rho_0^2\,L^4D\int d\boldsymbol{\Omega}_1\int d\boldsymbol{\Omega}_2\,f'(\cos\theta_1)\,f'(\cos\theta_2)\,\mathbf{R}_{\alpha\alpha}\,\boldsymbol{\Omega}_{1x}\,\boldsymbol{\Omega}_{2x} \tag{5.60}$$

where $f'(\cos\theta_i)$ is the derivative of the orientational distribution function with respect to its argument, and the average orientational correlation is represented by

$$\mathbf{R}_{\alpha\beta} \approx -\int d\mathbf{r}\;r^\alpha\,r^\beta\,C^2(\mathbf{r};\boldsymbol{\Omega}_1,\boldsymbol{\Omega}_2)\,/\,L^4\,D \tag{5.61}$$

$\mathbf{R}_{\alpha\beta}$ can be evaluated from the known analytic expression for $C^{(2)}$.

These expressions were evaluated in the low density limit by Poniewierski and Stecki[34] using variational approximation for $f(\cos\theta)$. For the particular cases of very long and narrow molecules, Lee and Meyer[39] evaluated elastic moduli in the same limit of low-density without variational approximations. Lee[36] calculated the elastic constants beyond Onsager's limit with a prescription for DPCF as

$$C^{(2)}(\mathbf{r},\mathbf{\Omega}_1,\mathbf{\Omega}_2) \approx \alpha(\eta)\, M_{HSC}(\mathbf{r},\mathbf{\Omega}_1,\mathbf{\Omega}_2) \quad (5.62)$$

where M_{HSC} is the Mayer function of the HSC and $\alpha(\eta)$ is given by

$$\alpha(\eta) = (4-\eta)/4(1-\eta)^4 \quad (5.63)$$

with the property $\alpha(0) = 1$. Equation (5.62) has not been obtained from the functional differentiation of a free-energy functional model, but it is constructed directly on some physical arguments. Lee found that K_1 (or K_2) is weakly dependent on the degree of ordering, while K_3 diverges to infinity at perfect order. Above a certain packing fraction, a HSC system undergoes a transition from the nematic phase to the smectic phase, for example, the system with $x_0 = 6$ forms the stable smectic phase at the scaled critical packing fraction η^* $(= \eta/\eta_{max}) \sim 0.6$; $\eta_{max} = \frac{\pi}{2\sqrt{3}}\left(\frac{L/D + 2/3}{L/D + \sqrt{2/3}}\right)$. For small values of x_0., the higher-order packing entropy effects provided by $\alpha(\eta)$ become more dominant, and therefore, all three elastic constants increase rapidly with η (or scaled concentration). As x_0 increases the predictions of this model approach the low-density approximation[33]

Somoza and Tarazona[37a] used an effective DCF for evaluating elastic constants of HSC system from eq. (5.60). The main idea is to evaluate the free-energy of a system of hard bodies with the help of a reference system of parallel hard ellipsoids (PHE) which may be obtained by a direct mapping of the hard sphere model. The shape of reference PHE is taken to reflect both the molecular shape and the local angular distribution. The model[37a] writes the free-energy of HSC as

$$A[\rho] = A_{id}[\rho] + \int d\mathbf{r}_1 \int d\mathbf{\Omega}_1\, \rho(\mathbf{r}_1,\mathbf{\Omega}_1)\, \Delta\psi_{PHE}\,(\overline{\rho}(\mathbf{r}_1))$$

$$\left[\frac{\tilde{M}_{HSC}(\mathbf{r}_1,\mathbf{\Omega}_1)}{\tilde{M}_{PHE}(\mathbf{r}_1)}\right] \quad (5.64)$$

where $\Delta\psi_{PHE}$ is the interaction contribution to the free-energy per molecule in the PHE system which is evaluated at an average density $\overline{\rho}(\mathbf{r}_1)$ and $\tilde{M}_{HSC}$ and $\tilde{M}_{PHE}$ are the Mayer functions of the HSC and the reference PHE system, averaged with the distribution function

$$\tilde{M}_{HSC}(\mathbf{r}_1,\mathbf{\Omega}_1) = \int d\mathbf{r}_2\, d\mathbf{\Omega}_2\, M_{HSC}(\mathbf{r}_1-\mathbf{r}_2,\mathbf{\Omega}_1,\mathbf{\Omega}_2)\,\rho(\mathbf{r}_2,\mathbf{\Omega}_2) \quad (5.65)$$

$$\tilde{M}_{PHE}(\mathbf{r}_1) = \int d\mathbf{r}_2\, d\mathbf{\Omega}_2\, M_{PHE}(\mathbf{r}_1 - \mathbf{r}_2)\, \rho(\mathbf{r}_2, \mathbf{\Omega}_2) \tag{5.66}$$

The functional differentiation of eq. (5.64) gives a fairly complex expression for DPCF,

$$\begin{aligned} C^{(2)}(\mathbf{r}_1 - \mathbf{r}_2, \mathbf{\Omega}_1, \mathbf{\Omega}_2) = & \frac{\Delta\psi_{PHE}(\eta)}{4\eta} M_{HSC}(\mathbf{r}_1 - \mathbf{r}_2, \mathbf{\Omega}_1, \mathbf{\Omega}_2) + \tilde{M}_{HSC}(\mathbf{\Omega}_1) \\ & H_1(\mathbf{r}_1 - \mathbf{r}_2, \mathbf{\Omega}_2) + \tilde{M}_{HSC}(\mathbf{\Omega}_2)\, H_1(\mathbf{r}_1 - \mathbf{r}_2, \mathbf{\Omega}_1) \\ & + \int d\mathbf{r}_3\, d\mathbf{\Omega}_3\, \rho(\mathbf{r}_3, \mathbf{\Omega}_3)[H_1(\mathbf{r}_3 - \mathbf{r}_1, \mathbf{\Omega}_1)\, M_{HSC}(\mathbf{r}_3 - \mathbf{r}_2, \mathbf{\Omega}_3, \mathbf{\Omega}_2) \\ & + H_1(\mathbf{r}_3 - \mathbf{r}_2, \mathbf{\Omega}_2)\, M_{HSC}(\mathbf{r}_3 - \mathbf{r}_1, \mathbf{\Omega}_3, \mathbf{\Omega}_1) \\ & + \int d\mathbf{r}_3\, d\mathbf{\Omega}_3\, \rho(\mathbf{r}_3, \mathbf{\Omega}_3)\, \tilde{M}_{HSC}(\mathbf{r}_2, \mathbf{\Omega}_3)\, H_2(\mathbf{r}_1, \mathbf{r}_2, \mathbf{r}_3, \mathbf{\Omega}_1, \mathbf{\Omega}_2) \end{aligned} \tag{5.67}$$

where

$$\Delta\psi_{PHE}(\eta) = \eta \frac{4 - 3\eta}{(1-\eta)^2} \tag{5.68a}$$

$$H_1(\mathbf{r}_1 - \mathbf{r}_2, \mathbf{\Omega}_2) = -\frac{\delta}{\delta\rho(\mathbf{r}_2, \mathbf{\Omega}_2)} \left[\frac{\Delta\psi_{PHE}(\rho(\mathbf{r}_1))}{\tilde{M}_{PHE}(\rho(\mathbf{r}_2, \mathbf{\Omega}_2))} \right] \tag{5.68b}$$

$$H_2(\mathbf{r}_1, \mathbf{r}_2, \mathbf{r}_3, \mathbf{\Omega}_1, \mathbf{\Omega}_2) = -\frac{\delta^2}{\delta\rho(\mathbf{r}_1, \mathbf{\Omega}_1)\, \delta\rho(\mathbf{r}_2, \mathbf{\Omega}_2)} \left[\frac{\Delta\psi_{PHE}(\rho(\mathbf{r}_3))}{\tilde{M}_{PHE}(\rho(\mathbf{r}_3, \mathbf{\Omega}_3))} \right] \tag{5.68c}$$

The functional derivatives (5.68b) and (5.68c) were evaluated for the homogeneous system. However, in the calculation a very simple form for the effective DPCF was adopted by Somoza and Tarazona[37a],

$$C^{(2)}_{eff}(\mathbf{r}_1 - \mathbf{r}_2, \mathbf{\Omega}_1, \mathbf{\Omega}_2) = \left(\frac{\Delta\psi(\eta)}{4\eta} \right) M_{HSC}(\mathbf{r}_1 - \mathbf{r}_2, \mathbf{\Omega}_1, \mathbf{\Omega}_2) \tag{5.69}$$

Two sets of calculations were performed by Somoza and Tarazona[37a]; in the first set eq. (5.69) has been used, whereas in the second set the full expression (5.67) was adopted for the DPCF. In the first case the elastic constants were evaluated directly from the Lee's work[36] by simple scaling by a factor $\Delta\psi(\eta)/4\eta\ \alpha(\eta)$. A comparison between different results is made in Table 5.1. It can be seen that the results of neither of the two calculations[36,37a] are in very good agreement with the simulation work[29]. The results obtained by Lee[36] using a truncated relation seems to be in better agreement with the simulation, probably due to a fortuitous cancellation of two independent sources of error. An improvement between the results of simulation[29] and Somoza and Tarazona[37a] was observed when the dependences of K_1 and K_2 on the order parameter $\overline{P}_2$ were considered and K_i's were scaled with $\overline{P}_2^{\,2}$. However, the situation in case of K_3 is not at all satisfactory.

Table 5.1. The elastic constants of a system of HSC with L/D = 5. K_i^* (in units of $k_B T/D$) are the reduced elastic constants, and η^* be the packing fraction scaled with the close-packing value $\pi\sqrt{2}\,D^3\left(1+\frac{3}{2}L/D\right) \Big/ \left[6\left(1+\frac{3}{2}\right)^{1/2} L/D\right]$.

	η^*	$\overline{P}_2$	K_1^*	K_2^*	K_3^*
Lee's work[36]	0.566	0.905	2.985	0.995	38.67
Somoza and Tarazona work[37a] using eq. (5.69)	0.566	0.905	0.533	0.178	6.905
Somoza and Tarazona work[37a] using eq. (5.67)	0.569	0.908	1.380	0.646	7.202
Simulation work[29]	0.569	0.910	2.06 ± 0.26	1.00 ± 0.09	0.67 ± 0.14

In an another work, Somoza and Tarazona[37b] derived the microscopic expression for the K_i's by following the lines of gradient expansion for simple fluids[59]. It was assumed that the continuum elastic energy should be valid for any smooth change of nematic director and the corresponding deformation should be not only smooth but also small everywhere, so that

$$\hat{\mathbf{n}}(\mathbf{r}) = \hat{\mathbf{n}} + \sum_{\mathbf{q}} \hat{\mathbf{n}}(\mathbf{q})\, e^{-i\mathbf{q}\cdot\mathbf{r}} \tag{5.70}$$

Further, the molecular distribution function in the deformed state is a simple local rotation of the equilibrium distribution

$$\rho(\mathbf{r},\mathbf{\Omega}) = \rho_0(\hat{\mathbf{n}}(\mathbf{r})\cdot\mathbf{\Omega}) \tag{5.71}$$

which together with eq. (5.70) gives

$$\Delta\rho_0(\mathbf{r},\mathbf{\Omega}) = \rho_0(\hat{\mathbf{n}}(\mathbf{r})\cdot\mathbf{\Omega}) - \rho_0(\hat{\mathbf{n}}\cdot\mathbf{\Omega}) \tag{5.72}$$

The functional Taylor expansion of free-energy performed up to the second order in $\Delta\rho_0$ gives the expression for elastic energy as quadratic in $\Delta\rho_0$,

$$\frac{2\Delta A_e}{V} = \sum_{\mathbf{q}} \int d\mathbf{\Omega}_1 \int d\mathbf{\Omega}_2\, C^{(2)}(\mathbf{q},\mathbf{\Omega}_1,\mathbf{\Omega}_2)\, \rho_0'(\hat{\mathbf{n}}\cdot\mathbf{\Omega}_1)\, \rho_0'(\hat{\mathbf{n}}\cdot\mathbf{\Omega}_2)\mathbf{\Omega}_1\cdot\hat{\mathbf{n}}(\mathbf{q})\mathbf{\Omega}_2\cdot\hat{\mathbf{n}}(-\mathbf{q}) \tag{5.73}$$

where

$$C^{(2)}(\mathbf{q},\mathbf{\Omega}_1,\mathbf{\Omega}_2) = \int d\mathbf{r}\; C^{(2)}(\mathbf{r},\mathbf{\Omega}_1,\mathbf{\Omega}_2)\, e^{i\mathbf{q}.\mathbf{r}} \tag{5.74}$$

and ρ_0' is the derivative of the equilibrium distribution function $\rho_0(\cos\theta)$ with respect to its argument $\cos\theta$.

If $\hat{\mathbf{n}}(\mathbf{q})$ can be taken to be nonzero only for the values of $\mathbf{q}$ that are small compared to the molecular size, $C^{(2)}(\mathbf{q},\mathbf{\Omega}_1,\mathbf{\Omega}_2)$ may be expanded, as a function of $\mathbf{q}$, around $\mathbf{q} = 0$,

$$C^{(2)}(\mathbf{q},\mathbf{\Omega}_1,\mathbf{\Omega}_2) = C^{(2)}(0,\mathbf{\Omega}_1,\mathbf{\Omega}_2) + \left.\frac{\partial C^{(2)}}{\partial q_m}\right|_{q=0} q_m + \frac{1}{2}\left.\frac{\partial^2 C^{(2)}}{\partial q_m\,\partial q_n}\right|_{q=0} q_m q_n + 0(q^3) + \cdots \tag{5.75}$$

As the first two terms do not contribute to the eq. (5.73), the elastic energy reads

$$\Delta A_e \approx \frac{1}{4} V \sum_q \int d\mathbf{\Omega}_1 \int d\mathbf{\Omega}_2 \left.\frac{\partial^2 C^{(2)}(\mathbf{q},\mathbf{\Omega}_1,\mathbf{\Omega}_2)}{\partial q_m\,\partial q_n}\right|_{q=0} q_m q_n \; \rho_0'(\hat{\mathbf{n}}\cdot\mathbf{\Omega}_1)\,\rho_0'(\hat{\mathbf{n}}\cdot\mathbf{\Omega}_2)\,\mathbf{\Omega}_i\,\hat{\mathbf{n}}_i(\mathbf{q})\,\mathbf{\Omega}_j\hat{\mathbf{n}}_j(\mathbf{q}) \tag{5.76}$$

Comparing eq. (5.76) with (5.14) and using the usual relationship between the Fourier transform derivatives and the real space moments, the Poniewierski and Stecki[34] expression for the elastic moduli was recovered. The modifications induced by the relaxation of the molecular distribution function in the deformed system have also been considered by these authors[37b]. Although they could demonstrate that this effect may lower the elastic energy and hence reduce the values of the elastic constants, its contribution for a simple model with reasonable molecular asymmetry was found to be quite small and can be neglected within the considered accuracy of the direct correlation function.

Adopting a perturbation scheme, the influence of vdW-type of molecular interactions on the K_i using eq. (5.48) were studied[60-62] by taking the following forms of $u(\mathbf{x}_1,\mathbf{x}_2)$: (i) $u(\mathbf{x}_1,\mathbf{x}_2)$ is described by the interaction energy of a reference HER system superposed with an attractive interaction (dispersion, induction); (ii) $u(\mathbf{x}_1,\mathbf{x}_2)$ is represented by the Berne and Pechukas Gaussian overlap (BP) model (eq. 5.44)).

The influence of dispersion interactions on the K_i was studied[60] for the system of prolate as well as oblate ellipsoids. Numerical calculations were performed for a range of hard core sizes and shapes and dispersional strengths and anisotropies. These authors found that the contribution of long-range dispersion interaction to the

K_i's is always positive. While the values of $K_{i,dis}$ decreases with x_0 in the prolate system, it increases with $1/x_0$ in case of the oblate molecules. This trend is reversed from the one observed in case of HER system[57]. For the prolate as well as oblate system $K_{i,dis}$ are insensitive to the magnitude of the anisotropic dispersion interaction. But it is quite sensitive to the magnitude of the isotropic dispersion interaction. The influence of the dispersion interactions on the K_i is determined not only by the angle-dependent part of the forces but, and more significantly, by the coupling between the isotropic part and the anisotropic hard-core repulsion, through the pair correlation function. It was also found that for prolate molecules the long-range dispersion interaction decreases the values of the ratio K_3/K_1 which is found to be increasing with x_0. Adopting a similar method Singh and Rajesh[61] studied the influence of the induction interaction on the K_i. They represented the reference interaction by the BP model and the perturbation potential by the dipole-induced dipole (induction) interaction. The contribution of induction interaction $K_{i,ind}$ was found to be positive for i = 1 and negative for i = 2, 3 in case of prolate ellipsoids. For the oblate molecules it is positive for i = 1 and 3 and negative for i = 2. The theory was applied[62] to evaluate K_i for the BP Gaussian overlap model. It has been found that the softness in the repulsive core has profound effect on the values of elastic constants and on its dependences on x_0.

5.1.3 Unified Molecular Theory for the Elastic Constants of Ordered Phases

In the previous subsection, we discussed the application of density functional theory for the elastic constants of nematic liquid crystals. The results obtained for the elastic constants of N_u phase were discussed for a number of model systems using approximate forms for the pair correlation functions of the medium. The results reported are supposed to be not accurate because of the use of approximate forms of DPCF. In this section, we discuss, in brief, a unified molecular theory, as developed by Singh, Singh and Rajesh[56] (referred to as SSR theory), for the deformation free-energy of ordered molecular phases (liquid crystals, plastic crystals and crystalline solids). This theory is based on the weighted density functional formalism[63] (WDFT) and writes exact expressions for the elastic constants of ordered phases in terms of integrals which involve spherical harmonic coefficients of DPCF of an effective isotropic system.

Neglecting all the terms involving $C^{(n)}(\rho_f)$ with $n \geq 3$ and using the symmetry of the system the deformation free-energy can be written as

$$\beta \Delta A_e[\rho] \approx -\frac{1}{2}\int d\mathbf{x}_1 \int d\mathbf{x}_2 \, [\rho_e(\mathbf{x}_1)\rho_e(\mathbf{x}_2) - \rho_0(\mathbf{x}_1)\rho_0(\mathbf{x}_2)]\, C^{(2)}(\mathbf{x}_1,\mathbf{x}_2;\rho_f) \tag{5.77}$$

In WDFT, which we follow in SSR theory, ρ_f is replaced by a weighted density, $\overline{\rho}[\rho]$, of the reference system.

$$\overline{\rho}[\rho] = \left(\frac{1}{\rho_0 V}\right) \int d\mathbf{x}_1 \int d\mathbf{x}_2\, \rho(\mathbf{x}_1)\, \rho(\mathbf{x}_2)\, \omega(\mathbf{x}_1, \mathbf{x}_2, \overline{\rho}) \tag{5.78}$$

where ω is a weight factor. $\overline{\rho}[\rho]$ is viewed here as a functional of $\rho(\mathbf{x})$ and ω must satisfy the normalization condition.
Requiring that ω satisfy

$$-C^{(2)}(\mathbf{x}_1, \mathbf{x}_2; \rho_u) = \lim_{\rho \to \rho_u} \frac{\delta^2(\beta \Delta A)}{\delta\rho(\mathbf{x}_1)\, \delta\rho(\mathbf{x}_2)} \tag{5.79}$$

exactly, one obtains

$$\omega(\mathbf{x}_1, \mathbf{x}_2; \overline{\rho}) = -(1/2\Delta a'(\overline{\rho}))\, [\beta^{-1} C^{(2)}(\mathbf{x}_1, \mathbf{x}_2; \overline{\rho}) + (1/V)\, \overline{\rho} \Delta a''(\overline{\rho})] \tag{5.80}$$

where $\Delta a(\overline{\rho})$ is the excess energy per particle and the primes on it denote derivatives with respect to the density. Thus ρ_f in eq. (5.77) is replaced by $\overline{\rho}$.

The singlet distribution functions of the undeformed state can be expressed as

$$\rho_0(\mathbf{x}_i) = \rho_0 \sum_{G} \sum_{\ell,m,n} Q_{\ell mn}(\mathbf{G}) \exp[i\mathbf{G}\cdot\mathbf{r}]\, D^{\ell}_{m,n}(\hat{\mathbf{z}}\cdot\hat{\mathbf{e}}_1) \tag{5.81}$$

Here the director $\hat{\mathbf{n}}(\mathbf{r})$ is assumed to point along the Z-axis.

In the limit of long-wavelength distortion the reciprocal lattice vectors $\mathbf{G}_e$ of the strained structure are related to $\mathbf{G}$ of the unstrained structure as

$$\mathbf{G}_e = (\mathbf{I} + \delta\mathbf{G})^{-1} \cdot \mathbf{G} \tag{5.82}$$

where $\delta\mathbf{G}$ is the strain matrix which governs the change in position. Thus for the deformed state

$$\rho_e(\mathbf{x}_i) = \rho_0 \sum_{G} \sum_{\ell,m,n} Q_{\ell mn}(\mathbf{G}) \exp[i\mathbf{G}_e\cdot\mathbf{r}]\, D^{\ell}_{m,n}(\hat{\mathbf{n}}(\mathbf{r}_2)\cdot\hat{\mathbf{e}}_2) \tag{5.83}$$

All the angles without subscript e refer to a space-fixed frame whose origin is located at $\mathbf{r}_1$.

In writing eqs. (5.81) and (5.83) it is assumed that the molecule 1 is located at the origin with the principal director $\hat{\mathbf{n}}(\mathbf{r}_1)$ pointing in the direction of SF Z-axis and molecule 2 is at a distance $\mathbf{r}_{12}$ from the origin where $\hat{\mathbf{n}}(\mathbf{r}_2)$ represents the direction of principal local director (Fig. 5.1). Using the rotational properties of generalized spherical harmonics, SSR theory writes

$$\begin{aligned} D^{\ell_2}_{m_2,n_2}(\boldsymbol{\Omega}_{2e}) &\equiv D^{\ell_2}_{m_2,n_2}(\hat{\mathbf{n}}(\mathbf{r}_2)\cdot\hat{\mathbf{e}}_2) \\ &= \sum_{m} D^{\ell_2}_{m,m_2}(\Delta\chi(\mathbf{r}))\, D^{\ell_2}_{m_2,n_2}(\boldsymbol{\Omega}_2) \end{aligned} \tag{5.84}$$

where $\chi(\mathbf{r})$ is the angle between the principal directors at $\mathbf{r}_1$ and $\mathbf{r}_2$ and is assumed to be small ($\sim \Delta\chi(\mathbf{r})$).

Expanding $C^{(2)}$ in terms of generalized spherical harmonics and performing angular integrations over $\mathbf{r}_1$, Ω_1 and Ω_2 the elastic free-energy density was obtained[56] as

$$\frac{1}{V}\beta\Delta A_e[\rho] = -\frac{1}{2}\rho_0^2 \sum_{\ell_1,\ell_2,\ell} \sum_{m_1,m_2,m,m'} \sum_{n_1,n_2} \sum_{G} [(2\ell_1+1)(2\ell_2+1)]^{-1}$$

$$C_g(\ell_1\ell_2\ell,\ m_1mm')\, Q_{\ell_1m_1n_1}(G)\, Q_{\ell_2m_2n_2}(-G) \int d\mathbf{r}$$

$$[\exp(i\mathbf{G}_e.\mathbf{r})\, D^{\ell_2}_{mm_2}(\Delta\chi(\mathbf{r})) - \exp(i\mathbf{G}\cdot\mathbf{r})]\, Y^*_{\ell m'}(\hat{\mathbf{r}})\, C(\ell_1\ell_2\ell;\ n_1n_2;r) \tag{5.85}$$

where $C(\ell_1\ell_2\ell;\ n_1n_2;\ r)$ are the spherical harmonic coefficients of DPCF. Equation (5.85) is the principal result of the SSR theory which can be used to derive the exact expressions for the elastic constants of ordered phases.

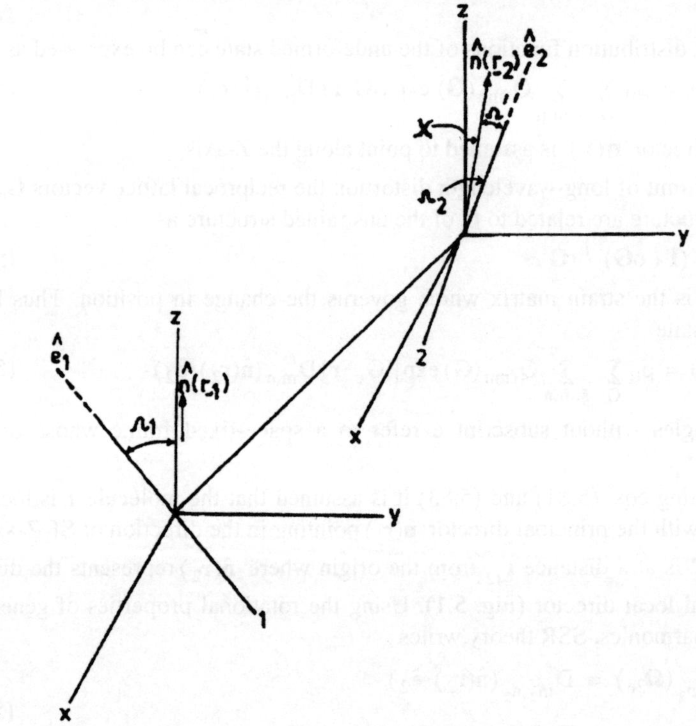

Fig. 5.1. Definition of angular variables for two linear molecules.

5.1.3.1 Application to uniaxial nematic phase

Exploiting the symmetry properties of N_u phase, eq. (5.85) reduces to

$$\frac{1}{V}\,\beta\,\Delta A_e[\rho] = -\frac{1}{2}\,\rho_n^2 \sum_{\ell_1,\ell_2,\ell}{}' \sum_m \left[\frac{(2\ell_1+1)\,(2\ell_2+1)}{16\pi^2}\right]^{1/2} \overline{P}_{\ell_1}\,\overline{P}_{\ell_2}$$

$$C_g(\ell_1\ell_2\ell,\ \mathrm{omm}) \int dr \left[\left(\frac{4\pi}{2\ell_2+1}\right)^{1/2} Y_{\ell_2 m}(\Delta\chi(\mathbf{r})) - 1\right] Y^*_{\ell m}(\hat{\mathbf{r}})\, C_{\ell_1\ell_2\ell}(r) \tag{5.86}$$

where ρ_n is the nematic number density. If the variation of $\hat{\mathbf{n}}(\mathbf{r}_2)$ is confined in a plane (say XZ plane),

$$Y_{\ell_2 m}(\Delta\chi(r)) = Y_{\ell_2 m}(\Delta\theta(r),\ 0)$$

and assuming $\Delta\theta$ to be small the spherical harmonics were expanded in ascending powers of $\Delta\theta$

$$Y_{\ell_2 m}(\Delta\theta(r),0) = \left(\frac{2\ell_2+1}{4\pi}\right)^{1/2} \delta_{m,0} + b_{\ell_2 1}(\delta_{m,1} - \delta_{m,\underline{1}})\,\Delta\theta(r)$$

$$+\frac{1}{2}[b_{\ell_2 0}\,\delta_{m,0} + b_{\ell_2 2}(\delta_{m,2} + \delta_{m,\underline{2}})\,\Delta\theta^2(\mathbf{r}) + \cdots \tag{5.87}$$

where

$$\Delta\theta(r) = \begin{pmatrix} -\,xz \\ 0 \\ xz \end{pmatrix}; \qquad \Delta\theta^2(r) = \begin{pmatrix} x^2 \\ y^2 \\ z^2 \end{pmatrix}$$

and

$$b_{\ell_2 0} = -\frac{1}{2}\,\ell_2(\ell_2+1)\left(\frac{2\ell_2+1}{4\pi}\right)^{1/2}$$

$$b_{\ell_2 1} = -b_{\ell_2 \underline{1}} = -\frac{1}{2}\ell_2(\ell_2+1)\left[\frac{2\ell_2+1}{4\pi}\,\frac{(\ell_2-1)!}{(\ell_2+1)!}\right]^{1/2}$$

$$b_{\ell_2 2} = b_{\ell_2 \underline{2}} = \frac{1}{4}(\ell_2-1)\,\ell_2\,(\ell_2+1)(\ell_2+2)\left[\frac{2\ell_2+1}{4\pi}\,\frac{(\ell_2-2)!}{(\ell_2+2)!}\right]^{1/2}$$

Following expressions were obtained[56] for $K_i'^s$ of N_u phase composed of cylindrically symmetric molecules

$$K_i = \sum_{\ell_1,\ell_2}{}' K_i(\ell_1,\ell_2) \tag{5.88}$$

where

$$K_1(\ell_1,\ell_2) = -\frac{1}{3\sqrt{5}}\,\rho_n^2\,(2\ell_1+1)^{1/2}\,\overline{P}_{\ell_1}\,\overline{P}_{\ell_2}\left\{\frac{1}{2}\sqrt{5}\,b_{\ell_2 0}\,C_g(\ell_1\ell_2 0,\,000)\,J_{\ell_1\ell_2 0} + \left[-\frac{1}{2}b_{\ell_2 0}\,C_g(\ell_1\ell_2 2,\,000) + \sqrt{6}\,b_{\ell_2 1}\,C_g(\ell_1\ell_2 2,\,011) + \sqrt{\frac{3}{2}}\,b_{\ell_2 2}\,C_g(\ell_1\ell_2 2,\,022)\right]J_{\ell_1\ell_2 2}\right\} \tag{5.89}$$

$$K_2(\ell_1,\ell_2) = -\frac{1}{6}\,\rho_n^2\,(2\ell_1+1)^{1/2}\,\overline{P}_{\ell_1}\,\overline{P}_{\ell_2}\left\{b_{\ell_2 0}\,C_g(\ell_1\ell_2 0,000)\,J_{\ell_1\ell_2 0} - \frac{1}{\sqrt{5}}\left[b_{\ell_2 0}\,C_g(\ell_1\ell_2 2,000) + \sqrt{6}\,b_{\ell_2 2}\,C_g(\ell_1\ell_2 2,022)\right]J_{\ell_1\ell_2 2}\right\} \tag{5.90}$$

and

$$K_3(\ell_1,\ell_2) = -\frac{1}{3\sqrt{5}}\,\rho_n^2\,(2\ell_1+1)^{1/2}\,\overline{P}_{\ell_1}\,\overline{P}_{\ell_2}\left\{\frac{1}{2}\sqrt{5}\,b_{\ell_2 0}\,C_g(\ell_1\ell_2 0,\,000)\,J_{\ell_1\ell_2 0} + \left[b_{\ell_2 0}\,C_g(\ell_1\ell_2 2,000) - \sqrt{6}\,b_{\ell_2 1}\,C_g(\ell_1\ell_2 2,011)\right]J_{\ell_1\ell_2 2}\right\} \tag{5.91}$$

with the structure factor given by

$$J_{\ell_1\ell_2\ell} = \int r^4\,dr\,C_{\ell_1\ell_2\ell}(r) \tag{5.92}$$

The numerical calculations were performed[56,64] for a model system of molecules having prolate ellipsoidal symmetry and interacting via a pair potential

$$u(\mathbf{r},\mathbf{\Omega}_1,\mathbf{\Omega}_2) = (u_{HER} + u_{dd} + u_{dq} + u_{qq} + u_{dis})(\mathbf{r},\mathbf{\Omega}_1,\mathbf{\Omega}_2) \tag{5.93}$$

where the subscripts dd, dq, qq and dis indicate, respectively, the interactions arising due to the dipole-dipole, dipole-quadrupole, quadrupole-quadrupole and dispersion interactions. The explicit form of these interactions are given elsewhere[8,66]. The dispersion interaction was described by Gay and Berne potential[67].

The evaluation of K_i's, eqs. (5.88)-(5.91), requires the values of density, order parameters and the structural parameters as a function of density and x_0. The values of spherical harmonics $C_{\ell_1\ell_2\ell}(r)$ of the DPCF for the potential model (5.93) are needed for calculating $J_{\ell_1\ell_2\ell}(r)$. $C_{\ell_1\ell_2\ell}(r)$ can be obtained[66] by solving the Ornstein-Zernike (OZ) equation using the Percus-Yevick (PY) closure relation. As this evaluation is difficult, in a calculation only a finite number of the spherical harmonic coefficients for any orientation-dependent function can be handled. It was found[56] that the inclusion of all the harmonics up to indices $\ell_1 = \ell_2 = 4$ makes the series fully convergent. Since the C-harmonic coefficients contributing to the free-energy of uniaxial nematics of axial molecules have even ℓ_1 and ℓ_2 indices, all the numerical results were calculated for the 14 C-harmonics C_{000}, C_{200}, C_{220}, C_{221}, C_{222}, C_{400}, C_{420}, C_{421}, C_{422}, C_{440}, C_{441}, C_{442}, C_{443} and C_{444}. Thus only those interactions will contribute to the C-harmonic coefficients which have nonvanishing potential $u^{\ell_1\ell_2 m}$-harmonics (in the body-fixed frame) for the even values of ℓ_1 and ℓ_2. From this following conclusions were derived[56],

(i) Since in case of the dd and dq interactions the nonvanishing $u^{\ell_1\ell_2 m}$-harmonics are u_{dd}^{110}, u_{dd}^{111}, u_{dq}^{120} and u_{dq}^{121}, these interactions do not contribute to the free-energy (and hence to the elastic constants) of the uniaxial phase.

(ii) For the qq interaction the nonvanishing $u^{\ell_1\ell_2 m}$-harmonics are u_{qq}^{220}, u_{qq}^{221} and u_{qq}^{222} and hence qq interaction contribute to the elastic constants of the ordered phase.

(iii) In case of HER and dispersion interactions u^{000}, u^{200}, u^{220}, u^{221}, u^{222}, u^{400}, u^{420}, u^{421}, u^{422}, u^{440}, u^{441}, u^{442}, u^{443} and u^{444} are the non-vanishing harmonics and so these interactions contribute.

The C-harmonics were evaluated[56,64] for x = 3.0, 3.25, 3.5 and 4.0. The values of $J_{\ell_1\ell_2\ell}$ were calculated as a function of reduced density $\rho_n^*(= \rho_n d_0^3)$. It has been observed that the contributions of HER interaction dominates and the series (5.88) rapidly converges as ℓ_1 and ℓ_2 increase. At low density all the interactions contribute almost equally but at higher densities (close and above NI transition ρ_n^* = 0.28 in a HER system) the contribution of the repulsive HER interaction increases rapidly. In the density range of the N_u phase the contribution of dispersion and quadrupole interactions are small as compared to the HER interaction. The elastic constants were calculated[64] for a range of x_0, T, ρ_n and molecular parameters. The results obtained for the HER system are in good agreement with that of Allen and Frenkel simulation[29] results. The ratios of the HER contributions are higher than the experimental values and the inclusion of dispersion and quadrupole interactions reduces the values of K_2/K_1 and K_3/K_1. The temperature dependences of K_i's and their ratios were studied and the results were found in accordance with the experiment[68-70].

5.1.3.2 Application to biaxial nematic phase

Exploiting the symmetry of the phase and of the constituent molecules, we have derived[27] relations for the 12 elastic constants (eq. (5.19)) of the N_b phase from eq.

(5.85). In this derivation the main task is to compare the orientation of director triad $(\ell, \hat{\mathbf{m}}, \hat{\mathbf{n}})$ at a point **R** from the origin with its orientation at some neighbouring point **R + r** (Fig. 5.2). Since the orientation of a director triad can be uniquely specified by three angles, the director components were expressed in terms of three Eulerian angles. The continuum relation, eq. (5.19), was also expressed[27] in terms of the director components.

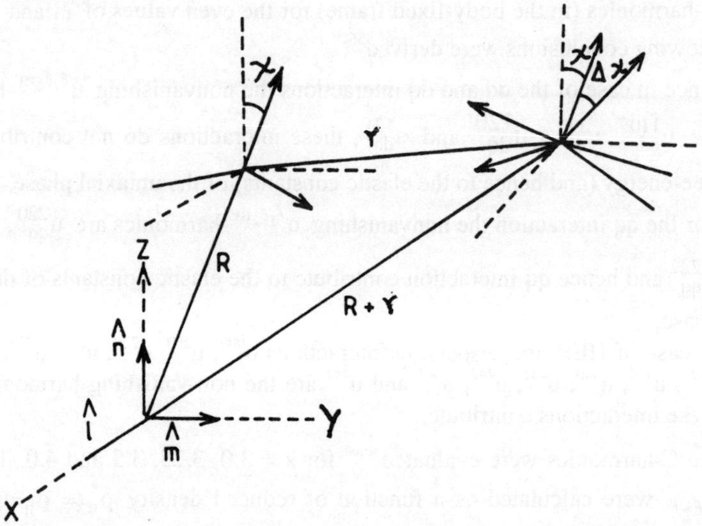

Fig. 5.2. Deviation angle for director triad at **R + r** with respect to director triad at **R**.

The rotation matrix $D^{\ell_2}_{mm_2}(\Delta\chi(\mathbf{r}))$ (eq. (5.85) with no positional correlation) which describes the rotation of director triad at **R + r** with respect to the director triad at **R** can be expressed as

$$D^{\ell_2}_{mm_2}(\Delta\chi(\mathbf{r})) = e^{im\alpha'}\, d^{\ell_2}_{mm_2}(\beta')\, e^{im_2\gamma'} \tag{5.94}$$

Here α', β' and γ are the Euler angles.

Assuming $\xi' = \alpha' + \gamma'$, eq. (5.94) was written as[27]

$$D^{\ell_2}_{mm_2}(\Delta\chi(\mathbf{r}_{12})) = e^{i(m-m_2)\alpha'}\, d^{\ell_2}_{mm_2}(\beta')\, e^{im_2\xi'} \,. \tag{5.95}$$

Since β' is small (weak deformation), we expand $d^{\ell_2}_{mm_2}(\beta')$ in terms of the powers of β' :

$$d^{\ell_2}_{mm_2}(\beta') = d^{\ell_2}_{mm_2}(0) + \beta' \left.\frac{\partial d^{\ell_2}_{mm_2}(\beta')}{\partial \beta'}\right|_{\beta'=0} + \frac{1}{2}\beta'^2 \left.\frac{\partial^2 d^{\ell_2}_{mm_2}(\beta')}{\partial \beta'^2}\right|_{\beta'=0} + \ldots \quad (5.96)$$

Retaining up to the second-order term in eq. (5.95) and performing the angular integrations, the elastic free-energy density was written in terms of the director components. Comparing this relation with the continuum relation in terms of director components, expressions were derived for the twelve elastic constants of biaxial nematic phase (see Refs. 8, 27). The evaluation of these elastic constants requires the values of structural parameters

$$J^{n_1 n_2}_{\ell_1 \ell_2 \ell} = \int r^4 \, dr \, C(\ell_1 \ell_2 \ell, n_1 n_2, r) \quad (5.97)$$

Since the evaluation of structural parameters is extremely difficult, we preferred to obtain order-of-magnitude contributions arising due to different ordering of elastic constants. A system of rigid molecules possessing three mutually orthogonal mirror planes with inversion symmetry through their intersection, e.g., ellipsoids with three different axes or spheroplatelets was considered. The constituent molecules were assumed to have the same symmetry as that of the phase. Such a system can be characterized by four order parameters $\overline{P}_2$, $\overline{\eta}_2$, $\overline{\mu}_2$ and $\overline{\tau}_2$ defined by eq. (2.40). In terms of these order parameters the first term of the elastic constants corresponding to $\ell_1 = \ell_2 = 2$ were evaluated which involve structural parameters J^{22}_{220}, J^{00}_{222}, J^{02}_{222} and J^{22}_{222} (see Appendix E in Ref. 27). For the discussion purposes we give here the expression for $\beta K_{LL}(2,2)$:

$$\begin{aligned} \beta K_{LL}(2,2) = \left(\tfrac{4}{5}\pi\right)^{1/2} \rho_b^2 \Big\{ & \overline{P}_2^2 \left[\tfrac{1}{2} J^{00}_{220} + \left(\tfrac{2}{7}\right)^{1/2} J^{00}_{220}\right] \\ & + \overline{\eta}_2^2 \left[\tfrac{1}{2} J^{22}_{220} + \left(\tfrac{2}{7}\right)^{1/2} J^{22}_{222}\right] + \overline{P}_2 \overline{\eta}_2 \left[\tfrac{1}{2} J^{02}_{220} + \left(\tfrac{2}{7}\right)^{1/2} J^{02}_{222}\right] \\ & + \tfrac{1}{\sqrt{6}} \overline{P}_2 \overline{\mu}_2 \left[\tfrac{1}{2} J^{00}_{220} + \left(\tfrac{2}{7}\right)^{1/2} J^{00}_{222}\right] + \tfrac{1}{\sqrt{6}} \overline{P}_2 \overline{\tau}_2 \left[\tfrac{1}{2} J^{02}_{220} + \left(\tfrac{2}{7}\right)^{1/2} J^{02}_{222}\right] \\ & + \tfrac{1}{\sqrt{56}} \left[\overline{\mu}_2^2 J^{00}_{222} + \overline{\tau}_2^2 J^{22}_{222}\right] \Big\} \end{aligned} \quad (5.98)$$

ρ_b is the number density of the N_b phase. Taking the following values of the order parameters and the structural parameters;

$\overline{\eta}_2 / \overline{P}_2 \approx \overline{\mu}_2 / \overline{P}_2 \sim 0.1$, $\qquad \overline{\tau}_2 / \overline{P}_2 \sim 0.01$,

$J^{22}_{22i} / J^{00}_{22i} \sim 0.20$, $\qquad J^{02}_{22i} / J^{00}_{22i} \approx (J^{22}_{22i} / J^{00}_{22i})^{1/2} \sim 0.45$, with i = 0 or 2, the numerical estimate of all the 12 elastic constants were made. It was observed from eq. (5.98) that the first term involving $\overline{p}_2^2$ is equal to K_2 (2,2) of uniaxial nematic phase[56]. The next two terms containing $\overline{\eta}_2^2$ and $\overline{p}_2\,\overline{\eta}_2$ are due to the breaking of axial symmetry of molecules and their contributions are 5% of that of the first term. The remaining three terms are due to the biaxial ordering of the phase. The combined contribution of these terms is of the order of 5% or less. Similar conclusion was derived for K_{MM} (2,2) also. Thus, while the elastic constants associated with twists equal to that of K_2 (~ 10^{-6} dyn), the twist elastic constant of the uniaxial phase, the magnitude of K_{NN} which represents the twist about the $\hat{\mathbf{n}}$ axis is about 3 or 4 orders of magnitude smaller than K_2. The constants K_{LM} and K_{ML} which represent, respectively, the splay and bend of $\hat{\ell}$ in the $(\hat{\ell}_0, \hat{\mathbf{m}}_0)$ plane, are very small compared to K_2 or K_3 of uniaxial nematic phase. The contribution of terms arising due to ordering giving rise to biaxiality in the phase is also small. Among the three C-constants associated with mixed modes of deformation it was found that C_{LM} is about ten times larger than the other two C_{MN} and C_{NL}. Since for a realistic pair potential model J_{222}^{00} is negative we concluded that while C_{LM} (2,2) and C_{MN}(2,2) are positive, C_{NL}(2,2) is negative. Thus, as argued by Kini and Chandrasekhar[26] the determination of the signs of the C's is experimentally difficult. A study of the deformation for different fields should yield an estimate of the magnitudes of the C's. However, the conoscopic observations[26] can determine the sign of the C-constants.

From the above estimate it was concluded that the three constants, viz., K_{NN}, K_{LM} and K_{ML}, which are associated with deformation in the $(\hat{\ell}_0, \hat{\mathbf{m}}_0)$ plane (the principal director $\hat{\mathbf{n}}$ is perpendicular to this plane in the underformed state), are of the order of 10^{-9} or 10^{-10} dyn, which is three or four orders of magnitude smaller than the values of the constants found in the N_u phase. The constant C_{MN}, associated with the mode of deformation representing simultaneous splay of $\hat{\ell}$ in the $(\hat{\mathbf{m}}_0, \hat{\ell}_0)$ plane and a bend of $\hat{\mathbf{n}}$ in the $(\hat{\mathbf{n}}_0, \hat{\ell}_0)$ plane, and the constant C_{NL}, which is associated with the simultaneous splay of $\hat{\mathbf{m}}$ in the $(\hat{\mathbf{n}}_0, \hat{\mathbf{m}}_0)$ plane and a bend of $\hat{\ell}$ in the $(\hat{\ell}_0, \hat{\mathbf{m}}_0)$ plane, are about one order of magnitude smaller than C_{LM} which is associated with simultaneous splay of $\hat{\mathbf{n}}$ in the $(\hat{\ell}_0, \hat{\mathbf{n}}_0)$ plane and a bend of $\hat{\mathbf{m}}$ in the $(\hat{\mathbf{m}}_0, \hat{\mathbf{n}}_0)$ plane. The seven constants, viz., K_{LL}, K_{MM}, K_{LN}, K_{NM}, K_{MN}, K_{NL} and C_{LM} have nearly equal value and are of the order of the values found in the uniaxial phase. Further, the effect of biaxial ordering and, also, the effect of departure from axial molecular symmetry on the value of the elastic constants are small (almost negligible in view of the large experimental error bars). However, these orderings give rise to several modes of deformation.

5.2 DYNAMICAL PROPERTIES OF NEMATICS

In this section we shall describe, in brief, the basic aspects of "nematodynamics".

5.2.1 Fluid Dynamics of Nematics

Nematics are viscous fluids, i.e., their flow gives rise to energy dissipation due to internal friction. As compared to the isotropic liquids the nematic flow regimes are more complex and more difficult to study experimentally. The translational motions are coupled to inner, orientational motions of the molecules. In most cases, the flow disturbs the orientational alignments. Also a flow in the nematic may be induced due to a change in the alignment by the application of external fields. It is very difficult to measure these effects quantitatively (see Ref. 12). From a theoretical view point, it is difficult to incorporate the coupling between orientation and flow. Using a macroscopic approach, based on classical mechanics, Ericksen[5], Leslie[7] and Parodi[71] (hereafter referred to as ELP) have analysed most of the existing data. A microscopic approach, based on a study of correlation functions, has been developed by several other workers[72]. However, in content the two approaches are identical. We discuss here a slightly amended version[12] of ELP theory.

The macroscopic motion of nematics can be adequately described by treating the nematic fluid as a continuous medium. A dynamical situation must be specified by two fields; namely a velocity field $\mathbf{v}(\mathbf{r})$ giving the flow of matter and the director $\hat{\mathbf{n}}(\mathbf{r})$ describing the local state of alignment. The main task is to write the equations of motion for the variables $\mathbf{v}$ and $\hat{\mathbf{n}}$ in the limit of slow space variations and low frequencies. For simplicity, we consider isothermal processes (no thermal gradients), and hence two kinds of dissipative losses - conventional viscosity effects and losses associated with a rotation of the optic axis with respect to the background fluid. Let a magnetic field $\mathbf{H}$ be applied which is uniform in space. We ignore the surface effects and consider situations of strong anchoring. Following closely the approach of de Groot and Mazur[73] for isotropic fluids, the free-energy stored in the nematic can be expressed as

$$\int \left\{ \frac{1}{2}\rho v^2 + F_0 + F_d + F_m \right\} d^3\mathbf{r}$$

The first term represents kinetic energy (ρ is the density), F_0 the internal free-energy, F_d the distortion free-energy and F_m gives the coupling between the director and an applied magnetic field $\mathbf{H}$.

For an isothermal process, the dissipation $T\dot{\Sigma}$ is equal to the decrease in stored free-energy

$$T\dot{\Sigma} = -\frac{d}{dt}\int \left\{ \frac{1}{2}\rho v^2 + F_0 + F_d + F_m \right\} d^3\mathbf{r} \tag{5.99}$$

The kinetic energy term can be derived from an equation for the local acceleration. The acceleration equation for the fluid reads

$$\rho \frac{dv_\beta}{dt} = \partial_\alpha \sigma_{\alpha\beta} \tag{5.100}$$

where $\sigma_{\alpha\beta}$ is called the stress tensor. Equation (5.100) does not define $\sigma_{\alpha\beta}$ uniquely; changes of the type $\sigma_{\alpha\beta} \rightarrow \sigma_{\alpha\beta} + g_{\alpha\beta}$ leave the acceleration unchanged, provided that $\partial_\alpha g_{\alpha\beta} = 0$.

Substitution of eq. (5.100) into eq. (5.99) and then integration by parts yields

$$-\frac{d}{dt} \int \left(\frac{1}{2}\rho v^2\right) d^3\mathbf{r} = \int \sigma_{\alpha\beta}\, \partial_\alpha\, v_\beta\, d^3\mathbf{r} \tag{5.101}$$

The contribution of $F_0 + F_d + F_m$ to the entropy source is given by (see Ref. 12)

$$-\frac{d}{dt} \int (F_0 + F_d + F_m)\, d^3\mathbf{r} = \int (-\sigma^e_{\alpha\beta}\, \partial_\alpha\, v_\beta + \mathbf{h}\cdot\dot{\hat{\mathbf{n}}})\, d^3\mathbf{r} \tag{5.102}$$

where σ^e is the Ericksen stress

$$\sigma^e_{\alpha\beta} = \sigma^d_{\alpha\beta} - p\,\delta_{\alpha\beta} \tag{5.103}$$

$\mathbf{h}$ is the molecular field

$$h_\alpha = \partial_\gamma \pi_{\gamma\alpha} - \frac{\partial F_d}{\partial n_\alpha} + \gamma_a (\hat{\mathbf{n}}\cdot\mathbf{H}) H_\alpha \tag{5.104}$$

and $\dot{\hat{\mathbf{n}}}$ represents the change of $\hat{\mathbf{n}}$ per unit time as experienced by a moving molecule

$$\dot{\hat{\mathbf{n}}} = \frac{d\hat{\mathbf{n}}}{dt} = \frac{\partial \hat{\mathbf{n}}}{\partial t} + (\mathbf{v}\cdot\nabla)\hat{\mathbf{n}} \tag{5.105}$$

Here $\boldsymbol{\sigma}^d$ is the distortion stress tensor and p is the pressure which plays the role of a Lagrange multiplier. In eqs. (5.101) and (5.102) the surface terms have been ignored.

In eq. (5.104) $\pi_{\gamma\alpha}$ represents

$$\pi_{\gamma\alpha} = \frac{\delta F_d}{\delta(\partial_\gamma\, n_\alpha)} = \frac{\delta F_d}{\delta g_{\gamma\alpha}} \tag{5.106}$$

Adding eqs. (5.101) and (5.102), we obtain

$$T\dot{\Sigma} = \int (\sigma'_{\alpha\beta}\, \partial_\alpha\, v_\beta + \mathbf{h}\cdot\dot{\hat{\mathbf{n}}})\, d^3\mathbf{r} \tag{5.107}$$

where $\sigma'_{\alpha\beta}$ is known as the viscous stress

$$\sigma'_{\alpha\beta} = \sigma_{\alpha\beta} - \sigma^{e}_{\alpha\beta} \tag{5.108}$$

In general the tensor $\sigma'_{\alpha\beta}$ is not symmetric. Let us introduce a vector $\mathbf{\Gamma}$

$$\Gamma_z = -\sigma'_{xy} + \sigma'_{yx}\text{, etc.} \tag{5.109}$$

and derive a relation for it. The total torque applied to the nematic is given by

$$\dot{\mathbf{L}} = \frac{d}{dt}\int (\mathbf{r} \times \rho\, \mathbf{v})\, d^3\mathbf{r} \tag{5.110}$$

In the present case $\dot{\mathbf{L}}$ contains three contributions : magnetic torques

$$\int (\mathbf{M} \times \mathbf{H})\, d^3\mathbf{r}$$

where $\mathbf{M}(\mathbf{r})$ is the total magnetization in the nematic; torques due to external stresses ($\boldsymbol{\sigma}$) acting on the sample boundary

$$\int \mathbf{r} \times (d\mathbf{s} : \boldsymbol{\sigma})$$

where $d\mathbf{s}$ is the surface element vector on the boundary and $(d\mathbf{s} : \boldsymbol{\sigma})_\alpha = ds_\beta\, \sigma_{\beta\alpha}$, and torques on the director at the boundary, which for the equilibrium situations are

$$\int \hat{\mathbf{n}} \times (d\mathbf{s} : \boldsymbol{\pi}).$$

Thus

$$\frac{d}{dt}\int(\mathbf{r} \times \rho\mathbf{v})\, d^3\mathbf{r} = \int (\mathbf{M} \times \mathbf{H})\, d^3\mathbf{r} + \int \{\mathbf{r} \times (d\mathbf{s} : \boldsymbol{\sigma}) + \hat{\mathbf{n}} \times (d\mathbf{s} : \boldsymbol{\pi})\} \tag{5.111}$$

Using eq. (5.100) and integrating by parts, one obtains

$$\int (\sigma_{yx} - \sigma_{xy})\, d^3\mathbf{r} = \int (\mathbf{M}\times\mathbf{H})_z\, d^3\mathbf{r} + \int \{\hat{\mathbf{n}}\times(d\mathbf{s}:\boldsymbol{\pi})\}_z \tag{5.112}$$

Now let $\boldsymbol{\sigma} = \boldsymbol{\sigma}^e + \boldsymbol{\sigma}'$ the contribution due $\boldsymbol{\sigma}^e$ is given[12] by

$$\int (\sigma^e_{yx} - \sigma^e_{xy})\, d^3\mathbf{r} = \int \{\mathbf{M} \times \mathbf{H} - \hat{\mathbf{n}} \times \mathbf{h}\}_z\, d^3\mathbf{r} + \int \{\hat{\mathbf{n}}\times(d\mathbf{s}:\boldsymbol{\pi})\}_z \tag{5.113}$$

The contribution from $\boldsymbol{\sigma}'$ is simply Γ_z. With these results, eq. (5.112) gives

$$\int \{\mathbf{\Gamma} - \hat{\mathbf{n}} \times \mathbf{h}\}\, d^3\mathbf{r} = 0 \tag{5.114}$$

This leads to

$$\mathbf{\Gamma} = \hat{\mathbf{n}} \times \mathbf{h} \tag{5.115}$$

Thus it is concluded that $\mathbf{\Gamma}$, the torque (per unit volume) exerted by the internal degrees of freedom on the flow, is non-zero only out of equilibrium as $\hat{\mathbf{n}}$ is co-linear to $\mathbf{h}$ at equilibrium.

For the convenience the velocity gradient tensor $\partial_\alpha v_\beta$ is separated into a symmetric part

$$A_{\alpha\beta} = \frac{1}{2}(\partial_\alpha v_\beta + \partial_\beta v_\alpha) \tag{5.116}$$

and an antisymmetric part, associated with the vector

$$\boldsymbol{\omega} = \frac{1}{2}(\boldsymbol{\nabla} \times \mathbf{v}) \qquad (5.117)$$

With these definitions eq. (5.107) reads

$$T\dot{\Sigma} = \int \{\mathbf{A}:\boldsymbol{\sigma}^s - \boldsymbol{\Gamma}\cdot\boldsymbol{\omega} + \mathbf{h}\cdot\dot{\hat{\mathbf{n}}}\}\, d^3\mathbf{r} \qquad (5.118)$$

where $\boldsymbol{\sigma}^s$ is the symmetric part of $\boldsymbol{\sigma}'$. Now using eq. (5.115), one obtains

$$T\dot{\Sigma} = \int \{\mathbf{A} : \boldsymbol{\sigma}^s + \mathbf{h}\cdot\mathbf{N}\}\, d^3\mathbf{r} \qquad (5.119)$$

where the vector **N** represents the rate of change of the director with respect to the background fluid,

$$\mathbf{N} = \dot{\hat{\mathbf{n}}} - \boldsymbol{\omega} \times \hat{\mathbf{n}} \qquad (5.120)$$

Equation (5.119) is the fundamental equation of 'nematodynamics' showing two kinds of dissipation : dissipation by shear flow and dissipation by rotation of the optic axis.

5.2.1.1 The laws of friction

In case of irreversible processes, the each contribution to the entropy source is usually written as the product of a 'flux' by the conjugate 'force'. Now starting from the entropy source (5.119), we can define as fluxes the quantity $A_{\alpha\beta}$ and N_α and as force the quantities $\sigma^s_{\alpha\beta}$ and h_α. The tensor **A** has six independent components, and the vector **N** has three components. For the eq. (5.119), we can make a choice

σ^s_{xx} is the force conjugate to A_{xx},

$2\sigma^s_{xy}$ is the force conjugate to A_{xy}, etc. (5.121)

h_α is the force conjugate to N_α.

Now one can write down a set of phenomenological equations expressing the force in terms of the flux (or vice versa). In the limit of slow motion (hydrodynamic limit), it is assumed that the fluxes are weak on the molecular scale. Hence the forces will be linear functions of the fluxes,

$$\sigma^s_{\alpha\beta} = L_{\alpha\beta\gamma\delta}\, A_{\gamma\delta} + M_{\alpha\beta\gamma}\, N_\gamma \qquad (5.122)$$

$$h_\gamma = M'_{\alpha\beta\gamma}\, A_{\alpha\beta} + P_{\gamma\delta}\, N_\delta \qquad (5.123)$$

Here all the coefficients **L**, **M**, **M′** and **P** have the dimension of a viscosity. However, in view of Onsager theorem[73], the matrices **M** and **M′** are identical

$$M_{\alpha\beta\gamma} = M'_{\alpha\beta\gamma} \qquad (5.124)$$

Since the matrices **L**, **M** and **P** must be compatible with the uniaxial $D_{\infty h}$ symmetry, they must be defined in terms of the vector $\hat{\mathbf{n}}$ only. All the measurable properties should remain invariant under the change $\hat{\mathbf{n}} \to -\hat{\mathbf{n}}$. Also in such a change $\mathbf{A}$ and $\boldsymbol{\sigma}^s$ are invariant, while $\mathbf{N}$ and $\mathbf{h}$ are odd. In view of these requirements eqs. (5.122) and (5.123) must read

$$\begin{aligned}\sigma^s_{\alpha\beta} = {} & \rho_1 \delta_{\alpha\beta} A_{\mu\mu} + \rho_2 n_\alpha n_\beta A_{\mu\mu} + \rho_3 \gamma_{\alpha\beta} n_\gamma n_\mu A_{\gamma\mu} \\ & + \alpha_1 n_\alpha n_\beta n_\mu n_\rho A_{\mu\rho} + \alpha_4 A_{\alpha\beta} + \frac{1}{2}(\alpha_5 + \alpha_6) \\ & \times (n_\alpha A_{\mu\beta} + n_\beta A_{\mu\alpha}) n_\mu + \frac{1}{2}\gamma_2 (n_\alpha N_\beta + n_\beta N_\alpha)\end{aligned} \tag{5.125}$$

$$h_\mu = \gamma'_2 n_\alpha A_{\alpha\mu} + \gamma_1 N_\mu \tag{5.126}$$

Again all the coefficients have the dimension of a viscosity. In view of relation (5.124)

$$\gamma'_2 = \gamma_2 \tag{5.127}$$

Thus the equation of nematic dynamics (eqs. (5.125) and (5.126)), in the general case, involves eight independent coefficients. However, if we consider only motions that are very slow as compared to sound waves, the fluid can be treated as incompressible. In case the density of the fluid is constant, the trace of the A tensor vanishes

$$A_{\mu\mu} = \operatorname{div} \mathbf{v} = 0 \tag{5.128}$$

Thus the ρ_1 and ρ_2 terms in eq. (5.125) drop out. The ρ_3 term does not contribute to the entropy source (5.119). So it is convenient to write eq. (5.125) as an equation for the complete viscous stress $\boldsymbol{\sigma}'$

$$\begin{aligned}\sigma'_{\alpha\beta} = {} & \alpha_1 n_\alpha n_\beta n_\mu n_\rho A_{\mu\rho} + \alpha_4 A_{\alpha\beta} + \alpha_5 n_\alpha n_\mu A_{\mu\beta} \\ & + \alpha_6 n_\beta n_\mu A_{\mu\alpha} + \alpha_2 n_\alpha N_\beta + \alpha_3 n_\beta N_\alpha\end{aligned} \tag{5.129}$$

$$h_\mu = \gamma_1 N_\mu + \gamma_2 n_\alpha A_{\alpha\mu} \tag{5.130}$$

with

$$\begin{aligned}\gamma_1 &= \alpha_3 - \alpha_2 \\ \gamma_2 &= \alpha_3 + \alpha_2 = \alpha_6 - \alpha_5\end{aligned} \tag{5.131}$$

The coefficients α_i's are known as the Leslie coefficients. The six α_i's related with eq. (5.129) were first derived by Parodi[71]. Thus the dynamics of incompressible nematic involves five independent coefficients having the dimension of a viscosity.

It is also possible to derive the equations for nematodynamics by making a different choice, i.e., $\sigma^s_{\alpha\beta}$ and N_α as fluxes, while $A_{\alpha\beta}$ and h_α the forces. This choice, first introduced by Harvard group[72], gives very compact formulae for the study of small motions in a nematic single crystal.

Considering the small amplitude motion and stresses only to first-order in n_x and n_y, one obtains

$$T\dot{\Sigma} = \sigma^s_{\alpha\beta}\, A_{\alpha\beta} + h_x\, N_x + h_y\, N_y \tag{5.132}$$

In writing this equation the component of **h** along $\hat{\mathbf{n}}$ has been chosen arbitrarily, so h_2 vanishes identically. Now the issue is to write relations for $\boldsymbol{\sigma}^s$ and **N** in terms of **A** and **h**. For an incompressible fluid,

$$\sigma^s_{\alpha\beta} = 2\nu_2\, A_{\alpha\beta} + 2(\nu_3 - \nu_2)\,(A_{\alpha\mu}\, n^0_\mu\, n^0_\beta + A_{\beta\mu}\, n^0_\mu\, n^0_\alpha)$$

$$+\, 2(\nu_1 + \nu_2 - 2\nu_3)\, n^0_\alpha\, n^0_\beta\, n_\mu\, n_\rho\, A_{\mu\rho} - \frac{1}{2}\, b\,(n^0_\alpha\, h_\beta + n^0_\beta\, h_\alpha) \tag{5.133}$$

$$N_i = h_i/\gamma_1 + b\, A_{iz} \qquad (i = x,y) \tag{5.134}$$

The parameters ν_1, ν_2 and ν_3 have the dimensions of viscosity and b is a dimensionless number. The components of $\boldsymbol{\sigma}^s$ are

$$\sigma^s_{xx} = 2\nu_2\, A_{xx}$$

$$\sigma^s_{xy} = 2\nu_2\, A_{xy}$$

$$\sigma^s_{xz} = 2\nu_3\, A_{xz} - \frac{1}{2}\, b\, h_x \tag{5.135}$$

$$\sigma^s_{zz} = 2\nu_1\, A_{zz}$$

$$\sigma^s_{yz} = 2\nu_3\, A_{yz} - \frac{1}{2}\, b\, h_y$$

In fact, eqs. (5.133) and (5.134) must be identical in content with eqs. (5.129) and (5.130). On comparing the two sets of equations, we get

$$2\nu_2 = \alpha_4 \tag{5.136a}$$

$$2\nu_3 = \alpha_4 + \alpha_6 + \alpha_3\, b = \alpha_4 + \alpha_5 + \alpha_2\, b \tag{5.136b}$$

$$2\nu_1 = \alpha_1 + \alpha_4 + \alpha_5 + \alpha_6 \tag{5.136c}$$

and

$$b = -\gamma_2/\gamma_1 \tag{5.136d}$$

The entropy source may be expressed in terms of the fluxes **A** and h as

$$T\dot{\Sigma} = 2\nu_1\, A^2_{zz} + 4\nu_2(A^2_{zx} + A^2_{zy}) + 2\nu_3(A^2_{xx} + A^2_{yy}) + (h^2_x + h^2_y)/\gamma_1 \tag{5.137}$$

As seen the dissipation does not involve b.

The equation giving the bulk force (per unit volume) in a nematic with distortions and flow can be defined from eq. (5.100),

$$\rho \frac{dv_\beta}{dt} \cong \partial_\alpha \left[\sigma^e_{\alpha\beta} + \sigma^s_{\alpha\beta} + \frac{1}{2}(n_\beta h_\alpha - n_\alpha h_\beta)\right] \tag{5.138}$$

In the limit of small motion the Ericksen stress $\boldsymbol{\sigma}^e$ drops out completely. The last term in eq. (5.138) represents the antisymmetric part of $\boldsymbol{\sigma}'$. Finally, the bulk force is represented by

$$\rho \frac{dv_\beta}{dt} \cong \partial_\alpha \sigma^s_{\alpha\beta} + \frac{1}{2}(n^0_\beta \partial_\alpha h_\alpha - n^0_\alpha \partial_\alpha h_\beta) \tag{5.139}$$

Here to first order in n_x, n_y, we have replaced $n_\beta h_\alpha$ by $n^0_\beta h_\alpha$, etc. Making the use of the fact that the stress tensor giving this bulk force is not unique, the Harvard group[72] has constructed a symmetric stress tensor $\sigma^H_{\alpha\beta}$ which yields the same set of bulk forces provided that no external torques are present in the bulk. The Harvard stress $\boldsymbol{\sigma}^H$ is defined by

$$\sigma^H_{\alpha\beta} = \sigma^s_{\alpha\beta} + \frac{1}{2}\{-n^0_\mu \partial_\mu \pi_{\alpha\beta} - n^0_\mu \partial_\mu \pi_{\beta\alpha} + n^0_\alpha \partial_\mu \pi_{\beta\mu} + n^0_\beta \partial_\mu \pi_{\alpha\mu}\} \tag{5.140}$$

The corresponding force is

$$\partial_\alpha \sigma^H_{\alpha\beta} = \partial_\alpha \sigma^s_{\alpha\beta} + \frac{1}{2}\{-n^0_\mu \partial_\mu h_\beta - n^0_\mu \partial_\mu \partial_\alpha \pi_{\beta\alpha} + n^0_\alpha \partial_\alpha \partial_\mu \pi_{\beta\mu} + n^0_\beta \partial_\mu h_\mu\} \tag{5.141}$$

The two eqns. (5.139) and (5.141) are identical.

5.2.2 Molecular Motions

The dynamics of nematic fluids is very poorly understood on the molecular scale. Since the experimental data are time averages, the relevant information on the dynamics of the molecules can only be obtained from experimental methods involving short time scales. Here we discuss some selected experiments which may be relevant concerning the molecular dynamics.

5.2.2.1 Dielectric relaxation

We consider the behaviour of the dielectric permittivity in the presence of changing field. The main difference between the static and dynamic permittivity is due to the process of reorientation of the permanent dipole moments in the changing field. In such a process a definite time interval is involved which can be seen after the removal of a static field; the orientational polarization decays exponentially to its equilibrium value with characteristic time constant τ. In changing fields this reorientation process leads to a time lag between the direction of the average orientations of the dipole moments and the field. The effect can be seen at frequencies of about τ^{-1}. At higher frequencies the time lag is such that the orientation polarization hardly contributes to the permittivity because the correlation between the orientational polarization and the variations of the field is lost. The residual permittivity ϵ_∞ is caused by the induced polarization only. The informations obtained from dielectric relaxation cover time scales from 10^{-8} s to the order of 10^{-3} s.

In the case of isotropic liquids the dielectric relaxation can be described in terms of a complex permittivity

$$\epsilon(\omega) = \epsilon'(\omega) - i\epsilon''(\omega)$$

where ω denotes the angular frequency of the applied alternating field. In Debye's classical theory $\epsilon(\omega)$ is expressed as

$$\epsilon(\omega) = \epsilon_\infty + \frac{\epsilon - \epsilon_\infty}{1 + i\omega\tau} \tag{5.142}$$

where $\epsilon = \epsilon(0)$ and $\epsilon_\infty = \epsilon(\infty)$ represent the static and high-frequency permittivity, respectively. Equation (5.142) leads directly to

$$\omega\tau = \frac{\epsilon''(\omega)}{\epsilon'(\omega) - \epsilon_\infty} \tag{5.143a}$$

$$\frac{(\epsilon - \epsilon_\infty)[\epsilon'(\omega) - \epsilon_\infty]}{[\epsilon'(\omega) - \epsilon_\infty]^2 + [\epsilon''(\omega)]^2} = 1 \tag{5.143b}$$

Consequently according to eq. (5.143b) a plot of $\epsilon'(\omega)$ vs $\epsilon''(\omega)$ must give a semicircle. For nematics eq. (5.143b) can be checked for $\epsilon_\parallel(\omega)$ and $\epsilon_\perp(\omega)$. It has been observed[74] that only for $\epsilon_\parallel(\omega)$ a true semicircle is found. In other cases deviations have been observed; thus the relaxation to the equilibrium configuration cannot be described by a single relaxation time.

The influence of permanent dipoles, (μ_ℓ and μ_t) on the behaviour of $\epsilon_\parallel$ and $\epsilon_\perp$ can be understood by the following equations

$$\epsilon_\parallel(\omega) - \epsilon_{\parallel,\infty} \sim \left\langle \mu_\parallel^2 \right\rangle = \frac{1}{3}[\mu_\ell^2(1 + 2\overline{P}_2) + \mu_t^2(1 - \overline{P}_2)] \tag{5.144a}$$

$$\epsilon_\perp(\omega) - \epsilon_{\perp,\infty} \sim \left\langle \mu_\perp^2 \right\rangle = \frac{1}{3}[\mu_\ell^2(1 - \overline{P}_2) + \frac{1}{2}\mu_t^2(2 + \overline{P}_2)] \tag{5.144b}$$

The behaviour of μ_ℓ depends on the mutual direction of the director and the applied field. If they are parallel, the reorientation process is considerably affected by the nematic order leading to relatively large values of τ which can easily be observed. When the director and the field are mutually perpendicular, keeping the angle between long axis and the director fixed, the reorientation of μ_ℓ can be accomplished by a rotation over π. The τ for such process may be expected to decrease with increasing nematic order. However, in general, this is difficult to observe because for a given $\overline{P}_2$ the contribution of μ_ℓ to $\in_\perp$ is small. The μ_t can be reoriented by a rotation about the long-axis. Such a rotation, in first approximation, does not depend on the nematic order. As nematic order increases the importance of μ_t increases for $\in_\perp$ but decreases for $\in_\parallel$. Thus the dielectric relaxation of $\in_\parallel$ and $\in_\perp$ can be ascribed, respectively, to the relaxation processes of μ_ℓ and μ_t. The main difference between the dielectric relaxation in the isotropic and the nematic phase must be associated with the reorientation of μ_ℓ.

5.2.2.2 Magnetic resonance and relaxation

As discussed in Sec. 2.3.1 magnetic resonance technique is used to measure the order parameters $\overline{P}_2$, $\overline{P}_4$. In the nmr technique the idea is to measure the line splitting $\Delta\nu$ caused by the nuclear spin-spin interaction. Here $\Delta\nu$ is considered to be an average value. This means that the correlation time τ must satisfy the relation

$$\tau\ \Delta\nu << 1$$

The nmr values of order parameter $\overline{P}_2$ are quite similar to those as obtained using the method of electron paramagnetic resonance (EPR) of spin-probes dissolved in liquid crystals. The time scale of nuclear-spin relaxation is of the order of 10^{-6} s where that of EPR is 10^{-8} s. Thus the supplementary averaging involved in the nmr measurement does not appear to have any influence. This suggests a time scale of the orientational fluctuations < 10^{-8} s and any effect of reorientation in the time range 10^{-8} s to 10^{-6} s must have a small amplitude. However, the measured values of $\overline{P}_2$ require large amplitude reorientations so that the fluctuations of the long molecular axis must be quite rapid. In addition to orientational fluctuations, the conformational changes within the molecules also may be relevant. Information about these changes can be derived from deuteron magnetic resonance (dmr) study; for example, the dmr spectra of alkyl chains indicate isomeric rotations about the C-C bonds. These rotations are fast on the time scale of dmr. Since the relevant time scales of these conformational changes are very small, the concept of an average conformation of the chains may be quite relevant.

The study of nuclear magnetic relaxation effects gives valuable information about the dynamics of the molecules. Since the nuclear spin relaxation is caused due to the molecular motions through the modulation of the interactions involving the nuclear spins, the longitudinal relaxation rate $1/T_1$ is very relevant. In case of isotropic liquids, the influence of the molecular motions on the relaxation process can be described in terms of one single correlation time τ_C, and hence the dependence of relaxation rate on the nuclear frequency ν_n is simple. For nematics the situation is very different; the frequency dependence of $1/T_1$ is more complex and cannot be described in terms of one correlation time τ_C. Several processes may contribute to $1/T_1$. Restricting to the ideal situation with rigid molecules and relaxation by intramolecular couplings only, one can think of contributions due to three classes of molecular motions :

(i) rotations about the long molecular axis,
(ii) large-angle rotations around a short molecular axis,
(iii) small-amplitude oscillations of the long axis around its average orientation.

For the process (i) the short correlation time is $\tau_1 \sim 10^{-10}$ s ($\nu_n \tau_1 << 1$). Process (ii) is rare with a long correlation time. It can contribute to the frequency dependence of $1/T_1$. Process (iii) cannot be described by one correlation time. Such motions can be analysed by the Leslie equations. For nuclear relaxation, the most important fluctuations are those for which the correlation time and the nuclear period should not differ too much $\eta / kq^2 \sim \nu_n^{-1}$. This corresponds to the wavelength

$$\lambda = \frac{2\pi}{q} = 2\pi\left(\frac{K}{\eta\,\nu_n}\right)^{1/2} \tag{5.145}$$

where η is a typical viscosity and K a Frank constant. Taking typical values $\nu_n = 10^7$, $K = 10^{-6}$ and $\eta = 10^{-1}$, one finds $\lambda = 600$ Å. Thus λ is significantly larger than the molecular length a, and the problem can be discussed in terms of the continuum theory. This has been done by several workers[74–76] and qualitatively, the resulting contribution to the nuclear relaxation may be expressed as

$$\frac{1}{T_1} \sim (\gamma\, H_L)^2\, \frac{k_B T\, \overline{P}_2^{\,2}}{K}\left(\frac{K}{\eta}+D\right)^{-\frac{1}{2}} \nu_n^{-1/2} \tag{5.146}$$

where γ is the nuclear gyromagnetic factor, H_L the local field and D is a translational self-diffusion coefficient. In case of PAA the experimental relaxation rates can be fitted by a power law

$$\frac{1}{T_1} = A + B\,\nu_n^{-1/2} \tag{5.147}$$

Here it appears that the superposition of two relaxation processes occurs; one (A) due to the local motions with a short correlation time and the other ($B\,\nu_n^{-1/2}$) is of Pincus type. The experiments which have been performed at ultra low frequencies using the $T_1\,\rho$ technique[77] show that the Pincus process is dominant.

5.2.2.3 Acoustic relaxation

Acoustic waves can be used to study the relaxation processes in the liquids, if the time scale is of the order of 10^{-7} s. The case of nematics is more complex, and the results are poorly understood.

At very low frequencies (< 1 MHz), longitudinal waves propagate in a nematic with a velocity C_0 which is independent of direction and depends only on the bulk rigidity coefficient

$$E_0 = -V\left(\frac{\partial p}{\partial V}\right)_{\text{adiabatic}}$$

$$C_0 = (E_0/\rho)^{1/2}$$

As usual, experimentally, C_0 is mainly a decreasing function of temperature. However, a dip near the NI transition point is observed.

At finite frequencies, the sound velocity is slightly modified and becomes dependent on the angle between the optic axis and the direction of propagation,

$$C(\theta) = C(0)\,[1 - \Delta \sin^2\theta]$$

where Δ is typically of order 10^{-3}, and is positive. At temperatures well below T_{NI}, Δ increases with frequency; near T_{NI} becomes essentially independent of frequency.

For the frequency in the megacycle range the attenuation α is much more angular dependent than the velocity C. Usually α can be fitted by the relation

$$\alpha(\theta) = \alpha(0)\,[1 - \delta \sin^2\theta]$$

where $\delta \sim 10^{-1}$, and is positive. There is of course a correlation between the attenuation α and the velocity C. In the isotropic liquids this correlation is of limited use but it is more useful in the case of anisotropic liquids.

5.2.2.4 Translational motions

a) Self-diffusion

Self-diffusion coefficients $D_\parallel$ and $D_\perp$ can be measured by various techniques using radioactive tracers, inelastic scattering of neutrons, studies on nuclear spin precession in a magnetic field gradient, etc. The results, using radioactive tracers, for PAA at 398K are $D_\parallel = 4\times10^{-6}$ cm^2 s^{-1}, $D_\perp = 3.3 \times 10^{-6}$ cm^2 s^{-1}. In case of inelastic scattering of neutrons use is made of the large, incoherent scattering due to the protons of the molecule. Corresponding to a monoenergetic ingoing beam, the energy width of the ingoing beam is given by

$$\Delta\omega_q = D_\parallel q_z^2 + D_\perp q_\perp^2$$

This is valid only if qa << 1. Such a small widths are difficult to measure accurately. The application of the method using nuclear spin precession in a magnetic field gradient to nematic liquids is difficult because the spin-spin relaxation time T_2 is short. However, by eliminating most of the dipolar interactions responsible for T_2, $D_\parallel$ has been measured[78] for MBBA.

b) Diffusion of a solute

Often it is more convenient to measure $D_\parallel$ and $D_\perp$ for a solute in the nematic phase than for the nematic molecules themselves. For example, if in a small region of the sample a dye is injected at time t = 0, a study of the spatial spread in coloration at later times t is enough to measure D. These studies can provide some useful information about the solvent-solute interactions. A macroscopic conformational change in a nematic can also be studied.

c) Mobility of the charge carriers

In the organic fluids it is not easy to measure the mobility μ of the charge carriers. In addition, convective motions caused by the electric field give rise to an apparent mobility which is much larger than the intrinsic μ. However, some information has been derived[12] by two methods : studies on transient regime in ultrapure specimens with electrodes suppressing all injection, and studies on a.c. instabilities in the limit of very small 'chevrons'.

The analysis of instabilities[79] was based on Ohm's law $\mathbf{J} = \sigma \mathbf{E}$. However, in the limit of rapid spatial variations, a diffusion term must be included

$$\mathbf{J} = \sigma \mathbf{E} - D_c \nabla q \tag{5.148}$$

where

$$D_c = \frac{k_B T \sigma}{ne^2} = \frac{k_B T}{e} \mu \tag{5.149}$$

Taking the divergence of **J** and using Poisson's equation to eliminate **E**, we obtain

$$\nabla \cdot \mathbf{J} = \frac{4\pi\sigma}{\in} q (1 + k^2 r_D^2) \tag{5.150}$$

where the length r_D is defined by

$$r_D = \left(\frac{D_C \in}{4\pi\sigma} \right)^{1/2} = \left(\frac{k_B T \in}{4\pi\, ne^2} \right)^{1/2} \tag{5.151}$$

Here r_D is nothing else but the Debye-Hückel screening radius, associated with the mobile carriers. Equation (5.151) shows that diffusion begins to play an important role in the charge balance when $k\, r_D \sim 1$, r_D turns out to be typically of order 1 μm. Knowing r_D one can obtain the carrier density n, and with n and measured σ, the mobility μ can be derived.

REFERENCES

1. C.W. Oseen, Trans Faraday Soc. **29**, 883 (1933).
2. H. Zöcher, Trans Faraday Soc. **29**, 945 (1933).
3. F.C. Frank, Disc. Faraday Soc. **25**, 19 (1958).

4. A. Anzelius, Uppsala Univ. Arsskr., Mat. Och. Naturvet 1 (1933).

5. J.L. Ericksen, Arch. Rational Mech. Anal. **4**, 231 (1960); Phys. Fluids **9**, 1205 (1966).

6. J.L. Ericksen, Trans. Soc. Rheol. **5**, 23 (1961).

7. F.M. Leslie, Quart. J. Mech. Appl. Math. **19**, 357 (1966); Arch. Ration. Mech. Analysis 28, 265 (1968); Adv. in Liq. Cryst. Vol. 4, ed., G.H. Brown (Academic Press, New York, 1979) p. 1.

8. S. Singh, Phys. Rep. **277**, 283 (1996).

9. W.H. de Jeu, Physical Properties of Liquid Crystalline Materials (Gordon and Breach, London, 1980).

10. J. Nehring, and A. Saupe, J. Chem. Phys. **54**, 337 (1971); 56, 5527 (1972).

11. S. Chandrasekhar, Liquid Crystals (Cambridge Univ. Press, 1977, 1992).

12. P.G. de Gennes, and J. Prost, The Physics of Liquid Crystals (Clarendon Press, Oxford, 1993).

13. L.D. Landau, and E.M. Lifshitz, Theory of Elasticity, Section 10 (Pergamon Press, London, 1959).

14. M. Klëman, Points, Lines and Walls (Wiley, New York, 1983).

15. A. Saupe, J. Chem. Phys. **75**, 5118 (1981).

16. M. Liu, Phys. Rev. **A 24**, 2720 (1981).

17. H. Brand, and H. Pleiner, Phys. Rev. **A24**, 2777 (1981).

18. E.A. Jacobsen, and J. Swift, Mol. Cryst. Liquid Cryst. **78**, 311 (1981).

19. G.E. Volovik, and E.I. Kats, Zh. Eksp. Teor. Fiz **81**, 240 (1981).

20. W.M. Saslow, Phys. Rev. **A25**, 3350 (1982).

21. U.D. Kini, Mol. Cryst. Liquid Cryst. **108**, 71 (1984).

22. E. Govers, and G. Vertogen, Phys. Rev. **A30**, 1998 (1984).

23. A. Chaure, J. Engg. Sci. 23, 797 (1985).

24. J.L. Ericksen, in, Adv in Liq. Cryst. Vol. 2, ed., G.H. Brown (Academic Press, New York, 1976).

25. L.M. Blinov, Electro-optical and Magneto-optical Properties of Liquid Crystals (Wiley, New York, 1983).

26. U.D. Kini, and S. Chandrasekhar, Physica **A156**, 364 (1989).

27. Y. Singh, K. Rajesh, V.J. Menon, and S. Singh, Phys. Rev. **E49**, 501 (1994).

28. (a) Th. W. Ruijgrok, and K. Sokalski, Physica **A111**, 627 (1982); (b) E.T. Kats, Sov. Phys. Usp. **27**, 42 (1984).

29. M.P. Allen, and D. Frenkel, Phys. Rev. **A37**, 1813 (1988); Phys. Rev. **A42**, 3641 (1990).

30. P.G. de Gennes, Mol. Cryst. Liquid Cryst. **12**, 193 (1971).

31. P. Sheng, and E.B. Priestley, Landau-de-Gennes Theory of Liquid Crystals Phase Transitions, in, Introduction to Liquid Crystals, eds. E.B. Priestley, and P.J. Wojtowicz (Plenum, New York, 1974) Chapt. 10, p. 143.

32. R.G. Priest, Phys. Rev. **A7**, 720 (1973).

33. J.P. Straley, Phys. Rev. **A8**, 2181 (1973).

34. A. Poniewierski, and J. Stecki, Mol. Phys. **38**, 1931(1979).

35. T.E. Faber, Proc. R. Soc. London **A353**, 271 (1977).

36. S.D. Lee, Phys. Rev. **A39**, 3631 (1989).

37. A.M. Samoza, and P. Tarazona (a) Phys. Rev. **A40**, 6069 (1989); (b) Mol. Phys. **72**, 911 (1991).

38. Y. Singh, and K. Singh, Phys. Rev. **A33**, 3481 (1986).

39. S.D. Lee, and R.B. Meyer, J.Chem. Phys. **84**, 3443 (1986).

40. G. Lasher, J. Chem. Phys. **53**, 4141 (1970).

41. T.E. Faber, Proc. R. Soc. London **A353**, 247 (1977); **A375**, 579 (1981).

42. J. Stecki, and A . Poniewierski, Mol. Phys. **41**, 1451 (1980).

43. H. Kimura, M. Hosino, and H. Nakano, Mol. Cryst. Liquid Cryst. **74**, 55 (1981).

44. W.M. Gelbart, and A. Ben Shaul, J. Chem. Phys. **77**, 916 (1982).

45. W.M. Gelbart, and B.A. Baron, J. Chem. Phys. **66**, 207 (1977).

46. M.A. Cotter, The vander Waals Approach to Nematic Liquid Crystals, in, The Molecular Physics of Liquid Crystals, eds., G.W. Gray, and G.R. Luckhurst (Plenum Press, New York, 1979) Chapt. 8, p. 181.

47. B. Barboy, and W.M. Gelbart, J. Chem. Phys. **71**, 3053 (1979); J. Stat. Phys. **22**, 709 (1980).

48. B.W. van der Meer, F. Postma, A. J. Dekker, and W.H. de Jeu, Mol. Phys. **45**, 1227 (1982).

49. A.V. Zakharov, Physica **A175**, 327 (1991).

50. E.T. Brook-Levinson, and A.V. Zakharov, J. Engg. Phys. **54**, 262 (1988).

51. L.A Rott, Statistical Theory of Molecular Systems (Nauka, Moscow, 1979) p. 280, (In Russian).

52. V.V. Belov, and E.T. Brook-Levinson, Phys. Lett. **A80**, 314 (1980).

53. J.R. McColl, and C.S. Shih, Phys. Rev. Lett. **29**, 85 (1972).

54. R. Eppenga, and D. Frenkel, Mol. Phys. **52**, 1303 (1984).

55. Y. Singh, Phys. Rep. **207**, 351 (1991); Phys. Rev. **A30**, 583 (1984).

56. Y. Singh, S. Singh, and K. Rajesh, Phys. Rev. **A45**, 974 (1992) (Referred to as SSR theory).

57. Y. Singh, and K. Singh, Phys. Rev. **A33**, 3481 (1986).

58. A.M. Somoza, and P. Tarazona, Phys. Rev. Lett. **61**, 2566 (1988); J. Chem. Phys. 91, 517 (1989).

59. J.K. Percus, The Pair Distribution Function in Classical Statistical Mechanics, in, 'The Equilibrium Theory of Classical Fluids, eds., H.L. Frisch and J.L. Lebowitz (Benjamin, New York, 1964), Section II, p. 33.

60. K. Singh, and Y. Singh, Phys. Rev. **A34**, 548 (1986).

61. S. Singh, and K. Rajesh, Mol. Cryst. Liquid Cryst. **200**, 133 (1991).

62. K. Singh, and Y. Singh, Phys. Rev. **A35**, 3535 (1987).

63. A.R. Denton, and N.W. Ashcroft, Phys. Rev. **A39**, 4701 (1989).

64. S. Singh, Liquid Crystals **20**, 797 (1996).

65. T.K. Lahiri, K. Rajesh, and S. Singh, Liquid Crystals **22**, 575 (1997).

66. C.G. Gray, and K.E. Gibbins, Theory of Molecular Fluids, Vol. I : Fundamentals (Clarendon Press, Oxford, 1984).

67. J.G. Gay, and B.J. Berne, J. Chem. Phys. **74**, 3316 (1981).

68. A. Saupe, Z. Naturforsch **a15**, 815 (1960).

69. W.H.de Jeu, W.A.P. Claassen, and A.M.J. Spruijt, Mol. Cryst. Liquid Cryst. 37, 269 (1976); W.H. de Jeu, and W.A.P. Claassen, J. Chem. Phys. 67, 3705 (1977); ibid 68, 102 (1978).

70. N.V. Madhusudana, and R. Pratibha, Mol. Cryst. Liquid Cryst. **89**, 249 (1982); P.P. Karat, and N.V. Madhusudana, Mol. Cryst. Liquid Cryst. **55**, 119 (1979).

71. O. Parodi, J. Phys. (Paris) **31**, 581 (1970)

72. D. Forster, T. Lubensky, P. Martin, J. Swift, and P. Pershan, Phys. Rev. Lett. **26**, 1016(1971); P.C. Martin, O. Parodi, and P.J. Pershan, Phys. Rev. **A6**, 2401 (1972); M.J. Stephen, Phys. Rev. **A2**, 1558 (1970); F. Jähnig, and M. Schmidt, Ann. Phys. **71**, 129 (1971).

73. S. de Groot, and P. Mazur, Non-equilibrium Thermodynamics (North-Holland, Amsterdam, 1962; Dover publications, New York, 1984).

74. P. Pincus, Solid State Commun **7**, 415 (1969).

75. J.W. Doane, and D.L. Johnson, Chem. Phys. Lett. **6**, 291 (1970).

76. T. Lubensky, Phys, Rev. **A2**, 2497 (1970).

77. W. Doane, C.E. Tarr, and M.A. Nickerson, Phys. Rev. Lett. **33**, 620 (1974).

78. R. Blinc, J. Pirš , and I. Zupančič , Phys. Rev. Lett. **30**, 546 (1973).

79. Y. Galerne, G. Durand, and M. Veyssie, Phys. Rev. **A6**, 484 (1972).

6

Smectic Liquid Crystals

This chapter is devoted to the smectic mesophases that are characterized by both the orientational order of the long-molecular axes and a reduced positional order. As discussed in chapter 1 (sec. 1.4), smectics have stratified structures with a well defined interlayer spacing; within each stratification a variety of molecular arrangements are possible. With the established layer structure, two kinds of phases, orthogonal and tilted, can be distinguished. In the orthogonal phases the director is perpendicular to the layers or equivalently parallel to the direction of the density wave. Tilted smectic phases are characterized by an angle between the director and normal to the smectic planes. The term "Smectic" was coined by G. Friedel to designate certain mesophases observed in soaps and lipid-water mixtures. He recognized only one kind of smectic – now known as smectic A (S_A). A number of smectic phases has now been identified giving rise to different macroscopic textures (see chapter 11) which are recognized by optical observation. A natural classification of these phases should be made on the basis of their symmetry properties. However, their notation is according to the order of discovery of the different phases and bears no relation with the symmetry and molecular packing.

6.1 SYMMETRY AND CHARACTERISTICS OF SMECTIC PHASES

A smectic is defined by its periodicity in one direction of space, and by its point group symmetry (see secs. 1.3 and 1.4 and Tables 1.1–1.3). On the basis of point group symmetry an infinite number of smectic phases can be possible. However, only a finite number of smectics[1–5] are known. The simplest is the smectic A (S_A) which is a stack of disordered two dimensional liquids ($D_{\infty h}$ Symmetry). The hexatic smectics are characterized by a D_{6h} point group symmetry. Smectics with C_{2h} symmetry are characterized by stacks of liquid crystal layers in which the molecules are, on the average, tilted with respect to the layers. In case of liquid crystals composed of chiral molecules the symmetry drops to C_2 and the dipolar order is allowed along the two-fold axes. The corresponding phases are ferroelectrics ; they shall be described in chapter 9.

Some major structural characteristics of smectic phases are given in Table 6.1. This table also illustrates the natural textures[3,6] that are commonly observed when the phase is formed directly from the nematic. The different kinds of ordering existing in the layers lead to the following three smectic types : Smectics with liquid layers, smectics with bond-orientational order and smectics with ordered layers. We shall describe each case separately in the following sections.

Table 6.1. Characteristics of nonchiral smectics.

Mesophase	Optical properties	Textures: Homogeneous/ planar/ alignment	Textures: Homeotropic/ orthogonal alignment	Structures
S_A	u(+) uniaxial positive	focal-conic (fan shaped) homogeneous polygonal defects	extinct black	Liquid-like layers with the molecules upright on the average, negligible in-plane and interlayer positional correlations
S_C	b biaxial positive	focal-conic broken	schlieren 4–brushes	Liquid-like layers with positional correlations between the layers as in S_A, but the molecular axes inclined to the layer normal; random arrangement within the layers
S_B	u(+) uniaxial positive	focal-conic	extinct black	Long-range 3D six-fold bond-orientational order; short-range in-plane positional correlation, negligible interlayer positional correlation
$S_{B(cry)}$	u(+) uniaxial positive	mosaic	extinct black	3D crystal, hexagonal lattice with upright molecules; weak interlayer forces

S_F	b biaxial positive	mosaic	schlieren mosaic	Tilted and lamellar hexatic phase; C-centred monoclinic (a > b) lattice with in-plane short-range positional correlation, weak or no interlayer positional correlation, pseudohexagonal molecular packing
S_I	b biaxial positive	focal conic broken	schlieren	Tilted and lamellar hexatic phase, C-centred monoclinic (b > a) lattice, slightly greater in-plane correlation than S_F
S_E	u(+) uniaxial positive	mosaic	shadowy mosaic	3D crystal, molecular axes orthogonal to the layers, orthorhombic lattice with interlayer herringbone arrangement of the molecules

6.1.1. Smectics With Liquid Layers

There are two kinds of smectic phase with two-dimensional liquid like layers, smectic A (S_A) and smectic C (S_C) (Fig. 6.1). In case the S_C phase is not chiral, they can be distinguished by the fact that smectics A are optically uniaxial and smectics C biaxial.

In the S_A phase the molecules are perpendicular to the layers and their centers of gravity show no long-range order within the layers. There is a strong correlation of positions between the layers. In the basic state of the crystal the layers are planar, but they can easily slip over each other and the molecules can diffuse readily within the layers. The director coincides with the layer normal and the directions $\hat{\mathbf{n}}$ and $-\hat{\mathbf{n}}$ are still equivalent. This explains the uniaxiality of the optical properties. A liquid-like layered structure gives the so-called focal-conic textures. Some times the focal-conic units are arranged in a polygon type superstructure. Apart from the focal-conic texture, S_A also exists in the homeotropic texture and in that case it cannot be distinguished optically from a homeotropic nematic texture.

The structure of the S_C phase (Fig. 6.1b) is similar to that of the S_A phase except that the long-axes of the molecules are tilted with respect to the planes of the layers (Fig. 6.2). Thus the optical character is positive biaxial. As a result of tilt angle the layer spacing in S_C phase is less than that of the corresponding S_A phase. The tilt directions of the various layers are correlated in the same way. The tilt direction can be described by a $\hat{\mathbf{c}}$ -director that lies in the layers and more or less remains uniform over the macroscopic distances. The states $\hat{\mathbf{n}}$ and $-\hat{\mathbf{n}}$ are still equivalent but $\hat{\mathbf{c}}$ and $-\hat{\mathbf{c}}$ are not equivalent. The tilt angle gives a natural order parameter

$$\chi = \omega \exp(i\,\phi) \tag{6.1}$$

that distinguishes S_C from S_A. The azimuthal angle ϕ gives the direction of $\hat{\mathbf{c}}$ in which the tilt takes place (Fig. 6.2). The tilt angle ω is temperature dependent and shows quite different values for the various compounds. Due to the tilted structure together with the focal conic textures additional textures are possible in the S_C phase. Non-uniformity of the $\hat{\mathbf{c}}$-director leads to schlieren textures which are somewhat more disturbed than those of nematics. The schlieren texture of the N_u phase possesses 2-brush singularities, whereas 4-brush singularities are observed in the S_C phase. Thus, the schlieren defects can determine whether or not the phase is N_u or S_C.

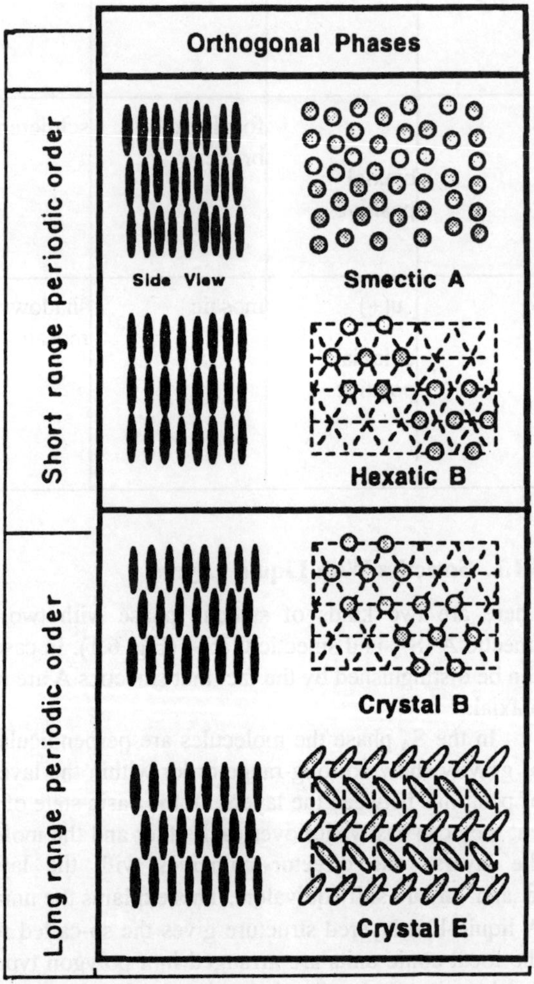

Fig. 6.1a. The structures of orthogonal smectic phases. The cross-sectional areas are shown as circles and ellipses. The dotted lines are the lattice sites for long-range order.

6.1.2. Smectics With Bond-Orientational Order

The existence of an intermediate phase between the 2D solid and liquid was predicted[7–9], on theoretical grounds, by several workers. In this phase, known as the hexatic phase, there exists short-range in-plane positional order, but quasi long-ranged, six fold bond-orientational order. This kind of order does not violet the fluid character of the layer. Several modifications of hexatic phase now exist (S_B, S_F, S_I) in which the molecules are arranged in a hexagonal manner.

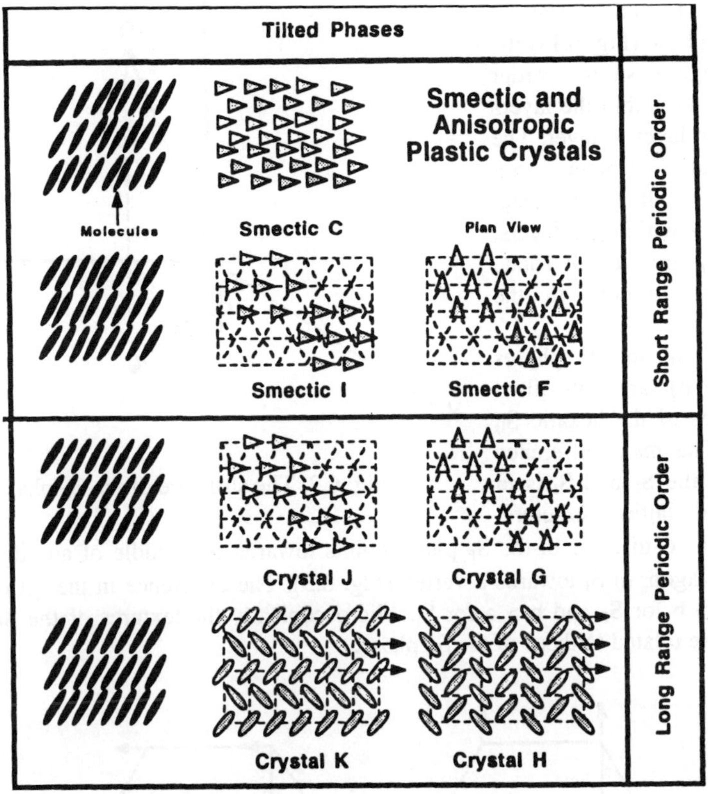

Fig. 6.1b. The structures of tilted smectic phases. The triangle denotes the tilt direction of the molecules.

The concept of bond-orientational order provides a distinction between the hexatic S_B and the crystalline S_B phases. In the hexatic S_B (S_B) phase (Fig. 6.1a) the positional order is quasi long-range in the direction normal to the layers and short-range within the plane of the layers but with much longer correlation lengths ($\xi \approx$ 100 Å) than in the S_A phase ($\xi \approx 20$ Å). The existence of bond-orientational order implies that the in-plane crystallographic axes are maintained over large distances both parallel and perpendicular to the layers. The molecules are orthogonal with respect to the layer plane and packed with hexagonal symmetry, and the orientation of the six-fold symmetry axes is maintained with true long-range order. The disordered crystalline smectic phases are possible due to the presence of long-rang positional order in the plane perpendicular to the director. In the crystal S_B ($S_{B(cry)}$) phase (Fig. 6.1a) the long-range positional order exists both within the layers and from layer to layer. The molecular centers are packed with hexagonal symmetry within the layers, and they are free to rotate about the normal to the layers.

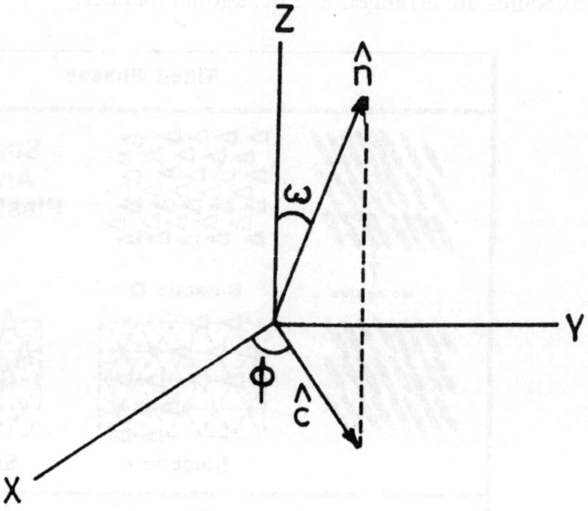

Fig. 6.2. The structure of the S_C phase

The S_F and S_I phases (Fig. 6.1b) are the tilted analogues of the hexatic S_B phase. The main difference between the S_F and S_I is in the local lattice structure. While the $\hat{c}$-director in the S_F phase points towards the middle of an edge of the local hexagon, in S_I towards a vertex (Fig. 6.3). The difference in the tilt direction gives a > b for S_F and b > a for S_I. Microscopically the textures of the S_F and S_I phases are related to those of the S_C phase.

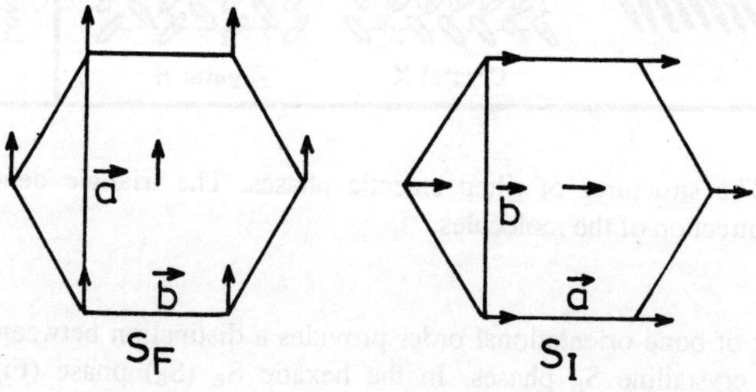

Fig. 6.3. Tilt directions for the S_F and S_I phases

6.1.3. Smectics With Ordered Layers

Direct evidence in favour of the three-dimensional positional order in smectics is obtained from the X-ray diffraction pattern. These phases (S_E, S_G, S_H, ...) are distinguished from the simple crystals by a large amount of disorder and from the ability of the molecules to reorient themselves. At the local level, there exists a herring-bone type of molecular packing. The tilted analogs of the crystal S_B phase are crystal S_G and S_J phases and of crystal S_E phase are crystal S_H and S_K (Fig 6.1b) phases. The tilt in the S_G phase is towards the side of the hexagon and towards the apex of the hexagon in the S_J phase. The molecules in both of these phases rotate freely about their long molecular axes. In the S_H phase the tilt is towards a next nearest neighbour similar to that of the S_G phase. In the S_K phase the tilt points towards a nearest neighbour, like the S_J phase. The rotation about the long molecular axis is restricted as in the S_E phase. In all of these phases no focal-conic textures are observed; mosaic textures are found consisting of various regions with a uniform director pattern. In addition, for the uniaxial phase homeotropic textures may be observed and schlieren textures in case of the biaxial phases.

6.2. STRUCTURE-PROPERTY RELATIONS

Some of the important issues related with the structure-property relationships of mesophases are discussed in sec. 4.3. We discuss, in brief, here how the structural units affect the generation of nonchiral smectic phase types and their physical properties. Since the smectic phases are lamellar in structure, the molecular structure required to generate a smectic phase must allow for the lateral intermolecular attractions. Consequently, the smectics are favoured by a symmetrical molecular structure. Any breaking of this symmetry or where the core is long relative to the overall molecular length may not facilitate the smectic formation and may help in the generation of the nematic phase. In addition, the use of lateral substituents tends to destabilise the lamellar packing and thus may facilitate the generation of nematic phase (compare compounds T6.2a and T6.5a).

The effects of core changes on the transition temperature of some smectogens are shown in Table 6.2. From a careful comparative study of the listed mesogens it is obvious that the core structure has considerable influence on the phase stability of smectic phases. When a core unit is narrow, linear and composed of similar compatible units, the mesogens tend to exhibit smectic phases rather than the nematic phase. The biphenyl core structure (compound T6.2a) exhibits mesophases with a strong smectic phase stability (the phase types are unidentified) . An aromatic core unit without a linking group (compound T6.2a) provides good lateral attractions and thus favours the smectic phase formation. In addition, for such a short core the end groups are reasonably long which stabilises the smectic phase stability by the mutual entanglement.

Table 6.2. Effects of core changes on the transition temperatures.

a. 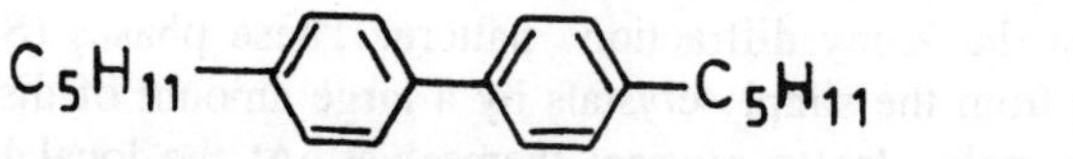

K 26 S 47.6 S 52.2 IL

b. 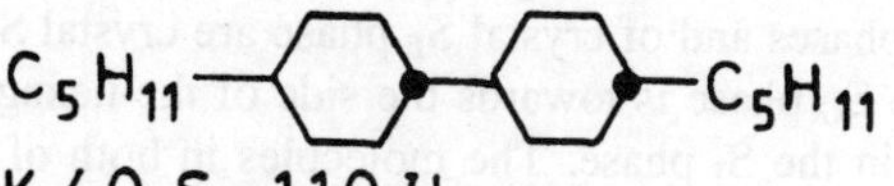

K 40 S_B 110 IL

c. 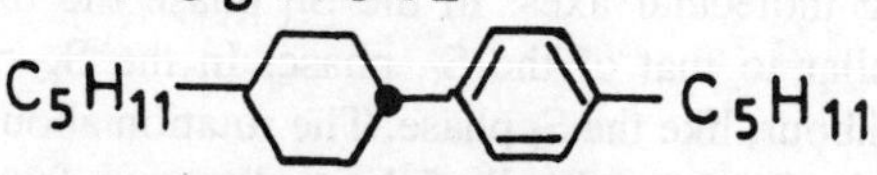

K –0.8 (S –8 N –5) IL

d. 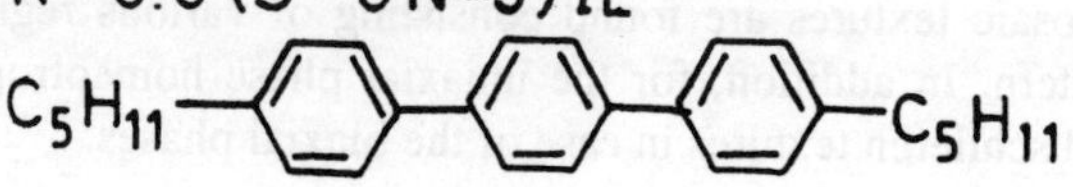

K 192 S_A 213 IL

e. 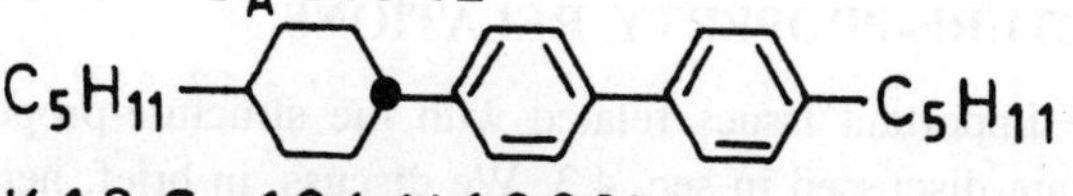

K 13 S_B 164 N 166 IL

f. 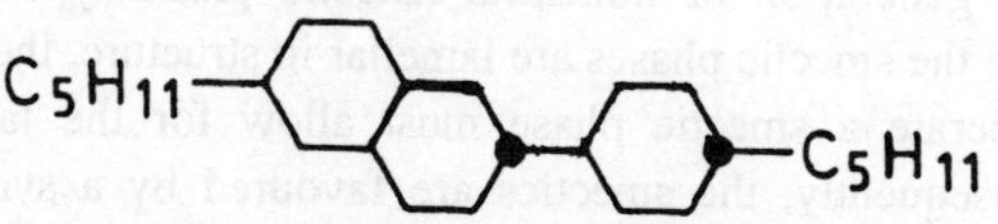

K 70 S_A 90.5 N 140.5 IL

g. 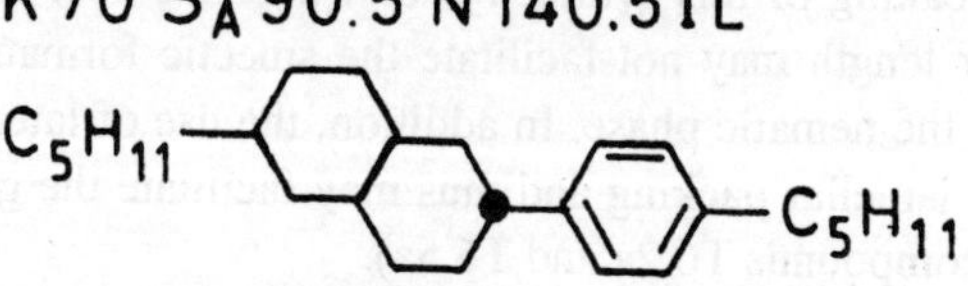

K 39.9 N 59.7 IL

For the similar reasons the analogous compound T6.2b with a alicyclic core is strongly smectic. The smectic phase stability of compound T6.2b is significantly higher as compared to the aromatic compound T6.2a. The smectic phase stability of compound T6.2c is considerably reduced as compared to both of the above compounds, because its core comprises a cyclohexane ring and a phenyl ring; the incompatibility of these two distinctly different core regions does not favour the lamellar packing.

The three ring systems have a higher length to breadth ratio and hence exhibit much higher liquid crystal phase stabilities than the two ring analogues. Compound T6.2d is a terphenyl and comprises of three aromatic rings which are structurally compatible and confers the strong smectic phase stability. The core of compound T6.2e has a cyclohexane ring and two phenyl rings. These core regions are expected to confer phase separation and hence poor smectic phase stability. Compound T6.2f has a completely saturated core that may be expected to pack well in a lamellar fashion exhibiting an S_A phase. However, the broad molecular structure reduces smectic phase stability and allows a large nematic range to be exhibited. When this broad core structure comprises saturated and aromatic regions (compound T6.2g), the smectic phase is not generated at all and the phase stability of nematic phase is also reduced.

The mesogens with two terminal alkyl or alkoxy chains tend to exhibit smectic phase types but the use of linking groups often disrupts the lamellar packing and facilitates the formation of nematic phase. Compound T6.2a, for example, exhibits strong smectic phase stability but the use of an ester ($-CO_2-$) group disrupts the lamellar packing and increases the core length relative to the terminal chains to generate a nematic phase (see Table 6.3, compound T6.3a). The use of an ester linking group in the cyclohexyl analogue (compound T6.3b) confers a high T_{NI} whereas the parent material (compound T6.2c) exhibits both the smectic and nematic phases of very low stability (see Table 6.2). The ester linking unit is a polar group that can favour lamellar packing and generate smectogens provided the end groups are sufficiently long. For example, although compound T6.3c is quite similar in structure to compound T6.3a, the terminal chains are both longer and incorporate a conjugative ether oxygen, and hence generate a smectic C phase. These structural changes confer a completely different mesogenic nature; the long chains stabilise a lamellar packing and generate a smectic phase with tilted molecules because of the arrangement of the lateral dipole units. Compounds T6.3d and T6.3e are wholly alicyclic. Compound T6.3d has a hydrocarbon dimethylene linkage that is compatible with the rings, and hence a high smectic phase stability results despite the use of short end chains. The use of methyleneoxy linkage (compound T6.3e) introduces incompatibiliy into the structure which destabilises the lamellar packing, reduces the melting point and smectic phase stability and exhibits a nematic phase of low stability.

Table 6.3. Effects of linking groups on the transition temperatures.

a.

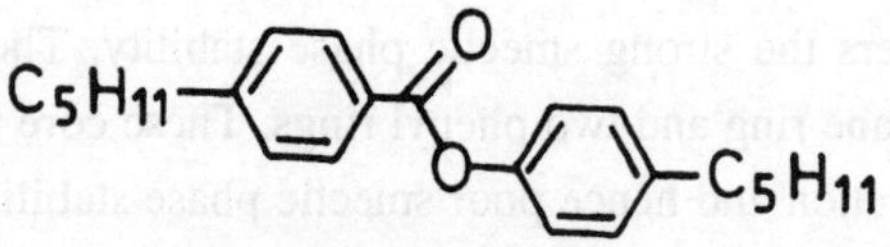

K 34.8 (N 26) IL

b

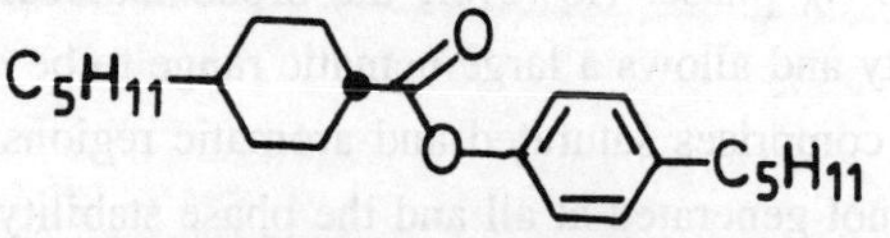

K 36 (S_A 29) N 48 IL

c.

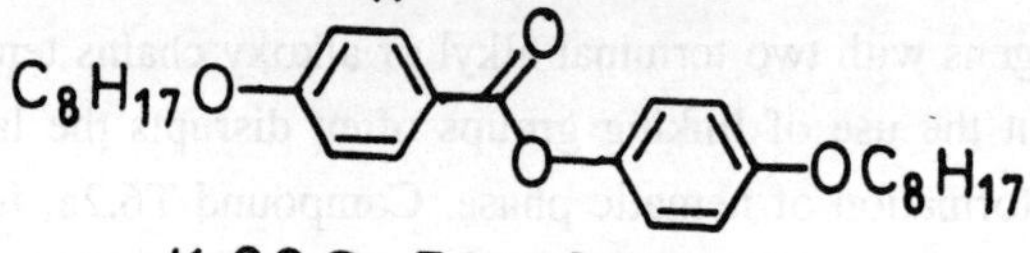

K 63 S_C 74 N 91 IL

d.

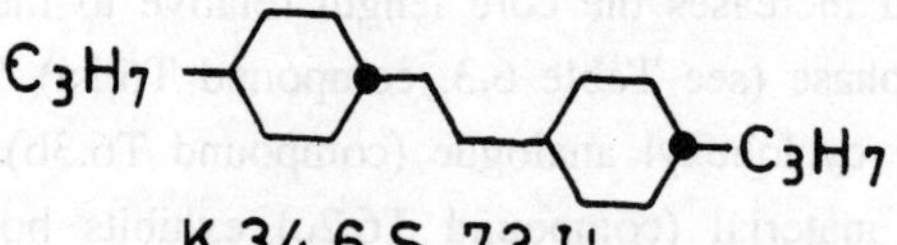

K 34.6 S 73 IL

e.

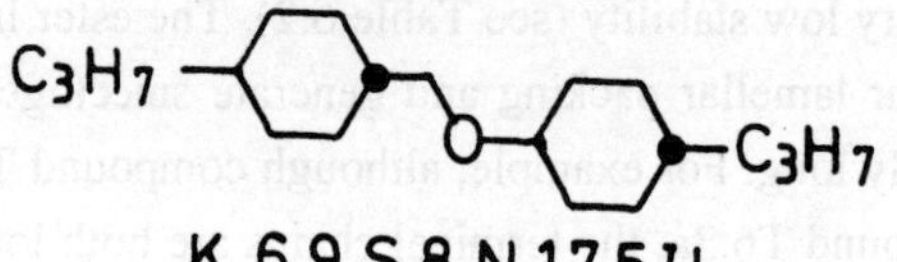

K 6.9 S 8 N 17.5 IL

The effects of the end chains on the smectic phase stability can be seen in Table 6.4. The phenylpyrimidine core with a terminal cyano group (compound T6.4a) exhibits a monotropic nematic phase. The use of a long alkyl chain and a long alkoxy chain in the phenylpyrimidine core unit (compounds T6.4b and T6.4c) increases the smectic phase stability considerably to the extent that the nematic phase is squeezed out (compound T6.4c). Compound T6.4b exhibits a S_C phase of lower phase stability and a nematic phase. The use of relatively longer alkyl chain length increases the smectic phase stability of the material considerably and eliminates the nematic phase. The influence of the length of end chains can also be seen in the last two compounds listed in Table 6.4.

Table 6.4. Effects of end groups on the transition temperatures.

a. 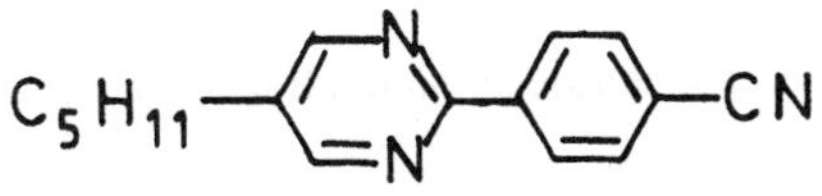

K 71 (N 52) IL

b 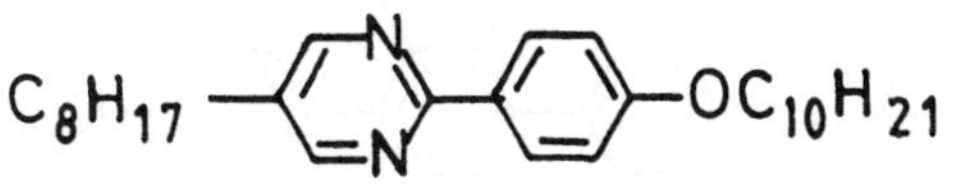

K 32 S_C 59.5 S_A 65.5N69.5 IL

c. 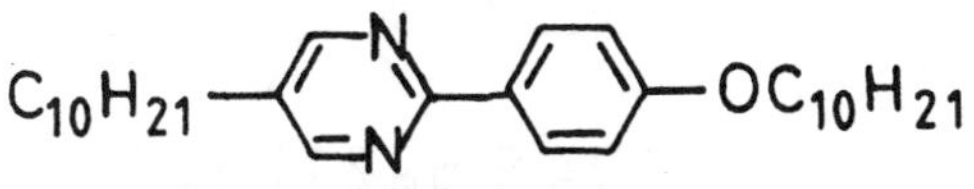

K 36 S_C 73 S_A 77 IL

d. 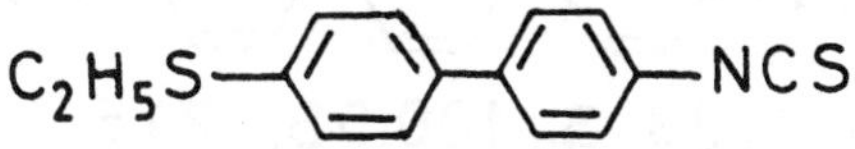

K 77.5 S_B 79 (N 43) IL

e. 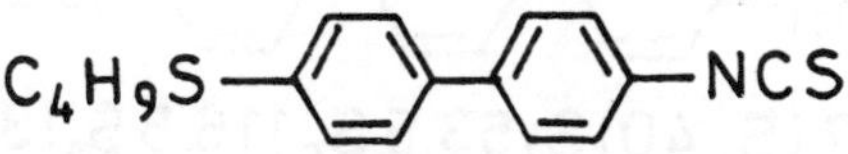

K 78.5 S_B 79 (N 44) IL

The influence of the lateral substituents on the liquid crystalline behaviour of smectogens is shown in Table 6.5. Compound T6.2d with short end chains exhibits smectic A phase but the use of fluoro substituents (compound T6.5a) eliminates all smectic character. The lateral fluoro substitution has been widely used to generate the smectogens that can be used as ferroelectric host materials[10]. The fluoro substituents provide a lateral dipole that can cause molecular tilting. As an example, the parent compound T6.5b exhibits strong smectic character but no tilted smectic phases. It can be seen that the use of one fluoro substituent considerably reduces the smectic character (see compounds T6.5c and T6.5d), but induces a variety of tilted smectic phases. In these compounds the influence of the position of the lateral substituents can be seen clearly. The use of two fluoro substituents in compound T

6.5e reduces the phase stability and eliminates ordered smectic phases. The two fluoro substituents in this arrangement give a strong lateral dipole and the compound exhibits only a S_C phase.

Table 6.5. Effects of lateral substituents on the transition temperatures.

a. 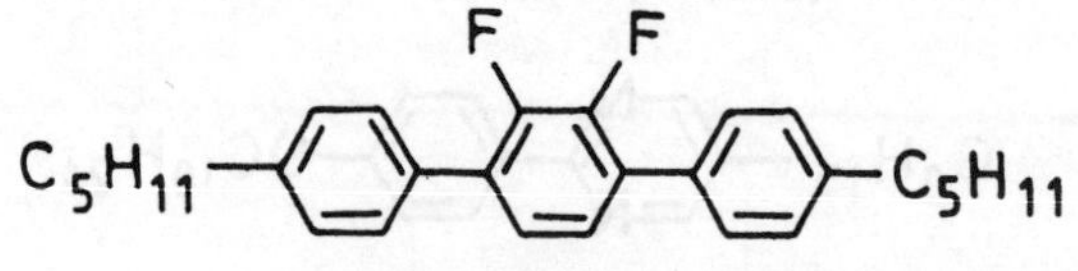

K 60 N 120 IL

b. 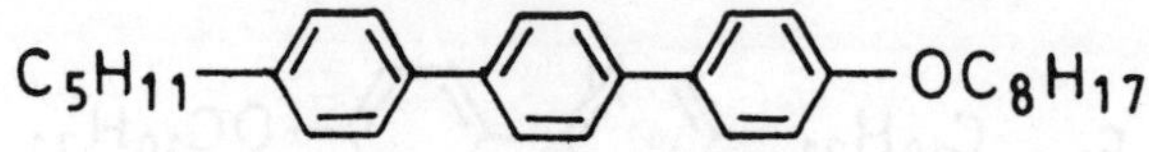

K 194.5 S_B 211 S_A 221.5 IL

c. 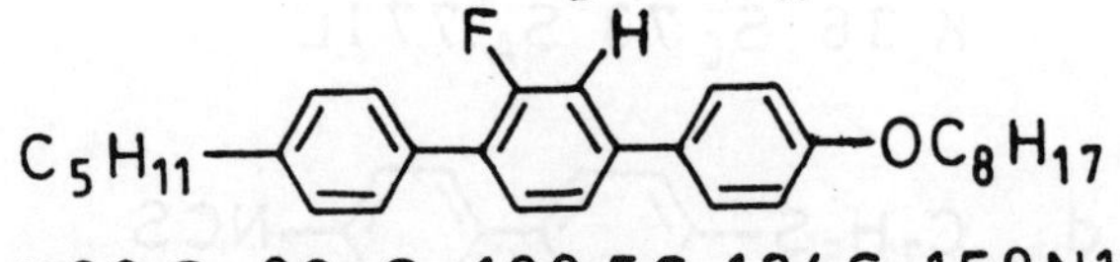

K 69 S_G 83 S_B 100.5 S_C 124 S_A 158 N 161 IL

d. 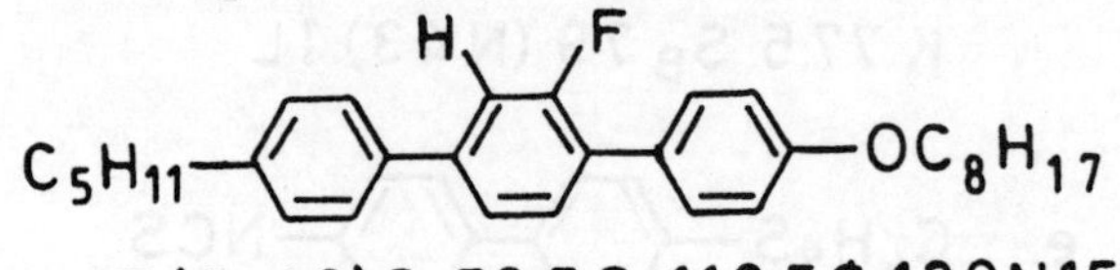

K 47 (S_J 40) S_I 53.5 S_C 116.5 S_A 130 N 155 IL

e. 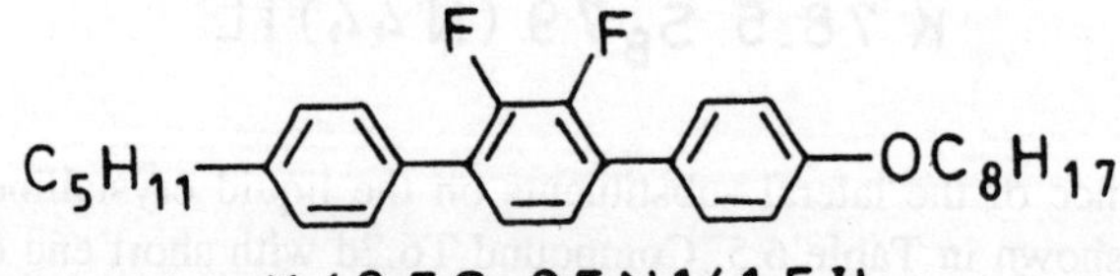

K 48.5 S_C 95 N 141.5 IL

6.3. SMECTIC A PHASE

6.3.1. Elasticity of Smectic A Phase

6.3.1.1. The basic equations

It has been indicated in sec. 6.1 that smectics have a liquid character along the plane of the layer and respond like solids to a force perpendicular to the layers. The smectic structure imposes certain restrictions on the kinds of deformation that can result in it. Consequently, the evaluation of elastic constants of smectics becomes more complicated in comparison to the nematic liquid crystals.

The basic idea of continuum theory for S_A phase was put forward by de Gennes[11]. Let us consider the S_A structure with parallel and equidistant layers. In the undeformed state the molecules are normal to the layers. Let some long-wavelength distortions be imposed on this initial state. The fluid character of the system introduces curvature in the director orientation, while due to partly solid like character the layers are dilated; the layers in some regions will be compressed and in others expanded. Let u(**r**) represent a small displacement of the layers normal to their planes (Fig. 6.4). We consider a weakly deformed medium, i.e., ∇u is small. Physically, this means that neither the layers are very much tilted from the (X,Y) plane nor strongly dilated. This assumption does not include certain interesting cases such as the focal-conic texture. Due to the layer dilation an additional term, $\frac{1}{2}B(du/dz)^2$, must be added to the Frank elastic free-energy density. The situation can be treated properly by introducing certain tilt angles as extra variable, in addition to the displacement variable u(**r**), in the theory[2].

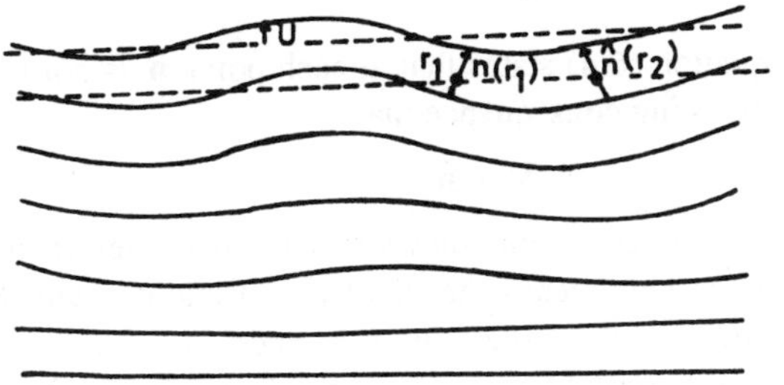

Fig. 6.4. One dimensional displacement variable for the S_A phase

In the deformed state the density ρ will usually differ from its equilibrium value ρ_0

$$\rho = \rho_0(1-\theta(\mathbf{r})) \tag{6.2}$$

On symmetry consideration θ(**r**), (the rotation angle of $\hat{\mathbf{n}}(\mathbf{r})$), is related to u as

$$\theta = m\frac{du}{dz} \tag{6.3}$$

where m is a dimensionless constant, characteristic of the material, which may be positive or negative. In case of dynamical effects θ(**r**) has to be treated as an independent variable because the equilibrium relation (6.2) breaks down at acoustic frequencies. For the static distortion of concern here, θ(**r**) will adjust itself such that the free-energy is minimized in a given u(**r**) configuration. In practice, all the experiments are performed at constant pressure and thus pressure plays no role in changing the density. Consequently, for long-wavelength static distortions θ(**r**) need not be considered as an independent variable. Similarly, the tilt of the molecules with respect to the layers is also not an independent variable here. It has been shown[2] that the first-order derivatives $\partial u/\partial x$, $\partial u/\partial y$ correspond to a simple rotation of the smectic system. This, in fact, cannot change the molecular orientations with respect to the layers. Thus the director $\hat{\mathbf{n}}$ will have the components,

$$n_x = -\frac{\partial u}{\partial x} << 1 \qquad (6.4)$$

$$n_y = -\frac{\partial u}{\partial y} << 1$$

Equation (6.4) states that at each point $\hat{\mathbf{n}}$ is normal to the layers which leads to the interesting consequence that

$$\hat{\mathbf{n}} \cdot \nabla \times \hat{\mathbf{n}} = 0$$

This constraint shows that the twist deformation becomes forbidden in the S_A phase. This means that the twist elastic constant K_2 must diverge in the S_A phase. Further, if the terms $\partial^2 u/\partial x\, \partial z$ and $\partial^2 u/\partial y\, \partial z$ are ignored with respect to $\partial u/\partial z$, the bend deformation also becomes forbidden in the smectic A. According to Oseen[12] the only distortions that must be accounted, in a first-order approximation for a description of S_A state, are specific undulations of the smectic layers. These undulations are required to be such that the interlayer distance is kept constant and the director remains normal to the layer. In the present formulation the Oseen's treatment requires that $\partial u/\partial z = 0$. Consequently, it imposes the following constraints on the allowed deformations

$$\nabla \times \hat{\mathbf{n}}(\mathbf{r}) = 0$$

This constraint states that the elastic constants K_2 and K_3 of a S_A phase are large enough to prevent the twist and bend deformations. Obviously, this is rigorously true for K_2 and to a very good approximation for K_3.

Let us discuss the issue by considering a few more details[2]. The variation of u in the neighbourhood of a reference point $\mathbf{r}$ can be expressed as

$$\delta u = \nabla \mathbf{u} \cdot \delta \mathbf{r} + \frac{1}{2}(\delta \mathbf{r} \cdot \nabla\nabla \mathrm{u}) \cdot \delta\, \mathbf{r} + 0(\delta r^3) \qquad (6.5)$$

This may be written as

$$\delta u = \partial_z u\, \delta z - \mathbf{n}_\perp \cdot \delta \mathbf{x}_\perp + \frac{1}{4}(\delta x, \delta y)\begin{pmatrix} \partial_x u + \partial_y u & 0 \\ 0 & \partial_x u + \partial_y u \end{pmatrix}\begin{pmatrix} \delta x \\ \delta y \end{pmatrix}$$

$$+ \frac{1}{2}(\delta x, \delta y)\begin{pmatrix} \frac{1}{2}(\partial_x u - \partial_y u) & \partial_x \partial_y u \\ \partial_x \partial_y u & \frac{1}{2}(\partial_y u - \partial_x u) \end{pmatrix}\begin{pmatrix} \delta x \\ \delta y \end{pmatrix}$$

$$+ \delta z\, \delta \mathbf{x}_\perp \cdot \partial_z \partial_{x_\perp} u + u \frac{1}{2} \delta z^2 \partial_z \partial_z u \qquad (6.6)$$

where

$$\partial_x u = \frac{\partial u}{\partial x}, \ \mathbf{n}_\perp = -\nabla_\perp u.$$

In this equation the first term corresponds to the dilation of the layers. Due to the rotational invariance the second term, which is associated with a mere tilt of the layers, cannot be included in the free-energy density. The third term possessing the symmetry of a vector in the Z-direction represents the splay deformation. The fourth term corresponds to a saddle-splay deformation. The fifth term has the symmetry of a vector in the (X,Y) plane and represents the bend deformation. The last term exhibits the symmetry of a vector along Z as the splay deformation and is interpreted as the rate of change of the compression along Z-axis.

The relation of the elastic free-energy density may be derived keeping in view that the free-energy must be invariant in the operations that leave smectic A unperturbed. Thus we must take into account the scalar square of each term separately, plus the product of the third and sixth terms.

The saddle-splay contribution to the free-energy is given by

$$\frac{1}{2}K\left[\left(\frac{\partial^2 u}{\partial x^2} - \frac{\partial^2 u}{\partial y^2}\right)^2 + 4\left(\frac{\partial^2 u}{\partial x \partial y}\right)^2\right]$$

$$= \frac{1}{2}K\left[\left(\frac{\partial^2 u}{\partial x^2} + \frac{\partial^2 u}{\partial y^2}\right)^2 + 4\left(\left(\frac{\partial^2 u}{\partial x \partial y}\right)^2 - \frac{\partial^2 u}{\partial x^2}\frac{\partial^2 u}{\partial y^2}\right)\right] \tag{6.7}$$

It is obvious that the first-term of the right hand side of this equation corresponds to the splay energy, while the second term can be recognized as a surface energy term. Thus it will not be included in the bulk equations. The bend energy differs from the product of the third and sixth terms of eq. (6.6) only through a surface term.

In view of the above provisos the elastic free energy of smectic A phase is written as

$$\Delta A_e = \int d\mathbf{r}\left[\frac{1}{2}K_1\left(\frac{\partial^2 u}{\partial x^2} + \frac{\partial^2 u}{\partial y^2}\right)^2 + \frac{1}{2}B\left(\frac{\partial u}{\partial z}\right)^2 + \frac{1}{2}K'\left(\frac{\partial^2 u}{\partial z^2}\right)^2 + \frac{1}{2}K''\frac{\partial^2 u}{\partial z^2}\left(\frac{\partial^2 u}{\partial x^2} + \frac{\partial^2 u}{\partial y^2}\right)\right] \tag{6.8}$$

The first-term which is a splay energy can be written as

$$\frac{1}{2}K_1\left(\frac{\partial^2 u}{\partial x^2} + \frac{\partial^2 u}{\partial y^2}\right) = \frac{1}{2}K_1(\boldsymbol{\nabla}\cdot\hat{\mathbf{n}})^2 \tag{6.9}$$

The second term in eq. (6.8) represents an elastic energy for the compression of the layers; B, the bulk or compressional elastic constant, has the dimension of an energy per unit volume.

The last two terms in eq. (6.8) are, in fact, unobservable in the limit of long-wavelength distortions. They contribute only if the displacement vector **u** varies along Z. But, in this case, they are dominated by the term $\frac{1}{2}B(\partial u/\partial z)^2$, which is of lower order in the space variations. Thus, the static elastic properties of S_A are described by two constants K_1 and B, and eq. (6.8) becomes

$$\Delta A_e = \int d\mathbf{r}\left[\frac{1}{2}K_1(\boldsymbol{\nabla}\cdot\hat{\mathbf{n}})^2 + \frac{1}{2}B\left(\frac{\partial u}{\partial z}\right)^2\right] \tag{6.10}$$

It is often convenient to associate a length parameter λ_A defined as

$$\lambda_A = (K_1/B)^{1/2} \tag{6.11}$$

Usually λ_A is comparable to the layer thickness. For example, for B ~ 4 × 10^6 dyne/cm^2 (i.e. quite elastic), and $K_1 \sim 10^{-6}$ dyne, we have $\lambda_A \sim 50$ Å. These values are of a reasonable order of magnitude in the middle of a domain to the existence of a smectic phase. In the vicinity of a transition towards the nematic, B coefficient must becomes zero and λ_A becomes infinite.

Let us consider the case of a planar smectic. Introduce the displacement u(x,y,z) of the layer situated at level z_0 before every deformation at the point (x,y) of this layer. The equation of this layer is given by

$$\Phi = z - u(x, y, z) = z_0 \tag{6.12}$$

where u varies slowly, and its derivatives expressing the deformation of the layers are small. To second order, the direction cosines of $\hat{\mathbf{n}}$ are written as

$$\hat{n} = \begin{cases} \dfrac{-\partial u}{\partial x}\left[1+\dfrac{\partial u}{\partial z}\right] \\[2ex] \dfrac{-\partial u}{\partial y}\left[1+\dfrac{\partial u}{\partial z}\right] \\[2ex] 1-\dfrac{1}{2}(\nabla_\perp u)^2 \end{cases} \tag{6.13}$$

where $$\nabla_\perp u = \begin{cases} \partial u / \partial x \\ \partial u / \partial y \\ 0 \end{cases} \tag{6.14}$$

It is obvious from eq. (6.13) that $\hat{\mathbf{n}} \cdot \nabla \times \hat{\mathbf{n}} = 0$. Thus, the layers are preserved as such by the deformation. On the other hand, $\hat{\mathbf{n}} \times \nabla \times \hat{\mathbf{n}} \neq 0$, but the role of K_3 has been seen to be negligible.

Let us introduce a dilation parameter defined as

$$\gamma = 1 - \frac{d}{d_0} = 1 - \frac{1}{|\nabla \Phi|} \sim \hat{\mathbf{n}} \cdot \nabla \mathbf{u} \tag{6.15}$$

where d_0 is the thickness of the undeformed layer and d is its thickness after deformation. To second order, γ can be written as

$$\gamma = \frac{du}{dz} - \frac{1}{2}(\nabla_\perp u)^2 \tag{6.16}$$

The free-energy per unit volume can be written as

$$f_{el} = \frac{1}{2} K_1 \left(\frac{\partial^2 u}{\partial x^2} + \frac{\partial^2 u}{\partial y^2} \right)^2 + \frac{1}{2} B \left(\frac{\partial u}{\partial z} \right)^2 + \frac{1}{2} K_3 \left[\left(\frac{\partial^2 u}{\partial x \partial z} \right)^2 + \left(\frac{\partial^2 u}{\partial y \partial z} \right)^2 \right] \tag{6.17}$$

The equation for the equilibrium can be obtained by minimizing free-energy. It reads

$$\frac{K_3}{B} \frac{\partial^2}{\partial z^2} \left(\frac{\partial^2}{\partial x^2} + \frac{\partial^2}{\partial y^2} \right) u + \lambda_A^2 \left(\frac{\partial^2}{\partial x^2} + \frac{\partial^2}{\partial y^2} \right)^2 u = \frac{\partial^2 u}{\partial z^2} \tag{6.18}$$

The magnetic or electric field effects can be included easily in eq. (6.10). The magnetic energy in a field parallel to the Z-direction can be written as

$$f_{mag} = \frac{1}{2} \chi_a H^2 \left[\left(\frac{\partial u}{\partial x} \right)^2 + \left(\frac{\partial u}{\partial y} \right)^2 \right]$$
$$= \text{const.} - \frac{1}{2} \chi_a (\hat{\mathbf{n}} \cdot \mathbf{H})^2 \tag{6.19}$$

Usually the diamagnetic anisotropy $\chi_a > 0$, and $\chi_a \sim 10^{-7}$ cgs units. The effects of an electric field are phenomenologically analogous, replacing **H** by **E** and χ_a by $\epsilon_a/4\pi$,

$$\chi_a (\hat{\mathbf{n}} \cdot \mathbf{H})^2 \rightarrow \frac{1}{4\pi} \epsilon_a (\hat{\mathbf{n}} \cdot \mathbf{E})^2$$

where ϵ_a is the dielectric anisotropy.

6.3.1.2. Transition induced by external forces : Helfrich-Hurault deformation

Let us consider first a homeotropically aligned smectic A between two glass plates. The unperturbed layers are parallel to the plane (X,Y) of the plates; the molecules are aligned along Z. If a magnetic field is applied parallel to the X-direction (Fig. 6.5) and $\chi_a > 0$, one may observe a Helfrich type of deformation above a critical field $\mathbf{H}_c$. The free-energy per unit volume reads

$$f = f_{el} + f_{mag} = \frac{1}{2}K_1(\nabla\cdot\hat{\mathbf{n}})^2 + \frac{1}{2}B\gamma^2 - \frac{1}{2}\chi_a(\hat{\mathbf{n}}\cdot\mathbf{H})^2 \tag{6.20}$$

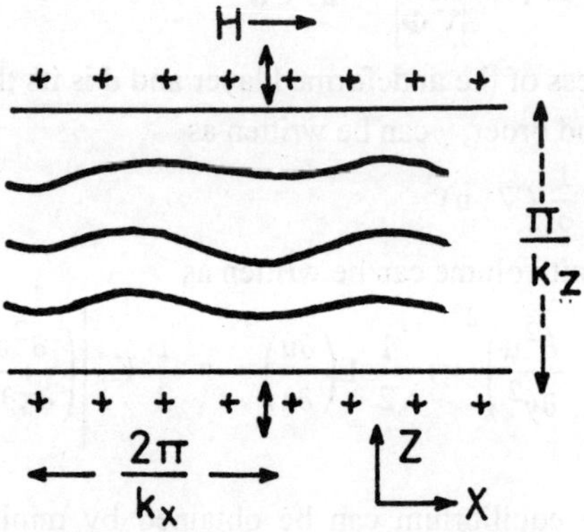

Fig. 6.5. The Helfrich-Hurault transition for S_A film subjected to a magnetic field. In practice, it is more convenient to apply a mechanical tension.

Let us assume that the layers undergo a periodic distortion along X (Fig. 6.5); in terms of displacements u, this corresponds to

$$u(x,z) = u_0(z)\cos k_x x \tag{6.21a}$$

where k_x is a certain wave vector. For small-amplitude distortions (just above threshold) $u_0(z)$ can be represented as a sine wave, vanishing both for z = 0 and z = D (sample thickness),

$$u_0(z) = u_0 \sin(k_z z) \tag{6.21b}$$

where $k_z = \pi/D$. The director now has components

$$\left.\begin{aligned} n_x &= -\frac{\partial u}{\partial x} = \theta_0 \sin(k_z z)\sin k_x x \\ n_y &= 0 \\ n_z &\approx 1 \end{aligned}\right\} \tag{6.21c}$$

where $\theta_0 = k_x u_0$. Evaluating the free-energy averaged over the sample thickness, one obtains

$$<f> = \frac{1}{8}\theta_0^2 B\left[\left(\frac{k_z}{k_x}\right)^2 + k_x^2 \lambda_A^2\right] - \frac{1}{8}\theta_0^2 \chi_a H^2 \tag{6.22}$$

When the overall coefficient of θ_0^2 is positive for all k_x values, the unperturbed arrangement is stable. Instability will occur first for that value of k_x which corresponds to

$$k_x^2 = k_z / \lambda_A = \frac{\pi}{\lambda_A D} \tag{6.23}$$

Thus the optimal wavelength of the deformation ($2\pi/k_x$) is equal to the geometric mean of the sample thickness and the microscopic length λ_A. At the critical field $\mathbf{H}_c$ the elastic and magnetic terms cancel exactly in eq. (6.22). We obtain

$$\chi_a H_c^2 = 2Bk_z \lambda_A = 2\pi\left(\frac{B\lambda_A}{D}\right) \tag{6.24}$$

This equation was derived by Hurault[13a] based on an earlier calculation of Helfrich[13b]. It shows that $\mathbf{H}_c$ is proportional to $\sqrt{1/D}$. However, in a conventional Frederiks transition $\mathbf{H}_c \approx 1/D$. Here the field decreases more slowly with the sample thickness. Equation (6.24) can be rewritten as

$$\chi_a H_c^2 = 2\pi \frac{K_1}{\lambda_A D} \tag{6.25}$$

Using the typical estimates $K_1 \sim 10^{-6}$ dyn, $\lambda_A = 20$ Å, D = 1mm, and $\chi_a \sim 10^{-7}$ cgs units, one gets $H_c \sim 20$ kG. Thus in order to observe the effect one needs to have very large fields or very large samples. Although the samples of this size have been prepared, the difficulty is that above the critical field the deformation amplitude θ_0 remains very small. The undulation of the layers is strongly limited due to the requirement of constant interlayer thickness.

A periodic undulation of the layers of the Helfrich-Hurault type, with a wavelength $2\pi/k_x$ proportional to the square root of the sample thickness D, can be achieved more easily by mechanical tension, i.e., by increasing the separation between the glass plates[14,15]. If no field is applied and u is assumed to be independent of y, eq. (6.8) reduces to

$$f_{el} = \frac{1}{2} B\gamma^2 + \frac{1}{2} K_1\left(\frac{\partial^2 u}{\partial x^2}\right)^2 \tag{6.26}$$

However, in the present case it is noted that the effective layer spacing along Z is altered due to the bending of the layers and therefore the linear form $\gamma = \partial u/\partial z$ must be improved. Considering a second order contribution to the layer dilation in that direction, we have

$$\gamma = \frac{\partial u}{\partial z} - \frac{1}{2}\left(\frac{\partial u}{\partial x}\right)^2 \tag{6.27}$$

The f_{el} is still of the form (6.26) with γ given by eq. (6.27). Let us split the u and γ into two terms

$$u = u_0 + u_1(x, z) \tag{6.28a}$$

$$\gamma = \gamma_0 + \gamma_1 \tag{6.28b}$$

Where u_1 describes the undulation of the layers and $u_0 = \gamma_0 z$ is the uniform displacement associated with the imposed strain γ_0 (>0). Of interest is the critical value for the onset of the undulation. γ_0 and u_0 are finite, but u_1 is infinitesimally small; so expanding the free-energy (6.26) upto second order u_1^2, we get

$$f_{el} = f_0 + B\gamma_0 \frac{\partial u_1}{\partial z} - \frac{1}{2}B\gamma_0\left(\frac{\partial u}{\partial x}\right)^2 + \frac{1}{2}K_1\left(\frac{\partial^2 u_1}{\partial x^2}\right)^2 \tag{6.29}$$

The second term gives 0 by integration over the Z variable (u_1 vanishes on both the plates.). The new term of interest is the third one, which is identical in form to the effect of a destabilizing magnetic field **H**. The correspondence is given by

$$-\frac{1}{2}\chi_a H^2\left(\frac{\partial u}{\partial x}\right)^2 = -B\gamma_0\left(\frac{\partial u}{\partial x}\right)^2 \tag{6.30}$$

The threshold value to the strain γ_{0c} is obtained from eqs. (6.24) and (6.30) as

$$\gamma_{oc} = 2\pi(\lambda_A/D) \tag{6.31}$$

This corresponds to a plate displacement $\Delta = 2\pi\lambda_A \sim 150$ Å, which can be easily realized in practice. These conclusions have been verified experimentally[14,16]. The plate separation was increased in a controlled manner by piezoelectric ceramics. Two transient bright spots in the laser diffracted beam were seen at the certain value of dilation; this confirmed the onset of a spatially periodic distortion above a critical strain. For CBOOA at 78^{0}C, the measurements gave $\lambda_A = 22 \pm 3$ Å and assuming $K_1 \sim 10^{-6}$ dyn, B was estimated to be 2×10^7 cgs units.

6.3.1.3. Correlations in the smectic A phase

Let us consider, in this section, the correlation between smectic layers in the S_A phase. This is important because the absence of true long-range correlation affects the phase transition behaviour involving S_A phase. This so-called Landau-Peierls instability is related to the mean-square fluctuation of the displacement of a given smectic layer. The relevant correlation function thus concerns the evaluation of $<u^2(r)>$. We write the Fourier transform u_q of the displacement u(r),

$$u_q = \int u(\mathbf{r}) \exp(i\,\mathbf{q}\cdot\mathbf{r}) d\mathbf{r} \tag{6.32}$$

In terms of u_q's, the free-energy density assumes the form

$$f = \frac{1}{2}\sum_q |u_q|^2\left[Bq_z^2 + K_1(q_\perp^2 + \xi^{-2})q_\perp^2\right] \tag{6.33}$$

where $q_\perp^2 = q_x^2 + q_y^2$ and $\xi = (K_1/\chi_a)^{\frac{1}{2}} H^{-1}$ is the magnetic coherence length. Application of equipartition theorem gives

$$<|u_q|^2> = \frac{k_B T}{Bq_z^2 + K_1(q_\perp^2 + \xi^{-2})q_\perp^2} = \frac{k_B T}{B(q_z^2 + \lambda_A^2(q_\perp^2 + \xi^{-2})q_\perp^2)} \tag{6.34}$$

Assuming the sample to be infinitely large, the mean square fluctuation is given by

$$<u^2(r)> = \frac{k_B T}{4\pi(BK_1)^{1/2}} \ell n\left(\frac{\xi}{d_0}\right) \tag{6.35}$$

where d_0 is the layer periodicity. It is obvious that as $\mathbf{H} \to 0$, $<u^2> \to \infty$, showing that such a structure can not be stable . In the absence of a magnetic field, for a sample thickness D,

$$<u^2(r)> = \frac{k_B T}{4\pi(BK_1)^{1/2}} \ell n\left(\frac{D}{d_0}\right) \tag{6.36}$$

Thus the layer fluctuations diverge logarithmically with the sample size. The coupling between the displacement field and the orientational field, which is represented by the term proportional to K_1, may be responsible for the logarithmic divergence. When this coupling is absent the mean square displacement diverges linearly with the length L. The logarithmic divergence causes the positional order of the S_A layering to be quasi long-range.

The displacement-displacement correlation function, in the harmonic approximation, may be defined as

$$G(r) = \exp\left[-\left(\frac{1}{2}q_z^2 <|u(r) - u(0)|^2>\right)\right] \tag{6.37}$$

From eq. (6.32), we get

$$<|u(r) - u(0)|^2> = \frac{1}{4\pi^3}\int <u_q^2>[1 - \exp(-i\,\mathbf{q}\cdot\mathbf{r})]dq \tag{6.38}$$

The Fourier transform of G(r) gives the intensity of scattering,

$$I \propto |q_z - q_0|^{-2+\eta}, \qquad q_\perp = 0 \tag{6.39a}$$

$$I \propto |q_\perp|^{-4+2\eta}, \qquad q_z = 0 \tag{6.39b}$$

where

$$q_0 = 2\pi/d_0$$

and

$$\eta = \frac{k_B T q_0^2}{8\pi (BK_1)^{1/2}} \tag{6.40}$$

Thus the sharp Bragg peak characteristic of the long-range ordered crystal lattice is replaced by strong thermal diffuse scattering with a power-law singularity[17]. This prediction has been confirmed by the high resolution X-ray measurements[18].

6.3.1.4. Molecular theory for the elasticity of S_A phase

In this section we consider the application of the unified molecular theory, developed by us[19,20a], (referred to as SSR theory ; see sec. 5.1.3), to study the elastic constants of S_A phase. As mentioned in sec. 5.1.3 this theory is based on the weighted density functional formalism[21] and writes the exact expressions for the elastic constants (K_1 and B) in terms of structural parameters, order parameters, etc.

For the S_A structure the elastic free-energy density (eq. (5.86)) assumes the form

$$\frac{1}{V}\beta\Delta A_e[\rho] = -\frac{1}{8\pi}\rho_s^2 \sum_{\ell_1,\ell_2,\ell}{}' \sum_m \sum_k Q_{k\ell_1} Q_{-k\ell 2} C_g(\ell_1 \ell_2 \ell; omm)$$
$$[(2\ell_1+1)(2\ell_2+1]^{-1/2} \int d\mathbf{r}_{12} \left[\left(\frac{4\pi}{2\ell_2+1}\right)^{1/2} \exp(i G_e z_{12})\right.$$
$$\left. Y_{\ell m}(\Delta\chi(\mathbf{r}_{12})) - \exp(iGz_{12})\right] C_{\ell_1\ell_2\ell}(\mathbf{r}_{12})\, Y^*_{\ell m}(\hat{\mathbf{r}}_{12}) \tag{6.41}$$

where $G = 2\pi k/d_0$ and $G_e = 2\pi k/d$, k is a positive or negative integer, z_{12} is the translational coordinate parallel to the layer normal, and $Q_{0\ell} = (2\ell+1)\overline{P}_\ell$, $Q_{k0} = \overline{\mu}_q$ and $Q_{k\ell} = (2\ell+1)\overline{\tau}_{k\ell}$ are, respectively, orientational, positional and mixed order parameters. In writing eq. (6.41) we have assumed molecules of the system to be cylindrically symmetric. Neglecting the coupling between the distortions caused by the curvature in director orientation and dilation in layer thickness, eq.(6.41) can be decomposed as[19, 20a]

$$\frac{\beta \Delta A_e^{(s)}[\rho]}{V} = -\frac{1}{8\pi}\rho_s^2 \sum_{\ell_1,\ell_2,\ell}{}' \sum_m \sum_k Q_{k\ell_1} Q_{-k\ell 2} C_g(\ell_1\ell_2\ell; omm)$$

$$[(2\ell_1+1)(2\ell_2+1]^{-1/2} \int d\mathbf{r}_{12}\, e^{iGz_{12}} \left[\left(\frac{4\pi}{2\ell_2+1}\right)^{1/2} Y_{\ell m}(\Delta\chi(\mathbf{r}_{12}))-1\right]$$

$$C_{\ell_1\ell_2\ell}(r_{12}) Y^*_{\ell m}(\hat{\mathbf{r}}_{12}) \tag{6.42}$$

and

$$\frac{\beta \Delta A_e^{(\ell)}[\rho]}{V} = -\frac{1}{8\pi}\rho_s^2 \sum_{\ell_1,\ell_2,\ell}{}' \sum_m \sum_k Q_{k\ell_1} Q_{-k\ell_2} C_g(\ell_1\ell_2\ell; omm)$$

$$[(2\ell_1+1)(2\ell_2+1]^{-1/2} \int d\mathbf{r}_{12}\left[\exp(iG_e z_{12}) - \exp(iGz_{12})\right]$$

$$C_{\ell_1\ell_2\ell}(r_{12}) Y^*_{\ell m}(\hat{\mathbf{r}}_{12}) \tag{6.43}$$

These eqs. (6.42) and (6.43) represent, respectively, the distortion free-energy arising due to curvature in the director orientation and dilation in layer thickness. ρ_s is the smectic number density.

Using the expansion

$$e^{iGz_{12}} = \sum_{\ell'} (i)^{\ell'} (2\ell'+1)\, j_{\ell'}(G r_{12}) P_{\ell'}(\cos\theta), \tag{6.44}$$

and comparing eq. (6.42) with eq. (6.8), the following expression, was obtained for the splay elastic constant of S_A phase

$$\beta K_1 = -\frac{1}{6}\rho_s^2 \sum_{\ell_1,\ell_2}{}' \sum_m Q_{k\ell_1} Q_{-k\ell_2} (2\ell_1+1)^{-1/2} (2\ell_2+1)^{-1} \sum_{\ell,\ell'} (i)^{\ell'}$$

$$(2\ell'+1)^{1/2} j^{\ell'}_{\ell_1\ell_2\ell}\{b_{\ell_2 0}\, C_g(\ell_1\ell_2\ell,000)\,\delta_{\ell\ell'} + (2\ell'+1)^{1/2}(2\ell+1)^{-\frac{1}{2}}$$

$$C_g(2\ell'\ell,000)[b_{\ell_2 0}\, C_g(\ell_1\ell_2\ell,000)\, C_g(2\ell'\ell,000) + 2\sqrt{6}\, b_{\ell_2 2}$$

$$C_g(\ell_1\ell_2\ell,011)\, C_g(2\ell'\ell,101) + \sqrt{6}\, b_{\ell_2 2}\, C_g(\ell_1\ell_2\ell,022)$$

$$C_g(2\ell'\ell,202)]\} \tag{6.45}$$

where $j_{\ell'}(Gr_{12})$ are spherical Bessel functions and

$$j^{\ell'}_{\ell_1\ell_2\ell} = \int dr_{12}\, r_{12}^4\, j_{\ell'}(G r_{12})\, C_{\ell_1\ell_2\ell}(r_{12}) \tag{6.46}$$

are the structural parameters.

Eq. (6.45) can be written as a double sum

$$\beta K_1 = \sum_{\ell_1}{}' \sum_{\ell_2}{}' \beta K_1(\ell_1, \ell_2) \tag{6.47}$$

Expanding eq. (6.43) in ascending powers of dilation parameter γ and comparing the terms associated with γ^2 (=$(du/dz)^2$) in eq. (6.8), the following expression for the bulk elastic constant was derived

$$\beta B = \frac{1}{12}(4\pi)^{-1/2} \rho_s^2 \sum_{\ell_1, \ell_2, \ell, \ell'}{}' \sum_{k} G^2 Q_{k\ell_1} Q_{-k\ell 2}[(2\ell_1+1)(2\ell_2+1)]^{-1/2}$$

$$(i)^{\ell'} j^{\ell'}_{\ell_1 \ell_2 \ell} (2\ell'+1)^{1/2} C_g(\ell_1 \ell_2 \ell, 000) \left[\delta_{\ell\ell''} + 2\left(\frac{2\ell'+1}{2\ell+1}\right)^{1/2} \right.$$

$$\left. C_g^2(2\ell'\ell;\ 000) \right] \tag{6..48}$$

This equation can be rewritten as

$$\beta B = \sum_{\ell_1}{}' \sum_{\ell_2}{}' \beta B(\ell_1, \ell_2) \tag{6.49}$$

The terms of the series (6.47) and (6.49) were evaluated for $0 < \ell_1, \ell_2 < 4$ and are given in Ref. 19.

The influence of the molecular interactions on the K_1 and B was investigated for a model system as described by eq. (5.93) and it was concluded that at the low densities all interactions (HER, dispersion and quadrupole) contribute almost equally but at higher densities the contribution of the HER interaction increases steeply and dominate over the contributions of other interactions. We have also obtained[19] the condition for the equilibrium layer separation; the interlayer spacing can be determined by minimizing free-energy of undistorted smectic phase with respect to G.

6.4 THE NEMATIC TO SMECTIC A (N S_A) TRANSITION

In spite of considerable research activity over the last three decades the nematic to smectic A phase transitions remain one of the principal unsolved problems[1, 2, 20b] in statistical physics of condensed matter. The uniaxial nematic to smectic A (NS_A) transition involves a rearrangement of the centers of the molecules; the centers are disordered in the nematic state and ordered in the smectic state. The ordering in the S_A phase is such that the molecular centers are, on average, arranged in equidistant planes with the interplanar spacing d_0 (say). Within these layers the molecules move at random, with the restriction that the director remains perpendicular to the smectic layers. Consequently, the density changes its behaviour at the S_AN transition from periodic to homogeneous. The one dimensional density wave characterizing the S_A phase is associated with the Landau-Peierls instability. In the thermodynamic limit there exists no long-range translational order in the S_A phase. Thus the theory of S_AN transition is a theory of one-dimensional melting. Since covering all aspects of this transition is not possible here, we shall discuss molecular models developed to describe the origin and details of S_A ordering and S_AN transition.

During the last 25 years, many high resolution heat capacity and X-ray studies have been devoted to the S_AN transition[22–36]. Most of the focus has been on the critical exponents in the hope of determining the universality class of this transition. The most extensively measured critical exponents are α, γ, $\nu_{\parallel}$ and $\nu_{\perp}$, where the last two correspond to the divergence of the correlation lengths parallel and perpendicular to the director, respectively. α and γ are, respectively, the critical exponents of the singular part of the specific heat and the susceptibility. In summarizing the data (see Table 6.6), it is convenient to specify the McMillan ratio T_{NA}/T_{NI}. At first, DSC seemed to indicate that most of the observed NS_A transitions were discontinuous, with only those below the tricritical value of 0.87 being continuous. Later work using adiabatic scanning calorimetry and AC calorimetry showed that pretransitional effect has been interpreted as latent heats in the DSC experiment, and that continuous behaviour occurred up to much higher values of the McMillan ratio. The critical exponents for the susceptibility and correlation lengths are measured from diffuse X-ray scattering experiments in the nematic phase[2,5]. For the McMillan ratio below about 0.93, these exponents are close to the 3D XY universality class, $\alpha = -0.007$, $\gamma = 1.316$, $\nu_{\parallel} = \nu_{\perp} = 0.669$. Above a ratio of roughly 0.93, they are no longer fairly constant but tend to approach their tricritical values, $\alpha = 0.5$, $\gamma = 1.0$, $\nu_{\parallel} = \nu_{\perp}$; $\nu_{\parallel} = 0.5$ as the McMillan ratio approaches unity. It is important to mention that the value of the McMillan ratio at the tricritical point is not universal and that the crossover from 3D XY values to tricritical values is not the same for different homologous series, but the trend in the critical exponents is very similar.

The theory of S_AN transition has received more attention[37–71] than any other smectic transition, partially due to the analogy with the superconducting normal transition. In spite of all these efforts, the situation remains very complicated with numerous questions unresolved. Frenkel and coworkers[72, 73] have performed the computer simulation studies on a system of hard spherocylinders to observe the S_AN transition. A list of the references to much of the theoretical and experimental work is given in an article by Garland and Nounesis[74a]. Longa[74b] has discussed the application of the Landau and molecular-field theories to the S_AN transition very well.

At the NS_A phase transition the continuous rotational symmetry of the nematic phase is spontaneously broken by the appearance of one-dimensional density wave in the S_A phase. Close to the transition the onset of quasi-smectic features in the nematic phase may lead to a drastic change in certain important properties (elastic coefficients, transport properties, cholesteric pitch, etc.). Original theories due to McMillan[45] and de Gennes[78] suggested that the NS_A transitions could be first or second order. The order of transition changes at a tricritical point (TCP)[75]. Alben[76] predicted a ^3He–^4He like TCP in binary liquid crystal mixtures. However, Halperine et. al.[77] argued that the NS_A transition can never be truly second order, which of course rules out the possibility of a TCP. This controversy has spurred experimental studies[22–36] which have shown that NS_A transition can indeed be continuous when measured to the dimensionless temperature $(T\text{-}T_{NA})/T_{NI}) \approx 10^{-5}$. Lelidis and Durand[79] predicted the field induced TCP in NS_A transition. Thus, in spite of early controversies, it is witnessed now that the NS_A transition is continuous, in the absence of special circumstances[26,80], in accordance with the suggestion of McMillan[45] and de Gennes[78a] for a specific models. The salient features of the NS_A transition have been documented well by several authors[4,81,82].

Table 6.6. Critical exponents measured for several NS_A systems (Ref. 117). Anisotropic scaling of the correlation lengths is observed.

	T_{NA}/T_{NI}	α	γ	$\nu_\perp$	$\nu_\parallel$	$\alpha+2\nu_\perp+\nu_\parallel$
T8	0.66	0.07	1.22	0.65	0.70	1.93
T7	0.71	0.05	1.22	0.61	0.69	2.04
60CB-80CB	0.895		1.75	0.73	0.95	
80PCBOB	0.897	−0.008	1.39	0.59	0.75	1.92
60CB-80CB	0.898		1.77	0.70	0.93	
7S5–8S5	0.908		1.21	0.67	0.64	
$\overline{7}$S5–$\overline{8}$S5	0.911		1.52	0.68	0.82	
80CB-$\overline{7}$S5	0.911	−0.55	1.58	0.78	0.90	1.92
60CB-80CB	0.920		1.61	0.61	0.81	
$\overline{7}$S5–$\overline{8}$S5	0.924		1,45	0.68	0.81	
407	0.926	−0.03	1.46	0.65	0.78	2.05
$\overline{8}$S5	0.936	0	1.53	0.68	0.83	2.19
CBOOA	0.940	0.15	1.30	0.62	0.70	2.09
408	0.958	0.15	1.31	0.57	0.70	1.99
8OCB	0.963	0.2	1.32	0.58	0.71	2.09
$\overline{9}$S5	0.967	0.22	1.31	0.57	0.71	2.07
8CB	0.977	0.31	1.26	0.51	0.67	2.00
$\overline{10}$S5	0.984	0.45	1.1	0.51	0.61	2.08
9CB	0.994	0.53	1.1	0.37	0.57	1.84
XY		−0.007	1.32	0.67	0.67	2.00
Tricritical		0.5	1.0	0.5	0.5	2.00

6.4.1 Phenomenological Description of the NS_A Transition

The Landau-de Gennes theory for the NI transition (sec. 4.4.1) can be extended to the NS_A transition. We start by defining an order parameter for the smectic A phase. The order parameter for this transition is $|\psi|$, the amplitude of the density wave describing the formation of layers in the smectic A phase. Since the difference between the values of $-|\psi|$ and $|\psi|$ only amounts to a shift of one half layer spacing in the location of all the layers, no change in the free-energy per unit volume results. Therefore, the expansion in terms of the powers of $|\psi|$ can only contain even power terms.

Let us begin with the simple case. The S_A phase is characterized by a periodic density modulation along the direction $\hat{\mathbf{z}}$ orthogonal to the layers[45,2,83].

$$\rho(r) = \rho(z) = \rho_0 + \sum_{i=1}^{\infty} \rho_i \cos(iq_s z - \Phi_i) \tag{6.50}$$

where ρ_1 represents the first harmonic of the density modulation and Φ an arbitrary phase. In a nematic $\rho_1 = 0$, whereas in S_A phase ρ_1 becomes the natural candidate for the order parameter. In the vicinity of the NS_A transition the free-energy per unit volume may be expanded in powers of ρ_1.

$$f_{S_A} = \frac{1}{2}\alpha_2 \rho_1^2 + \frac{1}{4}\alpha_4 \rho_1^4 + \ldots\ldots \tag{6.51}$$

At a certain temperature T_0, the coefficient $\alpha_2 \approx \alpha_0(T-T_0)$ vanishes. Above this temperature it is positive. The coefficient α_4 is always positive. Only by these considerations, a second order transition could be obtained $T_{AN} = T_0$. However, a number of complications are to be taken into the consideration. First, the influence of coupling between ρ_1 and S, the nematic order parameter, is to be accounted for. Because of this coupling the optimal value of S does not coincide with $S_0(T)$ ($\equiv \overline{P}_2$) obtained in the absence of smectic order.

Let,

$$\delta S = S - S_0(T), \tag{6.52}$$

measures the deviation of the alignment from its equilibrium value S_0.
The coupling term, to the lowest order, must have the from

$$f_{AN} = -C\rho_1^2\,\delta S \tag{6.53}$$

C is a positive constant. Further, to the free energy density must be added the nematic free-energy density f_N which is minimum for $\delta S = 0$.

$$f_N = f_N(S_0) + \frac{1}{2\chi}\delta S^2 \tag{6.54}$$

where $\chi(T)$ the response function is large in the vicinity of NI transition but is small for $T < T_{NI}$.

The free-energy density now reads

$$f = \frac{1}{2}\alpha_2\rho_1^2 + \alpha_4\rho_1^4 + \frac{1}{2\chi}\delta S^2 - C\rho_1^2\,\delta S \tag{6.55}$$

Clearly the appearance of the smectic order usually increases the average attraction between the molecules and hence reinforces the alignment.
Minimization with respect to δS gives

$$\delta S = \chi C\rho_1^2, \quad (>0) \tag{6.56}$$

and new coefficient appears with the fourth order term,

$$\alpha_4' = \alpha_4 - 2\chi C^2 \tag{6.57}$$

The sign of α_4' critically determines the order of transition. Clearly the sign of α_4' can change depending on whether χ is large or small, i.e., whether the nematic range is narrow or wide. When the nematic susceptibility is low (i.e. $T_0 << T_{NI}$), α_4' is positive and the NS_A transition is second-order at $T_{AN} \approx T_0$. If χ is large (i.e. $T_0 \approx T_{NI}$) α_4' is negative; the stability requires that ρ_1^6 term with positive coefficient must be added to eq. (6.55). In this case the NS_A transition is first-order at a temperature $T_{AN} = T_0 + \alpha_4'^2/2\alpha_0 V > T_0$. The point at which $\chi = \alpha_4/2C^2$ (i.e. $\alpha_4' = 0$) is a tricritical point on the NS_A line. Thus the change from a second-order to a first-order is induced by the coupling between ρ_1 and S.

It is well known that the layer fluctuations play an important role in smectics. Close to the NS_A transition, critical fluctuations of the smectic order parameter $\rho_1(\mathbf{r})$ are expected to be important too. The two fields $\rho_1(\mathbf{r})$ and $u(\mathbf{r})$ can be described by a complex order parameter, first introduced by de Gennes[78] and McMillan[37].

$$\psi(\mathbf{r}) = \rho_1(\mathbf{r})\,e^{i\Phi_1(\mathbf{r})} \tag{6.58}$$

The phase $\Phi_1(\mathbf{r})$ is simply related to the layer displacement $u(\mathbf{r})$ through

$$\Phi_1(\mathbf{r}) = -iq_s u(\mathbf{r}) \tag{6.58a}$$

where $q_s = 2\pi/d_0$. Now near the transition, the free energy density may be expanded in even powers of ψ. Since the spatially dependent order parameter has been defined, we must add to the free energy the gradient terms that express the tendency for the smectic to be homogenous. Keeping all these features in view, the general expansion for the free energy density can be written as

$$f = f_N(T,Q,\nabla Q) + f_{S_A}(T,\psi,\nabla\psi) + f_{AN}(Q,\psi,\nabla Q,\nabla\psi) + f_{ext}(Q,\psi,\nabla Q,\nabla\psi,P) \tag{6.59}$$

where $f_N(T,Q,\nabla Q)$ is the free-energy density of the nematic phase, $f_{S_A}(T,\psi,\nabla\psi)$ corresponds to the smectic A phase, f_{AN} is the contribution from the coupling between Q (nematic order parameter) and ψ and f_{ext} is the free- energy density associated with the coupling of the order parameters and the external perturbation.

$$f_{S_A} = \frac{1}{2}\alpha_2|\psi|^2 + \frac{1}{4}\alpha_4|\psi|^4 + \frac{1}{2}C_{\parallel}\left|\frac{\partial\psi}{\partial z}\right|^2 + \frac{1}{2}C_{\perp}\left(\left|\frac{\partial\psi}{\partial x}\right|^2 + \left|\frac{\partial\psi}{\partial y}\right|^2\right) \tag{6.60}$$

where $C_{\parallel} \neq C_{\perp}$ because of the nematic anisotropy. At the lowest order coupling term, we can write.

$$f_{AN} = \frac{1}{2}E_2Q|\psi|^2 + \frac{1}{2}E_4Q^2|\psi|^2 \tag{6.61}$$

E_2 and E_4 are coupling constants. E_2 is chosen negative to favour S_A phase when the nematic phase exists and $E_4 > 0$. Generally this term allows reentrant effects (see sec 6.7). Because of (Q,ψ) coupling the NS_A phase transition can be of second-order or first- order.

In writing the expansion (6.60) an implicit assumption has been made that the director $\hat{\mathbf{n}}$ is fixed in the Z-direction. In reality, $\hat{\mathbf{n}}$ fluctuates. So the gradient terms have to be taken in directions parallel and perpendicular to $\hat{\mathbf{n}}$. Owing to this only the $C_{\perp}$ term is modified. Using the notation

$\nabla_{\perp} = \left(\frac{\partial}{\partial x}, \frac{\partial}{\partial y}\right)$, it becomes

$$C_{\perp}|(\nabla_{\perp} - iq_s\delta\hat{n}_{\perp})\psi|^2 \tag{6.62}$$

where $\delta\hat{n}_{\perp} = \hat{\mathbf{n}} - \hat{\mathbf{z}}$. When the director fluctuation is taken into account the Frank-Oseen elastic contribution[20a] has to be added to the total free- energy.

A number of conclusions have been reached on this problem. The transition seems always to be weakly first-order in four dimensions due to the coupling between the smectic order parameter and the director fluctuation[77]. In three dimension, the behaviour on the low and high sides of the transition can be reversed from the 3D XY behaviour, i.e., an inverted 3D XY model[84,85]. A dislocation-loop melting theory, in which a divergence in the density of dislocation loops destroys the smectic order, yields anisotropic critical behaviour in the correlation length[86,87]. A noninverted behaviour has been observed in MC simulations[88]. A self-consistent one-loop theory employing intrinsically anisotropic coupling between the director fluctuations and the smectic order parameter predicts a gradual crossover from isotropic behaviour to strongly anisotropic behaviour[89,90]. Thus the NS_A transition is very complicated with many factors influencing the behaviour near the transition. Coupling between the smectic order parameter and the nematic order drives the transition towards tricritical behaviour, while coupling between the smectic order and nematic director fluctuations drives it into an anisotropic regime. While the McMillan ratio is a convenient indicator of the strength of both of these couplings, it is quite imprecise. Yet the general trends with McMillan ratio are clearly evident.

6.4.2 Mean-field Description of the NS_A Transition

McMillan[45] presented an elegant description of smectic A phase by extending the Maier-Saupe (MS) theory of the nematic phase to include the one-dimensional translational order of the S_A phase. A similar but somewhat more general treatment, based on the theory of melting[91], was proposed independently by Kobayashi[44]. Kobayashi developed a formalism of the translational and orientational ordering but due to the two particle distribution for the ordered phase the detailed numerical results for a realistic potential were hard to obtain. A number of other refinements and extensions[47–53] were proposed but the McMillan's model remains the simplest one for its computational convenience and comparison with experiments and explains all the qualitative features of the S_AN and S_AI transitions. A simpler treatment of the S_AN transition as compared to Kobayashi- McMillan approach has been given by Meyer and Lubensky (ML)[46]. The ML model limits the discussion of the S_AN transition to the case of an ideal orientational order, i.e, the temperature dependence of the orientation of the molecules is neglected.

We shall first describe the essential features of most widely used McMillan model. The physical idea behind the model is based on the structure of the molecules exhibiting S_A phase. The mesogenic materials forming the S_A phase typically have a central aromatic core and flexible alkyl chains at the two ends. The aromatic moieties have large polarizabilities and thus, the dispersion interaction energy is very strong between the cores. Consequently, this can lead to the formation of a layered arrangement as in the S_A phase if the alkyl chains are sufficiently long and serve to efficiently separate the aromatic cores in layers. McMillan assumed that the anisotropic interaction is short-ranged and can be expressed as

$$u_{12}(r_{12}, \cos\theta_{12}) = -\left(\frac{V_0}{N r_0^3 \pi^{3/2}}\right) \exp\left[-\left(\frac{r_{12}}{r_0}\right)^2\right] P_2(\cos\theta_{12}) \tag{6.63}$$

where r_0 is of the order of the length of the rigid section of the molecule, and the exponential term reflects the short-range character of the interaction.

The self-consistent one-particle potential that a test molecule would feel, retaining only the leading term in the Fourier expansion, can be written as

$$u_1(z, \cos\theta) = -V_0[\overline{P}_2 + \overline{\sigma}\alpha \cos(2\pi z / d_0)] P_2(\cos\theta) \tag{6.64}$$

where the McMillan parameter α is given by

$$\alpha = 2 \exp[-(\pi r_0 / d_0)^2] \tag{6.65}$$

$\overline{P}_2$ is the usual orientational order parameter and $\overline{\sigma}$ is an order parameter which couples translational and orientational orders. Equation (6.64) ensures that the energy is a minimum when the molecule is in the smectic layer with its axis along Z. The single-particle distribution function is given by

$$f_1(z, \cos\theta) = \exp[-\beta u_1(z, \cos\theta)] \tag{6.66}$$

Using this distribution function and the interaction potential (6.63) the one-body potential is recalculated

$$u_1(z_1, \cos\theta_1) \equiv \frac{N \int d^3 x_2\, d\Omega_2\, u_{12}(r_{12}, \cos\theta_{12})\, f(z_2, \cos\theta_2)}{\int d^3 x_2\, d\Omega_2\, f(z_2, \cos\theta_2)} \tag{6.67a}$$

$$= -V_0[P_2(\cos\theta_1) < P_2(\cos\theta_2) > + \alpha \cos(2\pi z_1 / d_0)$$
$$P_2(\cos\theta_1) \langle \cos(2\pi z_2 / d_0)\, P_2(\cos\theta_2) \rangle] \tag{6.67b}$$

Self-consistency of eqs. (6.64) and (6.67) demands that

$$\overline{P}_2 = \langle P_2(\cos\theta) \rangle \tag{6.68a}$$

$$\overline{\sigma} = \langle \cos(2\pi z / d_0)\, P_2(\cos\theta) \rangle \tag{6.68b}$$

Equation (6.68) must be solved self-consistently for both the order parameters,; the order parameter $\overline{P}_2$ defines the orientational order, exactly as in MS theory, and $\overline{\sigma}$ is a measure of the amplitude of the density wave describing the layered structure. These equations exhibit three types of solutions : (i) $\overline{P}_2 = \overline{\sigma} = 0$, no order, characteristic of the isotropic liquid phase, (ii) $\overline{P}_2 \neq 0$, $\overline{\sigma} = 0$, orientational order only, the theory reduces to the MS theory of the nematic phase; and (iii) $\overline{P}_2 \neq 0$, $\overline{\sigma} \neq 0$, orientational and translational order characteristic of the S_A phase.

The free-energy of the system is given by

$$A = U - T\Sigma \tag{6.69}$$

where the internal energy

$$U = -\frac{1}{2} N V_0 (\overline{P}_2^{\,2} + \alpha \overline{\sigma}^2) \tag{6.69a}$$

and the entropy is evaluated through

$$-T\Sigma = N V_0 (\overline{P}_2^{\,2} + \alpha \overline{\sigma}^{\,2}) - N k_B T\, \ell n \left[\frac{1}{d_0} \int_0^{d_0} dz \int_0^1 d(\cos\theta)\, f_1(z, \cos\theta) \right] \tag{6.69b}$$

At any temperature the phase with the lowest free energy is thermodynamically stable.

As is obvious two physical parameters V_0 and α enter the theory. V_0 determines the T_{NI} and fixes the temperature scale of the model, α, a dimensionless parameter, is a measure of the strength of layering interactions and can vary between 0 and 2. The interplanar distance is determined by the competition between the anisotropic forces which produce the smectic order and excluded volume effects. The smectic condensation energy is greater for larger values of α, that is for larger d_0. Experimentally the layer thickness is of the order of the molecular length. The parameter α increases with increasing chain length of the alkyl tails. Equations (6.68) were solved self consistently and the order parameters, entropy, specific heat as a function of temperature for several values of α were evaluated and the transition parameters were calculated. The main results of calculations are shown in Fig. 6.6. For $\alpha < 0.70$ and $T_{AN}/T_{NI} < 0.87$ the model predicts the second-order S_AN transition.

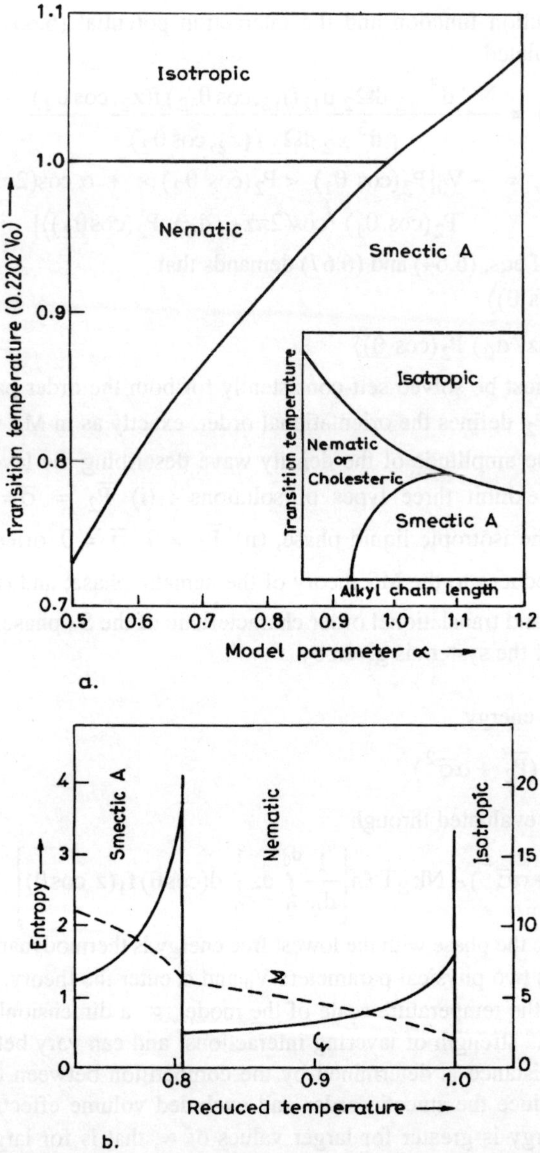

Fig. 6.6. (a) Phase diagram for the theoretical model parameter α. Inset : Typical phase diagram for homologous series of compounds showing transition temperatures versus length of the alkyl end-chains[45]; (b) The variation of entropy Σ and specific heat C_V with reduced temperature for $\alpha = 0.6$. A second order S_AN transition and a first order NI transition are exhibited by the model[45].

For $\propto > 0.98$, the S_A phase transforms directly into the isotropic phase, while for $\propto < 0.98$ there is a S_AN transition followed by a NI transition at higher temperature. Hence $\propto == 0.7$ and $T_{AN}/T_{NI} = 0.87$ correspond to a tricritical point at which the first-order transition terminates to a second-order transition. Figure 6.6(a) shows that the theoretical phase diagram broadly reflects the experimental trends in homologous series. However, it was found that the theoretical S_AN transition entropy is some what higher than the experimental values. McMillan made an attempt, in a later paper[37], to improve the agreement by using a modified pair potential

$$u_{12}(r_{12}, \cos\theta_{12}) = -(V_0 / N r_0^3 \pi^{3/2}) \exp\left(-\left(\frac{r_{12}}{r_0}\right)^2\right)(P_2(\cos\theta_{12}) + \delta) \tag{6.70}$$

The three parameters were fixed by the criterion that the theory has to fit the transition temperatures and entropies. The results were essentially the same as those obtained by the two parameter model (eq. 6.63) but some quantitative improvements were obtained. McMillan[45] himself suggested that if the excluded volume effect is included in eq. (6.63) it would favour $\hat{\mathbf{n}}$ to be parallel to $\hat{\mathbf{z}}$ Further, it was suggested by Priest[71] that a generalized rank-2 tensor model would give rise to additional terms which favour to be parallel to $\hat{\mathbf{q}}$. In most homologous series, the S_AN transition point T_{AN} reaches a maximum value for some chain length and tends to decrease for higher homologues.

In the Meyer-Lubensky model[46] the molecules are assumed to be perfectly oriented in a given direction (say the Z direction), whereas their centers of mass are situated on planes, parallel to the (X,Y) plane. Within the framework of the model (for the detail see Ref. 92) the nematic or isotropic phase is described by the uniform density ρ_0 only. The smectic phase is described by the infinite set of order parameters $\{\rho_n\}$,

$$\overline{\rho}_n = \langle \cos(nq_s z) \rangle$$

The main difference between the ML approach and the Kobayashi-McMillan approach concerns the nature of the smectic order parameter. The ML treatment considers only a pure translational order parameter $\overline{\rho}_1$, whereas the essential order parameter of the McMillan model is mixed order parameter $\overline{\sigma}$. In the limit of increasing orientational order $\overline{\sigma}$ approaches $\overline{\rho}_1$. Both the models stress the relevance of the range of the intermolecular potential with respect to the molecular length as far as the nature of the S_AN transition is concerned. The existence of a tricritical point is predicted in both of these models.

Woo and coworkers[61] have used a pairwise potential of the type.

$$u(1,2) = V_0 \exp\left[-\left(\frac{r_{12}}{r_0}\right)^2\right] [-\delta - P_2(\mathbf{\Omega}_1 \cdot \mathbf{\Omega}_2) + \varepsilon\{P_2(\mathbf{\Omega}_1 \cdot \hat{r}_{12}) + P_2(\mathbf{\Omega}_2 \cdot \hat{r}_{12})\}] \tag{6.71}$$

Four physical parameters V_0, δ, r_0 and $\in$ appear in the theory; δ is a measure of the interaction which gives rise to the translational order even in the absence of orientational order and $\in$ roughly accounts for the steric effects, which help to keep $\hat{\mathbf{n}}$ parallel to $\hat{\mathbf{q}}$. The four parameters of the model are adjusted to fit the measured transition temperature curves and approximate triple point. For each homologous series V_0 determines one point on the T_{NI} curve, while r_0 (or $\propto$) measures the length of the central section of the molecule. Both these parameters are chosen at the outset and there after held fixed. The remaining two parameters δ and $\in$ are adjusted to fit rest of the experimental phase diagram. The model predicted the phase diagrams similar to that of experimental one. Further the parameter $\in$ is sufficient to generate a reasonably good fit to the experimental phase diagram. The role of the end chains is merely to cause a larger interplanar spacing in the S_A phase and thus does not affect the model interaction. The characteristic feature of the S_A phase that the director prefers to be perpendicular to the smectic layers is incorporated in the model (6.71). The connection between $\in$ and the structure of molecules can explain the differences in phase transition properties between homologous series whose molecules are of similar structure but differ in the length of rigid section. Sokalski[62] has used a corner potential with Lennard-Jones type of interaction

$$u(1,2) = 4\in[(\sigma/r_{12})^m - (\sigma/r_{12})^n] \tag{6.72}$$

where σ depends on the orientations of the molecules. It was shown that the model (6.72) exhibits the smectic A, nematic and isotropic phases.

Kloczkowski and Stecki[48] considered a model system of hard spherocylinders superimposed with the attractive $1/r_{12}^6$ potential for the S_A phase. By using the stability analysis these authors have shown that this type of potential allows for the S_A phase formation. The influence of the attractive tail on the S_A phase formation was studied and it was found that the additional centre to centre r_{12}^{-6} potential produces an instability of the isotropic or nematic phases, towards the S_A phase formation. The calculations were performed for spherocylinders with length to diameter ratio 5 and 10 for different values of additional attractive potential parameters. This type of potential was also used by Nakagawa and Akahane[93] in their treatment of the S_A phase formation. These authors[52] also developed a McMillan type theory for binary mixtures including both intermolecular repulsions and attractions.

As discussed above the original version[45] of McMillan's theory contains just two order parameters, the orientational order parameter $\overline{P}_2$ and a mixed orientational-translational order parameter $\overline{\sigma}$. In this theory the occurrence of the mixed order parameter results in a strong correlation between the orientational and translational coordinates. It has been argued that this strong correlation may be responsible for the quantitative failure of the McMillan theory. In a later paper[37] McMillan introduced a purely translational order parameter $\overline{\tau}(= \langle \cos 2\pi z/d_0 \rangle)$ to improve the quantitative agreement. This third term has the effect of stabilizing the S_A phase and so allows a second-order S_AN transition to occur at a higher temperature by reducing the relative correlation between orientational and translational coordinates. In fact the basic structure of the theory remains unchanged. Despite success the introduction of third order parameter leads to the complication of the theoretical treatments in obtaining both analytic and numerical solutions. Katriel and Kventsel[47] have shown that this complication could be removed by writing the mixed order parameter as a product of the pure orientational and translational order parameters.

$$\overline{\sigma} = \langle P_2(\cos\theta) \rangle \langle \cos(2\pi z / d_0) \rangle = \overline{P}_2 \overline{\tau} \qquad (6.73)$$

This decoupling approximation has the additional advantage that the strength of correlation between the translational and orientational coordinates is readily controlled. The results obtained using this simplification (eq. (6.73)) exhibit the main features of the isotropic-nematic-smectic A transitions in a satisfactory manner. Kventsel et al.[53] made a quantitative comparison of the two theories, McMillan's type three order parameter theory[93], and Katriel and Kventsel[47] theory. These authors[53] have demonstrated the validity of the decoupling approximation for the mixed order parameter in both the models. For the decoupled model the phase diagram involving isotropic, nematic, smectic A and plastic crystal phases was constructed and the S_AN transition properties, which include the second order transition temperature and tricritical point, the transition entropy and the change in the order parameters, etc., were evaluated in detail. These properties have been found to agree well with the predictions of McMillan's theory for a wide range of strength parameters.

A lattice model for the S_A phase was proposed by Dowell[60a], and Ronis and Rosenblatt[60b]. Dowell investigated the effects of temperature, pressure, tail chain flexibility and tail chain length on the relative stabilities of the isotropic, nematic, smectic A and reentrant nematic phases (see sec. 6.7) using a lattice model[94] having only hard repulsions. The method provides a way to determine on an individual basis which of the molecular features are sufficient and /or necessary for the existence and relative stabilities of various phases. The thermodynamic and molecular ordering properties, such as S_A order parameter, core and tail intermolecular orientational order parameters, tail intramolecular orientational order parameter, density and entropy, were evaluated, in these phases and at the phase transitions. It has been shown[60a] that for the formation of S_A phase in real systems the presence of semiflexible tail chains of significant length is essential and that as the length of tail chain is shortened the S_A phase disappears. However, this conclusion is not always true. Further, the dipolar interactions are not necessary in this steric packing model to form S_A phase but do affect the stability ranges of temperature and pressure of the phases. It is known that the systems of hard particles, interacting through the excluded-volume effect, have played a fundamental role in the understanding of structural phase transitions in liquid cystals. Onsager made a major contribution to our understanding of nematic phase in a system of hard spherocylinders. Stroobants et. al.[95] provided evidence for the appearance of S_A order in the system of perfectly parallel hard spherocylinders from the simulation studies. Stimulated by the simulation results Mulder[66] developed a density functional calculation and Wen and Meyer[55] analysed about the origin of S_A phase in a parallel hard rod system. In this later work, these authors have described how the appearance of S_A order reduces the excluded volume of the nematic phase and have calculated the entropy for a system of parallel right circular cylinders. The arguement can be expressed quantitatively; there are three contributions to the entropy changes in going from the S_A order to the nematic order; a term describing directly the changes in long range order, a term due to the change of packing density within each layer, and a term due to the change in axial freedom of the motion of each rod. The first term has been expressed in terms of the distribution function of the centers of mass of the rods, and the second term in terms of 2D packing density. The third term is insignificant as compared to the other two terms. The calculation demonstrated that a second-order NS_A phase transition takes place as the packing fraction was raised to 0.202 times the value for the close packing. One of the limitations of this model is that the results have no dependence on the length to diameter ratio of the rods. Holyst and Poniewierski[57] studied the system of hard parallel cylinders in the framework of the smooth density approximation. Using a bifurcation analysis it was shown that apart from the nematic phase, smectic A, solid (or crystalline B) and columnar phases should also occur in this system. This approach fails to distinguish between a solid phase and a crystalline smectic B phase. Hosino et al.[54] had earlier found the persistence of the S_AN transition when hard cylinders were allowed to have three orthogonal orientations. Taylor et al.[59] constructed an excluded volume theory for the S_A and columnar phases of a system of hard spherocylinders by using the scaled particle theory in dimensions with full translational freedom, combined with a simple cell model to describe the positionally ordered dimensions. These authors obtained a phase diagram remarkably similar to that obtained from the Monte Carlo calculation[95]. The predicted smectic layer spacing at the nematic-smectic cross over is also almost identical to the MC results. The model suffers with a serious limitation that in it all transitions are required to be discontinuous. Consequently, the second-order nematic-smectic transition as demonstrated also by the MC results was found to be first-order.

A number of experimental works[96–102] show that the molecular biaxility has to be incorporated in the theoretical treatment in order to describe the structural and thermodynamic properties of smectics. The first attempt of such consideration was made by Averyanov and Primak[64] within the framework of McMillan model generalized to the biaxial molecules. These authors[64a] investigated the influence of smectic layering on the order parameters of solute molecules as well as on their positional orientational correlations. In a later paper[64b], the thermodynamics of S_AN phase transition in system of biaxial molecules was studied. The orientational-translational ordering of the molecule can be characterized by a set of 5 order parameters, $\overline{P}_2$, $\overline{\tau}$, $\overline{\sigma}$, and the other two defined as

$$\overline{G} = \frac{3}{2}\left\langle \sin^2\theta \cos 2\phi \right\rangle \equiv S_{x_1x_1} - S_{y_1y_1} \tag{6.74a}$$

$$\overline{K} = \frac{3}{2}\left\langle \sin^2\theta \cos 2\phi \cos(2\pi z/d_0) \right\rangle \tag{6.74b}$$

Taking into account these order parameters within the framework of McMillan model generalized to biaxial molecules the molecular pseudopotential was written as[64a]

$$\begin{aligned} u_1(\theta,\phi,z) = & -V_2\left[\overline{P}_2 + b_1\overline{G}\right]P_2(\cos\theta) + \alpha\left(\overline{\sigma} + b_1\overline{K}\right) \\ & P_2(\cos\theta)\cos(2\pi z/d_0) + \left(b_1\overline{P}_2 + b_2\overline{G}\right)\frac{3}{2}(\sin^2\theta\cos 2\phi) \\ & + \alpha\left(b_1\overline{\sigma} + b_2\overline{K}\right)\frac{3}{2}\sin^2\theta\cos 2\phi\cos(2\pi z/d_0) \\ & + \alpha\delta\,\overline{\tau}\cos(2\pi z/d_0) \end{aligned} \tag{6.75}$$

where

$$V_2 = u_{200} \exp\left[-\frac{a}{r_0^2}\right],$$

b_1 and b_2 are the biaxial parameters

$$b_1 = \frac{2u_{220}}{\sqrt{6}u_{200}} \quad \text{and} \quad b_2 = \frac{2u_{222}}{3u_{200}} \tag{6.76}$$

Here a is the average intermolecular distance within a smectic layer, r_0 is the effective interaction radius which is approximately equal to the length of the molecular core and $u_{2pq} = u_{2qp}$ are the expansion coefficients of the pairwise potential for the effective anisotropic intermolecular interactions. Thus in addition to three anisotropic parameters, (eq. (6.75)), V_2, b_1 and b_2 there appears a supplementary parameter, in the model using the relation $b_2 = b_1^2$. For the first-order S_AN transitions these four parameters can be evaluated from independent measurements of T_{NI}, T_{AN}, $\Delta\Sigma_{AN}$ and the correlation $\overline{G}(\overline{P}_2)$ in the nematic phase near T_{AN} or from the other set of experimental data. The signs of the parameters $\overline{G}$ and $\overline{K}$ coincide with the sign of b_1, so that changing the sign of b_1 does not influence the values of T_{NI}, T_{AN} and $\Delta\Sigma_{AN}$. Extensive numerical calculations were performed and it was found that the biaxiality modifies the phase diagram of the McMillan model. With increasing biaxiality parameter $b\left(=(3/2)^{\frac{1}{2}}b_1\right)$, T_{NI} increases but the dependence $T_{AN}(b)$ is weaker and more complex. For each value of δ (a parameter of McMillan theory) and b there is a critical value $\alpha_c(\delta,b)$ such that the inequality $T_{AN}(\delta,b) \geq T_{AN}(\delta,0)$ are valid for $\alpha \geq \alpha_c$. For each value of b there is a limited range of changes $0 < \delta \leq \delta_1(b)$, where $\alpha_{tcp}(\delta,b) \leq \alpha_c(\delta,b)$. With δ and α being constant the entropy change $\Delta\Sigma_{AN}$ decreases with increasing b. For sufficiently high values of δ and for α corresponding to a narrow nematic range increasing b can result in changing the S_AN transition from first order to second order. The parameter δ strengthens the layering tendency independent of orientational ordering and has to increase as the alkyl chain is lengthened at fixed length r_0 of the molecular aromatic core. It is expected that the anisotropic steric interactions do make a significant contribution to the effective value of b. The change $\Delta\overline{G}(T_{AN})$ and the dependence $\overline{G}(T)$ in the smectic phase are determined by the change of the order parameter $\overline{P}_2$ at T_{AN} and its dependence on the temperature. The character of the dependence of the derivative $d\overline{G}/dT$ on temperature changes qualitatively both for the first and second order S_AN transitions. It is important to mention here that the qualitative considerations, which are in agreement with known experiments, show that the maximum manifestation of the pseudopotential biaxiality should be expected for mesogens having sufficiently long end chains and narrow nematic ranges.

6.4.3 Application of Density Functional Theory to the NS$_A$ Transition

Lipkin and Oxtoby[103] were first to consider the application of the density functional approach to the mean-field theory of the isotropic-nematic-smectic A transition. This theory incorporates the steric effect through the pair correlation function and gives the result that $\mathbf{q}_s$ is parallel to $\hat{\mathbf{n}}$. These authors constructed a set of self consistent equations for the order parameters,

$$\delta_{\ell 0}\,\delta_{p0} + \mu_{p\ell} = \frac{2\ell+1}{4\pi V} \int d\Omega\, P_\ell(\cos\theta) \int d\mathbf{r}\, \exp(-i\mathbf{G}_{q_s}\cdot\mathbf{r})$$

$$\exp\left[\sum_{q_s} \exp(i\mathbf{G}_{q_s}\cdot\mathbf{r}) \sum_{\ell_1\ell_2} (2\ell_1+1)\, J(q_s,\ell_1,\ell_2)\,\mu_{q_s\ell_2}\, P_{\ell_1}(\cos\theta)\right] \tag{6.77}$$

where $\mu_{q_s\ell}$ are the order parameters of the theory and

$$J(q_s,\ell_1,\ell_2) = J(q_s,\ell_2,\ell_1)$$

$$= \frac{\rho_s}{(4\pi)^{3/2}} \sum_\ell G(\ell_1\ell_2\ell;00)\, i^\ell \left\{\frac{(2\ell+1)}{(2\ell_1+1)(2\ell_2+1)}\right\}^{\frac{1}{2}} C_{\ell_1,\ell_2\ell}(k_{q_s}) \tag{6.78}$$

with

$$C_{\ell_1,\ell_2\ell}\left(k_{q_s}\right) = 4\pi \int_0^\infty dr_{12}\, r_{12}^2\, j_\ell\left(k_{q_s} r_{12}\right) C_{\ell_1,\ell_2\ell}(r_{12}) \tag{6.79}$$

j_ℓ is a spherical Bessel function.

The free-energy difference between isotropic fluid and the ordered state was written as

$$\beta\Delta A = \rho_s V\left[-\mu_{00} + J(0,00)\mu_{00} + \frac{1}{2} \sum_{q_s\ell_1\ell_2}{}' J(q_s,\ell_1,\ell_2)\right] \mu_{q_{s\ell_1}}\, \mu_{q_{s\ell_2}} \tag{6.80}$$

where the summation excludes the term with $q_s = \ell_1 = \ell_2 = 0$. Equation (6.80) is a direct generalization of the work of Sluckin and Shukla[104] to the molecular system. A self consistent solution of eqs. (6.77) and (6.80) relate the phase diagram of the isotropic, nematic and S_A phases to the direct correlation function of the isotropic liquid. Truncating the order parameter expansion at the lowest term ($q_s = 0, \pm 1$, $\ell = 0, 2$) and relating the term μ_{00} to the density change between isotropic liquid and liquid crystal a connection with the McMillan theory was established. With these approximations the excess free energy was reformulated as

$$\frac{\beta \Delta A}{N} = \frac{1}{2} J(0,22)\mu_{02}^2 + J(1,00)\mu_{10}^2 + J(1,22)\mu_{12}^2$$

$$+ 2J(1,02)\mu_{10}\,\mu_{12} - \ell n\left(\frac{1}{d_0}\right)\int_0^{d_0} dz \int_0^1 d(\cos\theta)$$

$$\times \exp[5J(0,22)\mu_{02}\,P_2(\cos\theta) + \cos(2\pi z/d_0)\{2J(1,00)\,\mu_{10}$$

$$+ 2J(1,20)\mu_{12} + [10J(1,20)\mu_{10} + 10J(1,22)\mu_{12}]\,P_2(\cos\theta)\}] \qquad (6.81)$$

Equation (6.81) has the same form as obtained in the McMillan theory if the following identifications are made,

$$J(0,22) \rightarrow \frac{1}{25}\beta V_0$$

$$J(1,22) \rightarrow \frac{1}{50}\beta V_0 \alpha$$

$$J(1,00) \rightarrow \frac{1}{2}\beta \alpha \delta V_0$$

$$J(1,20) \rightarrow 0$$

Thus here an extra cross term between μ_{10} and μ_{12} appears which is absent in the McMillan theory of smectic A.

The density functional theory has been used to study the S_AN transition in a system of parallel hard spherocylinders[105,66–68] as well as system with orientational degrees of freedom[69]. Mahato et al.[70] developed a density wave theory which involves the direct correlation function of ellipsoids of revolution and showed that the S_A phase is metastable with respect to the bcc crystal in such a system. Mulder[66] made an attempt to locate the NS_A transition in a system of perfectly aligned hard spherocylinders (PAHSC) by using a bifurcation analysis of the free-energy functional in the second virial coefficient approximation and obtained a second-order transition of mean field type towards a smectic phase. He also studied the effect of higher order terms in the density expansion on the location of the bifurcation point by considering the influence of the third and fourth order terms. The values of critical packing fraction and wavelength are in good agreement with the simulation values[95] as compared to the results obtained by Hosino et al.[54] by using the method of symmetry breaking potential in the second virial coefficient approximation,

$$\eta^*_M \approx 0.37, \quad \eta^*_{MC} \approx 0.36, \quad \eta^*_H = 0.729$$

$$\lambda^*_M \approx 1.34, \quad \lambda^*_{MC} \approx 1.27, \quad \lambda^*_H = 1.414.$$

A more systematic calculation was reported by Somoza and Tarazona[68,105]. These authors constructed a free-energy functional for a system of parallel hard spherocylinders[68] as well as for a system of hard bodies with arbitrary shape and orientational distribution[105]. A free-energy functional was constructed[68] by generalizing the Lee's functional scaling method[106] of a hard sphere reference system to the parallel hard ellipsoids (PHE) and then to use the later as the reference system for the real hard bodies (HB). The interaction part of the free-energy functional expressed as

$$\Delta A[\rho(\mathbf{r},\Omega)] = \int d\mathbf{r} \int d\Omega\, \rho(\mathbf{r},\Omega)\, \Delta\psi_{PHE}[\overline{\rho}(\mathrm{r})]$$

$$\times \frac{\int d\mathbf{r}' \int d\Omega' \rho(\mathbf{r}',\Omega') M_{HB}(\mathbf{r}-\mathbf{r}',\Omega,\Omega')}{\int d\mathbf{r}' \rho(\mathrm{r}') M_{PHE}(\mathrm{r}-\mathrm{r}')} \qquad (6.82)$$

was evaluated at the effective density $\overline{\rho}(\mathrm{r})$. Here M_{PHE} and M_{HB} are the respective Mayer functions, which give the second virial coefficients by integration over all the variables. A criterion has to be defined for choosing the reference PHE. It should reflect both the molecular shape and the orientational distribution function. An empirical rule was proposed[68,105] on the basis of the tensor of inertia of the HB, $\langle \mathbf{I}^{HB}(\Omega) \rangle$, averaged over the orientations with the function $\rho(\mathbf{r}, \Omega)$,

$$\langle \mathbf{I}^{HB} \rangle = \int d\Omega\, \rho(\mathbf{r},\Omega)\, \mathbf{I}^{HB}(\Omega) / \rho(\mathbf{r}) \qquad (6.83)$$

The length of the PHE along the principal axes are taken so that the eigenvalues of its inertia tensor are proportional to the corresponding eigenvalues of the hard body tensor of inertia :

$$\frac{I_1^{PHE}}{\langle I_1^{HB}\rangle} = \frac{I_2^{PHE}}{\langle I_2^{HB}\rangle} = \frac{I_3^{PHE}}{\langle I_3^{HB}\rangle} \tag{6.84}$$

These equations together with the equal volume condition $V_{PHE} = V_{HB}$, fully specify the PHE. Equation (6.82) may be regarded as a way to study the general HB system as a perturbation from the PHE system, for which the direct mapping onto HS can be used. The density profiles and the free-energy of S_A phase were evaluated numerically and a second order NS_A phase transition was observed. The treatment was extended[105] to a system of HB with arbitrary shape and orientational distribution. The phase diagram was evaluated for a system of parallel hard spherocylinders, and its dependence only on the ratio L/d_0 was observed. The use of different volumes for the PHSC and PHE as to give the same close packing density together with Eq. (6.84) gave the best results for the molecules of intermediate elongation ($2 \le L/d_0 \le 5$). The use of equal volume for the reference and real molecules and eq. (6.84) underestimates the critical density by about 10%. The continuous character of the NS_A phase transition for the PHSC provided an easy way to calculate the phase diagram based on the analysis of the stability of the nematic against a density modulation. In order to achieve this, Somoza and Tarazona[105] calculated the direct correlation function for the homogeneous phase from the second functional derivative of eq. (6.82). From the Fourier transform of this direct correlation function $C(q_s)$, the structure factor $S(q_s)$ was obtained which, contrary to the case of simple fluid depends not only on the modulus of the wave vector (q_s) but also on its direction (relative to the nematic director). The numerical results show that the inclusion of the contribution to the direct correlation function coming from the real HB Mayer function changes this qualitatively because it induces a nontrivial dependence of $C_{HB}(q_s)$ with the direction of the wave vector. A stable S_A phase is observed (see Fig. 6.7) in an interval of densities between the nematic and the crystal phases. This provides an insight into how the slight difference in the shape between the PHE and the PHSC may lead to such dramatic changes in the phase diagram.

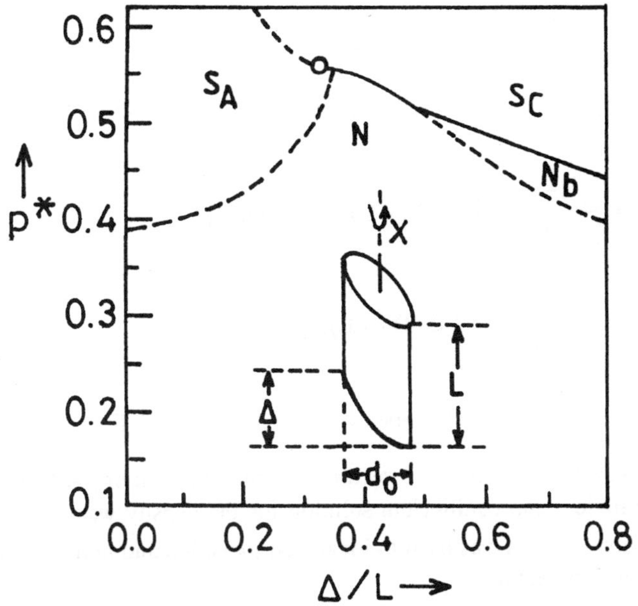

Fig. 6.7. Phase diagram for a system of parallel hard oblique cylinders. N, S_A, S_C and N_b phases may appear. The dashed lines are continuous transitions and the full lines first-order transitions. The circle separates first and second order phase transitions between the S_A and S_C phases.

Poniewierski and Holyst[69] used density functional theory to study the NI and NS_A phase transitions in systems of hard spherocylinders of length L and diameter D_0 with full translational and orientational degrees of freedom. The free-energy functional was constructed in the spirit of the smooth density approximation (SDA) as developed by Tarazona[107] and also by Curtin and Ashcroft[108] for the inhomogeneous hard sphere fluids. Both phase transitions, NI and NS_A were observed in a wide range of length-to-width ratio $x_0 = (=L/D_0)$ and the results tend to Onsager's limit of $x_0 \to \infty$. In an another paper these authors[67] applied the SDA theory to study the NS_A transition for the system of perfectly aligned hard spherocylinders. The free-energy functional was constructed as

$$\frac{\beta A}{N} = \ell n(\rho \wedge^3) - 1 + \frac{1}{2\pi} \int_0^{2\pi} \varphi(\xi)\, \ell n\, \varphi(\xi)\, d\xi + \frac{1}{2\pi} \int_0^{2\pi} \varphi(\xi)\, \Delta\psi(\rho\overline{\varphi}(\xi))\, d\xi \quad (6.85)$$

where

$$\overline{\varphi}(\xi) = 1 + \sum_{n=1}^{\infty} \overline{\mu}_n \omega(n\mathbf{k}) \cos(n\xi) \quad (6.86)$$

and

$$\omega(n\mathbf{k}) = \left[\frac{4\pi}{(n\mathbf{k})^3}\{\sin[n\mathbf{k}(D_0 + L)] - \sin(n\mathbf{k}L)\}\right.$$

$$\left. - \frac{4\pi Do}{(n\mathbf{k})^2}\cos[n\mathbf{k}(D_0 + L)]\right]/v_0 \quad (6.87)$$

For two parallel spherocylinders $v_0 = 2\pi LD_0^2 + \frac{4}{3}\pi D_0^3$, $\overline{\mu}_n$ is the nth order parameter and **k** the smectic wave vector. The minimization of βA/N with respect to $\overline{\mu}_n$ and **k** provides the equilibrium solution for the density wave of the S_A phase. The bifurcation analysis was used to locate the transition and numerical calculations were performed by taking

$$\Delta\psi(\rho) = k_B T \frac{\eta(4-3\eta)}{(1-\eta)^2} \quad (6.88)$$

Using bifurcation analysis the transition density ρ^* at which the nematic solution becomes unstable with respect to the perturbations of the symmetry of the S_A phase was obtained and the following asymptotic relation for the difference between the free energies of the smectic and nematic phases was arrived at.

$$\Delta A_{S_A N} \sim -(\rho_s - \rho^*)^2$$

The variation of the transition density ρ^* / ρ_{cp} (ρ_{cp} is the close packing density) with L/D_0 was studied. It was found that ρ_s^* / ρ_{cp} ranges from 0.31 for $x_0 \to \infty$ to 0.41 for $x_0 = 0.5$ which agrees with the simulation results[95] within 25%. Further, the NS_A transition occurs for all the values of x_0 whereas in simulations NS_A transition is not observed for $x_0 \leq 0.25$.

6.5 POLYMORPHISM IN S_A PHASE

Mesogenic molecules with long aromatic cores and strongly polar head groups exhibit a rich smectic A polymorphism[109–114]. Since the discovery of first

$S_{A_1} - S_{A_2}$ (monolayer-bilayer) transition by Sigaud et. al.[109] in 1979, seven different smectic phases have been identified in pure compounds or in binary mixtures. Extensive experimental studies over the past two decades lead to a considerable amount of knowledge about the structures, phase diagrams and physical properties of these phases. A variety of structures arises from the asymmetry of the molecules; in addition to the usual phases exhibited by the symmetric and nonpolar molecules, the long-range organization of the position of the polar heads generates a number of additional phases[1,2,117] (Fig. 6.8).

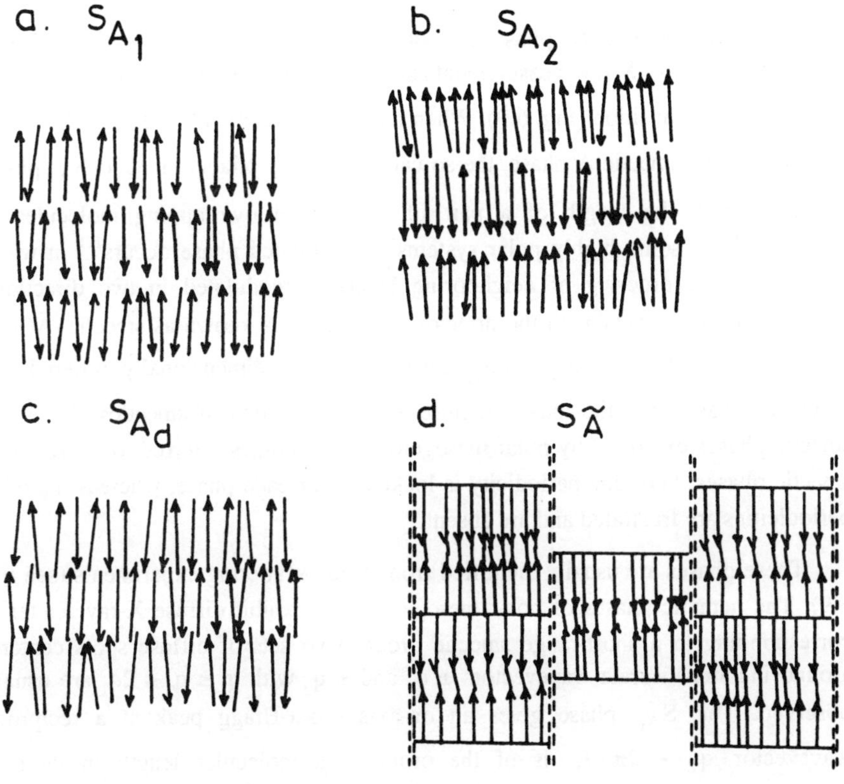

Fig. 6.8. Schematic representation of the four types of S_A phases formed by polar molecules : monolayer S_{A_1}, bilayer S_{A_2}, partially bilayer S_{A_d} and antiphase $S_{\tilde{A}}$.

The first evidence[109] of the S_A polymorphism was found in the DB5/TBBA (DB5 ≡ 4-pentyl-phenyl 4-cyanobenzoyloxy benzoate, TBBA ≡ Terephtal-bis-4n-butylaniline) binary phase diagram that was detected by calorimetry, but could not be observed optically. X-ray studies confirmed that this was a transition between two forms of the S_A phase; the higher temperature phase is called the monolayer (S_{A_1}) phase and the lower temperature one the bilayer (S_{A_2}) phase. When in a S_A (or S_C) phase the dipoles are randomly oriented, the asymmetry can be forgotten and a monolayer S_{A_1} (or S_{C_1}) phase is obtained. An antiferroelectric ordering of pairs of layers generates the bilayer S_{A_2} (or S_{C_2}) phase. Although the S_{A_1} phase is not different from the usual S_A phase, the subscript 1 indicates the potential ability to exhibit S_{A_2} phase. X-ray measurements show the doubling of the lattice spacing at the $S_{A_1} - S_{A_2}$ transition (Fig. 6.9). When DB5 is replaced with DB7 (i.e. C_5H_{11} by C_7H_{15}), a third smectic A phase, the so called partially bilayer smectic S_{A_d} phase, identified with the length of a pair of antiparallel overlapping molecules, is exhibited[110]. In some highly polar systems, a modulated phase occurs[111] in which antiferroelectric ordering of neighboring layers is modulated in two directions, parallel and perpendicular to the director. This phase is known as smectic $\tilde{A}(S_{\tilde{A}})$ phase (Fig. 6.8d). The $S_{\tilde{A}}$ and $S_{\tilde{C}}$ phases are two dimensionally ordered like columnar phases, but the local structure is similar to that of smectics. All of the smectic phases exhibited by polar mesogens are sometimes referred to as frustrated smectic phases; here one periodicity is locked in for each phase, whereas the other periodicities are frustrated and are absent.

These phases are usually identified from X-ray scattering experiments (see Fig. 6.9). The nematic phases of polar mesogens often exhibit diffuse X-ray scattering corresponding to a short-range smectic order. Two sets of diffuse spots centered around incommensurate wavevectors $\pm q_1$ and $\pm q_2$ with $q_1 \le q_2 \le 2q_1$ are usually observed. The S_{A_1} phase gives an on-axis quasi-Bragg peak at a reciprocal wavevector $q_2 \sim 2\pi/\ell$, ℓ is of the order of a molecular length in its most extended configuration. The S_{A_2} phase produces a quasi-Bragg peak at a reciprocal wavevector of π/ℓ, i.e., the period is twice that of the molecule. The S_{A_d} phase has a peak at a reciprocal wavevector $q_1 \approx 2\pi/\ell'$ where $\ell < \ell' < 2\ell$. All these peaks are on-axis. The $S_{\tilde{A}}$ phase shows weak, diffuse off-axis peaks. In fact, these spots can be seen in the S_{A_1} phase, near the transition to the $S_{\tilde{A}}$ phase[112]. The S_C phases corresponding to these S_A phases produce similar X-ray reflections, but the tilt of the director relative to the layer normal shifts the peaks off-axis. X-ray studies on highly polar smectic mesogens have also revealed the existence of phases with more than one periodicity. In the incommensurate smectic A phase, S_{Ainc} , peaks indicating both S_{A_d} and S_{A_2} ordering are present[113]. However, some experiments have shown[112,115,116] that these phases are probably broad two-phase regions due to extremely slow conversion of one phase to the other. The experimental phase diagrams (see Refs. 1 and 117) of polar materials are usually drawn in axes temperature-pressure or temperature-concentration in binary mixtures. All the phase structures as described above could not be observed in one single material, but most of the phase diagrams fit in a common topology; N_u, S_{A_1}, S_{A_2} and S_{A_d} form the generic phase diagram of polar mesogens. If the biaxial phases $S_{\tilde{A}}$ and $S_{\tilde{C}}$ are present, they occupy position between S_{A_1} and S_{A_d}. The incommensurate structures are observed in triangular region surrounded by S_{A_1}, S_{A_2} and S_{A_d} .

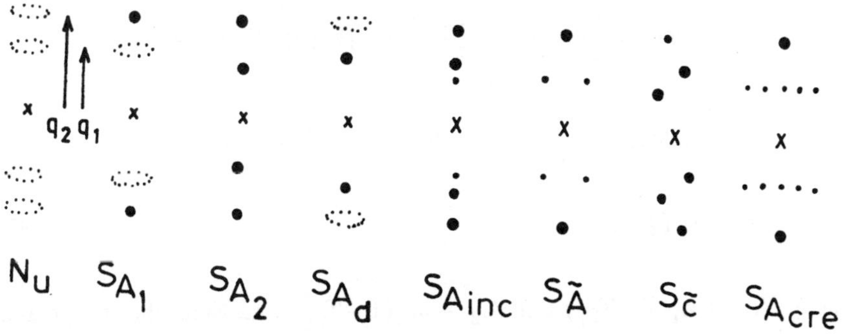

Fig. 6.9. Typical X-ray scattering patterns for polar S_A phase.

6.5.1 Frustrated Smectics Model

Prost[118] showed that the properties and structures of frustrated smectics can be described by two coupled order parameters. The first one $\rho(\mathbf{r})$ measures the mass density wave familiar in traditional S_A phases[2], and the second $P_z(\mathbf{r})$, the polarization wave, describes long- range head-to-tail correlations of asymmetric molecules along the Z-axis (normal to the smectic layers). In order to describe the appearance of modulated order two complex fields ψ_1 and ψ_2 are introduced :

$$P_z(\mathbf{r}) = \mathrm{Re}\,(\psi_1(\mathbf{r})) \tag{6.89a}$$

$$\rho(\mathbf{r}) = \mathrm{Re}\,(\psi_2(\mathbf{r})) \tag{6.89b}$$

with

$$\psi_1(\mathbf{r}) = |\psi_1|\, e^{i(\mathbf{Q}_1\cdot\mathbf{r} + \phi_1)} \tag{6.90a}$$

$$\psi_2(\mathbf{r}) = |\psi_2|\, e^{i(\mathbf{Q}_2\cdot\mathbf{r} + \phi_2)} \tag{6.90b}$$

The rule of the game is to build the phenomenological Landau free-energy density expansion with ψ_1 and ψ_2; the expansion up to fourth order reads[117].

$$\begin{aligned} f = {} & \frac{1}{2}A_1|\psi_1|^2 + \frac{1}{2}D_1\left|(\Delta + q_1^2)\psi_1\right|^2 + \frac{1}{2}C_1|\boldsymbol{\nabla}_\perp\,\psi_1|^2 \\ & + \frac{1}{2}U_1|\psi_1|^4 + \frac{1}{2}A_2|\psi_2|^2 + \frac{1}{2}D_2\left|(\Delta + q_2^2)\psi_2\right|^2 + \frac{1}{2}C_2|\boldsymbol{\nabla}_\perp\,\psi_2|^2 \\ & + \frac{1}{2}U_2|\psi_2|^4 + \frac{1}{2}U_{12}|\psi_1|^2|\psi_2|^2 - \omega\,\mathrm{Re}\,(\psi_1^2\,\psi_2^*) \\ & + \omega'\,\mathrm{Re}\,(\psi_1\,\psi_2^*) \end{aligned} \tag{6.91}$$

where $A_1 = a_1(T - T_{C_1})$ and $A_2 = a_2(T - T_{C_2})$ measure the temperatures from the non-interacting mean-field transition temperatures T_{C_1} and T_{C_2} of the fields ψ_1 and ψ_2. $\boldsymbol{\nabla}_\perp$ is the in-layer gradient operator. The terms in D_1 and D_2 favour $Q_1^2 = q_1^2$ and $Q_2^2 = q_2^2$, respectively. The coupling term $\omega\,\mathrm{Re}\,(\psi_1^2\,\psi_2^*)$ favours lock in condition $2\mathbf{Q}_1 = \mathbf{Q}_2$ which is appropriate to the case $\ell' = 2\ell$. The last linear coupling term $\omega'\,\mathrm{Re}\,(\psi_1\,\psi_2^*)$ dominates in the strong overlapping limit $\ell' = \ell$. It is impossible to satisfy simultaneously all these tendencies (incommensurability and lock-in) and hence the name frustrated smectics[118] is used.

Minimization of free-energy density (6.91) with respect to the smectic amplitudes $|\psi_1|$ and $|\psi_2|$ and the wavevectors $\mathbf{Q}_1$ and $\mathbf{Q}_2$ gives a mean-field theory for this model with very rich phase structures[117–119]. The following phases are possible :

$|\psi_1| = |\psi_2| = 0$; nematic phase, the director $\hat{\mathbf{n}}$ defines the Z-axis.

$|\psi_1| = 0, |\psi_2| \neq 0$ and $\mathbf{Q}_2 = q_2$ defines the S_{A_1} phase.

$|\psi_1| \neq 0, |\psi_2| \neq 0$ and $\mathbf{Q}_2 = 2\mathbf{Q}_1$ defines the S_{A_2} phase.

$|\psi_1| \neq 0, |\psi_2| \neq 0$ and at least $\mathbf{Q}_1$ non collinear with $\mathbf{Q}_2$ defines the biaxial layered structures.

$|\psi_1| \neq 0, |\psi_2| \neq 0$ and modulated phases defines the uniaxial modulated structures.

The mean-field analysis of the Prost's model (6.91) was carried out by Barois[117] by reducing the number of external parameters through an appropriate rescaling of the variables. It has been found that the model exhibits very rich phase structures depending on the relative values of ℓ and ℓ', and the incommensurability parameter $\tilde{Z} \sim (q_1^2 - q_2^2/4)\omega$. The incommensurability parameter turns out to be a relevant quantity and basically controls the frustration. The elastic coefficients of tilt γ_1 and γ_2 control the appearance of the biaxial phases $S_{\tilde{A}}$ and $S_{\tilde{C}}$. Certain interesting features were noticed from the calculated[117] phase diagrams (Fig. 6.10). For a small $\tilde{Z}$ and the symmetric fourth order coefficients in the expansion, the calculated phase diagram (Fig. 6.10a) is similar to the very first diagram calculated by Prost[118] in which incommensurability was not considered. The phase diagram (Fig. 6.10a) is characterized by a second-order NS_{A_1} line terminating to a mean-field critical end point Q where the N, S_{A_1} and S_{A_2} phases meet. A second order NS_{A_2} line terminates at a tricritical point P. The NS_{A_2} line QP is first-order and continues into the smectic region as a first-order $S_{A_1}S_{A_2}$ line. Beyond a tricritical point R, the $S_{A_1}S_{A_2}$ line is second-order. Due to the coupling term ω, the phase of the bilayer order parameter ψ_1 is locked in the monolayer smectic S_{A_1} so that its amplitude only is critical at the $S_{A_1}S_{A_2}$ transition. It is therefore expected to be in the Ising universality class[118,120]. For higher values of $\tilde{Z}$ and/or asymmetric fourth order terms[121], a new phase boundary separating two S_{A_2} phases appears (Fig. 6.10b).

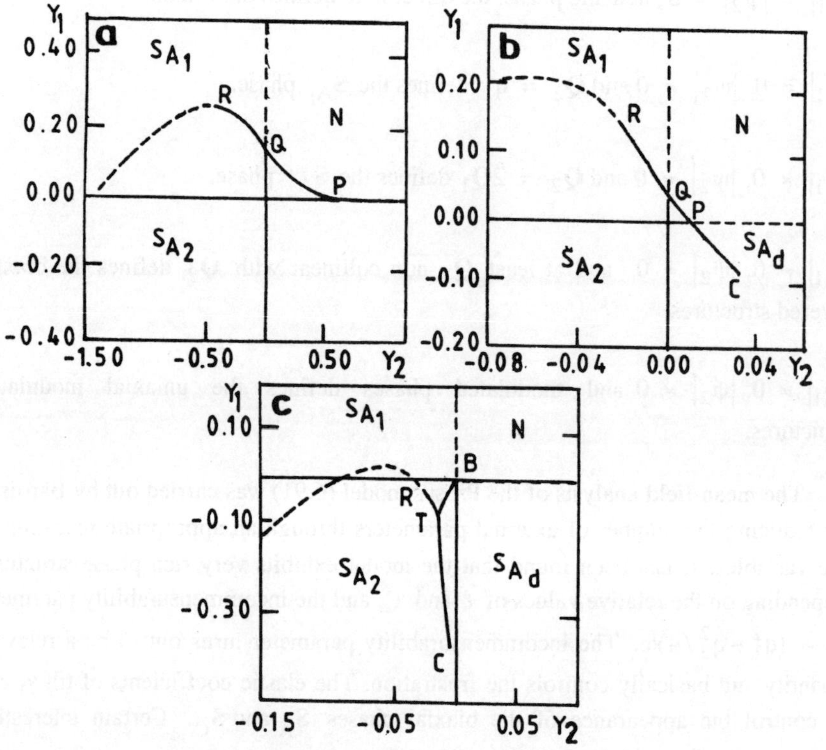

Fig. 6.10. Calculated phase diagrams in mean-field for (a) weak, (b) medium and (c) strong incommensurability parameter $\tilde{Z}$ (Ref. 117).

The two S_{A_2} phases are distinguished by different values of the amplitudes of order parameters; one of the phases is identified as the partially bilayer S_{A_d} phase. The new phase boundary is tangent to the first order $N S_{A_2}$ line in P and terminates at a critical point C where the jump in wave vector goes to zero. The new critical point C does not belong to the Ising universality class[122]. When the incommensurability parameter $\tilde{Z}$ is further increased, a new $S_{A_1} S_{A_d}$ line appears (Fig. 6.10c) terminating at a mean-field bicritical point B where the N, S_{A_1} and S_{A_d} phases meet and a S_{A_1}, S_{A_2} and S_{A_d} coexisting triple point T is obtained.

In case of a weak/medium incommensurability parameter a biaxial domain opens up in case the elastic coefficient $\gamma_1 < 2\tilde{Z}$. The $S_{A_2} S_{\tilde{A}}$ transition is always first-order, whereas the $S_{A_1} S_{\tilde{A}}$ line is second-order or first-order within mean-field with a tricritical point. It meets the first-order $S_{A_1} S_{A_2}$ line at a triple point. At the higher values of $\tilde{Z}$, the biaxial domain reaches the S_{A_d} region. The tilted antiphase $S_{\tilde{C}}$ is stable in-between S_{A_d} and $S_{\tilde{A}}$ so that two new lines, a first-order $S_{\tilde{C}} S_{A_d}$ and a second-order $S_{\tilde{A}} S_{\tilde{C}}$ are expected.

In summary to this section, it may be concluded that the mean-field analysis of the Prost's model of frustrated smectics makes it possible to explain most of the experimental observations on polar smectics. However, it fails to describe correctly those situations where fluctuations play an important role (these points are not covered here).

6.6 SMECTIC C PHASE

6.6.1 Elasticity of Smectic C Phase

To construct a continuum theory for the S_C phase, we have to take into account two distinct types of degrees of freedom[1,2,20a]; the orientational fluctuations of the director about the layer normal (Z-axis) and the distortions of the layers themselves. Expressions for the former was given by Saupe[123] but the complete theory including both the contributions and also the coupling between the two was worked out by Orsay group[124]. Let us choose a cartesian coordinate system such that the projection of the mean molecular direction on the basal (X,Y) plane is along the Z-axis. It is convenient to use simultaneously the three rotation angles as,

$$\Omega_x = \frac{\partial u}{\partial y}; \qquad \Omega_y = -\frac{\partial u}{\partial x}; \qquad \Omega_z \,. \tag{6.92}$$

Here Ω_x and Ω_y represent, respectively, rotations about the X-and Y-axes. Equation (6.92) implies that

$$\frac{\partial \Omega_x}{\partial x} + \frac{\partial \Omega_y}{\partial y} = 0 \tag{6.93}$$

Since uniform rotations do not change the free-energy, ΔA_e must be a function of $\mathbf{\nabla\Omega}$. If the unperturbed interlayer distance has its equilibrium value, the terms linear in $\mathbf{\nabla\Omega}$ and $\partial u/\partial z$ do not contribute to the ΔA_e. With all these provisos, the free-energy density of elastic distortion may be expressed as

$$\frac{1}{V}\Delta A_e = \frac{1}{2}A\left(\frac{\partial \Omega_x}{\partial x}\right)^2 + \frac{1}{2}A_{12}\left(\frac{\partial \Omega_y}{\partial x}\right)^2 + \frac{1}{2}A_{21}\left(\frac{\partial \Omega_x}{\partial y}\right)^2$$

$$+\frac{1}{2}B_1\left(\frac{\partial\Omega_z}{\partial x}\right)^2 + \frac{1}{2}B_2\left(\frac{\partial\Omega_z}{\partial y}\right)^2 + \frac{1}{2}B_3\left(\frac{\partial\Omega_z}{\partial z}\right)^2$$

$$+ B_{13}\left(\frac{\partial\Omega_z}{\partial x}\frac{\partial\Omega_z}{\partial z}\right) + C_1\left(\frac{\partial\Omega_x}{\partial x}\frac{\partial\Omega_z}{\partial x}\right)$$

$$+ C_2\left(\frac{\partial\Omega_x}{\partial y}\frac{\partial\Omega_z}{\partial y}\right) + \frac{1}{2}B\left(\frac{\partial u}{\partial z}\right)^2 \quad (6.94)$$

Here the terms A, A_{12}, A_{21} describe curvature distortion of the smectic planes and are analogous (for a system of monoclinic symmetry) to the splay terms occurring in S_A. The terms B_1, B_2, B_3, B_{13}, introduced by Saupe[123], represent the director distortions when the smectic planes are unperturbed. C_1, C_2 describe the coupling between the above types of distortions. The B term is associated, as in S_A, with the compression of the layers. The Saupe coefficients (B_1, B_2, B_3, B_{13}), the A coefficients (A, A_{12}, A_{21}) and C coefficients (C_1, C_2) have the dimensions of energy per unit length and are approximately of the same order of magnitude (~ 10^{-6} dyn) as that of the N_u elastic constants. The B coefficient has the dimension of energy per unit volume as in S_A. The fluctuations of the director are evidently related to the fluctuations in $\mathbf{\Omega}$. Equation (6.94) and the equipartition theorem give for a general wavevector **q**

$$<|\Omega_z(q)|^2> = \frac{k_B T}{B_1 q_x^2 + B_2 q_y^2 + B_3 q_z^2 + 2B_{13} q_x q_z} \quad (6.95)$$

From the light scattering studies on monodomain samples of the S_C phase, it has been confirmed[125] that the intensity of scattering arising from the director fluctuations in the vertical (scattering) plane (the k_e, k'_e configuration, e refers to extraordinary) is extremely weak. It is quite large due to fluctuations normal to the scattering plane (the k_e, k'_0 or k_0, k'_e configuration). For the k_e, k'_0 configuration,

$$(B_1 \cos^2\theta + B_2 \sin^2\theta + 2B_{13}\sin\theta\cos\theta)q^2 I = B(\theta)\, q^2 I = \text{constant} \quad (6.96)$$

where $\tan\theta = q_z/q_x$ and I is the intensity. A plot of $q_x\sqrt{I}$ versus $q_z\sqrt{I}$ should give an ellipse which has been verified experimentally.

6.6.2 Phase Transitions Involving S_C Phase

This section covers the smectic A to smectic C (S_AS_C) transition and the NS_AS_C multicritical point. Only a brief summary is presented.

6.6.2.1 Smectic A – smectic C transitions

The S_AS_C transitions is characterized by the onset of a tilt in a one dimensional layered matrix. A number of experiments[2] have been carried out using a variety of techniques (calorimetry, optical, X-ray scattering, ESR, NMR, neutron scattering, etc.) which confirmed the earlier prediction by Taylor et al[126] that the S_AS_C transition is continuous. Many of these experiments, however, probably do not reliably obtain the asymptotic behaviour near the transition. The critical fluctuations for the S_AS_C transitions are those in the molecular tilt. Most of these transitions show mean-field behaviour for the same reason as superconductors do; the intrinsic coherence lengths are long and reduce the fluctuations until very close to T_{AC}. The mean-field value of the critical exponent β characterizing the temperature dependence of the tilt angle in the S_C phase was clearly shown. A second order S_AS_C transition in a 3D space has been discussed by Kats and Lebedav[127] using renormazation group technique.

For a given set of layers, to describe the smectic C order, one must specify the magnitude ω of the tilt angle, and also the azimuthal direction of tilt, specified by an angle ϕ (see Fig. 6.2). The tilt angle ω of the molecules can point in any azimuthal direction; there are thus two independent components : $\omega_x = \omega \cos\phi$ and $\omega_y = \omega \sin\phi$. Thus due to the azimuthal degeneracy of the tilt angle one can use a complex order parameter ψ_1 (= $\omega \exp(i\phi)$) to describe the transition. If one neglects the coupling to the layers undulation mode, a remarkable analogy is seen with the superfluid helium. An overall change of the phase ϕ does not modify the free-energy. This leads to the following possibilities :

The S_CS_A transition may be continuous. The specific heat should show[2] a singularity,

$$\delta C_p = \text{const.} + A^{\pm}|t|^{-\alpha}$$

Below T_{CA} the tilt angle should obey the law

$$\omega = \text{const}\,|t|^{\beta} \tag{6.97}$$

where $\alpha = -0.007$ (too small to measure), and with $\beta \approx 0.35$ and $t = (T-T_{CA})/T_{CA}$. Above T_{CA} the application of a magnetic field that is oblique with respect to the optic axis of the smectic phase induces a tilt angle

$$\omega = C_1 \frac{\chi_a H_x H_z}{k_B T_{CA}}\, t^{-\gamma} \tag{6.98}$$

where $\gamma \approx 1.33$ is the susceptibility critical exponent. The predicted tilt is small (about 10^{-2} rad for $t \approx 10^{-4}$ near room temperature) because the diamagnetic energy $\chi_a H^2$ is very weak as compared to k_BT_{CA}. Starting from the S_A phase and decreasing

T toward T_{CA}, one expects to observe the onset of a strong (depolarized) light scattering due to fluctuations in the tilt angle. However, the regime of critical fluctuations may lie outside of the experimental observation. The Ginzburg criterion indicates that for this transition a reduced temperature ($\sim 10^{-5}$) may be necessary to observe critical phenomena. The experimental data show mean-field Landau behaviour with a large six order term[128].

The application of Landau theory to the S_AS_C transition was first considered by de Gennes[129], who treated the tilt angle as the relevant order parameter. It has two components : the magnitude $|\omega|$ and the azimuthal angle ϕ. This shows an analogy with the superfluid-normal fluid transition. So as usual the free-energy density of S_C phase can be expressed as

$$f_C = A|\omega|^2 + \frac{1}{2}D|\omega|^4 + \frac{1}{3}E|\omega|^6 \tag{6.99}$$

The total free-energy involves the order parameter due to the density wave, $|\psi_1|$, and its coupling with the orientational order parameter **Q**. It has been shown[130], from an analysis of the specific heat anomaly near the second-order S_AS_C transition, that the 6th order term in eq. (6.99) is unusually important.

Several theories for the S_C phase were developed taking into account the specific features of the molecules. Wulf[131] proposed a steric model which considers, a zig-zag shape for the molecules and assumes that the tilted structure results due to freezing of free-rotation around the long-axis. It was shown that the steric effects in particular, the zig-zag gross shape of the smectogenic molecules, may be able to account for the second order S_AS_C phase transitions. In the S_C phase the biaxial order parameters play a primary role and may grow to values of the order of 10^{-1} McMillan[132] took note of the molecular transverse dipoles which are present in all the smectogenic compounds and developed a model for the S_AS_C transition in which the molecules rotate freely about their long axes in the S_A phase leading to the dipolar ordering. Incorporating the dipole-dipole interaction of the permanent molecular dipole moments the theory[132] predicts three orientationally ordered phases, one with the physical properties of the S_C phase (tilted director, optically biaxial, second-order S_AS_C phase transition), the second a two-dimensional ferroelectric, and the third a low temperature ordered phase which is both tilted and ferroelectric. If there are two oppositely oriented outboard dipoles which are away from the geometric center of the molecules, the medium will not exhibit ferroelectric properties. The presence of the two dipoles with longitudinal components also favours the tilting of molecules in the layers to minimise the energy of the oriented dipoles. However, NMR and other experiments clearly show that the molecules are practically freely rotating about their long-axes in the S_C phase in disagreement with the assumptions of both Wulf and McMillan. Cabib and Benguigui[133] made an attempt to overcome this problem by postulating a special type of molecular structure in which only the longitudinal components of two symmetrically placed outboard dipoles are effective. However, most of the mesogenic compounds do not have such a structure. Based on the argument that many compounds have only one outboard dipole moment Van der Meer and Vertogen[134] proposed that in such cases the tilting optimises the attractive energy due to dipole-induced dipole interaction. This interaction remains effective even if the molecules rotate freely. Priest[135] developed a model for the nematic, smectic A and smectic C phases in which the intermolecular interactions are characterised by the second-rank tensor quantities and are supposed to produce the orientational order of the molecules. The model predicts that the free rotation is possible for uniaxial molecules and that the extent of this rotation being hindered depends on the degree of biaxility in certain intrinsic molecular second-rank tensors. According to the model the uniaxial molecules can exhibit S_C state. The model predicts a second-order S_AS_C transition with two independent degrees of freedom having divergent fluctuations. With the decreasing temperature the tilt angle of S_C phase is predicted to grow continuously from zero and to asymptotically saturate at 49.1°. However, the model fails to identify the physical origin of the proposed interactions. The model considered by Matsushita[136] incorporates the excluded volume effects due to the hard rod features of the molecules which actually favour the S_A phase. If the mesogenic molecules are uniaxial these models do not give rise to a biaxial order parameter in the S_C phase. They describe only a tilted S_A phase rather than the S_C phase with its intrinsically biaxial symmetry. Goossens[137] has shown that none of the above models using attractive interactions are satisfactory. In these molecular models the tilt angle does not appear as a natural order parameter. Only in the models of Wulf[131] and McMillan[132] the S_C phase is characterized by new order parameters. But they are unsatisfactory because of stringent requirement that rotations about the long-axes of the molecules should be frozen out. A detailed calculation of the intermolecular interactions between molecules of ellipsoidal shape has been carried out by Goossens[138]. The attractive potential arising from the anisotropic dispersion energy and the permanent quadrupole moments has been considered. The mean-field potential depends strongly on the anisotropy of the excluded volume. It has been shown that the relative weights of the 3 terms proportional to $\langle \cos qz \rangle^2$, $\langle \cos (qz)\, p_2(\cos\theta) \rangle$ and $\langle \cos qz \rangle \langle \cos qz\, p_2(\cos\theta) \rangle$ depend on the x_0 ratio of the ellipsoidal molecules. Further, from the calculations for a tilted director, about which the orientational order is still considered to be uniaxial, it was shown that the extra mean-field energy contains a term proportional to $\sin^2 \omega$. The angle of tilt is proportional to a new order parameter which goes to zero at a temperature which can be identified with T_{AC}. Gießelmann and Zugenmaier[139] have developed a model for the S_AS_C phase transition which is, in principle, analogous to a ferromagnetic phase transition with spin number $s \rightarrow \infty$. Assuming a bilinear mean-field potential, the macroscopic tilt angle has been calculated by Boltzmann statistics as a thermal average of the molecular tilt. The calculation gives an equation of state for the S_C phase which is self-consistent field equation involving the Langevin function of a reduced tilt and a reduced temperature. An excellent agreement with the experimental results has been obtained.

6.6.2.2 The nematic-smectic A-smectic C (NS_AS_C) multicritical point

The existence of NS_AS_C multicritical point was shown independently by Sigaud et al[140] and Johnson et al[141]. It is the point of intersection of the NS_A, S_AS_C and NS_C phase boundaries in a thermodynamic plane, e.g., in the T-P or T-y (concentration) diagram. All three phase transitions are continuous in the vicinity of it and at the point itself the three phases are indistinguishable[142]. It has been observed in the T-y diagram of binary liquid crystal mixtures and also in the T-P diagram of a single component mesogenic material[140,143]. High resolution studies have been carried out[143,144] in both T-y and T-P planes and it is now accepted that the topology of the phase diagram in the vicinity of NS_AS_C multicritical point exhibits universal behaviour (Fig. 6.11). It has been found that the analysis of the phase boundaries gives identical exponents for both the T-y and T-P diagrams showing the universal behaviour of NS_AS_C point. Further, in the NS_A case only two components of the mass density wave exhibit large fluctuations near T_C, whereas at the NS_C transition "Skewed" cybotactic groups (i.e. S_C type fluctuations) are concentrated on two rings in the reciprocal space. Thus the natural order parameter has an infinite number of components. In the description of the NS_AS_C point one should be free to move from one type of fluctuations to the other.

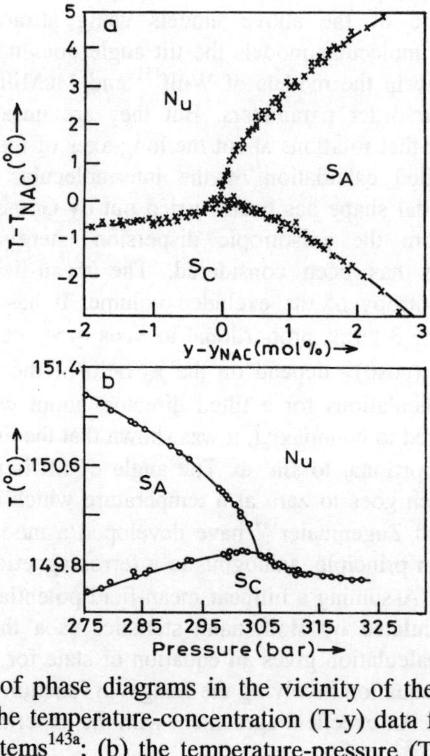

Fig. 6.11. Topology of phase diagrams in the vicinity of the NS_AS_C multicritical point (a) the temperature-concentration (T-y) data for four binary liquid crystal systems[143a]; (b) the temperature-pressure (T-P) data for a single component system[143b].

The phase boundaries obey simple power laws

$$T_{NA} - T_{NAC} = A_{NA} |X - X_{NAC}|^{\phi_1} + B(X - X_{NAC}) \quad (6.100a)$$

$$T_{NC} - T_{NAC} = A_{NC} |X - X_{NAC}|^{\phi_2} + B(X - X_{NAC}) \quad (6.100b)$$

$$T_{AC} - T_{NAC} = A_{NC} |X - X_{NAC}|^{\phi_3} + B(X - X_{NAC}) \quad (6.100c)$$

Here X is either the pressure or the concentration. All experimental data are consistent with $\phi_1 = \phi_2 \approx 0.57 \pm 0.03$ and $\phi_3 \approx 1.4$–1.7. The NS_A and NS_C lines have the same slope parallel to the temperature axis at the NS_AS_C point, whereas that of S_AS_C line is orthogonal. An alternative fit has been proposed by Anisimov[144] in which $\phi_1 = \phi_3 \approx 0.67 \pm 0.03$ and $\phi_2 \approx 0.87 \pm 0.04$. With this fit, all phase boundaries are tangent and parallel to the temperature axis at the NS_AS_C point which can only be fortuitous and approximate.

There have been several theoretical descriptions[41,130,145–148] on the NS_AS_C point. The framework of the Chen-Lubensky model[146] seems to correspond more closely to the experiment. It is in some way a generalization of de Gennes model[129] of the NS_A and NS_C transition. However, the Landau theory is not able to shed any light on the physical origin of the tilt. A simple model based on the quadrupolar nature of the molecules was proposed[149] which made use of the idea that a gradient in the scalar orientational order parameter $\overline{P}_2$ implies that of the quadrupole density, and hence leads to an order electric polarization of the medium[147]. The resulting dielectric self energy due to associated order electric polarization is given by

$$f_{oe} = \frac{1}{2}\alpha^2\overline{P}_2^2 |\psi_1|^2 \left(\cos^2\omega' - \frac{1}{3}\right)^2 \quad (6.101)$$

where α is related with the quadrupole moment of each molecule and the dielectric constant along the Z-axis, ω' the orientation of the principal axis of the quadrupole tensor of the medium with reference to the direction of the gradient in the scalar order parameter $\overline{P}_2$. Equation (6.101) will be minimised when $\cos^2\omega' = \frac{1}{3}$. This mechanism operates only in case of a layered structure, i.e., in the smectic phase. The total free-energy density is expressed as

$$f = \frac{1}{2} a(T - T_{AN})|\psi_1|^2 + \frac{1}{4} D|\psi_1|^4 - \frac{1}{4} C_1|\psi_1|^2\left(\cos^2\omega' - \frac{1}{3}\right)$$

$$+\frac{1}{2}\alpha^2|\psi_1|^2 \overline{P}_2^2\left(\cos^2\omega' - \frac{1}{3}\right)^2 \quad (6.102)$$

The relative stability of the S_C and S_A phases is determined by the ratio $\overline{P}_2(T)\alpha^2 / C_1$. As temperature decreases, $\overline{P}_2(T)$ increases. It has been found that the calculated α^2 / C_1 versus $(1-T/T_{NI})^{1/2}$ phase diagram looks similar to the experimental one close to the NS_AS_C point.

Grinstein and Toner[150] applied the renormalization group technique and considered the application of a dislocation loop model to the NS_AS_C multicritical point and made a striking prediction : Four, rather than three, phases meet at the point where S_AN and S_AS_C phase boundaries cross. A new, biaxial nematic, phase was found to intervene between the nematic and smectic C phases. It exhibits the orientational long-range order of the S_C phase, whereas the translational properties are those of the nematic. However, the layers have only short-range positional order. These four phases meet at a decoupled tetracritical point. This prediction is supported by a fluctuation corrected mean-field theory of Lubensky[151]. The high resolution specific heat measurements of Wen et al[152] have shown anomalous variations near the S_CN transition very close to the NS_AS_C point. They suggested that these anomalous variations may be due to the biaxial fluctuations.

6.7 REENTRANT PHASE TRANSITIONS (RPT) IN LIQUID CRYSTALS

An exception to the phase sequence rule (1.1) (chapter 1) and observed sequences (Tables 1.4–1.6) of phase transitions in liquid crystals was discovered, at atmospheric pressure, in the year 1975 by Cladis[153] in certain strongly polar materials. In a binary mixture of two cyano compounds, HBAB {p–[p–hexyloxybenzylidene)–amino] benzonitrile} and CBOOA[N–p–cyano-benzylidene–p–n–octyloxyaniline], over a range of compositions, the following sequence was observed on cooling the mixture from the isotropic phase :

$$IL \to N_u \to S_{A_d} \to N_R \to \text{Solid} \qquad (6.103)$$

where N_R stands for a second nematic, known as reentrant nematic, which appears at lower temperature. The smectic-A phase existing between two nematics was identified as the partially bilayer phase S_{A_d}. A similar result was reported by Cladis et al[154] in many other binary mixtures and in a pure compound COOB (4-cyano-4'-octyloxy biphenyl) at high pressures (see Fig. 6.12). These authors also observed that the pure compounds CBNA(N-p-cyanobenzylidene-p-nonylaniline), CBOA(N-p-cyanobenzylidene-p-octylaniline) and COB(4-cyano-4'-octylbiphenyl) did not show reentrant behaviour for pressures under 10 Kbar, although they exhibit bilayer smectic A phase.

X-ray and microscopic studies[155] show that the reentrant nematic phase is quite different from the classical (higher-temperature) nematic phase. The transition from smectic-A to reentrant nematic phase is reversible from the point of view of X-ray and optical results. The possibility that the reentrant nematic is actually a smectic-C phase was excluded[153,158] on the experimental grounds; N_R phase is uniaxial whereas S_C is biaxial. The defect structure of reentrant nematic observed in cylindrical geometry was identical to that observed for the classical nematic. For mixtures forming a N_R phase the birefringence[156] has been found to be continuous at both the transitions NS_A and S_AN_R. The magnitude of the increase at the NS_A transition point and the decrease on going from the S_A phase to the N_R phase has been found to depend on the length of smectic A range and tend to zero as this vanishes.

The above example of the phase transition (6.103) shows that the high symmetry phase may re-enter at lower temperature than low symmetry phase in a rather unexpected way. This kind of phenomenon is termed as "re-entrant phenomenon." A system is said to be exhibiting reentrant phase transition (RPT) if a monotonic variation of any thermodynamic field results in two (or more) phase transitions and finally attains a state which is macroscopically similar to the initial state or the system re-enters the original state. The phenomenon of RPT is intrinsically novel and the continued interest in this problem[157] is underlined by its discovery in amazingly diverse systems in addition to liquid crystals, e.g., binary gases[158–160], liquid mixtures[161–165], ferroelectrics[166], organometallic compounds[167], granular superconductors[168], gels[169], aqueous electrolytes[170], antiferromagnets[171] etc., The subject is nicely reviewed by Cladis[172] and very recently by the author[173] for liquid crystals, and by Narayan and Kumar[157] for multicomponent liquid mixtures.

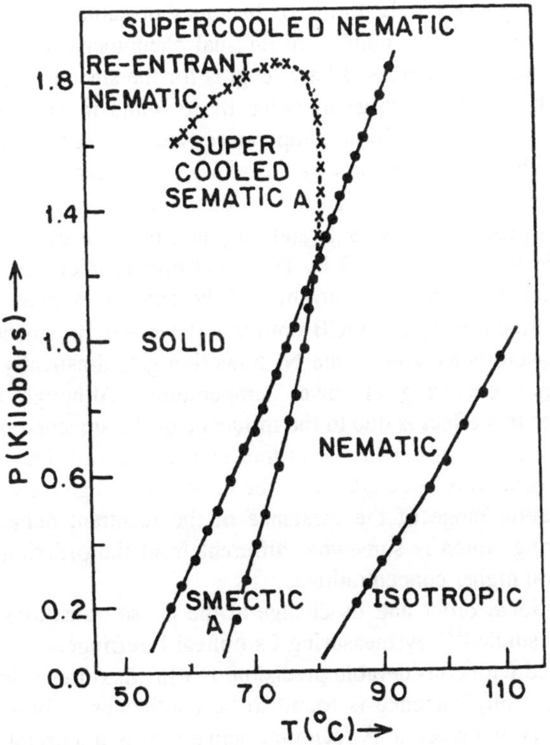

Fig. 6.12. PT phase diagram for COOB[154]. Data taken on the N_RS_A transition in the supercooled liquid are shown as crosses (X).

The discovery of reentrant behaviour[153] in liquid crystals has resulted in extensive studies[157,172–174], both experimental and theoretical, of this intricate kind of phenomena. Much of the experimental work has been concerned with the synthesis of the other reentrant systems, the establishment of their phase diagrams as a function of temperature, pressure and composition and the determination of orientational order and its change at the phase transitions. The main concern of theoretical work has been to understand the microscopic origin and the true nature of RPT. A few representative examples of RPT in liquid crystals are illustrated below.

6.7.1 Examples of Single Reentrance

There are quite a few studies[155,165,175–182] on the binary mixture of hexyloxycyanobiphenyl (6OCB) and 4'-n-octyloxy-4-cyanobiphenyl (8OCB) which exhibit nematic reentrance as a function of temperature, pressure and composition. Kortan et al[180] reported a high resolution X-ray study of the smectic A fluctuation in the nematic phases of this mixture with emphasis on the region of concentration when the smectic phase barely forms or becomes unstable before the correlation length has truely diverged. A number of unusual phenomena were observed in this region, e.g., dramatically increased bare lengths for the smectic correlation, critical exponents which may be as large as twice those found in pure 8OCB crossover effects, etc. The phase transition temperature was measured by observing the temperature at which the minimum in the width of q_1 scans occurred rather than treated as adjustable parameter. Figure 6.13 shows the phase diagram so obtained. The influence of pressure on the S_AN and NI phase boundaries were studied[178] as a function of mole fraction of 6OCB by DTA and optical microscopy. It was found that the maximum pressure of occurrence of the smectic A phase decreases with increasing mole fraction (y) of 6OCB until y $\approx$ 0.3 where no smectic phase exists. Further, the pressure behaviour of the NI transition gets drastically affected by the structural changes occurring at lower temperatures. Although they could not ascertain whether this effect is due to the influence of the smectic ordering or due to the presence of a reentrant nematic at farther temperatures or due to a combination of both, it was tentatively concluded that the classical or high temperature nematic, in the concentration range of the existence of the reentrant nematic, possesses a molecular ordering which is somewhat different from the ordering of the namatic phase occurring at higher concentrations.

The orientational order and its change at the phase transitions for the 6OCB/ 8OCB has been studied[156] by measuring its optical birefringence, a quantity which can be determined with considerable precision. For the mixtures exhibiting reentrant nematic phase the birefringence is found to be continuous at both the transitions; nematic-smectic A and smectic A-reentrant nematic. It is important to note that the birefringence in the smectic A phase is found to be greater than that obtained by an appropriate extrapolation from the results in the nematic phase. Contrary to it birefringence of the N_R phase is observed to be less than that determined by extrapolation from the smectic A phase. The magnitude of the increase in case of nematic-smectic A transition and the decrease on going from smectic A to reentrant nematic phase are found to depend on the range of stability of smectic A phase. Emsley et al[182] investigated in detail the orientational behaviour of this mixture using deuterium NMR spectroscopy, a technique which is able to provide information about the orientational order of the individual components of a mixture. More importantly this technique can determine the order parameters for the rigid subunits in a mesogenic molecules. This variation in the order parameter has been determined in the mixture 6 OCB/8 OCB as well as in the pure components as a function of temperature. The order parameters were found to undergo subtle changes at the transition from the smectic A to the reentrant nematic phase.

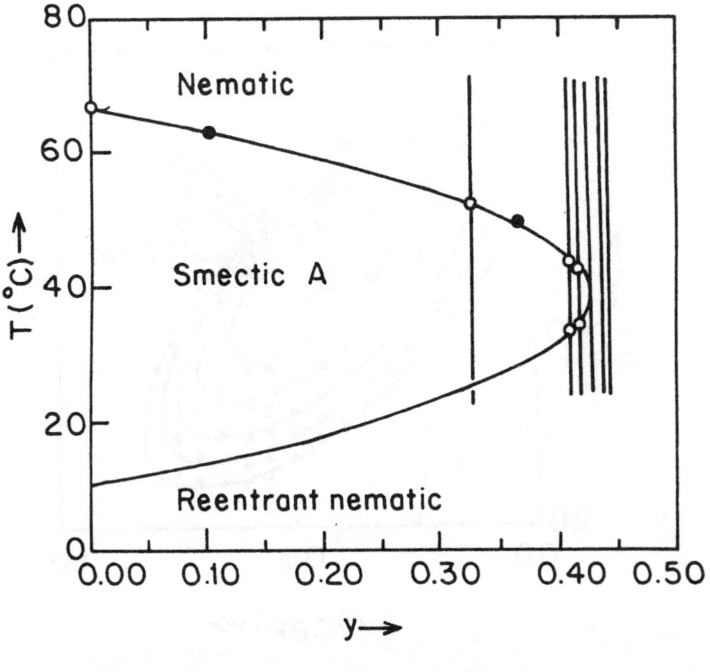

Fig. 6.13. The phase diagram for 6OCB/8OCB mixture[180]. y is the molecular ratio 6OCB : 8OCB. The open and solid circles represent, respectively, the data obtained by X-ray scattering and light scattering experiments. The vertical lines indicate the experimental path.

One of the important features of the RPT in liquid mixtures is the existence of a closed-loop phase diagram with upper and lower critical solution temperatures (T_u and T_L, respectively). A double critical point (DCP) results when the T_u and T_L are made to coincide. Such systems provide richer information as they permit a multitude of paths by which a critical point can be approached. A wide variety of phases can be obtained in these systems by mere variation of temperature, pressure, composition, additional components, etc.. Perhaps the first closed-loop existence curve was obtained by Hudson[183] for nicotine/water mixture while trying to crystallize nicotine from its aqueous solution. Since then the closed-loop phase diagrams have been reported in many systems[157]. Figure 6.14 shows a closed-loop nematic-smectic A phase boundary observed[184] in mixtures of $\bar{7}$ CBP (4-n-heptyloxy-4'-cyanobiphenyl) and a binary mixture of CBOOA and $\bar{8}$.0.5 (4-n-pentylphenyl-4-n-octyloxy-benzoate) as a function of temperature, pressure and composition. It can be seen that the pressure reduces the area of smectic A phase and the reentrance vanishes (i.e. a DCP is attained) at 145 bar.

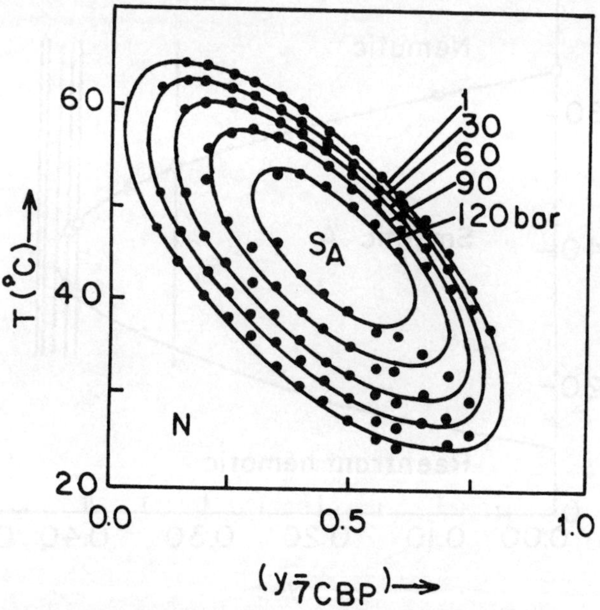

Fig. 6.14. T–y phase diagram of the system [8.0.5/CBOOA(y_{CBOOA} = 0.88)]/ $\bar{7}$ CBP for different pressures.

In principle, the reentrant phases can occur[60a,174,185] in nonpolar systems also. The experimental evidence in support of this prediction has been unveiled by the Halle group[186,187]. A nematic reentrant system of nonpolar compounds, with one component a side-chain LC polymer, has been reported[188] which exhibits this kind of inversion of symmetry. The polymer is an atactic polymethylsiloxane substituted with 4-undecyloxyphenyl ester of 4-methoxybenzoic acid ($P_{11,1}$) which follows the phase sequence.

$$S_C \xrightarrow{60^0C} S_A \xrightarrow{134^0C} IL \qquad (6.104)$$

The second component is a symmetrically substituted compound, the 4'-heptyloxybenzoate of 6-heptyloxynaphthyl-2, exhibiting the order of phase stability

$$S_C \xrightarrow{[64^0C]} N_u \xrightarrow{130^0C} IL \qquad (6.105)$$

Here $S_C \rightarrow N_u$ transition is monotropic.

The reentrant sequence (N_R-S_A-N_u-IL) has been observed[188] in a homologous mixture composed of 60% (weight) of the low molecular weight compound and 40% (weight) of polymer by using optical polarizing microscopy and X-ray analysis. On cooling the mixture the focal conic S_A texture transforms at 70^{0}C into a Schlieren twinkling nematic, which is similar to the high temperature nematic texture above 129^{0}C. This is connected on the X-ray pattern with a change from a resolution limited Bragg-spot to a diffuse scattering at small angles indicative of the loss of the layered structure at long range. It has been found[188] that the layer thickness in the S_A phase of the mixture is noticeably larger than both the molecular length of the mesogenic groups side chain of the polymer (33.5Å) and the length of the low molecular weight compound (37°C). Similar anomalies of periodicity have been observed in the pure polymer $P_{11,1}$ for which the layer thickness increases continuously with decreasing temperature[189].

The reentrant behaviour has also been observed in discotic liquid crystals. In a hexa-n-alkanoates of truxene for the higher homologues the following sequence is observed[190,191], on cooling,

$$IL \rightarrow D_h \rightarrow D_r \rightarrow N_D \rightarrow D_{Rh} \rightarrow \text{crystal} \qquad (6.106)$$

D_{Rh} is the reentrant hexagonal columnar phase. It has been conjectured that the truxene molecules are probably associated in pairs and that these pairs break up at higher temperatures and might be responsible for this extraordinary behaviour.

6.7.2 Examples of Multiple Reentrance

High resolution X-ray scattering and the heat capacity studies of the nematic-smectic A transitions in 4'-(4"-n-alkoxybenzyloxy)-4-cyanostilbene with the alkyl chain of length 7 (T7) and 8 (T8) were reported by Evans-Lutterodt et al.[192]. It was observed that the material T7 exhibits only a single nematic-smectic A_1 (S_{A_1}, monolayer) transition in which the S_{A_1} period is commensurate with the molecular length L whereas T8,

$$C_8H_{17}O - \phi - COO - \phi - CH = CH - \phi - CN$$

exhibits[192,193], with decreasing temperature, the double reentrance sequence

$$IL \rightarrow N_u \rightarrow S_{A_d} \rightarrow N_R \rightarrow S_{A_1} \rightarrow \text{Solid} \qquad (6.107)$$

Here ϕ represents a benzene ring and S_{A_d} period is incommensurate with $d \approx 1.2L$. It has been observed that in the reentrant nematic phase of T8, S_{A_d}, and S_{A_1} fluctuations are essentially independent; the S_{A_d} fluctuations change over from being S_A-like to S_C-like with decreasing temperature in the N_R phase.

Probably the most spectacular case of multiple reentrance has been observed[189,194,195] in the material 4-nonyloxyphenyl-4'-nitrobenzoyloxybenzoate (DB9ONO$_2$), with molecular structure[196]

$$C_9H_{19}O-\phi-OOC-\phi-OOC-\phi-NO_2$$

This compound exhibits three nematic, four smectic A and two smectic C phases in the following sequence, on cooling,

$$IL \rightarrow N_u \rightarrow S_{A_d} \rightarrow N_R \rightarrow S_{A_d} \rightarrow N_R \rightarrow S_{A_1} \rightarrow S_{\tilde{C}} \rightarrow S_{A_2} \rightarrow S_{C_2} \rightarrow \text{Solid} \qquad (6.108)$$

Here $S_{\tilde{C}}$ and S_{C_2} are two different forms of smectic C phase[196]. This unusual behaviour was rationalized in terms of the gradual cross-over from dominant S_{A_d} to dominant S_{A_1} fluctuations with decreasing temperature.

From the molecular point of view, only approximate qualitative explanations of the reentrant behaviour have been possible. As discussed above, the liquid crystal systems exhibiting reentrance consist of organic molecules usually with three or four aromatic rings with ester linkages and having polar cyano or nitro-end groups. Apart from pure compounds reentrant polymorphism, on cooling, have also been shown by binary mixtures of polar-polar, polar-nonpolar and nonpolar-nonpolar compounds. Further, as a homologous series is ascended the reentrant phase sequence is exhibited by the higher homologs which are neither very short nor very long. These experimental observations indicate that the dipolar forces play a crucial role in the reentrant polymorphism. However, the observation of single reentrance in a binary mixture of nonpolar compounds cannot be caused probably because of dipolar forces. Reviews of theories and experiments for the reentrant polymorphism are available[172,173,197,198] which present an account of the subject. No first principle understanding exists yet for the reentrance phenomena observed in condensed phase.

6.8 DYNAMICAL PROPERTIES OF SMECTICS

A nice description of dynamical properties of the smectics is given by de Gennes and Prost (see Ref. 2, chapter 8). Here we choose to present only the basic equations of 'smectodynamics' rather than to give a complete overview of the problem.

6.8.1 Basic Equations of 'Smectodynamics'

We have seen in chapters 4 and 5 that the description of nematics required the introduction of a director $\hat{\mathbf{n}}$, its relevance is directly linked to rotational invariance.

A homogeneous rotation of $\hat{\mathbf{n}}$ does not require any restoring torque and hence the rotated system has an infinite lifetime. Similarly, in a S_A phase, a uniform layer displacement **u** does not create any restoring force; it corresponds to a simple translation and the translated state has an infinite lifetime. $\hat{\mathbf{n}}$ and **u** are known as 'hydrodynamic' variables linked to broken rotational and translational symmetries. Here we consider only the small-amplitude (i.e. long-wavelength) distortions.

As we have seen in case of a nematic, the main task is to write down the equations for the entropy source. This leads to the introduction of a set of linearly related (in the lowest order approximation) fluxes and forces. Another central issue is to write a set of conservation (mass, momentum and energy) equations,

$$\frac{\partial\rho}{\partial t} + \partial_\alpha(\rho v_\alpha) = 0 \qquad (6.109a)$$

$$\frac{\partial}{\partial t}(\rho v_\alpha) - \partial_\beta\, \sigma_{\alpha\beta} = 0 \qquad (6.109b)$$

$$\frac{\partial \in}{\partial t} + \partial_\alpha\, j_\alpha = 0 \qquad (6.109c)$$

Equations (6.109b) and (6.109c) can be regarded as defining the stress tensor $\sigma_{\alpha\beta}$ and the energy flux j_α; $\in$ is the energy density.

Local equilibrium implies (per unit volume)

$$d\in = \mu d\rho + Td\Sigma + v_\alpha d(\rho v_\alpha) + \sigma^u_{\alpha\beta} d(\partial_\beta u_\alpha) - h_\alpha d\Omega_\alpha \qquad (6.110)$$

The first three terms of eq. (6.110) are usual terms common to any system, whereas the last terms correspond to the existence of translational and rotational order. Since the energy variations can result only from gradients in the displacement variable, the fourth term involves $d(\partial_\beta u_\alpha)$. On the same argument, for the angular variable Ω the correct expression would be

$$h_{\alpha\beta}\, d(\partial_\beta \Omega_\alpha) \qquad (6.111)$$

The equivalence of eqs. (6.110) and (6.111) requires that $h_\alpha = \partial_\beta h_{\alpha\beta}$. Here $\sigma^u_{\alpha\beta}$ and h_α are the fields conjugate to $\partial_\beta u_\alpha$ and Ω_α.

Following closely the approach of Martin et. al[199], and de Groot and Mazur[200] and using eqs. (6.109) and (6.110) (neglecting higher order and surface terms), the entropy production can be expressed as

$$T\frac{\partial\Sigma}{\partial t} = -\frac{\partial_\alpha T}{T}\left(j^{\in}_\alpha - (\in^0 + P)\, v_\alpha\right) + (\sigma_{\alpha\beta} + P\delta_{\alpha\beta})\partial_\beta v_\alpha$$
$$- \sigma^u_{\alpha\beta}\, \partial_\beta \dot{u}_\alpha + h_\alpha \dot{\Omega}_\alpha \qquad (6.112)$$

where

$$\dot{u}_\alpha = \partial u_\alpha/\partial t, \qquad \dot{\Omega}_\alpha = \partial\, \Omega_\alpha/\partial t.$$

Since $\dot{u}_\alpha$ and $\dot{\Omega}_\alpha$ each contain a reactive (i.e. reversible, non-dissipative) and a dissipative part, the eq. (6.112) still requires to be modified. The dissipative part is the one that is expressed in the 'flux-force' linear relations. The reversible part has time-reversal properties opposite to those of the dissipative part, and to lowest order, can be written as

$$\dot{u}_\alpha = \dot{u}^r_\alpha + \dot{u}^d_\alpha; \quad \dot{\Omega}_\alpha = \dot{\Omega}^r_\alpha + \dot{\Omega}^d_\alpha \tag{6.113}$$

Here superscripts r and d refer, respectively, to the reversible and dissipative parts. Since u_α and Ω_α are even under time reversal, $\dot{u}^r_\alpha$ and $\dot{\Omega}^r_\alpha$ have to be odd; only v_α satisfies this requirement. Thus

$$\dot{u}^r_\alpha = A_{\alpha\beta}\, v_\beta + A_{\alpha\beta\gamma}\, \partial_\gamma\, v_\beta + \ldots\ldots \tag{6.114a}$$

$$\dot{\Omega}^r_\alpha = B_{\alpha\beta}\, v_\beta + B_{\alpha\beta\gamma}\, \partial_\gamma\, v_\beta + \ldots\ldots \tag{6.114b}$$

where As and Bs are coefficients that characterize the phase that we describe. In view of Galilean invariance

$$A_{\alpha\beta} = \delta_{\alpha\beta}, \qquad B_{\alpha\beta} = 0$$

For systems having inversion symmetry, $A_{\alpha\beta\gamma} = 0$. Thus

$$\dot{u}^r_\alpha = v_\alpha$$

and the entropy source (6.112) can be written in terms of the dissipative fluxes

$$T\dot{\Sigma} = -\frac{\partial_\alpha T}{T}\, j^{\in d}_\alpha + \sigma^d_{\alpha\beta}\, \partial_\beta v_\alpha + \partial_\beta\, \sigma^u_{\alpha\beta}\, \dot{u}^d_\alpha + h_\alpha \dot{\Omega}^d_\alpha \tag{6.115}$$

where

$$j^{\in d}_\alpha = j^{\in}_\alpha - (\in^0 + P)v_\alpha \tag{6.116a}$$

$$\sigma^d_{\alpha\beta} = \sigma_{\alpha\beta} + P\,\delta_{\alpha\beta} - \sigma^u_{\alpha\beta} + B_{\gamma\alpha\beta}\, h_\gamma \tag{6.116b}$$

$$\dot{u}^d_\alpha = \dot{u}_\alpha - v_\alpha \tag{6.116c}$$

$$\dot{\Omega}^d_\alpha = \dot{\Omega}_\alpha - B_{\alpha\beta\gamma}\, \partial_\gamma\, v_\beta \tag{6.116d}$$

In eq. (6.115) the first term on the right-hand side represents the dissipation due to heat transport and the second term describes the friction effects. The third term, that corresponds to dissipation due to 'permeation' in the Helfrich sense, is non-zero when $\dot{u}_\alpha \neq v_\alpha$, i.e., when there is a mass transport through the structure. The fourth term describes the dissipation due to rotational motion of the angular variables (tilt direction, bond angle, nematic director).

It is important to mention that $\sigma^u_{\alpha\beta}$ is found to contribute to the reactive part of the stress tensor. In a solid, a columnar phase, or a smectic A it is, in fact, equal to that stress as expected. In S_C and $B_{B(hex)}$ phases there are angular contributions also. In the absence of an external field, **H** = 0, a symmetric stress tensor can be chosen through a suitable transformation.

The flux-force relations can be written as (see Ref. 2, chapter 8),

$$\sigma^d_{\alpha\beta} = \eta_{\alpha\beta\gamma\delta}\, \partial_\delta\, v_\gamma \,, \qquad (6.117a)$$

$$j^{\in d}_\alpha = K_{\alpha\beta}\, E_\beta + \mu_{\alpha\beta}\, g_\beta \,, \qquad (6.117b)$$

$$\dot{u}^d_\alpha = \dot{u}_\alpha - v_\alpha = \mu_{\alpha\beta} E_\beta + \lambda_{\alpha\beta}\, g_\beta \,, \qquad (6.117c)$$

$$\dot{\Omega}^d_\alpha = \dot{\Omega}_\alpha - B_{\alpha\beta\gamma}\, \partial_\gamma v_\beta = \nu_{\alpha\beta}\, h_\beta \,, \qquad (6.117d)$$

In these relations $j^{\in d}_\alpha$, $\sigma^d_{\alpha\beta}$, $\dot{u}^d_\alpha$ and $\dot{\Omega}^d_\alpha$ are chosen as fluxes, whereas $E_\alpha (= -\frac{\partial_\alpha T}{T})$, $\partial_\beta v_\alpha, g_\alpha (= \partial_\beta\, \sigma^u_{\alpha\beta})$ and h_α are the respective forces. $\eta_{\alpha\beta\gamma\delta}$ is the usual viscosity tensor containing 5 independent components in all uniaxial systems, 13 in S_C and 21 in a low symmetry crystal. $K_{\alpha\beta}$ is the thermal conductivity with 2 components in uniaxial systems, 4 in S_C and 6 for low symmetry crystal. $\mu_{\alpha\beta}$ describes a kind of Soret effect. $\lambda_{\alpha\beta}$ is the permeation tensor; it has one component in S_A, $S_{B(hex)}$, S_C and hexagonal columnar (D_h) phases, 2 in biaxial columnar (D_r,D_0) phases, and 6 in low symmetry crystal. $\nu^{-1}_{\alpha\beta}$ is the generalization of the twist viscosity of nematics. There are no dissipative cross-couplings between the stress and the g_α, E_α and h_α forces due to time-reversal symmetry. Conversely, velocity gradients are not coupled to $j^{\in d}_\alpha$, $\sigma^d_{\alpha\beta}$, $\dot{u}^d_\alpha$, $\dot{\Omega}^d_\alpha$ via dissipative terms. Equations (6.109), (6.116) and (6.117) describe the dynamical behaviour of the system, in the limit of small amplitude, if P, T, g_α, h_α are expressed as functions of the slow hydrodynamic variables. For the fast processes a convenient choice is ρ, Σ, v_α, u_α, Ω_α, while for the slow processes it is ρ, T, v_α, u_α, Ω_α.

A detail description of the specific cases (for example, S_A, D_h, $S_{B(hex)}$, S_C) and flow properties is given by de Gennes and Prost[2]; we shall not describe here.

6.9 COMPUTER SIMULATIONS OF PHASE TRANSITIONS IN LIQUID CRYSTALS

Computer simulations have played a key role in developing the understanding of phase transitions and critical phenomena in liquid crystals. A survey of the existing numerical studies of NI, NS_A, S_AS_B, S_AN_R transitions and transitions to the discotic phase is given in this section. The primary issues addressed in these simulation studies for model systems and the main results obtained from these are summarized[5,20,72,73,88, 95,201–249] here.

The models studied in numerical simulations of mesophase transitions may be classified into two categories called as molecular models and field models. Four broad classes of molecular models have been developed[240]; these are Lasher lattice model, hard-particle models, single site soft potentials (Gay-Berne model) and atomistic models. In the molecular models, the description is in terms of the molecules constituting the system and their interactions. The interactions included in these models must describe both translational and orientational orders. In reality, the mesogenic molecules are complicated objects consisting of rigid cores and flexible side chains. As a result, it is quite difficult to construct a model that provides a realistic description of the interaction between two such molecules. Even if one could construct a realistic model for the intermolecular interactions, it would be extremely difficult and time-consuming to simulate the properties of a liquid crystalline system. Owing to these reasons, the simulations are usually carried out for much simpler models in which the molecules are assumed to be simple non-spherical rigid objects such as rotational ellipsoids, spherocylinders, parallel plate cylinders, cut spheres, etc. The interaction between two such molecules is also assumed to have a simple form. Most of the simulations carried out for such models assume only a hard-core repulsion between two molecules[221], arising from the excluded volume interactions. A posteriori justification for the study of such models comes from the observation that these models do exhibit some of the phases that are found in real liquid crystals. The principle of universality provides another justification for such studies which states that the parameters characterizing the critical behaviour near a continuous phase transition do not depend on the microscopic details of the system and are determined by a few factors such as the dimensionality of space and the symmetry of the order parameter. Few attempts[222] have also been made to include in the simulation model the van der Waals interaction and dipolar interactions in an approximate way. The field models provide a coarse-grained description of the system in terms of an order parameter field appropriate for the ordered state under consideration. A model of this kind is defined by a Ginzburg-Landau free-energy expressed as a functional of the order parameter field. In some cases, due to symmetry or other considerations, terms coupling the order parameter to other non-ordering fields may have to be included in the free-energy functional. A typical example is that of de Gennes model[78] for the NS_A transition. It is important to note that unlike molecular models, a field model is specific only to the transition in which long-range order described by the particular order parameter field sets in and it cannot be used to describe other phase transitions which may be exhibited by the same physical system. The order parameter field appearing in models of this kind are continuous functions of the spatial coordinates. The results obtained from the numerical studies of such models can be compared directly with those obtained from the analytic calculations.

In computer simulation studies for the phase transitions in liquid crystals, the aim is to calculate the equilibrium thermodynamic properties of a classical system described by the appropriate Hamiltonian or free-energy functional. There are two categories of methods for carrying out such simulations : Monte Carlo (MC) methods[250] and molecular dynamics (MD) methods[251]. In the MC simulations, equilibrium properties are calculated as ensemble averages computed with the Boltzmann probability distribution function. In the MD simulations, the equilibrium properties are calculated as time averages computed over a part of a trajectory of the system in phase space; the relevant equations of motion are the force equations describing the center-of-mass motion and the torque equations describing the rotation of the molecules. The details of these methods are well documented[239,250,251].

Typical MD and MC simulations give results for very small systems containing of the order of a few thousand variables. Therefore, a proper analysis of the data obtained from such simulations is necessary for the extraction of meaningful information about the behaviour of macroscopic systems. A question which arises in the study of phase transitions is whether the transition is first-order or continuous. This question has been addressed in simulations by looking for some of the well known features (such as hysteresis and two-phase coexistence) of first-order phase transitions. However, in simulations of small system size it is difficult to distinguish a weakly first-order transition from a continuous one by employing these criteria. For detecting weakly first-order transitions from the simulation data a number of more efficient methods[252] are available.

6.9.1 Lebwohl-Lasher Model

The questions which have to be addressed in simulations of NI transitions are concerned with the (i) determination of the minimal characteristics of molecules and their interactions exhibiting nematic order and (ii) estimation of the magnitudes of the discontinuities shown by various thermodynamic functions (e.g. order parameter, density, internal energy and entropy) at the first-order transition in a 3D system. The first model used in the simulation of NI transition is the Lebwohl-Lasher model[201] which is the lattice version of the MS model of a nematic. It assumes that each site of a simple cubic lattice is occupied by a classical unit vector. The Hamiltonian for the model is defined as

$$H = -J \sum_{<ij>} P_2(\hat{\mathbf{e}}_i \cdot \hat{\mathbf{e}}_j) \qquad (6.118)$$

where the sum is over nearest-neighbour pairs of lattice sites and $J > 0$ measures the strength of the nematic coupling. Because the molecules are fixed on the sites of a lattice, translational motion is absent. The Monte Carlo (MC) simulation carried out by Lebwohl and Lasher[201] showed that this model exhibits a strongly first-order NI transition near $J/k_BT = 0.89$. The order parameter at the transition exhibits a jump from zero to about 0.33, while the transition entropy is very close to 1.09 J. This model has been intensively investigated using MC technique[202,204–207]. Further numerical work on similar models of the NI transition has been persued by others[203,204]. The accuracy of Lebwohl and Lasher's results were improved by Jensen et al.[203] by simulating the behaviour of same model on larger lattices. Luckhurst and Romano[204b] have carried out simulations of a system in which the molecules are not confined to a lattice. These authors assumed that the molecules interact via a MS type anisotropic potential with a Lennard-Jones-type distance dependence and found a weakly first-order NI transition. The influence of an external field on the thermodynamic behaviour of the Lebwohl-Lasher model was studied by Luckhurst et al.[204c]. The results obtained for the dependence of the order parameter and the internal energy on the value of the external field were in qualitative agreement with the predictions of mean-field theory. Zhang and co-workers[208,209] have simulated a simple cubic lattice with upto 28^3 sites. The free-energy function allows the determination of the limits of stability of the nematic and isotropic phases. Zhang et al.[208] estimated that these temperatures are within 5×10^{-4} reduced temperatures of T_{NI} which is in reasonable agreement with the experimental data.

The Lebwohl and Lasher model has also been extended to model nematics in porous media such as aerogels[210,211]. The Hamiltonian for the extended model is given by

$$H = -J\sum_{<ij>} P_2(\hat{\mathbf{e}}_i \cdot \hat{\mathbf{e}}_j) - D\sum_i P_2(\hat{\mathbf{e}}_i \cdot \hat{\mathbf{n}}_i)\ , \qquad (6.119)$$

where D is the strength of coupling to the random axis which mimics the local pore environment selecting a preferred direction for the nematic order within the pore. Cleaver and co-workers[210] studied this model for $D/J = 1$ using MC techniques on lattices upto 64^3 in size. These authors did not find any evidence for a first-order phase transition and their data are consistent with the absence of nematic long-range order at low temperatures.

6.9.2 Hard-Core Models

The first computer simulation of anisotropic hard-core model was carried out by Vieillard-Baron[212] in two dimensions. Frenkel and Mulder[72a] carried out a systematic study of the properties of a three-dimensional system of hard ellipsoids of revolution by using MC simulations. These authors studied the behaviour of this system for values of x_0 lying in the range between 3 and 1/3. The results show a first-order NI transition with a density change of about 2% at the transition only in the range $x_0 > 2.5$ or $x_0 < 0.4$. The phase diagram obtained is shown in Fig. 6.15.

In addition to the isotropic and nematic phases, they also found orientationally ordered and disordered (plastic) crystalline phases in the high-density region of the phase diagram. An interesting feature of the phase diagram is its approximate symmetry between oblate and prolate ellipsoids. However, the values found by Frenkel and Mulder[72a] appear to change for the larger systems studied[224]. Allen and Frenkel[217a] have observed for $x_0 = 3$ pretransitional nematic fluctuations in the isotropic phase. A number of authors[73,225,226] have performed numerical simulations of molecular models of two-dimensional nematics. Frenkel and Eppenga[73b] have carried out simulations to study the thermodynamics of a system of infinitely thin hard needles in two dimensions. They observed at high densities a stable nematic with algebraic decay of orientational correlations. The behaviour of a system of two-dimensional ellipses with aspect ratios 2, 4 and 6 have been simulated by Cuesta and

Frenkel[225]. For the aspect ratios 4 and 6, a stable nematic phase with a power-law decay of orientational correlations was found. While the NI transition in the system with the aspect ratio 6 was found to be continuous, the system with aspect ratio 4 was found to undergo a first-order transition. Denham et al.[226] have studied by MC simulations a two-dimensional version of the Lebwohl-Lasher model. The results obtained show a continuous NI transition with a power-law decay of orientational correlations.

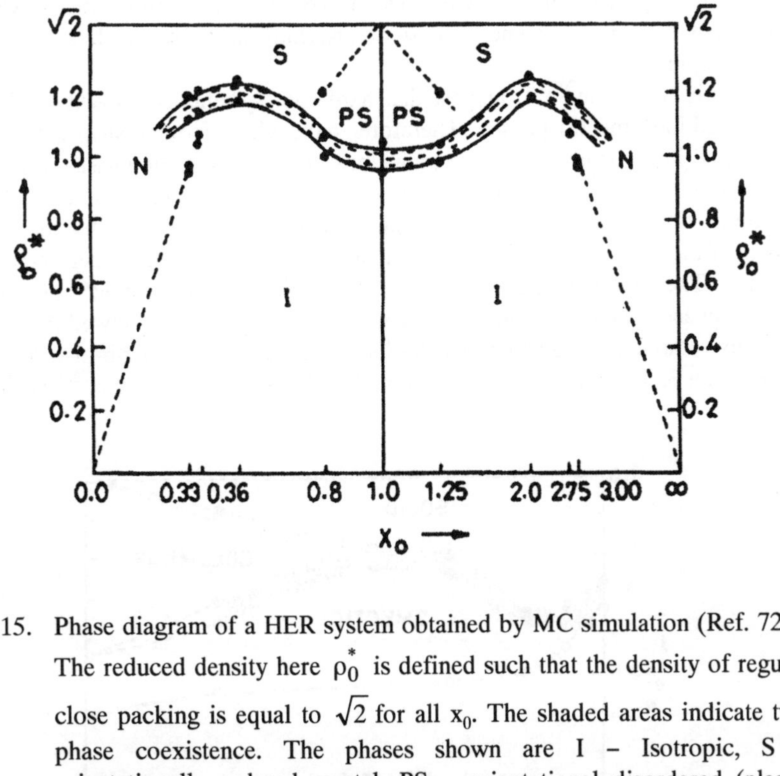

Fig. 6.15. Phase diagram of a HER system obtained by MC simulation (Ref. 72a). The reduced density here ρ_0^* is defined such that the density of regular close packing is equal to $\sqrt{2}$ for all x_0. The shaded areas indicate two phase coexistence. The phases shown are I – Isotropic, S – orientationally ordered crystal, PS – orientational disordered (plastic crystal) and N – nematic phase.

Allen[221a] has carried out a rough survey of the phase diagram of a biaxial phase which can be produced if the molecules are no longer modelled as ellipsoids of revolution, but rather spheroids with unequal semi-axes a, b and c. It has been found that the biaxial phase is stable only for a narrow range of the values of the semi-axes near the critical value $b/a \simeq \sqrt{10}$. Further, an approximate symmetry is observed under the transformation (a,b,c) ↔ (c.ca/b,a). The role of topological defects in the three-dimensional NI transition has been studied by Lammert et al.[227]. These authors considered a system whose Hamiltonian contains, in addition to the microscopic interactions producing nematic ordering, a new term which is related to the extra core energy for the line defects. A phase diagram was obtained in which the NI transition continues to be second order for large but finite values of the core energy of the line defects. Two interesting predictions have been made by this calculation. The first-order NI transition is recovered for a core-energy lower than a critical value. Second prediction is the existence of two different isotropic phases which are distinguished by the free-energy cost of inserting an additional line defects in a given sample.

Simple argument[205,218] suggest that hard-core systems composed of ellipsoids cannot exhibit smectic phases. Therefore, molecules of other shapes have to be considered in the studies of phase transitions involving smectic phases. Frenkel and co-workers[72c,95,218b,223b] have performed extensive MC and MD simulations to study the properties of a system of hard spherocylinder characterized by the different values of the ratio of L/d_0. The first simulation was carried out for perfectly aligned systems. The phase diagram is shown in fig. 6.16. It was found[95] that a stable smectic phase appears for $L/d_0 > 0.5$. The later work[72c,218b,223b] showed that a system of freely rotating spherocylinders exhibits a smectic phase if length-to-width ratio is higher than 3. The phase diagram obtained by Veerman and Frenkel[223b] is shown in Fig. 6.17.

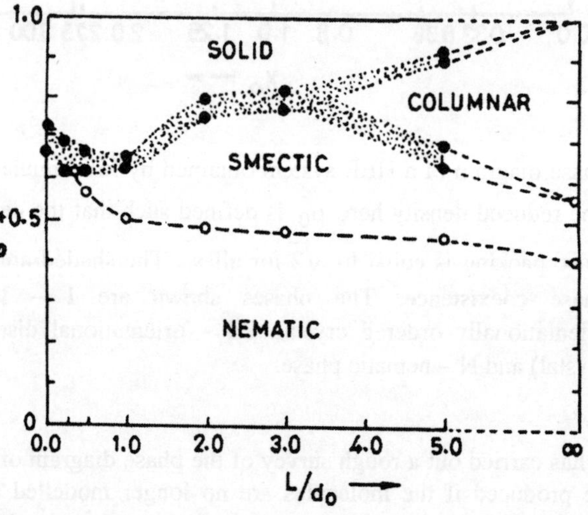

Fig. 6.16. Phase diagram of a system of hard parallel spherocylinder obtained from MC simulation studies (Ref. 95)

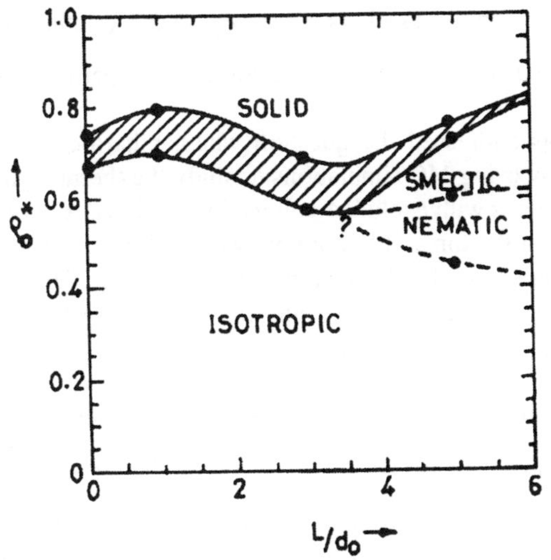

Fig. 6.17. Phase diagram of a system of freely rotating hard spherocylinders. The shaded area is the two-phase region separating the densities of the coexisting solid and fluid phases (Ref. 223b).

The NS_A transition for $L/d_0 = 5$ appears to be continuous in the simulation. However, due to the accuracy of the results a weakly first-order transition cannot be ruled out. The simulation data could not fully resolve the phase diagram near the point indicated by the question mark (?). In this region, the existence of two triple points is suggested, an isotropic-smectic-solid triple point at L/do slightly higher than 3 and an isotropic-nematic-smectic triple point at a higher values of L/d_0 but less than 5. Dasgupta[228] carried out a MC simulation of a discretized version of the de Gennes model, for understanding the nature of the NS_A transition. This model neglects the fluctuation of the magnitude of the smectic order parameter. No sign of a first-order transition was found in these simulations. The observed behaviour is qualitatively consistent with the predictions of finite-size scaling theory for a continuous phase transition. This simulation produces strong evidence for the de Gennes model.

Using an improved version of the frustrated spin gas model[253], Netz and Berker[222] carried out MC simulations and obtained phase diagrams which show the expected nematic re-entrance. With an appropriate choice of the model parameters, the simulations also show the existence of a S_C phase which arises due to a lock-in of molecular permeation and rotation. The onset of the S_C phase is detected by monitoring the tilt angle of the layer relative to the Z-axis. This work suggests a microscopic origin of the S_C phase. Further, the model exhibits some of the modulated smectic A phases (such as S_{A_1}, $S_{\tilde{A}_1}$ and S_{A_d} phases) and indicates the reason for their occurrence in terms of microscopic properties of the molecules and their interactions.

Frenkel[218c] has addressed the question whether hard-core models can exhibit columnar phases by using MC simulations to study the thermodynamic behaviour of a system of cut spheres with excluded volume interactions. A cut or truncated sphere is a sphere cut-off at the top and bottom by two parallel cuts. It is characterized by the ratio L/d_0 where L is the distance between the cuts. Veerman and Frenkel[223a] have studied the cases $L/d_0 = 0.1$, 0.2 and 0.3. For $L/d_0 = 0.1$, the system forms a nematic phase at the reduced density $\rho_0^* \simeq 0.33$ ($\rho_0^* = \rho_0/\rho_{cp}$, where ρ_{cp} is the density for close packing). At $\rho_0^* \simeq 0.5$ the nematic undergoes a strong first-order transition to a columnar phase with $\rho_0^* \simeq 0.53$. At the higher densities $\rho_0^* = 0.8$, a columnar-crystalline transition occurs. At $\rho_0^* = 0.5$ a transition to a cubatic phase having cubic symmetry but no translational order occurs. The nematic order parameter in this phase is zero. For $L/d_0 = 0.3$ both the nematic and cubatic phases are absent and a direct isotropic fluid-solid transition takes place. Frenkel[218c,220] simulated the thermodynamic behaviour for $L/d_0 = 0.1$ and 0.2. The system with $L/d_0 = 0.1$ was found to exhibit an NI transition at the reduced density $\rho_0^* = 0.335$ and a strongly first-order transition to a columnar phase at $\rho_0^* = 0.49$. The transition to crystalline phase occurs for $\rho_0^* > 0.8$. The system with $L/d_0 = 0.2$ did not exhibit any nematic ordering even at higher density of the fluid branch. For $\rho_0^* > 0.58$ strong evidence of cubatic phase was found.

Birgeneau and Litster[229] suggested that the S_B phase found in many liquid crystals may be a realization of a stacked hexatic phases which are characterized by short-positional order and quasi-long-range bond-orientational order. Evidence for the existence of local herring-bone packing of the molecules near the S_AS_B transition has been seen in the X-ray scattering studies[230]. Jiang et al.[231] have studied the thermodynamic behaviour of a model system defined by the reduced Hamiltonian

$$\beta H = \frac{-J_1}{T} \sum_{<ij>} \cos(\Psi_i - \Psi_j) - \frac{J_2}{T} \sum_{<ij>} \cos(\phi_i - \phi_j) - \frac{J_3}{T} \sum_{i} \cos(\Psi_i - 3\phi_i) \tag{6.120}$$

in two dimensions by MC simulation. In eq. (6.120) <(ij)> represents nearest-neighbour pairs of sites on a d-dimensional hypercubic lattice, the two angular variables, Ψ_i and ϕ_i, are located at each lattice site i ($-\pi \le \Psi_i, \phi_i \le \pi$) and the dimensionless coupling constants J_1, J_2 and J_3 are all positive. The values of J_1 and J_3 were kept fixed at 1.0 and 2.1, respectively, and the behaviour of the system as a function of temperature was simulated for different values of J_2. For $J_2 = 0.3$ results of simulations show two transitions. The transition at the higher temperature exhibits a rounded heat capacity peak in accordance with the expectation for a two-dimensional XY transition. At lower temperature the transition exhibits a sharp heat-capacity peak. For $J_2 = 1.4$ a single transition with a broad heat capacity peak is obtained. For intermediate values of J_2 (e.g. $J_2 = 0.85$ and 0.95) a single continuous transition with a sharp heat capacity anomaly is observed.

6.9.3 Gay-Berne Model

The Gay–Berne (GB) potential[254] is proving to be a valuable model for investigating the behaviour of liquid crystals[232–249] using computer simulation methods. It is one of the most useful single site or Corner[255] potentials developed from the Gaussian overlap model[256]. The GB potential was modelled to give the best fit to the pair potential for a molecule consisting of a linear array of four equidistant Lennard–Jones centers with separation of $z\sigma_0$ between the first and fourth sites. It is given by

$$u(\mathbf{r},\hat{e}_1,\hat{e}_2) = 4\in(\mathbf{r},\hat{e}_1,\hat{e}_2)\,(R^{-12} - R^{-6}) \tag{6.121}$$

where

$$R = (r - \sigma(\hat{\mathbf{r}},\hat{e}_1,\hat{e}_2) + \sigma_0)/\sigma_0 \tag{6.121a}$$

$$\sigma(\hat{\mathbf{r}},\hat{e}_1,\hat{e}_2) = \sigma_0\left[1-\chi\frac{(\hat{\mathbf{r}}\cdot\hat{e}_1)^2+(\hat{\mathbf{r}}\cdot\hat{e}_2)^2-2\chi(\hat{\mathbf{r}}\cdot\hat{e}_1)(\hat{\mathbf{r}}\cdot\hat{e}_2)(\hat{e}_1\cdot\hat{e}_2)}{1-\chi^2(\hat{e}_1\cdot\hat{e}_2)^2}\right]^{-1/2} \tag{6.121b}$$

$$\in(\hat{\mathbf{r}},\hat{e}_1,\hat{e}_2) = \in_0\in^{\nu}(\hat{e}_1,\hat{e}_2)\in_1^{\mu}(\hat{\mathbf{r}},\hat{e}_1,\hat{e}_2) \tag{6.121c}$$

with

$$\chi = \frac{(\sigma_e/\sigma_s)^2-1}{(\sigma_e/\sigma_s)^2+1} = \frac{x_0^2-1}{x_0^2+1} \tag{6.121d}$$

$$\in(\hat{e}_1,\hat{e}_2) = [1-\chi^2(\hat{e}_1\cdot\hat{e}_2)^2]^{-1/2} \tag{6.121e}$$

$$\in_1(\hat{\mathbf{r}},\hat{e}_1,\hat{e}_2) = 1-\chi'\left(\frac{(\hat{\mathbf{r}}\cdot\hat{e}_1)^2+(\hat{\mathbf{r}}\cdot\hat{e}_2)^2-2\chi'(\hat{\mathbf{r}}\cdot\hat{e}_1)(\hat{\mathbf{r}}\cdot\hat{e}_2)(\hat{e}_1\cdot\hat{e}_2)}{1-\chi'^2(\hat{e}_1\cdot\hat{e}_2)^2}\right) \tag{6.121f}$$

Here σ_e and σ_s are the separation of end-to-end and side-by-side molecules, respectively. The parameter χ' is determined by the ratio of well depths,

$$\chi' = \frac{(k')^{1/\mu} - 1}{(k')^{1/\mu} + 1} \tag{6.121g}$$

k′ is the ratio of well depths for end-to-end and side-by-side configurations.

There are many different parametrizations of the GB potential which exhibit mesogenic behaviour. In addition to two parameters (σ_0, ϵ_0) which scale, respectively, the distance and energy, the potential (6.121) contains four parameters (x_0, k′, μ and ν). The ratio of the end-to-end and side-by-side contact distance x_0 (= σ_e/σ_s) is related to the anisotropy of the repulsive force and it also determines the difference in the depth of the attractive well between the side-by-side and the cross configurations. The ability of a system to form an orientationally ordered state is influenced by these quantities. The parameter k′ determines the tendency of the system to form a smectic phase. Two other parameters μ and ν appearing as exponents in (6.121c) influence the nematic and smectic forming character of the anisotropic attractive forces in a more subtle manner. As is obvious, there is a seemingly endless variety of GB homologues, each differing by the value chosen for the four parameters (x_0, k′, μ and ν). The GB potential represented by the eq. (4.176) takes the values of μ and ν as 2 and 1, respectively. The best fit to the linear array of four Lennard–Jones centers gives ν = 1, μ = 2, x_0 = 3 and k′ = 5. The original parametrization of the GB potential (ν = 1, μ = 2), or modest variations on the parameters, have proved to be remarkably successful in the study of liquid crystalline behaviour. It has a rich phase behaviour with isotropic, nematic and smectic B phases having been clearly identified[232b,233,241]. In addition, there are claims to have observed a smectic A[233], a tilted smectic B[232b] and a rippled smectic B[243] phase. The properties of the phases, and the transitions between them also appear to be in reasonable agreement with those found for the real mesogens. Such success has contributed to its increasing use and further development; for example, these developments include the introduction of dipolar forces[244], flexibility[245], the construction of more complex particle shapes based on a collection of GB sites[246] and extension to biaxial particles[247].

De Miguel and coworkers[232,234] have performed a complete simulation of the GB potential with its original parametrization. These authors simulated 256 molecules using MD simulations in the canonical (NVT) ensemble. The phase diagram obtained is shown in Fig. 6.18. The NI transition is found to be first-order. It has not been possible to acertain whether the S_B phase is crystalline or hexatic because of the small system size considered in the simulation. The biaxial S_B(t) phase is a tilted version of the S_B phase. These authors[232a] also studied a purely repulsive GB potential, defined by subtracting the attractive part of the potential at a given relative molecular orientation, and observed the NI transition for fixed T at a slightly higher values of ρ_0^*. Further, the S_B and S_B(t) phases do not appear and the system remains nematic even at high densities. This signifies the role of the attractive part of the potential in stabilizing smectic phases. De Miguel and co-workers[234] have also studied the rotational and translational dynamics of the GB fluid in the isotropic and nematic phases. Various autocorrelation functions were studied. The behaviour of translational velocity autocorrelation function indicates that in the nematic the molecules diffuse along cylindrical cages whose long axes are parallel to the director.

Fig. 6.18. Phase diagram for the Gay-Berne fluid obtained from the MD simulation. The dashed lines are extrapolations of the data, $T^* = k_B T/\epsilon_0$ and $\rho_0^* = \rho_0 \sigma_0^2$ (Ref. 232b).

Luckhurst et al.[233] performed an extensive MD simulation of a reparametrized ($\mu = 1$, $\nu = 2$) GB potential in which the side by side configuration is supposed to be more stable relative to the cross and tee configurations, 256 particles were simulated using molecular dynamics. In addition to the isotropic, nematic, smectic B phases, a smectic A phase also appeared. A more systematic parametrization was carried out by Luckhurst and Simmonds[235], who constructed a site-site potential for p-terphenyl. This potential is not uniaxial, but the biaxiality was projected out and then the result was mapped onto the GB potential by examining various configurations of

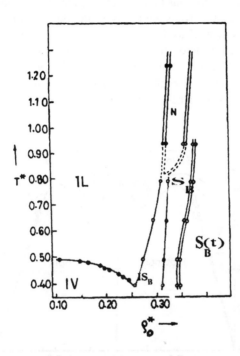

molecular pairs. This procedure produced the following values : $\mu = 0.8$, $\nu = 0.74$, $\sigma_0 = 3.65$ A, $\epsilon_0/k_B = 4302$ K, $x_0 = 4.4$ and $k' = 39.6$. Luckhurst and Simmonds[235] then simulated 256 particles interacting with this new potential and isotropic, nematic and smectic A phases were found. The nematic phase disappears when the reduced density is too low. These authors also compared the results of a number of GB simulations with those of a hard ellipsoid system (see Fig. 6.19). The good agreement shows that the NI transition is dominated by the excluded volume effects. On the other hand, the attractive potential dominates the formation of smectic phases as observed in GB system. Thus, it is concluded that hard ellipsoids do not exhibit smectic phases and hard spherocylinders do so only if length to width ratios exceed 4. The reparametrized GB potential $x_0 = 0.345$, $k' = 1/5$, $\nu = 2$, $\mu = 1$ has also been used to simulate a system of 256 discotic molecules by Emerson and coworkers[238]. A phase diagram was obtained consisting of isotropic, discotic-nematic and columnar phases.

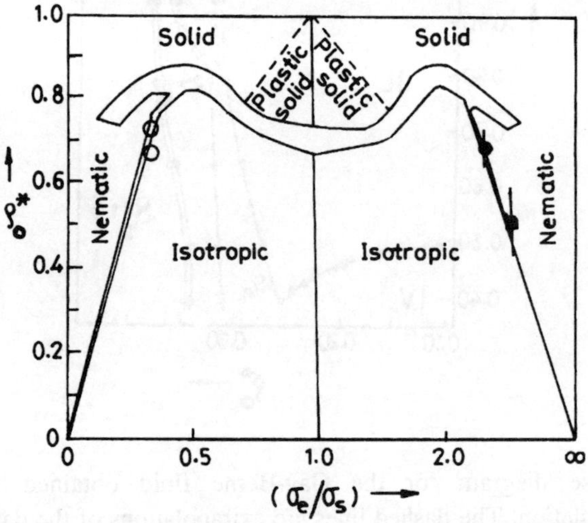

Fig. 6.19. Phase diagram of hard ellipsoids (Fig. 6.15) along with the data points resulting from simulations of the GB potential due to Luckhurst and Simmonds (Ref. 235). Vertical lines indicate the approximate density range where a NI transition was observed at constant density. The symbols ■, O and ● correspond to the different values of length to breadth ratio in the GB potential.

Recently Bates and Luckhurst[249] carried out a far more detailed study of GB particles interacting via a potential parametrized to reflect the anisotropic forces based on a fit to a realistic mesogenic molecule. They simulated 2000 molecules using canonical MC simulations with a new parametrization of the potential, x_0 =

4.4, $k' = 20$, $\mu = \nu = 1$. The behaviour of the phases and its pressure dependence, and the transitions between them have been investigated. At low pressures, the mesogen GB (4.4, 20, 1, 1) exhibits isotropic, smectic A and smectic B phases but, as the pressure is increased, a nematic phase also appears. The phase diagram determined for this mesogen is found to be in good agreement with those found for real mesogens. Reasonably accurate values of transition properties, with the exception of fractional volume change at the NI transition, are predicted by this parametrization of the potential. The simulated value of fractional volume change at T_{NI} is approximately ten times larger than that observed experimentally. These authors[249] also investigated the structures of the four phases in detail for a much larger system of 16000 particles using canonical MD simulations at state points taken from the phase diagram determined from the MC simulations. A wide range of singlet and pair distribution functions were evaluated together with orientational correlation coefficients, and for the smectic phases, a bond orientational correlation function. The results of these properties were used to identify the different phases. The essential difference between the S_A and S_B phases is revealed by the bond orientational correlation function which decays to zero for the S_A phase but not for the S_B phase.

Tsykalo[237] was the first to modify the Berne Pechukas potential to include chirality, and later Memmer and co-workers[236] modified the GB potential. The total potential consists of the GB term plus an additional chiral interaction :

$$u_{chiral} = 4g\varepsilon(\hat{r}_{12},\hat{e}_i,\hat{e}_j)\left(\frac{\sigma_0}{r_{12}-\sigma(\hat{r}_{12},\hat{e}_i,\hat{e}_j)+\sigma_0}\right)^7 (\hat{e}_i\cdot\hat{e}_j)(\hat{e}_i\times\hat{e}_j.\hat{r}_{12}) \qquad (6.122)$$

The parameter g measures the strength of the chiral coupling. Memmer and co-workers[236] simulated a system of 256 particles interacting with this new potential. The phase diagram was determined by adjusting the chiral coupling. A cholesteric phase was observed for $0.6 < g < 0.7$, while for $1.0 < g < 1.1$ a phase with geometric structure of blue phase appeared.

REFERENCES

1. S. Chandrasekhar, Liquid Crystals (Cambridge Univ. Press, Cambridge, 1992).
2. P.G. de Gennes, and J. Prost , The Physics of Liquid Crystals (Clarendon Press, Oxford, 1993).
3. G.W. Gray, and J.W. Goodby, Smectic Liquid Crystals; Textures and Structures (Leonard Hill, London, 1984).
4. G. Vertogen, and W.H. de Jeu, Thermotropic Liquid Crystals, Fundamentals (Springer-Verlag, 1988).

5. P.J. Collings, and J.S. Patel, eds., Handbook of Liquid Crystal Research (Oxford Univ. Press, Oxford, 1997).
6. S. Elston, and R. Sambles, eds., The Optics of Thermotropic Liquid Crystals (Taylor and Francis Ltd., London, 1998).
7. B.I. Halperin, and D.R. Nelson, Phys. Rev. Lett. **41**, 121 (1978).
8. D.R. Nelson, and B.I. Halperin, Phys. Rev. **B19**, 2457 (1979).
9. A.P. Young, Phys. Rev. **B19**, 1855 (1979).
10. P.J. Collings, and M. Hird, Introduction to Liquid Crystals : Chemistry and Physics (Taylor and Francis Ltd., London, 1997) Chapt. 6, p.111.
11. P.G. de Gennes, J. Phys. (Paris) Colloq. **30**, C4-65 (1969).
12. C.W. Oseen, Trans. Faraday Soc. **29**, 883 (1933).
13. (a) J.P. Hurault, J. Chem. Phys. **59**, 2086 (1973); (b) W. Helfrich, Appl. Phys. Lett. **17**, 531 (1970).
14. M. Delaye, R. Ribotta, and G. Durand, Phys. Lett. **A44**, 139 (1973).
15. N.A. Clark, and R.B. Meyer, App. Phys. Lett. **22**, 493 (1973).
16. G. Durand, in, Proceedings of the International Liquid Crystal Conference, Bangalore, Dec. 1973, Pramana Supplement **1**, 23 (1973).
17. (a) L. Gunther, Y. Imry, and J. Lajzerowicz, Phys. Rev. **A22**, 1733 (1980);
    (b) J.D. Litster, in, Proceedings of the Conference on Liquid Crystals of One- and Two-Dimensional Order, eds., W. Helfrich, and G. Heppke (Springer-Verlag, 1980) p. 65;
    (c) J. Als-Nielsen, Symmetries and Broken Symmetries in Condensed Matter Physics, ed., N. Boccara (IDSET-Paris, 1981) p. 107.
18. J. Als-Nielsen, J.D. Litster, R.J. Birgeneau, M. Kaplan, C.R. Safinya, A. Lindegaard-Andersen, and S. Mathiesen, Phys. Rev. **B22**, 312 (1980).
19. Y. Singh, S. Singh, and K. Rajesh, Phys. Rev. **A45**, 974 (1992), (Referred as SSR theory in the text).
20. S. Singh, (a) Phys. Rep. **277**, 283 (1996); (b) Phys. Rep. **324**, 107(2000).
21. (a) A.R. Denton, and N.W. Ashcroft, Phys. Rev. **A39**, 4701 (1989);
    (b) Y. Singh, Phys. Rep. **207**, 351 (1991).
22. W.L. McMillan, Phys. Rev. **A7**, 1419 (1973).
23. J. Als-Nielsen, R.J. Birgeneau, M. Kaplan, J.D. Litster, and C.R. Safinya, Phys. Rev. Lett. **39**, 352 (1977).
24. D. Brisbin, R. DeHoff, T.E. Lockhart, and D.L. Johnson, Phys. Rev. Lett. **43**, 1171 (1979),

25. R.J. Birgeneau, C.W. Garland, G.B. Kasting, and B.M. Ocko, Phys. Rev. **B24**, 2624 (1981).
26. H. Marynissen, J. Thoen, and W. Van Dael, Mol. Cryst. Liquid Cryst. **97**, (1983).
27. J.M. Viner, and C.C. Huang, Solid State Commun. **39**, 789 (1981).
28. B.M. Ocko, R.J. Birgeneau, J.D. Litster, and M.E. Neubert, Phys. Rev. Lett. **52**, 208 (1984).
29. K.K. Chan, P.S. Pershan, L.B. Sorensen, and F. Hardouin, Phys. Rev. Lett. **54**, 1694 (1985); Phys. Rev. **A34**, 1420 (1986).
30. K.K. Chan, M. Deutsch, B.M. Ocko, P.S. Pershan, and L.B. Sorensen, Phys. Rev. Lett. **54**, 920 (1985).
31. L.J. Martinez, A.R. Kortan, and R.J. Birgeneau, Phys. Rev. Lett. **56**, 2264 (1986).
32. S.B. Rananavare, V.G.K.M. Pisipati, and J.H. Freed, Chem. Phys. Lett. **140**, 255 (1987).
33. (a) C.W. Garland, G. Nouneiss, K.J. Stine, and G. Heppke, J. Phys. (Paris) **50**, 2291 (1989), (b) C.W. Garland, and G. Nouneiss, Phys. Rev. **E49**, 2964 (1984).
34. L. Chen, J.D. Brock, J. Huang, and S. Kumar, Phys. Rev. Lett. **67**, 2037 (1991).
35. G. Nouneiss, K.I. Blum, M.J. Young, C.W. Garland, and R.J. Birgeneau, Phys. Rev. **E47**, 1910 (1993).
36. L. Wu, M.J. Young, Y. Shao, C.W. Garland, R.J. Birgeneau, and G. Heppke, Phys. Rev. Lett. **72**, 376 (1994).
37. W.L. McMillan, Phys. Rev. **A6**, 936 (1972).
38. B.I. Halperin, and T.C. Lubensky, Solid State Comm. **14**, 997 (1974).
39. J.H. Chen, and T.C. Lubensky, Phys. Rev. **A14**, 1202 (1976).
40. P.E. Cladis, W.A. Saarloos, D.A. Huse, J.S. Patel, J.W. Goodby, and P.L. Finn, Phys. Rev. Lett. **62**, 1764 (1989).
41. J. Prost, Adv. Phys. **33**, 1(1984).
42. R.F. Bruinsma, and C.R. Safinya, Phys. Rev. **A43**, 5377 (1991).
43. P.J. Collings, Phase Structures and Transitions in Thermotropic Liquid Crystals, in, Handbook of Liquid Crystal Research, eds., P.J. Collings, and J.S. Patel (Oxford Univ. Press, New York, Oxford, 1997) Chapt. 4, p. 99.
44. K.K. Kobayashi, J. Phys. Soc. Japan **29**, 101 (1970); Mol. Cryst. Liquid Cryst. **13**, 137 (1971).
45. W.L. McMillan, Phys. Rev. **A4**, 1238 (1971).

46. R.B. Meyer, and T.C. Lubensky, Phys. Rev. **A14**, 2307 (1976).
47. J. Katriel, and G.F. Kventsel, Phys. Rev. **A28**, 3037 (1983); G.F. Kventsel, G.R. Luckhurst, and H.B. Zwdie, Mol. Phys. **56**, 589 (1985).
48. A. Kloczkowski, and J. Stecki, Mol. Phys. **55**, 689 (1985).
49. D.A. Badalyan, Sov. Phys. Crystallogr. **27**,10 (1982).
50. C.D. Mukherjee, B. Bagchi, T.R. Bose, D. Ghosh, M.K. Roy, and M. Saha Phys. Lett. **A92**, 403 (1982).
51. W. Wagner, Mol. Cryst. Liquid Cryst. **98**, 247 (1983).
52. M. Nakagawa, and T. Akahane, J. Phys. Soc. Japan **54**, 69 (1985).
53. G.F. Kventsel, G.R. Luckhurst, and H.B. Zewdie, Mol. Phys. **56**, 589 (1985).
54. M. Hosino, H. Nakano, and H. Kimura, J. Phys. Soc. Japan **46**, 740, 1709 (1979).
55. X. Wen, and R.B. Meyer, Phys. Rev. Lett. **59**, 1325 (1987).
56. N.V. Madhusudana, B.S. Srikanta, and M. Subramanya Raj Urs, Mol. Cryst. Liquid Cryst. **108**, 19 (1984).
57. R. Holyst, and A. Poniewierski, Mol. Phys. **71**, 561 (1990).
58. I.I. Ptichkin, Sov. Phys. Crystallogr. **36**, 134 (1991).
59. M.P. Taylor, R. Hentschke, and J. Herzfeld, Phys. Rev. Lett. **62**, 800 (1989).
60. (a) F. Dowell, Phys. Rev. **A28**, 3520, 3526 (1983), (b) C. Rosenblatt, and D. Ronis, Phys. Rev. **A23**, 305 (1981); D. Ronis, and C. Rosenblatt, Phys. Rev. **A21**, 1687 (1980).
61. (a) F.T. Lee, H.T. Tan, Y.M. Shih, and C.W. Woo, Phys. Rev. Lett.**31**,1117 (1973), (b) L. Senbetu, and C.W. Woo, Phys. Rev. **17A**, 1529 (1978).
62. K. Sokalski, Physica **A113**, 133 (1982).
63. C.D. Mukherjee, T.R. Bose, D. Ghosh, M.K. Roy, and M. Saha, Mol. Cryst. Liquid Cryst. **124**, 139 (1985).
64. E.M. Averyanov, and A.N. Primak, (a) Liquid Crystals **10**, 555 (1991); (b) ibid **13**, 139 (1993).
65. H. Wang, M.Y. Jin, R.C. Jarnagin, T.J. Bunning, W. Adams, B. Cull, Y. Shi, S. Kumar, and E.T. Samulski, Nature **238**, 244 (1996).
66. B. Mulder, Phys. Rev. **A35**, 3095 (1987).
67. R. Holyst, and A. Poniewierski, Phys. Rev. **A39**, 2742 (1989).
68. A.M. Somoza, and P. Tarazona, Phys. Rev. Lett. **61**, 2566 (1988).
69. A. Poniewierski, and R. Holyst, Phys. Rev. Lett. **61**, 2461 (1988).

70. M.C. Mahato, M. Raj Lakshmi, R. Pandit, and H.R. Krishnamurthy, Phys. Rev. **A38**, 1049 (1988).

71. R.G. Priest, Mol. Cryst. Liquid Cryst. **37**, 101 (1976).

72. (a) D. Frenkel, and B.M. Mulder, Mol. Phys. **55**, 1171 (1985), (b) D. Frenkel, B.M. Mulder, and J.P. McTague, Phys. Rev. Lett. **52**, 287 (1984), (c) D. Frenkel, H.N.W. Lekkerkerker, and A. Stroobants, Nature (London) **332**, 822 (1988).

73. (a) R. Eppenga, and D. Frenkel, Mol. Phys. **52**, 1303 (1984), (b) D. Frenkel, and R. Eppenga, Phys. Rev. **A31**, 1776 (1985).

74. (a) C.W. Garland, and G. Nounesis, Phys. Rev. **E49**, 2964 (1994), (b) L. Longa, J. Chem. Phys. **85**, 2974 (1986).

75. R.B. Griffiths, Phys. Rev. Lett. **24**, 715 (1970); M. Marynissen, J. Thoen, and W. Van Dael, Mol. Cryst. Liquid Cryst. **124**, 195 (1985).

76. R. Alben, Solid State Commun **13**, 1783 (1973).

77. B.I. Halperin, T.C. Lubensky, and S.K. Ma, Phys. Rev. Lett. **32**, 292 (1974).

78. P.G. de Gennes, (a) Solid State Commun. **10**, 753 (1972); (b) Phys. Lett. **A30**, 54 (1969).

79. I. Lelidis, and G. Durand, J. Phys. II (Paris) **6**, 1359 (1996).

80. M.A. Anisimov, V.P. Voronov, A.O. Kulkov, and F. Kholmurodov, J. Phys. (Paris) **46**, 2137 (1985).

81. M.A. Anisimov, Critical Phenomena in Liquids and Liquid Crystals (Gordon and Breach, 1991).

82. W.H. de Jeu, NATO School on Phase Transitions in Liquid Crystals, Erice, Italy (May, 1991).

83. P. Barois, Phase Transitions in Liquid Crystals : Introduction to Phase Transition Theories, in, Phase Transitions in Liquid Crystals, eds., S. Martellucci, and A.N. Chester (Plenum Press, New York, 1992) Chapt. 4, p. 41.

84. C. Dasgupta, and B.I. Halperin, Phys. Rev. Lett. **47**, 1556 (1981).

85. T.C. Lubensky, J. Chim. Phys. **80**, 31 (1983).

86. D.R. Nelson, and J. Toner, Phys. Rev. **B24**, 363 (1981).

87. J. Toner, Phys. Rev. **B26**, 462 (1982).

88. C. Dasgupta, Int. J. Mod. Phys. **B9**, 2219 (1995).

89. B. R. Patton, and B. S. Andereck, Phys. Rev. Lett. **69**, 1556 (1992).

90. B.S. Andereck, and B.R. Patton, Phys. Rev. **E49**, 1393 (1994).

91. J.G. Kirkwood, and E. Monroe, J. Chem. Phys. **9**, 514 (1941).

92. G. Vertogen, and W.H. de Jeu, Thermotropic Liquid Crystal, Fundamentals (Springer-Verlag, 1988) Chapt. 14, p. 283.

93. M. Nakagawa, and T. Akahane, J. Phys. Soc. Japan **53**, 1951 (1984).

94. (a) F. Dowell, and D.E. Martire, J. Chem. Phys. **68**, 1088 (1978), (b) F. Dowell, Phys. Rev. **A28**, 3520 (1983).

95. (a) A. Stroobants, H.N.W. Lekkerkerker, and D. Frenkel, Phys. Rev. **A36**, 2929 (1987); (b) Phys. Rev. Lett. **57**, 1452 (1986).

96. J.W. Emsley, R. Hashim, G.R. Luckhurst, G.N. Rumbles, and F.R. Viloria, Mol. Phys. **49**,1321 (1983).

97. J.W. Emsley, R. Hashim, G.R. Luckhurst, and G.N. Shilstone, Liquid Crystals **1**, 437 (1986).

98. J.W. Emsley, G.R. Luckhurst, G.N. Shilstone, and I. Sage, J. Chem. Soc. Faraday Soc. Trans. II **83**, 371 (1987).

99. D. Catalano, C. Forte, C.A. Veracini, J.W. Emsley, and G.N. Shilstone, Liquid Crystals **2**, 357 (1987).

100. E.M. Averyanov, and V.A. Gunyakov. Opt. Spectrosc. **66**, 72 (1989).

101. J.W. Emsley, G.R. Luckhurst, and H.S. Sachdev, Liquid Crystals **5**, 953 (1989).

102. M.P. Fontana, B. Rosi, N. Kirov, and I. Dozov, Phys. Rev. **A33**, 4132 (1986).

103. M.D. Lipkin, and D.W. Oxtoby, J. Chem. Phys. **79**, 1939 (1983).

104. T.J. Sluckin, and P. Shukla, J. Phys. **A16**, 1539 (1983).

105. A.M. Somoza, and P. Tarazona, J. Chem Phys. **91**, 517 (1989).

106. S.D. Lee, (a) J. Chem. Phys. **87**, 4972 (1987); (b) **89**, 7036 (1988).

107. P. Tarazona, (a) Mol. Phys. **52**, 81 (1984); (b) Phys. Rev. **A31**, 2672 (1985); erratum **A32**, 3148 (1985).

108. W.A. Curtin, and N.W. Ashcroft, Phys. Rev. **A32**, 2909 (1985).

109. (a) G. Sigaud, F. Hardouin, and M.F. Achard, Phys. Lett. **A72**, 24 (1979); (b)G. Sigaud, F. Hardouin, M.F. Achard, and H. Gasparoux, J. Phys. (Paris) Colloq. **40**, C3-356 (1979).

110. F. Hardouin, A.M. Levelut, M.F. Achard, and G. Sigaud, J. Chim. Phys. **80**, 53 (1983).

111. G. Sigaud, F. Hardouin, M.F. Achard, and A.M. Levelut, J. Phys. (Paris) **42**, 107 (1981).

112. M.J. Young. L. Wu, G. Nounesis, C.W. Garland, and R. J. Birgeneau, Phys. Rev. **E50**, 368 (1994).

113. R. Shashidhar, and B.R. Ratna, Liquid Crystals **5**, 421 (1989); B.R. Ratna, R. Shashidhar, and V.N. Raja, Phys. Rev. Lett. **55**, 1476 (1985).

114. A.M. Levelut, J. Phys. (Paris) Lett. **45**, 603 (1984).

115. S. Kumar, L. Chen, and V. Surendranath, Phys. Rev. Lett. **67**, 322 (1991).

116. P. Patel, S.S. Keast, M.E. Neubert, and S. Kumar, Phys. Rev. Lett. **69**, 301 (1992).

117. P. Barois, Phase Transitions in Liquid Crystals : Introduction to Phase Transition Theories, in, Phase Transitions in Liquid Crystals, eds., S. Martillucci, and A. N. Chester (Plenum Press, N.Y. 1992), Chapt. 4, p. 47.

118. J. Prost, J. Phys. (Paris) **40**, 581 (1979).

119. J. Prost, and P. Barois, J. Chim. Phys. **80**, 65 (1983).

120. J. Wang, and T.C. Lubensky, Phys. Rev. **A29**, 2210 (1984).

121. P. Barois, J. Prost, and T.C. Lubensky, J. Phys. (Paris) **46**, 391 (1985).

122. Y. Park, T.C. Lubensky, P. Barois, and J. Prost, Phys. Rev. **A37**, 2197 (1988).

123. A. Saupe, Mol. Cryst. Liquid Cryst. **7**, 59 (1969); J. Nehring, and A. Saupe, J. Chem. Soc. Faraday Trans. II **63**, 1 (1972).

124. Orsay Group on Liquid Crystals, Solid State Commun. **9**, 653 (1971).

125. Y. Galerne. J.L. Martinand, G. Durand, and M. Veyssie, Phys. Rev. Lett. **29**, 562 (1972).

126. T.R. Taylor, J.L. Fergason, and S.L. Arora, Phys. Rev. Lett. **24**, 359 (1970); **25**, 722 (1970).

127. E.I. Kats, and V.V. Lebedev, Physica, **A135**, 601 (1986).

128. C.C. Huang, and J.M. Viner, Phys. Rev. **A25**, 3385 (1982); R.J. Birgeneau, C.W. Garland, A.R. Kortan, J.D. Lister, M. Meichle, B.M. Ocko, C. Rosenblatt, L.J. Yu, and J.W. Goodby, Phys. Rev. **A27**, 1251 (1983).

129. P.G. de Gennes, (a) C.R. Acad. Sci. **B274**, 758 (1972); (b) Mol. Cryst. Liquid Cryst. **21**, 49 (1973).

130. C.C. Huang, and S.C. Lien, Phys. Rev. Lett. **47**, 1917 (1981).

131. A. Wulf, Phys. Rev. **A11**, 365 (1975).

132. W.L. McMillan, Phys. Rev. **A8**, 1921(1973).

133. D. Cabib, and L. Benguigui, J. Phys. (Paris) **38**, 419 (1977).

134. B.W. Van der Meer, and G. Vertogen, J. Phys. (Paris) **40**, C3-222 (1979).

135. R.G. Priest, J. Phys. (Paris) **36**, 437 (1975); J. Chem. Phys. **65**, 408 (1976).

136. M. Matsushita, J. Phys. Soc. Japan **50**, 1351 (1981).

137. W.J.A. Goossens, J. Phys. (Paris) **46**, 1411 (1985).

138. W.J.A. Goossens, Europhys. Lett. **3**, 341 (1987); Mol. Cryst. Liquid Cryst. **150**, 419 (1987).

139. F. Gieβelmann, and P. Zugenmaier, Phys. Rev. **E55**, 5613 (1997).

140. G. Sigaud, F. Hardouin, and M.F. Achard, Solid State Commun. **23**, 35 (1977).

141. D. Johnson, D. Allender, D. Dehoff, C. Maze, E. Oppenheim, and R. Reynolds, Phys. Rev. **B16**, 470 (1977).

142. D.L. Johnson, J. Chim. Physique **80**, 45 (1983).

143. (a) D. Brisbin, D.L. Johnson, H. Fellner, and M.E. Neubert, Phys. Rev. Lett. 50, 178 (1983), (b) R. Shashidhar, B.R. Ratna, and S. Krishna Prasad, Phys. Rev. Lett. **53**, 2141 (1984).

144. M.A. Anisimov, Mol. Cryst. Liquid Cryst. **162**, 1 (1988).

145. K.C. Chu, and W.L. McMillan, Phys. Rev. **A15**, 1181 (1977).

146. J. Chen, and T.C. Lubensky, Phys. Rev. **A14**, 1202 (1976).

147. J. Prost, and J.P. Marcerou, J. Phys. (Paris) **38**, 315 (1977).

148. L Benguigui, J. Phys. (Paris) Colloq. **40**, C3–419 (1979).

149. N.V. Madhusudana, Theories of Liquid Crystals, in, Liquid Crystals : Applications and Uses, Vol. 1, ed. B. Bahadur (World-Scientific, Singapore) Chapt. 2, p. 80.

150. G. Grinstein, and J. Toner, Phys. Rev. Lett. **51**, 2386 (1983).

151. T.C. Lubensky, Mol. Cryst. Liquid Cryst. **146**, 55 (1987).

152. X. Wen, C.W. Garland, and M.D. Wand, Phys. Rev. **A42**, 6087 (1990).

153. P.E. Cladis, (a) Phys. Rev. Lett. 35, **48** (1975); (b) Philos. Mag. **29**, 641 (1975).

154. P.E. Cladis, R.K. Bogardus, W.B. Daniels, and G.N. Taylor, Phys. Rev. Lett. **39**, 720 (1977).

155. D. Guillon, P.E. Cladis, and J. Stamatoff, Phys. Rev. Lett. **41**, 1598 (1978).

156. N.S. Chen, S.K. Hark, and J.T. Ho, Phys. Rev. **A24**, 2843 (1981).

157. T. Narayanan, and A. Kumar, Phys. Rep. **249**, 135 (1994).

158. A. Deerenberg, J.A. Schouten, and N.J. Trappeniers, Physica **A103**, 183 (1980).

159. J.A. Schouton, Phys. Rep. **172**, 33 (1989).

160. R.J. Tufeu, P.H. Keyes, and W.B. Daniels, Phys. Rev. Lett. **35**, 1004 (1975).

161. J.D. Cox, J. Chem. Soc. **4**, 4606 (1952).

162. T. Narayanan, A. Kumar, and E.S.R. Gopal, Phys. Lett. **A155**, 276 (1991).

163. R.G. Johnston, N.A. Clark, P. Wiltzius, and D.S. Cannel, Phys. Rev. Lett. **54**, 49 (1985).

164. V.P. Zaitsev, S.V. Krivokhizha, I.L. Fabelinskii, A. Tsitrovskii, L.L. Chaikov, E.V. Shvets, and P. Yani, Sov. Phys. JETP Lett. **43**, 112 (1986).

165. S.V. Krivokhizha, O.A. Dugovaya, I.L. Fabelinskii, L.L. Chaikov, A. Tsitrovskii, and P. Yani, Sov. Phys. JETP **76**, 62 (1993).

166. G.B. Kozlov, E.B. Kryukova, S.P. Lebedev, and A.A. Sobyanin, Sov. Phys. JETP **67**, 1689 (1988).

167. M.J. Naughton, R.V. Chamberlin, X. Yan, S.Y. Hsu. L.Y. Chiang, M.Ya. Azbel, and P.M. Chaikin, Phys. Rev. Lett. **61**, 621 (1988).

168. T.H. Lin, X.Y. Shao, M.K. Wu, P.H. Hor, X.C. Jin, C.W. Chu, N. Evans, and R. Bayuzick, Phys. Rev. **B29**, 1493 (1984).

169. S. Katayama, Y. Hirokawa, and T. Tanaka, Macromolecules **17**, 2649 (1984).

170. H. Glasbrenner, and H. Weingartner, J. Phys. Chem. **93**, 3378 (1989).

171. R.B. Griffiths, in, Critical Phenomena in Alloys, Magnets and Superconductors, eds., R.E. Mills, E. Ascher, and R.I. Jaffee (McGraw Hill, New York, 1971) p. 377.

172. P.E. Cladis, Mol. Cryst. Liquid Cryst. **165**, 85 (1988).

173. S. Singh, Phase Transitions **72**, 183 (2000).

174. L. Longa, and W.H. de Jeu, Phys. Rev. **A26**, 1632 (1982).

175. F.R. Bouchet, and P.E. Cladis, Mol. Cryst. Liquid Cryst. Lett. **64**, 97 (1980).

176. K.J. Lushington, G.B. Kasting, and C.W. Garland, Phys. Rev. **B22**, 2569 (1980).

177. R.Y. Dong, Mol. Cryst. Liquid Cryst. Lett. **64**, 205 (1981).

178. R. Shashidhar, H.D. Kleinhans, and G.M. Schneider, Mol. Cryst. Liquid Cryst. Lett. **72**, 119 (1981).

179. P.E. Cladis, D. Guillon, F.R. Bouchet, and P.L. Finn, Phys. Rev. **A23**, 2594 (1981).

180. A.R. Kortan, H.V. Känel, R.J. Birgeneau, and J.D. Litster, Phys. Rev. Lett. **47**, 1206 (1981).

181. N. Hafiz, N.A.P. Vaz, Z. Yaniv, D. Allender, and J.W. Doane, Phys. Lett. **A91**, 411 (1982).

182. J.W. Emsley, G.R. Luckhurst, P.J. Parsons, and B.A. Timimi, Mol. Phys. **56**, 767 (1985).

183. C.S. Hudson, Z. Phys. Chem. **47**, 113 (1904).

184. G.Illian, H. Kneppe, and F. Schneider, Ber. Bunsenges, Phys. Chem. **92**, 776 (1988).

185. F. Dowell, Phys. Rev. **A36**, 5046 (1987); (b) **A38**, 382 (1988).

186. G. Pelzl, S. Diele, I. Latif, W. Weissffog, and D. Demus. Cryst. Res. Technol **17**, K 78 (1982).

187. S. Diele, G. Pelzl, I. Latif, and D. Demus, Mol. Cryst. Liquid Cryst. Lett. **92**, 27 (1983).

188. G. Sigaud, F. Hardouin, M. Mauzac, and N.H. Tinh, Phys. Rev. **A33**, 789 (1986).

189. F. Hardouin, A.M. Levelut, M.F. Achard, and G. Sigaud, J. Chim. Phys. **80**, 53 (1983).

190. C. Destrade, J. Malthete, N.H. Tinh, and H. Gasparoux, Phys. Lett. **A78**, 82 (1980).

191. N.H. Tinh, J. Malthete, and C. Destrade, J. Phys. (Paris) Lett. **42**, L-417 (1981).

192. K.W. Evans-Lutterodt, J.W. Chung, B.M. Ocko, R.J. Birgeneau, C. Chiang, C.W. Garland, E. Chin, J. Goodby, and N.H. Tinh, Phys. Rev. **A36**, 1387 (1987).

193. F. Hardouin, G. Sigaud, M.F. Achard, and H. Gasparoux, Phys. Lett. **A71**, 347 (1979).

194. N.H. Tinh, J. Chim. Phys. **80**, 83 (1983).

195. R. Shashidhar, B.R. Ratna, V. Surendranath, V.N. Raja, S. Krishna Prasad, and C. Nagabhushana, J. Phys. (Paris) Lett. **46**, L445 (1985).

196. N.H. Tinh, F. Hardouin, C. Destrade, and A. M. Levelut, J. Phys. (Paris) Lett. **43**, L33 (1982).

197. S. Chandrasekhar, Proceedings of 10th International Liquid Crystal Conference, New York, 1984; Mol. Cryst. Liquid Cryst. **124**, 1 (1985)

198. A. Nayeem, and J.H. Freed, J. Phys. Chem. **93**, 65 (1989).

199. P. C. Martin, O. Parodi, and P.S. Pershan, Phys. Rev. **A6**, 2401 (1972).

200. S. de Groot, and P. Mazur, Non-equilibrium Thermodynamics (North-Holland, Amsterdam, 1962).

201. P.A. Lebwohl, and G. Lasher, Phys. Rev. **A6**, 426 (1972).

202. C. Zannoni, Computer simulations, in, The Molecular Physics of Liquid Crystals, eds., G.R. Luckhurst, and G.W. Gray (Academic Press, 1979) Chapt. 9, p. 191.

203. H.J.F. Jensen, G. Vertogen, and J.G.J. Ypma, Mol. Cryst. Liquid Cryst. **38**, 87 (1977).

204. (a) G.R. Luckhurst, and P. Simpson, Mol. Phys. **47**, 251 (1982);

(b) G.R. Luckhurst, and S. Romano, Proc. Roy. Soc. London **A373**, 111 (1980);

(c) G.R. Luckhurst, P. Simpson, and C. Zannoni, Chem. Phys. Lett. **78**, 429 (1981).

205. D. Frenkel, Computer Simulations of Phase Transitions in Liquid Crystals, in, Phase Transitions in Liquid Crystals, eds., S. Martellucci, and A.N. Chester (Plenum Press, N.Y., 1992) Chapt. 5, p. 67.

206. U. Fabri, and C. Zannoni, Mol. Phys. **84**, 424 (1986).

207. F. Biscarini, C. Zannoni, C. Chiccoli, and P. Pasini, Mol. Phys. **73**, 439 (1991).

208. Z. Zhang, O.G. Mouritsen, and M.J. Zuckermann, Phys. Rev. Lett. **69**, 1803 (1992).

209. Z. Zhang, M.J. Zuckermann, and O.G. Mouritsen, Mol. Phys. **80**, 1195 (1993).

210. D.J. Cleaver, D. Kralj, T.J. Sluckin, and M.P. Allen, in, Liquid Crystals in Complex Geometries Formed by Polymer and Porous Networks, eds., G.P. Crawford, S. Zumer, (Taylor and Francis, London, 1995), Chapt. 21, p. 467.

211. Y.Y. Goldschmidt, in Recent Progress in Random Magnets, ed., D.H. Ryan (World Scientific, Singapore, 1992) p. 151.

212. J. Vieillard-Baron, Mol. Phys. **28**, 809 (1974).

213. D.W. Rebertus, and K.M. Sando, J. Chem. Phys. **67**, 2585 (1977).

214. T. Boublik, I. Nezbeda, and O. Trnka, Czech. J. Phys. **B26**, 1081 (1976).

215. P.A. Monson, and M. Rigby, Mol. Phys. **35**¸1337 (1978).

216. I. Nezbeda, and T. Baublik, Czech. J. Phys. **B28**, 353 (1978).

217. (a) M.P. Allen and D. Frenkel, Phys. Rev. Lett. **58**, 1748 (1987);

(b) M.P. Allen, D. Frenkel, and J. Talbot, Computer Phys. Rep. **9**, 301 (1989).

218. D. Frenkel, (a) Mol. Phys. 60, 1 (1987); (b) J. Phys. Chem. **92**, 3280 (1988); (c) Liquid Crystals **5**, 929 (1989).

219. (a) A. Perera, P.G. Kusalik, and G.N. Patey, J. Chem. Phys. **87**, 1295 (1987); **89**, 5969 (1988);

(b) A. Perera, and G.N. Patey, J. Chem. Phys. **89**, 5861 (1988); **91**, 3045 (1989)l

(c) J. Talbot, A. Perera, and G.N. Patey, Mol. Phys. **70**, 285 (1990).

220. D. Frenkel, in, Liquids, Freezing and Glass Transitions eds., J.P. Hansen and D. Levesque (North Holland, 1991) p. 691.

221. M.P. Allen (a) Liquid Crystals **8**, 499 (1990); (b) Philos. Trans. Roy. Soc. London **A344**, 232 (1993).

222. R.R. Netz, and A.N. Berker, Phys. Rev. Lett. **68**, 333 (1992).

223. J.A.C. Veerman, and D. Frenkel, (a) Phys. Rev. **A45**, 5632 (1992); (b) ibid **A41**, 3237 (1990).

224. G.J. Zarragoicoechea, D. Levesque, and J.J. Weis, Mol. Phys. **75**, 989 (1992).

225. J.A. Cuesta, and D. Frenkel, Phys. Rev. **42A**, 2126 (1990).

226. J.Y. Denham, G.R. Luckhurst, C. Zannoni, and J.W. Lewis, Mol. Cryst. Liquid Cryst. **60**, 185 (1980).

227. P.E. Lammert, D.S. Rokhsar, and J. Toner, Phys. Rev. Lett. **70**, 1650 (1993).

228. C. Dasgupta, Phys. Rev. Lett. **55**, 1771 (1985); J. Phys. (Paris) **48**, 957 (1987).

229. R.J. Birgeneau, and J.D. Litster, J. Phys. Lett. **39**, L399 (1978).

230. R. Pindak, D.E. Moncton, S.C. Davey, and J.W. Goodby, Phys. Rev. Lett. **46**, 1135 (1981).

231. I.M. Jiang, S.N. Huang, J.Y. Ko, T. Stoebe, A.J. Jin, and C.C. Huang, Phys. Rev. **E48**, 3240 (1993).

232. (a) E. De Miguel, L.F. Rull, M.K. Chalam, K.E. Gubbins, and F.V. Swol, Mol. Phys. **72**, 593 (1991);

(b) E. De Miguel, L.F. Rull, M.K. Chalam, and K.E. Gubbins, Mol. Phys. **74**, 405 (1991).

233. G.R. Luckhurst, R.A. Stephens, and R.W. Phippen, Liquid Crystals **8**, 451 (1990).

234. E. De Miguel, L.F. Rull, and K.E. Gubbins, Phys. Rev. **A45**, 3813 (1992).

235. G.R. Luckhurst, and P.S.J. Simmonds, Mol. Phys. **80**, 233 (1993).

236. R. Memmer, H.G. Kuball, and A. Schonhofer, (a) Liquid Crystals **15**, 345 (1993); (b) Ber. Bunsen-Ges. Phys. Chem. **97**, 1193 (1993).

237. A.L. Tsykalo, Mol. Cryst. Liquid Cryst. **129**, 409 (1985).

238. A.P.J. Emerson, G.R. Luckhurst, and S.G. Whatling, Mol. Phys. **82**, 113 (1994).

239. P. Pasini, and C. Zannoni, eds., Advances in the Computer Simulation of Liquid Crystals (Kluwer Academic Publishers, Dordrecht Netherlands, 2000).

240. M.P. Allen, in, Observation, Prediction and Simulation of Phase Transitions in Complex Fluids, eds., M. Baus, L.F. Rull, and J.P. Ryckaert (Kluwer Academic Publishers, Dordrecht Netherlands, 1995).

241. D.J. Adams, G.R. Luckhurst, and R.W. Phippen, Mol. Phys. **61**, 1575 (1987).

242. A.P.J. Emerson, R. Hashim, and G.R. Luckhurst, Mol. Phys. **76**, 241 (1992).

243. R. Hashim, G.R. Luckhurst, and S. Romano, J. Chem. Soc. Faraday Trans. **91**, 2141 (1995).

244. K. Satoh, S. Mita, and S. Kondo, Liquid Crystals **20**, 757 (1996).

245. G. La Penna, D. Catalino, and C.A. Veracini, J. Chem. Phys. **105**, 7097 (1996).

246. M.P. Neal, A.J. Parker, and C.M. Care, Mol. Phys. **91**, 603 (1997).

247. D.J. Cleaver, C.M. Care, M.P. Allen, and M.P. Neal, Phys. Rev. **E54**, 559 (1996).

248. M.P. Allen, J.T. Brown, and M.A. Warren, J. Phys. Condens. Matter **8**, 9433 (1996).

249. M.A. Bates, and G.R. Luckhurst, J. Chem. Phys. **110**, 7087 (1999).

250. For a review of Monte Carlo methods, see K. Binder, and D.W. Heermann, Monte Carlo Simulation in Statistical Physics (Springer-Verlag, 1988);

K. Binder, ed., The Monte Carlo Methods in Condensed Matter Physics (Springer–Verlag, 1991).

251. For a review of molecular dynamics methods, see M.P. Allen, and D.J. Tildesley, Computer Simulation of Liquids (Clarendon, Oxford, 1987).

252. J. Lee and J.M. Kosterlitz, Phys. Rev. Lett. **65**, 137 (1990); Phys. Rev. **B43**, 3265 (1991), and references cited therein.

253. A.N. Berker, and J.S. Walker, Phys. Rev. Lett. **47**, 1469 (1981).

254. J.G. Gay, and B.J. Berne, J. Chem. Phys. **74**, 3316 (1981).

255. J. Corner, Proc. R. Soc. London Ser. **A192**, 275 (1948).

256. B.J. Berne, and P. Pechukas, J. Chem. Phys. **56**, 4213 (1972).

7

Liquid Crystals of Disc-Like Molecules

Since the early investigations of Lehmann[1] and others[2–4] it was recognized that mesomorphism could only be exhibited by the molecules of rod-like or lath-like structure. Exception to this accepted principle was first observed in 1977 when Chandrasekhar et. al.[5] discovered liquid crystalline behaviour in pure compounds consisting of simple disc-like molecules, viz., benzene-hexa-n-alkanoates (see Table 7.1, compound T7.1a). From a careful thermodynamic, optical and X-ray studies they showed that the structure of mesophase has translational periodicity in two dimensions and liquid like disorder in the third. The word "discotic" was proposed by Billard et.al.[6] to describe the disc-like molecules as well as the resulting mesophases. For details of mesophases of discotic molecules we refer to references[7–16].

The discotic molecules exhibit two distinct classes of phases, the discotic nematic and the columnar. A twisted nematic phase with the helical axis normal to the director has also been found. A mesophase with smectic like (lamellar) structure has also been reported[17] but the precise layer structure is not fully understood. A number of variants of columnar structure have been found (see sec. 7.2).

The existence of mesophases of disc-shaped molecules was theoretically predicted[18] even before the discovery[5] of nematic discotics. In fact, it is interesting to note that theory has been ahead of experiment in case of two dimensional ordered systems. In 1930s their stability was stressed[19] and a complete classification scheme was developed[20] according to group theory. In 1924 Vorländer[21] considered the possibility of the existence of mesophases in systems with flat star-like or cross-shaped molecules. However, his attempts to observe liquid crystalline behaviour in these systems failed. After 1960 anisotropic mesophases made up of flat polyaromatic molecules with nematic texture characteristics were found[22,23], but these phases were generated at high temperatures which decomposed the materials and so the analysis was not possible in only detail.

Some typical discotic molecules are shown in Table 1.7 (chapter 1). Generally, they have more or less flat cores with 3, 4, 6, 8 or even 9 long-chain substituents, commonly with ester or ether (but rarely more complex) linkage groups. Available experimental evidence indicates that the presence of side chains is crucial to the generation of mesophases of discotic molecules. However, discotic mesomorphism has been observed even in substances without long-flexible substituents (compound T7.1b). In this table we have also listed "pyramidic" compounds (T 7.1c) which no longer have a flat core but a cone shaped central moiety. These compounds, also called as "cone-shaped" or "bowlic"[24,25] are of special interest. For generating a dense packing in these structures, the up and down symmetry of the ordinary discotic molecules is no longer preserved. Therefore compounds of this type are suspicious to generate ferroelectric phases or anomalously high electrical conductivity.

Table 7.1. (a) BHn-alkanoates; (b) scyllo-inosithehexa-acetate; (c) hexa-substituted tribenzocyclononene.

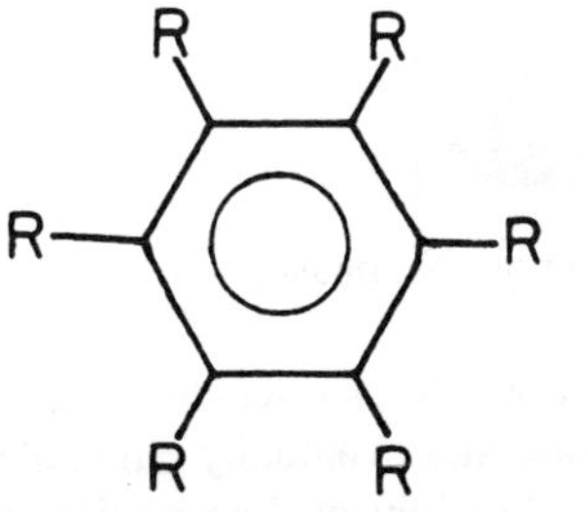

(a) R = n-C_4H_9-COO to n-C_9H_{19}-COO

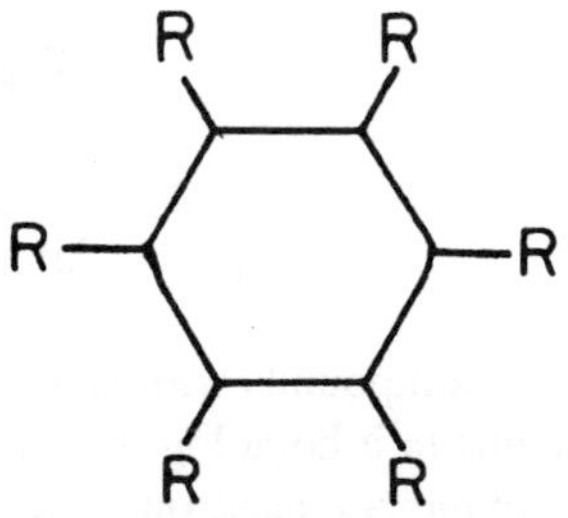

(b) R = CH_3-COO

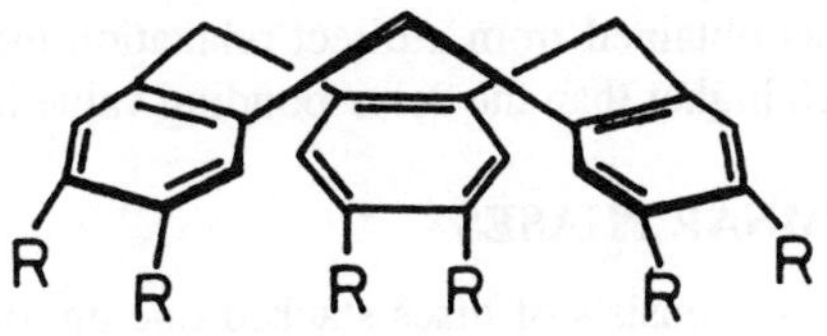

(c) R = $C_9H_{19}COO^-$

7.1 THE DISCOTIC NEMATIC PHASE

The most simple discotic phase is the discotic nematic (designated N_D); it is the least ordered and the least viscous. As shown in Fig. 7.1, the discs are anisotropic and are able to assemble so that their short axes, on the average, preferentially orient along the director. There is no long-range positional correlation. The symmetry of two kinds of nematics (the calamitic uniaxial nematic, N_u and the discotic nematic, N_D) is the same, and identical types of defects – the schlieren texture, and umbilics, etc., (see chapter 11) are observed in both the cases. However, the two are completely different and immiscible. While the calamitic nematic (N_u) is diamagnetically and optically positive, the N_D phase is negative. The dielectric anisotropy $\Delta\epsilon$ of both the phases can be either positive or negative.

Fig. 7.1. Arrangement of molecules within the N_D phase.

Compounds with shorter chains exhibit nematic phases. In fact, lengthening the chains in a homologous series narrows the nematic phase and eventually eliminates it. In the N_D phase only few quantitative measurements of the physical properties are available, but what has been done reveals that the values of elastic constants for the discotic and calamitic systems are of the same order. Using Frederiks technique the splay and bend constants have been measured, but the measurement of twist constant could not be possible. The negative diamagnetic anisotropy makes it very difficult to measure these constants by the Frederiks method.

The hydrodynamic equations of the calamitic nematic are applicable to the N_D phase as well. There are six viscosity coefficients (or Leslie coefficients) which reduce to five if Onsager's reciprocal relations are used. The effective value of the viscosity of N_D, as obtained from a direct relaxation measurement[26], shows that its magnitude is much higher than the corresponding value for the calamitic nematic.

7.2 THE COLUMNAR PHASES

The columnar phase consists of discs stacked one on top of the other aperiodically to form liquid-like columns; the different columns constituting a two dimensional lattice (Fig. 1.6 (c-e)). A number of variants of this structure have been identified – D_{ho}, D_{hd}, D_{rd} and $D_{ob.d}$. These phases arise because of the different symmetry classes of the two dimensional lattice of columns and the order or the disorder of the molecular stacking within the columns (see Table 7.2). The liquid crystal community prefer the levels D_h, D_r, D_{ob} (D refers to the discoid shape of the molecules and h, r, ob stand, respectively, for the hexagonal, rectangular and oblique symmetry groups); the second index (Table 7.2) refers to the range of positional correlations along the columnar axis (o stands for ordered, and d for disordered).

Table 7.2. Columnar mesophases of discotic molecules.

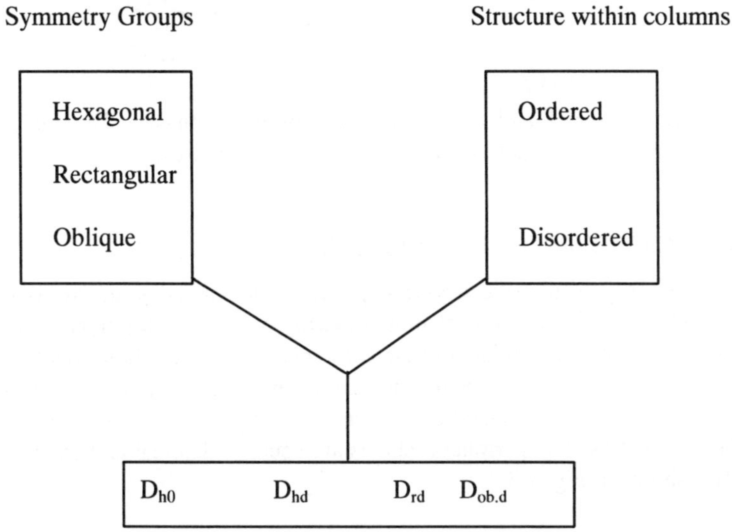

The high temperature hexagonal columnar phase D_h possesses a hexagonal lattice of columns with the molecules tilted within the column (Fig. 7.2a). There is no order in the azimuthal direction of the tilt. The lattice distance from a molecule to its nearest neighbour is equal irrespective of the direction (i.e., a=b). The difference between D_{ho} and D_{hd} phases is not sharp. Figure 7.2b shows the structure of the rectangular columnar phase (D_r); as yet only rectangular phases with the disordered molecular arrangement within the columns (D_{rd}) have been identified. Depending upon the space group four variants of rectangular structure have been identified and Fig. 7.2b shows a typical example. In these phases the columns occupy a rectangular lattice but the short axis of the molecules in each column is tilted away from the column axis. The lattice distance from a molecule to its nearest neighbour is directional (i.e., a≠b). The tilt direction of the columns is a herringbone arrangement as shown in the figure. An oblique phase is characterized by an average tilt of the molecules with respect to a plane perpendicular to the column axis.

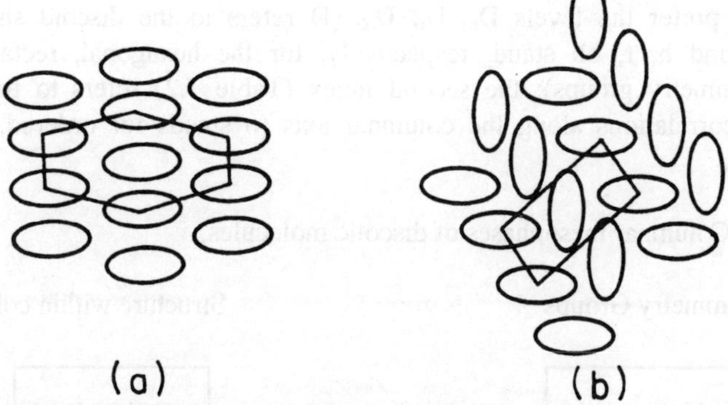

Fig. 7.2. (a) hexagonal and (b) rectangular columnar phases. The molecules are tilted with respect to the column axis in both phases.

7.3 STRUCTURE-PROPERTY RELATIONSHIP

The vast majority of discotic materials possess a molecular structure as shown[8] in Fig. 7.3. The discotic central core unit is usually benzene or a polyaromatic such as triphenylene or phthalocyanine, but the columnar phases have been synthesized with alicyclic cores such as cyclohexane and carbohydrate. The triphenylene and benzene cores usually have six peripheral chains whereas the phthalocyanine cores have eight peripheral moieties. The peripheral chains are usually identical and chosen from the examples shown in Fig. 7.3.

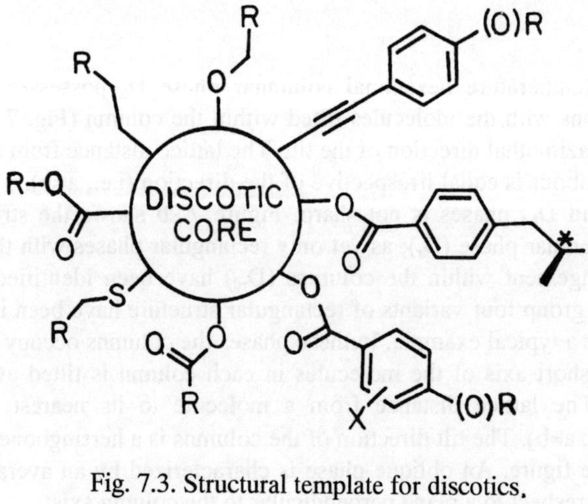

Fig. 7.3. Structural template for discotics.

7.3.1 Benzene Based Compounds

Most of the discotic mesogens generate columnar phases; some of these also exhibit the discotic nematic (N_D) phase. Very few materials generate only the N_D phase. The hexaalkanoyloxy benzenes (for example, compound T 7.3a, see Table 7.3) were the first discotic materials to exhibit the D_{rd} phase. Compound T 7.3b forms D_{rd} phase only when intermediate peripheral chain lengths are used; the hexa (alkoxyphenyl) benzenes have six directly linked benzene rings to a central benzene ring that forms a highly conjugated central core. Since both of these compounds (T 7.3a,b) have only small planar areas of the core, they show relatively low mesomorphic tendencies.

Table 7.3. Benzene based compounds; (a) hexaalkanoyloxy benzenes, (b) hexa (alkoxyphenyl) benzenes.

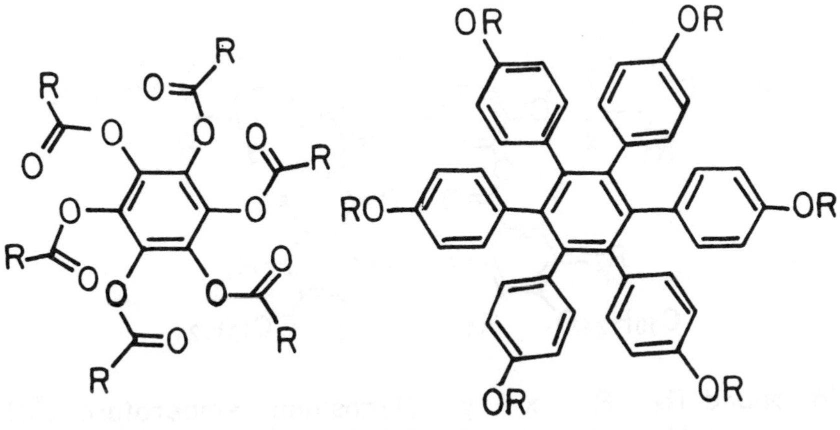

(a) $R = C_5H_{11}$ (b)

K 68.3 D_{rd} 86.0 IL K 68.0 D_{rd} 97.0 IL

7.3.2 Triphenylene-Core Based Compounds

The triphenylene core is composed of three benzene rings that give a planar aromatic unit; accordingly six peripheral moities can be symmetrically linked with the core. Since in these compounds core becomes much larger than benzene, the mesomorphic tendency is much higher. Table 7.4 shows the phase sequences (transition temperatures) of some triphenylene based compounds. As can be seen among all the compounds listed, the parent material (T7.4a) shows a much higher mesomorphic behaviour and exhibits a D_{rd} phase at the lower temperature and a N_D

phase at higher temperature. The influence of steric effects on the mesomorphic properties can be seen from the Table 7.4 (compound T7.4b-e); these compounds differ from each other in the presence of lateral methyl substituents in the peripheral units. It can be seen that the position and number of CH_3 units affect the liquid crystalline character significantly.

Table 7.4. Effect of the peripheral units on the transition temperatures of some triphenylene.

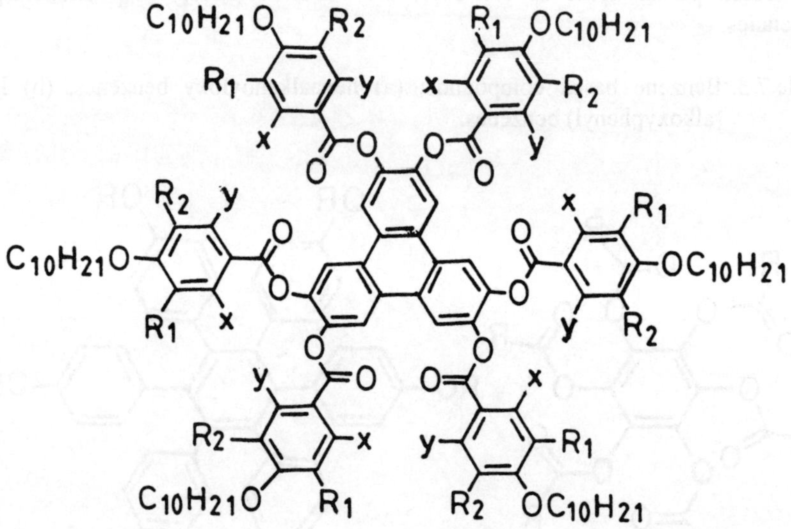

Compound	R_1	R_2	x	y	Transition Temperatures (°C)
a	H	H	H	H	K142 D_{rd} 191 N_D 212 IL
b	CH_3	H	H	H	K102 D_{hd} 127 N_D 192 IL
c	H	H	CH_3	H	K107 N_D 162 IL
d	CH_3	CH_3	H	H	K157 D_{hd} 167 N_D 182 IL
e	H	H	CH_3	CH_3	K108 N_D 134 IL

7.3.3 Truxene-Core Based Compounds

The truxene core consists of four benzene rings; the three radial rings are linked symmetrically to the central ring in two ways : first through a conjugative single bond, and second by a methylene spacer. Thus the compounds based on the hexa-substituted truxene core exhibit very high mesomorphic character.

7.4 PHASE TRANSITIONS IN DISCOTIC NEMATIC AND COLUMNAR PHASES

Transitions between the columnar (D) and discotic nematic (N_D) phases have been observed in few compounds. Majority of discotic mesogens show a phase sequence solid-columnar-isotropic. Available data on the discotic homologous series show that the lower members of the series generate only the N_D phase, the next few members exhibit both D and N_D phases, and the higher members exhibit transition directly from D to the isotropic phase. A number of experimental studies have been carried out on the structure and physical properties of discotic columnar phases. In contrast the first experimental data on the orientational order of N_D phase has been reported by Aver'yanov[27]. However, the results obtained[27] are at odds with the results of computer simulations for rigid discotic molecules (see below).

Attempts were made by several authors[28–30] to give a qualitative description of the D-N_D-IL transitions by extending McMillan's mean-field model of smectic A so that the density wave is periodic in two dimensions. Theoretical calculations were done by Feldkamp et. al.[29] for the D_h phase and by Chandrasekhar et. al.[30] for the general case of D_r phase. The hexagonal phase can be described by a superposition of three density waves with wavevectors[14,29]

$$\mathbf{A} = \frac{4\pi}{\sqrt{3}\,a}\hat{j}$$

$$\mathbf{B} = \frac{4\pi}{\sqrt{3}\,a}\left(\frac{\sqrt{3}}{2}\hat{i} - \frac{1}{2}\hat{j}\right)$$

$$\mathbf{C} = \mathbf{A} + \mathbf{B}$$

where a is lattice constant. The single particle potential depends on both the orientation of the short axis of the molecule and the position of its center of mass; it can be expressed in the mean-field approximation as

$$U_1(x, y, \cos\theta) = -\ U_0 P_2(\cos\theta)\{\overline{P}_2 + \alpha\,\overline{\sigma}[\cos(\mathbf{A}\cdot\mathbf{r}) + \cos(\mathbf{B}\cdot\mathbf{r}) + \cos(\mathbf{C}\cdot\mathbf{r})]\} \qquad (7.1)$$

In eq. (7.1) only the leading terms in the Fourier expansion have been considered. Here U_0 is the interaction energy that determines the nematic-isotropic transition, α is the McMillan parameter given by

$$2\exp\left[-(2\pi r_0 / \sqrt{3}a)^2\right]$$

where r_0 represents the range of interaction that is of the order of the size of the aromatic core, and the mixed (orientational-translational) order parameter $\overline{\sigma}$ is given by

$$\overline{\sigma} = \frac{1}{3} < [\cos(\mathbf{A}\cdot\mathbf{r}) + \cos(\mathbf{B}\cdot\mathbf{r}) + \cos(\mathbf{C}\cdot\mathbf{r})]\, P_2(\cos\theta) > \qquad (7.2)$$

The form of the potential (7.1) ensures that the energy of the molecule is a minimum when the disc is centred in the column with its plane normal to the Z-axis.
The free-energy can now be evaluated from the relation

$$\frac{\beta A}{N} = \frac{1}{2}\beta U_0(\overline{P}_2^{\,2} + \alpha\overline{\sigma}^{\,2}) - \ell n\left[\frac{2}{\sqrt{3}\,a^2}\int dx\,dy \int_0^1 d(\cos\theta)\, \exp(-\beta U_1)\right] \tag{7.3}$$

Three possible solutions result from the calculations

(i) $\overline{P}_2 = \overline{\sigma} = 0$, IL phase

(ii) $\overline{P}_2 \neq 0,\ \overline{\sigma} = 0$, N_D phase

(iii) $\overline{P}_2 \neq 0,\ \overline{\sigma} \neq 0$, D_h phase

For the given values of α and T, the minimization of free-energy determines which phase is stable. It has been found that the temperature range of the discotic nematic phase decreases with increasing α and for $\alpha > 0.64$ direct transition from a columnar phase to an isotropic phase is observed; further, the columnar–discotic nematic transition is first-order.

Some qualitative improvements in the phase diagram for a homologous series have been obtained by Ghose et. al.[31] by including more of Fourier components in the potential and using variational principle to solve the equations. A simpler mean field version of this theory giving essentially similar results has also been presented by these authors[32].

It has been shown by Chandrasekhar et. al.[30] that the nature of the phase diagram depends on the axial ratio b/a for the rectangular lattice. When b/a is only slightly different from $\sqrt{3}$ the phase diagram is similar to that for the hexagonal structure. As the asymmetry of the lattice is increased, one of the density wave disappears before the other to generate a smectic (or lamellar) phase which, in turn, transforms to the N_D phase at a higher temperature.

Theoretical models have been proposed [33–35] to describe transitions between different types of columnar phases. The influence of molecular interactions and short–range orientational order on the discotic nematic-isotropic transition properties has been investigated by us[36]. A remarkable symmetry between phase transition properties of calamitic (N_u phase) and discotic (N_D phase) molecules has been found. This is in agreement with the computer simulation[37] and other[38] results.

Frenkel and coworkers[39–42] have shown that the main features of the phase diagrams of discotic molecules could be reproduced by a system of cut-spheres. The cut-sphere is a hard convex body (see Fig. 7.4) that is defined as follows : Consider a sphere with diameter D and remove those parts of the sphere that are more than ½L above or below the equatorial plane; L is the height of the particle in the direction perpendicular to the equatorial plane. What remains is a disc-like object with flat caps and spherical rims. The aspect ratio L/D determines the flatness of the cut-sphere. From the point of view of numerical calculations, the cut-sphere have many advantages over other models for the discotic molecules such as short cylinder[46] and oblate spherocylinder[47]. They have a finite excluded volume even in case of perfect alignment. An assembly of aligned cut-spheres cannot be mapped on a hard sphere system. The test for the overlap between two cut-spheres can be carried out in a finite number of steps.

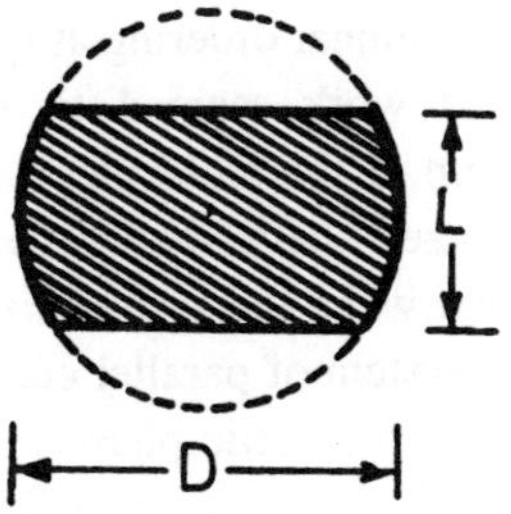

Fig. 7.4. Side view of a cut sphere with diameter D and thickness L.

For arbitrary L/D ratio, the second virial coefficient of cut-spheres is given by[39a,48]

$$B_2 = \frac{\pi D^3}{6}\left(\cos\theta_M\left[1 + \frac{1}{2}\sin^2\theta_M\right] + 3\left[\cos\theta_M + \frac{1}{2}\theta_M\sin\theta_M\right]\right.$$
$$\left.\times\left[\cos\theta_M + \frac{1}{2}\sin\theta_M\right]\right) \tag{7.4}$$

where $\theta_M \equiv \arccos(L/D)$. At high densities, cut spheres can be stacked in a regular close-packed lattice. The packing fraction of this close-packed structure is given by

$$\eta_{cp} = \frac{\pi}{6}(3-(L/D)^2)^{1/2} \tag{7.5}$$

For L/D = 1 (hard sphere), eq. (7.5) gives the well known hard sphere result $\eta_{cp} = \pi/\sqrt{18}$, while for $L/D \to 0$ (flat, cylindrical platelets), the 2D hard-disk value $\eta_{cp} = \pi/\sqrt{12}$ is recovered. The volume of a single cut-sphere is

$$v_0 = \frac{\pi}{12}D^3(L/D)[3 - (L/D)^2] \tag{7.6}$$

and so the number density at regular close packing is given by

$$\rho_{cp} = \frac{2(D/L)}{[3 - (L/D)^2]^{1/2}} \tag{7.7}$$

A number of simulation studies on a system of cut-spheres with L/D = 0.1, 0.2 and 0.3 over a range of densities are available[39–43]. Surprisingly, it turned out that the systems with different values of aspect ratio behave completely differently. The simulation studies[44,45] has also been performed to investigate the influence of attractive interactions (Polar and Gay-Berne potentials) on the phase behaviour of discogens.

For a system of 256 freely rotating cut-spheres with aspect ratio L/D = 0.1 Frenkel[39] identified a phase sequence isotropic-discotic nematic-columnar-solid. Apparently no smectic phase was identified. Although the transition densities with precision could not be located, it was found[39] that a discotic nematic phase exists at packing fraction $\eta \approx 0.32$ and columnar ordering at $\eta \approx 0.57$. The columnar ordering disappears at $\eta \approx 0.41$. This work seemed to indicate that constraining the orientational degrees of freedom of the molecules can modify the phase behaviour appreciably. Veerman and Frenkel[41] have shown that small systems (N ~ 100) can stabilize phases which disappear when the system size is increased. Azzouz et. al.[43] have shown that a constrained system of parallel cut-spheres transforms from a low density discotic nematic phase to a smectic phase in the density range $\eta \approx 0.4$ and that this latter phase exists upto density $\eta \approx 0.566$. These authors[43] considered 2500 parallel cut-spheres with L/D = 0.1 in a cubic box with Z-axis parallel to the symmetry axis of the cut-spheres. For $\eta < 0.4$ the initial configuration was assumed to be a perfect bcc lattice, while for $\eta = 0.4$–0.57 the initial configuration was that of a close-packed structure with fcc like hexagonal symmetry within the layers parallel to the X Y plane. The number of planes in the X, Y and Z directions was $N_x = 6$, $N_y = 7$, and $N_z = 59$ corresponding to a total of 2478 particles in an orthorhombic (but nearly cubic) simulation box of sides L_x, L_y and L_z. The close-packed structure was then uniformly expanded in all three directions in order to obtain a given lower density. The a priori trial moves were adjusted to give an acceptance ratio of 50%. The usual criterion for the overlap of the cut-spheres were adopted and the structure of system was simulated by evaluating various correlation functions. The compressibility factor was calculated over a range of packing fraction $\eta \approx 0.30$–0.566. The data points show the existence of a low-density discotic nematic phase and a high-density smectic phase. The transition between two phases occurs in the vicinity of $\eta \approx 0.40$. In the density region considered no tendency of the formation of columnar ordering was observed. These results show that hindering the rotational motion drastically modifies the phase diagram of the cut-sphere system.

Veerman and Frenkel[42] examined whether or not columnar ordering is stable as the system size is increased. They performed MC simulations on a system of hard cut-spheres with aspect ratio L/D = 0.1, 0.2 and 0.3 and found that the phase behaviour is strongly dependent on the ratio L/D of the particles. In all cases they performed both expansion runs, starting from the close-packed crystal, and the compression runs, starting from the dilute gas phase, or in some cases, some other well-equilibrated-state point. For L/D = 0.1, systems of 256 and 288 particles in the isotropic and discotic nematic phases were considered, while 576 particles were taken in the columnar and crystalline phases. Most of the simulations in case of L/D = 0.2 and 0.3 were performed on larger systems N = 1728 and 2048 for L/D = 0.2, and N = 2048 for L/D = 0.3. The systems were allowed to equilibrate at constant volume after energy expansion or compression. During this equilibration, the simulation box was allowed to change its shape, in order to relieve possible stresses in the system. In the low density simulations (isotropic and discotic nematic phases) the shape of the box was always kept fixed. In all the MC production runs, the hexagonal-close-packed structure was assumed as the initial configuration with the crystal unit cell containing four cut-spheres with their axes parallel to the Z-axis.

The first MC simulations of cut-spheres for L/D = 0.1 by Frenkel[39] provided strong evidence that the system can form four distinct phases : isotropic, discotic nematic, columnar and solid. The range of stability of these phases was estimated by Veerman and Frenkel[42]. They found that a stable columnar phase is indeed formed in systems of sufficiently flat cut-spheres. For L/D = 0.1 the columnar phase exists with a discotic nematic phase. The transitions between the discotic nematic and columnar phases is strongly first-order. No evidence for any effect of the system size on the stability of columnar phase was found. It proved very difficult to estimate accurately the location of the columnar-solid phase transition on the basis of the simulation data. For the system size studied, the main problem was that the crystal-columnar transition appears to be continuous. For this transition $\rho^* = 0.80$ was identified as the lower limit. The density of the discotic nematic phase at the isotropic-discotic nematic transition was estimated as $\rho^*_{nem} = 0.335 \pm 0.005$, while the isotropic transition density was found to be $\rho^*_{iso} = 0.330 \pm 0.005$.

The first numerical study of cut-spheres for L/D = 0.2 was reported by Frenkel[39b]. This work provided the evidence for the existence of a novel phase with cubic orientational order.

Veerman and Frenkel[42] performed a more detailed study of the different phases of the cut-spheres system with L/D = 0.2. They found that the discotic nematic phase does not exist; a columnar phase, and a "cubatic" phase with extended cubic orientational order and no translational order were found. However, the transition densities could not be located accurately. The cubatic phase is thermodynamically stable in a narrow density range (see Fig. 7.5). The existence of phases with cubic orientational order has been predicted by Nelson and Toner[49] in different context. The Nelson-Toner theory has shown that the cubic-bond orientational order without translational order can occur in a solid phase containing free dislocation loops. For L/D = 0.3, it was found that the system has an isotropic, a solid and probably a cubatic phase. However, the cubatic phase is not thermodynamically stable. The formation of a columnar phase was not observed.

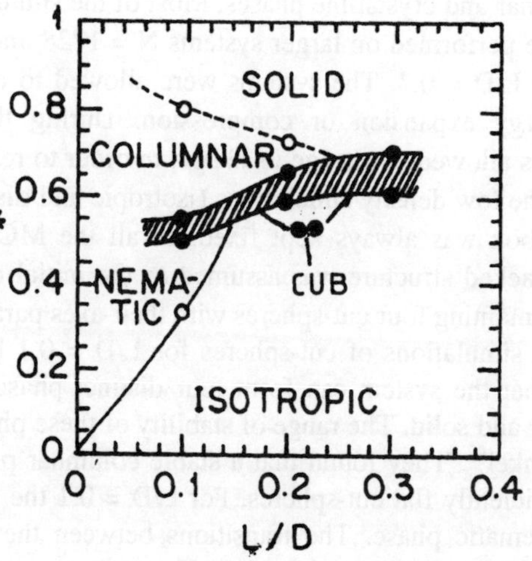

Fig. 7.5. Phase diagram of cut-spheres with aspect ratio L/D between 0.1 and 0.3 (Ref. 42).

The MC simulation of polar discotic molecules has been performed[44] in order to study the effect of dipolar interactions on the phase behaviour of a system of cut-spheres with L/D = 0.1. The computations were done with a cubic simulation cell for a system size of N = 480 molecules. Although small values of permanent moments $\mu^* [= (\mu^2 / k_B TD^2)^{1/2}] \approx 0.1 - 0.4$ were taken, but they correspond to very strong dipolar couplings between neighbouring cut-spheres because their distance can be as small as 0.1D. The main conclusions of this work can be summarized as follows : in the isotropic and discotic nematic phases the thermodynamic properties and structural order are not modified qualitatively by the dipolar interactions. The density domain of the stable isotropic phase seems to have the same extension for nonpolar and polar cut-spheres up to values of $\mu^* \leq 0.5$. In the discotic nematic phase, the domain of stability appears to be reduced in favour of that of the columnar phase. In the columnar phase the dipole-dipole interaction can induce the polarization of the columns which are arranged with antiferroelectric order. When the orientation of the dipole moment breaks the axial symmetry of the molecules, a rectangular columnar phase with molecules tilted with respect to the columnar axis appears to be the stable dense liquid crystal phase.

Emerson et al.[45] performed the molecular dynamics simulations for a system of molecules interacting with Gay-Berne potential (eq. (6.121)). With a simple parametrization ($x_0 = 0.345$, $k' = 1/5$, $\nu = 2$, $\mu = 1$) of the Gay-Berne potential, they have constructed a model discogen which exhibits isotropic, discotic nematic and columnar phases. At low density the symmetry of the columnar phase is D_{ho}, but this forms a rectangular phase at higher densities. The system contains 256 discotic particles in a cubic box with the usual boundary conditions and minimum image summation. The simulations were started from an α-f.c.c. lattice at the trial density $\rho^*(= N\sigma_0^3 / V)$ of 2.7. It was found that at this density the initial lattice was slow to disorder, and so the much lower density $\rho^* = 1.8$ was employed to facilitate melting. At the temperature T^* $(= k_BT/\epsilon_0) = 4.0$ the crystal melted rapidly to form an isotropic liquid. The system was slowly compressed in stages by increasing the density in steps of 0.3 and reequilibrated at the same temperature until the density has reached 2.7. At this state point, the $\overline{P}_2$ was found to be 0.94. The temperature was then raised in steps of 1.0 unit until $T^* = 14.0$ at which $\overline{P}_2$ had dropped to a value consistent with that of isotropic phase. The scaled density was then increased to 3.0 and the system reequilibrated at $T^* = 16.0$; this was also found to be orientationally disordered. The system was then slowly cooled by reducing the temperature until a state point with T^* close to 0.5 was obtained. From the available simulation data following conclusions were derived : By probing the structures of the ordered phases with the aid of positional distribution functions a phase sequence columnar (low T^*) → discotic nematic → isotropic (high T^*) was found. The radial distribution function $g(r^*)$ is very structured, indicating a high degree of translational order. The first peak must be the result of stacking of discs in a single column, but the particles are not close-packed. The second peak occurs at a separation ($0.9\ \sigma_0$) less than σ_0, so the columns should be interdigitated. The formation of ordered columns was monitored via the longitudinal correlation function $g_{\|}(r^*)$ which provides information on the positional ordering in a direction parallel to the director. The data reveals that at the higher temperature (N_D phase) there is very little structure and the function decays to unity very quickly. There must be correlations at shorter separations from particles within neighbouring columns, and two discs within one column are staggered with respect to those in an adjacent one. This situation is confirmed if one examines snapshots of a sample configuration taken from a production run in the columnar phase (see Fig. 7.6). It can be seen that the packing of the columns of discs, instead of hexagonal ordering, forms a quasi-rectangular ordering of the columns. From the parallel projection shown in Fig. 7.6, it can be observed that not only are the discs in one column staggered with respect to those in the neighbouring columns but also they overlap to a small extent. Consequently, the columns with significant interdigitation would not readily form a hexagonal structure and would prefer instead to pack in a square or rectangular manner. Figure 7.7 shows the situation at a lower density ($\rho^* = 2.5$) and temperature ($T^* = 2.0$). At this density a columnar phase is also observed. The snapshots taken shows that the columns are now packed in a hexagonal array. From the simulations data of perpendicular correlation function $g_{\perp}(r_{\perp}^*)$ it was noted that for the columnar phases at $\rho^* = 3.0$, $T^* = 2.5$ the first peak is centred at about $r_{\perp}^* = 0.8$; thus the separation between the centers of mass of the discs in neighbouring columns is significantly less than the diameter of one disc. At the state point $\rho^* = 2.5$, $T^* = 2.0$ the first peak is centred almost exactly at a separation close to 1.7 which is consistent with a second shell of neighbours in a hexagonal array. Thus, it can be concluded that at low densities the columns of the discs pack in the usual hexagonal way, while at higher densities, when the columns are interdigitated, hexagonal packing is no longer possible and instead a rectangular structure is formed.

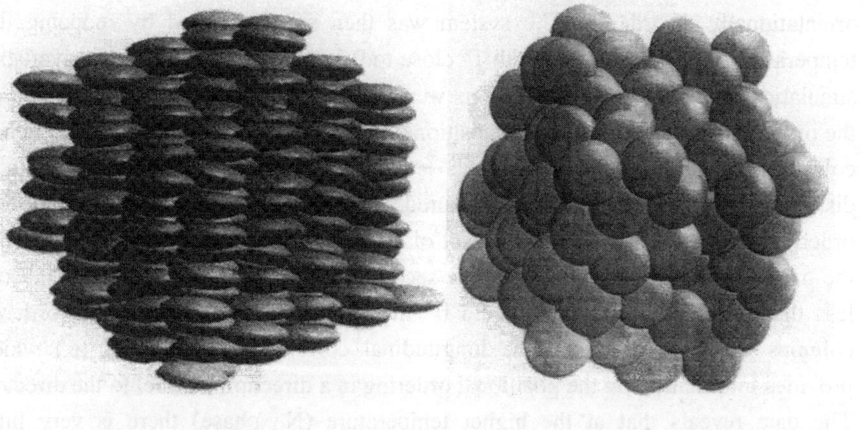

(a) (b)

Fig. 7.6. Snapshots taken (Ref. 45) at the state point $\rho^* = 3.0$ and $T^* = 2.5$; (a) a view seen in direction perpendicular to the director and (b) the molecules are viewed looking down the director.

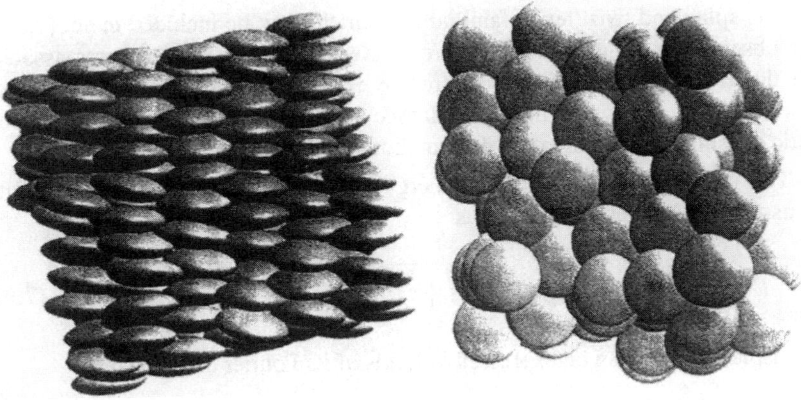

(a) (b)

Fig. 7.7. Snapshots taken (Ref. 45) at the state point $\rho^* = 2.5$ and $T^* = 2.0$. As in Fig. 7.6, views are shown in direction (a) perpendicular and (b) parallel to the director.

7.5 CONTINUUM DESCRIPTION OF COLUMNAR PHASES

7.5.1 Uniaxial Columnar Phases

The derivation of the distortion free-energy for small values of the derivatives of the two dimensional vector field $\mathbf{u}(\mathbf{r})$ follows much the same lines as adopted for the smectics. Let us assume that the liquid-like columns are along the Z-axis and that the two dimensional lattice (hexagonal) is parallel to the XY plane. In such a structure two basic deformations may result : (i) the curvature deformation (or bending) of the columns without distortion of lattice, and (ii) the dilation of lattice without columnar curvature. A term associated with the coupling between the two types of deformation can also be considered. However, it has been shown by Kleman and Oswald[50] that the coupling term only rescales the bend elastic constant of the columns. Hence the distortion free-energy density may be written as[28,51]

$$f = \frac{1}{2}B\left(\frac{\partial u_x}{\partial x} + \frac{\partial u_y}{\partial y}\right)^2 + \frac{1}{2}C\left[\left(\frac{\partial u_x}{\partial x} - \frac{\partial u_y}{\partial y}\right)^2 + \left(\frac{\partial u_x}{\partial y} + \frac{\partial u_y}{\partial x}\right)^2\right]$$

$$+ \frac{1}{2}K_3\left[\left(\frac{\partial^2 u_x}{\partial z^2}\right)^2 + \left(\frac{\partial^2 u_y}{\partial z^2}\right)^2\right] \qquad (7.8)$$

Both the splay and twist terms familiar in nematics can be included in eq. (7.8), but their observation would always be pre-empted by the first-order terms associated with the B and C. Here B and C are the elastic constants corresponding to the deformation of the two-dimensional lattice in its own plane and K_3 is the bend elastic constant for the curvature deformation of the columns.

The B and C have dimensions energy length^{-3}, while K_3 has energy length^{-1}; two associated lengths are defined as

$$\lambda_B = \left(\frac{K_3}{B}\right)^{1/2}, \qquad \lambda_C = \left(\frac{K_3}{C}\right)^{1/2} \tag{7.9}$$

The displacement u can be expressed in terms of its Fourier components,

$$u(\mathbf{r}) = \sum_q u(\mathbf{q}) \exp(i\mathbf{q}\cdot\mathbf{r}) \tag{7.10}$$

Substituting (7.10) in (7.8), one obtains, in the harmonic approximation

$$f = \frac{1}{2}\sum_q (B_0 q_\perp^2 + k_0 q_z^4) < u_q^2 > \tag{7.11}$$

where $B_0 = B+2C$, $k_0 = 2K_3$ and $q_\perp^2 = q_x^2 + q_y^2$. From the equipartition theorem

$$< u_q^2 > = k_B T / (B_0 q_\perp^2 + k_0 q_z^4) \tag{7.12}$$

The mean square displacement at any lattice point is given by

$$< u^2 > = \sum_q < u_q^2 > = \frac{1}{(2\pi)^3}\int < u_q^2 > dq$$

$$= \frac{k_B T}{(2\pi)^3}\int_{2\pi/L}^{2\pi/d}\int_{2\pi/L'}^{\infty}\frac{2\pi q_\perp dq_\perp dq_z}{B_0 q_\perp^2 + k_0 q_z^4} \tag{7.13}$$

where L is the linear dimension of the lattice in the XY plane with d its periodicity and L′ is the length of the columns.

In case $L' >> L$,

$$< u^2 > = [k_B T / 4B_0(\lambda d)^{1/2}]\,[1-(d/L)^{1/2}] \tag{7.14}$$

where $\lambda = (k_0/B_0)^{1/2}$. The structure is thus stable as $L \rightarrow \infty$.

7.5.2 Biaxial Columnar Phases

The distortion free–energy for the biaxial columnar phases can be expressed as[52]

$$f = \frac{1}{2}B\left(\frac{\partial u_x}{\partial x}\right)^2 + \frac{1}{2}B'\left(\frac{\partial u_y}{\partial y}\right)^2 + \frac{1}{2}C\left(\frac{\partial u_x}{\partial y} + \frac{\partial u_y}{\partial x}\right)^2$$

$$+ D\frac{\partial u_x}{\partial x}\frac{\partial u_y}{\partial y} + K\left(\frac{\partial^2 u_x}{\partial z^2}\right)^2 + K'\left(\frac{\partial^2 u_y}{\partial z^2}\right)^2 \qquad (7.15)$$

Equation (7.15) holds for the rectangular columnar phase as well as for the oblique rectangular columnar phase. In case of oblique phase, expressions such as $\partial u_x/\partial x$, $\partial u_x/\partial z$ can be written without violating the symmetry requirements. However, since $\partial u_x/\partial z$ refers to a mere rotation, it does not change the energy.

REFERENCES

1. O. Lehmann, Z. Physikal. Chem. **4**, 462 (1889).
2. L. Gattermann, and A. Ritschke, Ber. Deut. Chem. Ges. **23**, 1738 (1890).
3. D. Vörlander, Kristallinisch Flussige Substanzen (Stuttgart, Enke, 1908); Ber. Deut. Chem. Ges. **39**, 803 (1906).
4. G. Friedel, Ann. Physique **18**, 273 (1922).
5. S. Chandrasekhar, B.K. Sadashiva, and K.A. Suresh, Pramana **9**, 471 (1977).
6. J. Billard, J.C. Dubois, N.H. Tinh, and A. Zann, Nouveau J. Chim. **2**, 535 (1978).
7. S. Chandrasekhar, Liquid Crystals (2nd edition, Cambridge Univ. Press, 1992) Chapt. 6, p. 388.
8. P.J. Collings, and M. Hird, Introduction to Liquid Crystals : Chemistry and Physics (Taylor and Francis Ltd., London, 1997) Chapt. 4, p. 79.
9. S. Chandrasekhar, (a) Adv. in Liq. Cryst. Vol. 5, ed., G.H. Brown (Acad. Press, New York, 1982) p. 47; (b) Phil. Trans. Roy. Soc. London **A309**, 93 (1983); (c) Liquid Crystals **14**, 3 (1993); (d) in, Handbook of Liquid Crystals, eds., D. Demus, J. Goodby, G.W. Gray, H.W. Spiess, and V. Vill (Weinheim : Wiley – VCH) Chapt. VIII.
10. C. Destrade, N.H. Tinh, H. Gasparoux, J. Malthete, and A.M. Levelut, Mol. Cryst. Liquid Cryst. **71**, 111 (1981).
11. A.M. Levelut, J. Chim. Phys. **88**, 149 (1983).
12. C. Destrade, P. Foucher, H. Gasparoux, and N.H. Tinh, Mol. Cryst. Liquid Cryst. **106**, 121 (1984).
13. H.P. Hinov, Mol. Cryst. Liquid Cryst. **136**, 221 (1986).

14. S. Chandrasekhar, and G.S. Ranganath, Rep. Prog. in Phys. **53**, 57 (1990).
15. P. Hindmarsh, M.J. Watson, M. Hird, and J.W. Goodby, J. Materials Chem. **14**, 2111 (1995).
16. S. Chandrasekhar, and S. Krishna Prasad, Contemp. Phys. **40**, 237 (1999).
17. A.M. Giroud - Godquin, and J. Billard, Mol. Cryst. Liquid Cryst. **66**, 147 (1981).
18. L.K. Runnels, and C. Colvin, in, Liquid Crystals Vol. 3, eds., G.H. Brown, and M.M. Labes (Gordon and Breach Science Publishers, New York, 1972) p. 299.
19. R.E. Peierls, Annales de 'lInstitut Henri Poincare' **5**, 177 (1935); L.D. Landau, Phys. Z. Sowjet Union **2**, 26 (1937).
20. E. Alexander, and K. Herrmann, Z. Kristallographie **69**, 285 (1928); **70**, 328 (1929).
21. D. Vorländer, Chemische Kristallographie der Flüssigkeiten, Akadem. Verlagsges (Leipzig 1924).
22. J.D. Brooks, and G.H. Taylor, Carbon **3**, 185 (1965).
23. J.E. Zimmer, and J.L. White, Mol. Cryst. Liquid Cryst. **38**, 177 (1977).
24. J. Malthete, and A. Collet, Nouv. J. Chem. **9**, 151 (1985.
25. H. Zimmerman, R. Poupko, Z. Luz, and J. Billard, Z. Naturforsch. **a40**, 149 (1985).
26. B. Mourey, J.N. Perbet, M. Hareng, and S. Le Berre, Mol. Cryst. Liquid Cryst. **84**, 193 (1982);

    G. Heppke, H. Kitzerow, F. Oestreicher, S. Quentel, and A. Ranft, Mol. Cryst. Liquid Cryst. Lett. **6**, 71 (1988).
27. E.M. Aver'yanov, Pis'ma Zh. Eksp. Teor. Fiz. **61**, 796 (1995).
28. E.I. Kats, Sov. Phys. JETP **48**, 916 (1978).
29. G.E. Feldkamp, M.A. Handschy, and N.A. Clark, Phys. Lett. **A85**, 359 (1981).
30. S. Chandrasekhar, K.L. Savithramma, and N.V. Mudhusudana, ACS Symposium on Ordered Fluids and Liquid Crystals Vol. 4, eds., J.F. Johnson and A.C. Griffin (Plenum Press, New York, 1982) p. 299.
31. D. Ghose, T.R. Bose, C.D. Mukherjee, M.K. Roy, and M. Saha, Mol. Cryst. Liquid Cryst. **138**, 379 (1986).
32. D. Ghose, T.R. Bose, M.K. Roy, C.D. Mukherjee, and M. Saha, Mol. Cryst. Liquid Cryst. **154**, 119 (1988).
33. D. Ghose, T.R. Bose, M.K. Roy, M. Saha, and C.D. Mukherjee, Mol. Cryst. Liquid Cryst. **173**, 17 (1989).

34. E.I. Kats, and M.I. Monastyrsky, J. Phys. (Paris) **45**, 709 (1984).

35. Y.F. Sun, and J. Swift, Phys. Rev. **A33**, 2735, 2740 (1984).

36. (a) K. Singh, U.P. Singh, and S. Singh, Liquid Crystals **3**, 617 (1988); (b) T.K. Lahiri, K. Singh, and S. Singh, Liquid Crystals **27**, 431 (2000).

37. D. Frenkel, B.M. Mulder, and J.P. McTague, Phys. Rev. Lett. **52**, 287 (1984).

38 U.P. Singh, and Y. Singh, Phys. Rev. **A33**, 2725 (1986).

39. D. Frenkel, (a) Liquid Crystals **5**, 929 (1989); (b) in proceedings of the Les Houches Summer School on : Liquids, Freezing and Glass Transition, eds, J.P. Hansen, D. Levesque, and J. Jin Justin (North Holland, 1991) p. 691.

40. D. Frenkel, Computer Simulations of Phase Transitions in Liquid Crystals, in, Phase Transitions in Liquid Crystals, eds., S. Martellucci, and A.N. Chester (Plenum Press, 1992) Chapt. 5, p. 67.

41. J.A.C. Veerman, and D. Frenkel, Phys. Rev. **A43**, 4334 (1991).

42. J.A.C. Veerman, and D. Frenkel, Phys. Rev. **A45**, 5632 (1992).

43. H. Azzouz, J.M. Caillol, D. Levesque, and J.J. Weis, J. Chem. Phys. **96**, 4551 (1992).

44. G.J. Zarragoicoechea, D. Levesque, and J.J. Weis, Mol. Phys. **78**, 1475 (1993).

45. A.P.J. Emerson, G.R. Luckhurst, and S.G. Whatling, Mol. Phys. **82**, 113 (1994).

46. M.C. Duro, J.A. Martin–Pereda, and L.M. Sese, Phys. Rev. **A37**, 284 (1988).

47. M. Wojcik, and K.E. Gubbins, Mol. Phys. **53**, 397 (1984).

48. T. Boublik, and I. Nezbeda, Coll. Czech. Chem. Commun. **51**, 2301 (1986).

49. D.R. Nelson, and J. Toner, Phys. Rev. **B24,** 363 (1981).

50. M. Kleman, and P. Oswald, J. Phys. (Paris) **43**, 655 (1982).

51. J. Prost, and N.A. Clark, in, Proceedings of the International conference on Liquid Crystals, Bangalore, Dec. 1979; ed., S. Chandrasekhar (Heyden, London, 1980) p. 53.

52. H. Brand, and H. Pleiner, Phys. Rev. **A24**, 2777 (1981).

8

Polymer Liquid Crystals

8.1 INTRODUCTION

Polymers are long-chain molecules (or macromolecules) formed by the repetition of certain basic units or segments known as monomers. Polymers having identical, repeating monomer units are called "homopolymers", whereas those formed from more than one polymer types are called "copolymers". In a copolymer the monomers may be arranged in a random sequence, in an alternating sequence, or may be grouped in blocks. Polymers invariably consist of a mixture of chain lengths, the size and distribution of which is independent of the structural units present and the manner of synthesis. The average size of each chain is often referred to as the degree of polymerization ($\overline{DP}$), while the distribution of the chain sizes as the polydispersity. A polydispersity of one (monodisperse) denotes that all of the polymer chains are identical in size. Overall, the properties of all polymers are different and highly dependent upon the manner of synthesis; some are soft, while others are very hard and strong.

The low molar mass mesogens can be used as monomers in the synthesis of liquid crystal polymers[1–21]. There are two aspects to the synthesis of macromolecule mesogens, the conventional synthesis to generate the monomeric unit(s) and then the polymerisation reaction that yields the desired liquid crystal polymer (LCP). It is the combination of polymer-specific properties, together with the properties specific to the liquid crystalline phase that has led to a multitude of new prospectives; this has made possible a wide range of applications of liquid crystal polymers with excellent properties. However, it must be noted that the subject of LCPs is much more involved than such a simple combination of two areas of science. Consequently, the understanding, at the molecular level, about their behaviour is much more worse as compared to the liquid crystal systems. The key issue centres around the questions – to what extent the ordered liquid crystal state influences the kinetics, polymerisation, stereochemistry and rheological behaviour of the polymers. What is the effect of the spacer unit, mesogenic unit, polymer backbone, distribution of chain sizes, flexibility of different units, etc., on the mesomorphic behaviour. Some of these questions will be examined in this chapter. We shall discuss about the various types of liquid crystal polymers, and some of their important properties.

8.2 TYPES OF LIQUID CRYSTAL POLYMERS

It is well known that low molar mass mesogens can generate liquid crystalline phases if the molecules are approximately anisotropic in shape. This suggests that the concept of calamitic and discotic molecules can be extended to the macromolecules. On the basis of molecular anisotropy two types of structures are possible[5,11,12,16–21] (see chapter 1, Fig. 1.5). In the first type the macromolecule as a whole has a more or less rigid form, whereas in the second kind only the monomer unit possesses a mesogenic structure, connected to the rest of the molecule via a flexible spacer.

On the basis of the location of mesogenic group, that is, depending on whether the mesogenic group is inserted within the main chain, or as side group, the LCPs can be divided mainly into two kinds (see Fig. 1.5) : main-chain liquid crystal polymers (MCLCPs) and side-chain liquid crystal polymers (SCLCPs). A third kind can also be generated by inserting the mesogenic units both within the main-chain and as side groups; this class is known as combined liquid crystal polymers (CLCPs). Polymers with more complex artitectures are also possible[21–23]. The mesogenic units used in the generation of LCPs can be rod-like, disc-like, amphiphilic, etc. In addition to linear polymer structure, polymers with many other artitectures, such as, cyclic[24,25], hyperbranched[26–28], dendrimeric[29,30], crosslinked[31,32], etc., have also been synthesized. Consequently, LCPs exhibit thermotropic and lyotropic mesophases as displayed by the thermotropic and lyotropic low molar mass liquid crystals.

8.3 TYPES OF MONOMERIC UNITS

The low molar mass liquid crystals, and the main-chain and side-chain liquid crystal polymers have been synthesized by employing the rigid rod-like mesogenic units. The flexible rod-like units, or rod-like units based on conformational isomerism have also been used in the synthesis of main-chain and side-chain liquid crystal polymers. Therefore, based on the difference between the rotational energy barrier of different configurational isomers or conformers of calamitic mesogens, Percec and Zuber[33] suggested the classification of rod-like mesogenic groups into three categories : rigid, semirigid or semiflexible, and flexible. The rigid rod-like groups are rigid units with rigid shape, like for example, diphenylacetylene, oligo–p–phenylene, benzoxazole, etc., and configurational isomeric units that require a high rotational energy barrier or activation energy, like for example, stilbene (ΔEa = 42.8 Kcal/mol). The semirigid or semiflexible rod-like mesogens are conformationally flexible and behave like configurational isomers with a medium rotational energy barrier; classic examples are aromatic amides and esters (ΔEa = 14.4 Kcal/mol and

7.1 Kcal/mol, respectively). The flexible rod-like mesogens are conformationally flexible groups with rotational energy barrier in the range ΔEa = 2.8–3.7 Kcal/mol; examples are phenylbenzylether (ΔEa = 3.87 Kcal/mol) and 1,2-diphenylethane (ΔEa = 3.61 Kcal/mol) based molecules.

Liquid crystal polymers exhibit the same mesophases as exhibited by the low molar mass liquid crystals. But the identification of polymeric mesophases is usually far more difficult in comparison with the low molar mass mesogens. The nematic phase can be readily characterized but the smectic phases often remain uncharacterized. On cooling, many liquid crystal polymers, like conventional polymers, exhibit a glass transition temperature (T_g). T_g is defined as the temperature at which the material becomes less rigid and more rubbery. At the transition into the polymer glass the orientational order is frozen in without any change. However, such a transition is difficult to define accurately.

8.4 MAIN CHAIN LIQUID CRYSTAL POLYMERS (MCLCPs)

Main chain liquid crystal Polymers consist of repeating anisotropic monomer (mesogenic) units, which form a long chain, as shown in Fig. 1.5 (chapter 1). The monomer unit must be anisotropic and bifunctional (one function at each end) to enable the polymerisation and the generation of liquid crystal phases. When the mesogenic units are directly linked, a rigid (and intractable) structure is produced for the polymer chain. If the linking units are long and flexible, a semiflexible polymeric structure results. The overall flexibility of the polymer chain can be influenced with the introduction of flexible spacers (e.g. alkyl chains) between the functional groups and the rigid mesogenic units. Both the structural composition and the degree of flexibility determine the liquid crystalline properties of MCLCPs.

It was observed in 1956 that a polypeptide formed a mesophase in solution. However, systematic studies on MCLCPs did not begin until the 1970s, and the first MCLCP was reported[34] in 1975. In the same year de Gennes[35] independently advanced the concept of the flexible spacr. Several work dealing the development of MCLCPs are available[3,20,21,34–40] in the literature.

Unlike low molar mass mesogens, the MCLCPs have chain units which vary in size. The distribution of chain sizes is responsible for the existence of wide-ranges of melting to a liquid crystal phase and wide ranges over which the polymer clears to the isotropic liquid. A schematic view of the uniaxial distribution of semiflexible polymer chains in the nematic and smectic phases is shown in Fig. 8.1. The rigid structure of different chain sizes of a rigid polymer finds it most inconvenient to generate a layered structure, and so usually exhibits the nematic phase. On the other hand, the polymers having flexible units between the mesogenic moieties can easily generate layered structures. The flexible spacer tends to help the layer-like packing and so longer the spacer units the greater the smectic tendency.

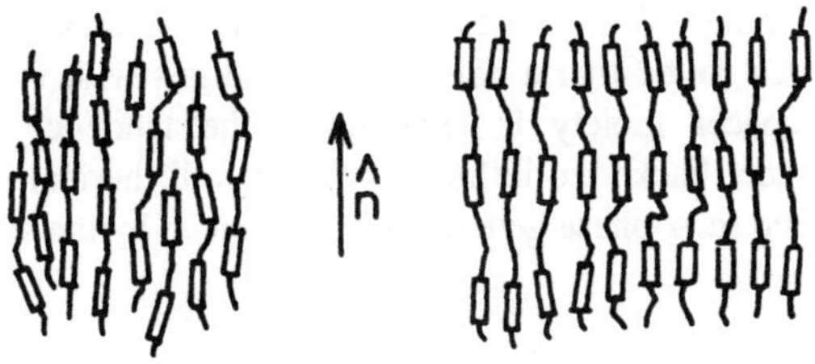

a. b.

Fig. 8.1. Schematic representation of (a) nematic and (b) smectic phases exhibited by MCLCPs.

8.5 SIDE CHAIN LIQUID CRYSTAL POLYMERS (SCLCPs)

Mesomorphic phase behaviour of SCLCPs, with respect to structure-properties relationship, is analogous to that of a low molar mass liquid crystals. The fixed mesogenic units must be attached to the polymer backbone through a flexible spacer usually at least 3–4 atoms long (for example, alkyl, alkoxyl or siloxane chains). There are two aspects of the description of the phase behaviour of SCLCPs. First, no rigid segments are present in the polymer main chain. The chain segments can rotate around the various bonds freely. This gives a tendency towards a statistically distributed chain conformation. The second aspect concerns with the role of mesogenic side units towards orientational order; these units restrict the conformational freedom of the main chain. The most crucial aspect is the union between the polymer main chain and the mesogenic side units. When mesogenic units are directly attached to the polymer main chain, the tendency towards a statistically distributed chain conformation hinders any anisotropic orientation of the side unit. Secondly, the importance of this effect is increased by the steric hindrance of the relatively voluminous structure that further decreases any orientational order. The influence of these two effects can be decreased by the use of flexible spacers in-between the polymer main chain and the mesogenic side groups. Due to the introduction of flexible spacers the motion of polymer main chain are decoupled to certain extent from the possible orientational order of the side groups. This concept is outlined in Fig. 8.2. Warner[41] predicted that the conformation of the polymer backbone should be distorted in the liquid crystal phase. In addition, both the X-ray scattering[42–45] and small angle neutron scattering[46–49] experiments have shown that the statistical random-coil conformation of the polymer backbone is slightly distorted in the nematic phase and highly distorted in the smectic phase.

If the above concept of decoupling is correct, the backbone should not influence the nature and thermal stabilities of the generated polymer mesophases. However, it has been seen that the backbone does affect the mesomorphic behviour of SCLCPs.

In fact, the spacer moiety does not totally decouple the mesogenic unit from the backbone, it has been found that a part of spacer aligns with the mesogenic units. However, as the spacer moiety is lengthened, the tendency of decoupling is increased. The polymer backbone influences the overall thermal stability, while the nature of least ordered mesophase generated is influenced by the spacer length.

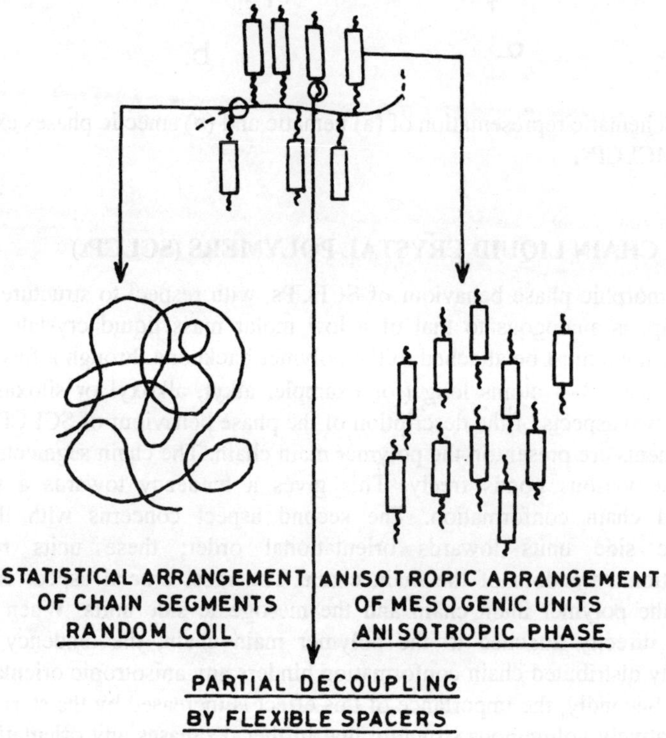

Fig. 8.2. Schematic representation of SCLCPs showing the necessity of decoupling the mesogenic units and the polymer backbone via flexible spacer (Ref. 21).

Over the past few decades a large number of SCLCPs have been synthesized. This could be possible because of the combination of the vast number of mesogenic units available and the many different backbone types possible (e.g., siloxanes, acrylates, methacrylates, ethylenes, epoxides). Both the nematic and smectic phases (Fig. 8.3) have been generated. The SCLCPs exhibiting a reentrant nematic phase have also been generated[50–54]. All these polymers, exhibiting a phase sequence

IL–N_u–S_{A_d}–N_R–, are based on mesogenic units that contain a cyano group, five or six atoms in the flexible spacer, and a polyacrylate or polyvinyl ether backbone.

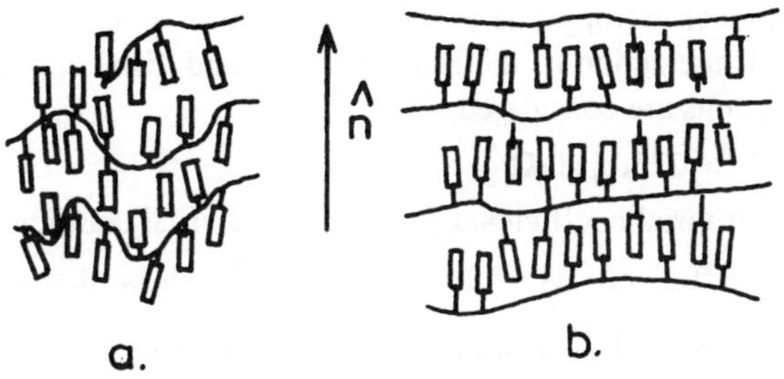

Fig. 8.3. Schematic representation of types of SCLCPs (a) nematic phase, and (b) smectic phase.

8.6 PHASE TRANSITIONS AND PHASE DIAGRAMS IN LC POLYMERS

Rigid rod polymers are thermally intractable, i.e., they chemically degrade at temperatures below their melting points. As a result these polymers must be solubilized in order to exhibit mesophase formation. Much of the academic activities, in previous years, have been focused on thermotropic semiflexible polymers. This section is devoted to the description of the salient features of some of the work done on the phase transitions in LC polymers.

8.6.1. Nematic Order in Polymer Solution

We discuss here one of the simplest systems which exhibit isotropic-nematic phase transition.

A. An athermal solution of long rigid rods

Assuming that the liquid crystalline order arises from purely steric causes, Onsager[55] proposed the first molecular theory of nematic ordering for an athermal solution of cylindrical, long, rigid rods of length L and diameter d_0 ($L >> d_0$). This system can be considered as a model for a system of rigid chain macromolecules with so insignificant flexibility that it can not be manifested in the length L. The basic steps of Onsager method consists of the following. Consider a system of N rods distributed in volume V such that their concentration is c = N/V and the volume fraction of rods in the solution is $\varphi = \pi c L d_0^2 / 4$. The free-energy of the solution of rods is written as

$$A = Nk_BT[\ell n\, c + \int f(\hat{e})\, \ell n\,[4\pi f(\Omega)]d\Omega \; + \frac{1}{2}\, c \int f(\hat{e}_1)\, f(\hat{e}_2)\, B_2(\Omega_{12})\, d\Omega_1\, d\Omega_2] \tag{8.1}$$

The first term is due to the translation motion of the rods, the second describes the losses of orientational entropy due to liquid crystalline order, and the third term is the free-energy of interaction of the rods in the second virial approximation. When only steric interactions are considered,

$$B_2(\Omega_{12}) = 2L^2 d_0 \sin\Omega_{12} \tag{8.2}$$

Simple estimates of virial coefficients, giving second virial coefficient $B_2 \sim L^2 d_0$ and the third virial coefficient $B_3 \sim L^3 d_0^3 \ell n(L/d_0)^8$, show that the second virial approximation ($cB_2 >> c^2 B_3$) is valid under the condition $c << 1/Ld_0^2$ or $\varphi << 1$. In the limit $L >> d_0$ a liquid–crystalline transition in the solution of rods occurs precisely at $\varphi << 1$.

For each concentration the free-energy of the system must be a minimum; the resulting minima correspond to possible phases. Above a certain concentration such a minimum is obtained by a state in which part of the system is isotropic and the other part nematic (with different concentration c_i^* and c_a^*; respectively; $c^* = (L/d_0)\varphi$ is a dimensionless concentration). It turns out that for low concentrations of the rods in solution ($c^* << 1$), the anisotropic (nematic)-isotropic (liquid) transition is first-order. When $c^* < c_i^*$ the solution is isotropic, when $c^* > c_a^*$ it is anisotropic, and when $c_i^* < c^* < c_a^*$ the solution separates into isotropic and anisotropic phases. The Onsager trial function

$$f_{on}(\hat{e}) = \alpha \cosh(\alpha \cos\theta)/4\pi \sinh\alpha \tag{8.3}$$

gives the following results

$$c_i^* = 3.340, \quad c_a^* = 4.486, \quad \alpha = 18.58, \quad \overline{P}_{2NI} = 0.848 \tag{8.4}$$

Here $\propto$ is the variational parameter.

Use of Gaussian distribution function[56]

$$f_G(\hat{e}) \sim N(\alpha)\, \exp(-\frac{1}{2}\, \alpha\theta^2), \qquad 0 \le \theta \le \pi/2$$

$$\sim N(\alpha)\, \exp(-\frac{1}{2}\alpha(\pi-\theta)^2), \qquad \pi/2 \le \theta \le \pi \tag{8.5}$$

leads to the following results

$$c_i^* = 3.45, \quad c_a^* = 5.12, \quad \alpha = 33.4, \quad \overline{P}_{2NI} = 0.91 \tag{8.6}$$

$N(\alpha)$ is a normalization constant. These results show that the coexisting concentrations and nematic order parameter depend critically on the exact form of the distribution function employed in the calculations.

The integral equation arising from the exact minimization of the free-energy can be solved numerically to a high degree of accuracy. Adopting this method, the following values were obtained[57]

$$c_i^* = 3.290, \qquad c_a^* = 4.191, \qquad \overline{P}_{2NI} = 0.792 \tag{8.7}$$

The results (8.4) and (8.7) show that the use of variational method leads to a very small error (~ 5%) in determining the characteristics of the mesophase transitions. The Onsager method has been generalized to the case of high concentrations also. The details are summarized in other reviews[8,13]. Here at this stage we would like to mention that the Onsager approach can be applied for describing solutions of any arbitrary concentration.

Another approach for the above problem was developed by Flory[58]. Using the lattice model of polymer solutions, Flory replaced flexible chains with rigid rods and demonstrated the formation of an ordered phase above a critical volume fraction of rods that depend on the rod aspect ratio ($x_0 = L/d_0$). The liquid crystal phase transition was studied[59,60] using a modified variant of Flory theory and for an athermal solution ($x_0 >> 1$) the results are

$$c_i^* = 7.89, \qquad c_a^* = 11.57, \qquad \overline{P}_{2NI} = 0.92 \tag{8.8}$$

Upon comparing the results (8.4) and (8.8) we can conclude that the results of lattice model although qualitatively correct differ quantitatively rather strongly from the exact results (in the limit $x_0 >> 1$) of Onsager.

A.1 : Polydispersity : In an experimental situation, most often the polymer solutions prove to be polydisperse. That is, they are composed of macromolecules of differing masses. This may have strong influence on the isotropic-nematic phase transition. The influence of polydispersity on a nematic transition has been studied by using both the Onsager method[57,61–64] and the Flory theory[65–68]. The qualitative results of these studies coincide.

Let us consider a simple bidisperse system, composed of particles of equal diameter d_0 but different lengths L_1 and L_2. Let φ_1 and φ_2 be the volume fractions in the solution of rods of the first and second type, respectively, and let

$$\varphi = \varphi_1 + \varphi_2 \qquad \text{and} \qquad q = L_2 / L_1 > 1.$$

z, the weight fraction of the long rods, is given by

$$z = \varphi_2 / \varphi$$

and $\overline{L^w}$ the weight average length of the rods is

$$\overline{L^w} = L_1(1-z) + zL_2 \tag{8.9}$$

Just like a monodisperse system, the nematic transitions occurs at $\varphi \sim d_0/\overline{L^w}$. Using the Onsager method, computer simulation was performed[57] for q = 2 to obtain the phase diagram of the system. Numerical calculations have also been performed for two systems with length ratio = 2 and 5. From these results the following features emerge[70] (see Fig. 8.4).

(i) The polydispersity leads to a small decrease in the lower (isotropic) boundary of the separation region and simultaneously to a strong increase in its upper (anisotropic) boundary. The longer rods preferentially go into the anisotropic phase.

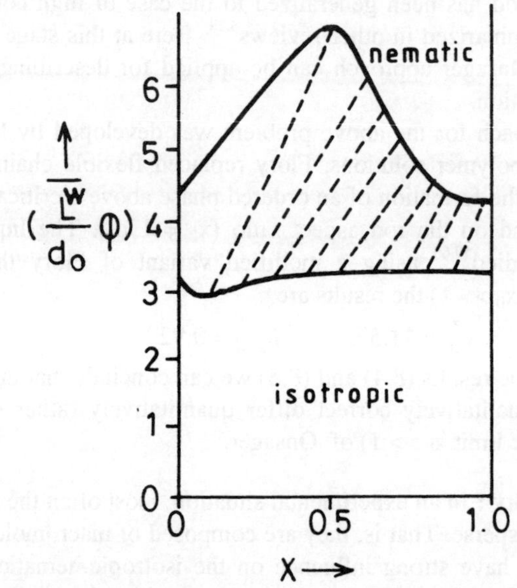

Fig. 8.4. Volume fraction scaled by the weight-averaged length against mole fraction of the longer rods for a bidisperse mixture of rods ($L_2/L_1 = 2$) (Ref. 57, 69). The broken lines indicate in which two phases compositions falling in the two-phase region will divide.

(ii) The relative width of the separation region, i.e., the concentration difference between isotropic and nematic phase may be much larger as compared to the monodisperse case.

(iii) A weak variation in $\overline{L^w}\,\varphi/d_0$ with mole fraction of the longer rods is found. This implies that the molecular weight dependence of the bifurcation density is a good indication for the onset of phase separation.

(iv) The mean-order parameter,

$$\overline{P}^M_{2NI} = (1-z)\overline{P}_{2,1} + z\overline{P}_{2,2}$$

is found to be appreciably higher (≈ 0.92) than in the monodisperse case (≈ 0.79).

For the length ratio q = 5 similar, but more pronounced, results were obtained with two additional features :

(a) For certain compositions a phase sequence of isotropic-nematic-reentrant isotropic-reentrant nematic has been observed as a function of concentration.

(b) For some compositions the system shows a biphasic region where the system separates into two different nematic phases, or even a three-phase region in which an isotropic phase coexists with two nematic phases.

B. Athermal Solutions of Partially Flexible Polymers

The real rigid chain macromolecules always have a certain finite flexibility. The macromolecules may differ with respect to the mechanism of the flexibility of the polymer chain. The simplest kind belongs to the freely linked chain, which amounts to a sequence of hinged rigid rods of length L and diameter d, with L >> d (Fig. 8.5a). The orientation of each successive rod, in the equilibrium state, is random and is independent of the orientation of previous ones. Consequently, the mean-square distance between the ends of the chain $\langle R^2 \rangle$ is given by

$$\left\langle R^2 \right\rangle_0 = L\ell, \qquad \ell >> L \tag{8.10}$$

where ℓ is the total contour length of the chain.

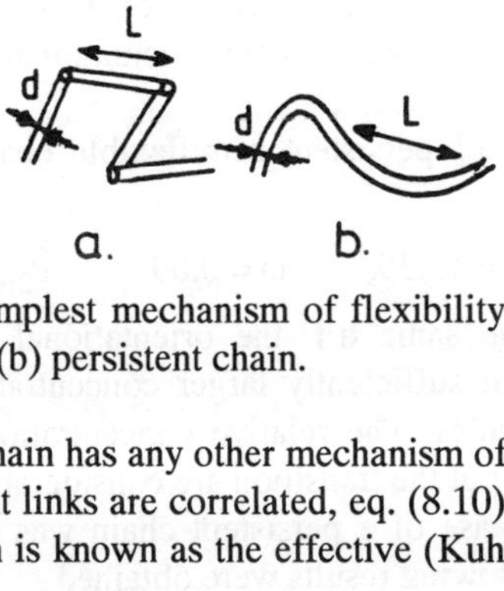

Fig. 8.5. The simplest mechanism of flexibility (a) freely linked chain, (b) persistent chain.

In case the polymer chain has any other mechanism of flexibility, for example, if the orientations of adjacent links are correlated, eq. (8.10) is still satisfied but with a renormalized length which is known as the effective (Kuhn) segment of the polymer chain.

In a persistent mechanism of flexibility, the flexibility arises from the accumulated effect of small oscillations in the valence angles. A persistent macromolecule can be represented in the form of a homogeneous cylindrical elastic filament of diameter d (Fig. 8.5b). The elasticity of the filament is such that it can be substantially bent only on scales of the order of L. Most of the macromolecules belong to this class of polymeric objects. On the basis of the relationship between ℓ and L, rigid chain macromolecules can belong to one of the following three fundamental classes : (a) if L >> ℓ >> d, the flexibility of the polymer chain can be neglected; this refers to the case of limiting rigid-chain macromolecules (or rigid rods); (b) if $\ell >> L >> d$, the rigid-chain macromolecule includes many Kuhn segments; this refers to the case of semiflexible macromolecule which stays in the state of a random coil and, (c) $\ell \sim L$; in real experiment this kind of macromolecule is found rather often.

Using the continuum approach (the Onsager method) the transition of an athermal solution of partially flexible polymer chains to an anisotropic phase has been studied[71–74] extensively. In full analogy with the Onsager method, these studies[71,72] led to the following conclusions : For the model shown in Fig. 8.5, the orientational ordering of the athermal solution shows the features of a first-order phase transition and occurs at low concentrations of the polymer in solution. When $\varphi < \varphi_i$, the solution is homogeneous and isotropic; if $\varphi > \varphi_a$ it is homogeneous and anisotropic, and in case of $\varphi_i < \varphi < \varphi_a$, it separates into isotropic and nematic phases, with $\varphi_i \sim \varphi_a \sim d/L << 1$. For an athermal solution of freely linked semiflexible chains[71],

$$c_i^* = 3.25, \quad c_a^* = 4.86, \quad \omega = \frac{c_a^*}{c_i^*} - 1 = 0.5, \quad \overline{P}_{2NI} = 0.87 \qquad (8.11)$$

A comparison of the results (8.11), (8.4) and (8.7) shows that hinge linking of the rods in long chains leads only to quite insignificant changes in the characteristics of the isotropic-nematic transition. The region of phase separation is some what expanded, while the order parameter of the orientationally ordered phase is slightly increased.

For an athermal solution of persistent semiflexible chains following results were obtained[72] :

$$c_i^* = 10.48, \quad c_a^* = 11.39, \quad \omega = 0.09 \quad \overline{P}_{2NI} = 0.49 \qquad (8.12)$$

It is obvious that for the same d/L the orientational ordering in a solution of persistent chains occurs at sufficiently larger concentrations than in a solution of freely linked macromolecules. The relative concentration jump of the polymer as well as the order parameter at the transition are considerably smaller.

The phase separation in case of a persistent chain was calculated in the Gaussian approximation and the following results were obtained

$$c_i^* = 7.77, \quad c_a^* = 9.71, \quad \propto = 12.34 \quad \overline{P}_{2NI} = 0.759 \qquad (8.13)$$

A somewhat better result was obtained[56] by employing the Onsager trial function

$$c_i^* = 5.41, \quad c_a^* = 6{,}197, \quad \propto = 6.502 \quad \overline{P}_{2NI} = 0.61 \qquad (8.14)$$

Solving the nonlinear integro-differential equation, as obtained from the minimization of free-energy, Vroege and Odijk[75] obtained the results

$$c_i^* = 5.124, \quad c_a^* = 5.509, \quad \overline{P}_{2NI} = 0.4617 \qquad (8.15)$$

The conformations of the semiflexible macromolecules in the liquid crystalline phase depend on the mechanism of the flexibility. A very important conformational charcteristic is the mean square of the projection $\mathbf{R}_z$ of the segment joining the ends of the chain on the director direction (Z-axis), $\langle \mathbf{R}_z^2 \rangle$,

$$\langle \mathbf{R}_z^2 \rangle = \chi_0 \langle \mathbf{R}^2 \rangle_0 \tag{8.16}$$

Here $\langle \mathbf{R}^2 \rangle_0 = \ell L$ is the mean square of the distance between the ends of the chain in the isotropic phase and χ_0 is the susceptibility of the system to an external orienting field. It has been found that in the nematic solution the freely–linked chains are somewhat extended in the direction of the axis of orientational order. However, the magnitude of $\langle \mathbf{R}_z^2 \rangle$ increases by a factor of not more than three as compared to its value in the isotropic phase. In the case of persistent flexibility, the behaviour of susceptibility is completely different. For this case the susceptibility χ_0, and hence $\langle \mathbf{R}_z^2 \rangle$, sharply increases according to an exponential law upon increasing the concentration of the nematic solution. In other words, the macromolecules are strongly stretched out along the director. This effect may be referred to as stiffening of persistent macromolecules in the liquid crystalline state. Other flexibility mechanisms[76,77] have also been considered to address the problem of orientational ordering in solutions of polymer chains.

C. Non-athermal polymer solutions

In case of athermal polymer solutions, the nematic ordering exists at the low concentrations of the rigid chain polymers. In calculating the effect of attractive forces on the transition properties, one faces some problems arising due to the following aspects; the separated anisotropic phase can be very concentrated and the Onsager's second virial approximation is inapplicable for treating the system. Efforts were made[78,79] to analyse the role of attractive forces of the links in nematic ordering of a solution of rigid-chain polymers. Since the treatment in these studies was based to some degree on the Flory lattice approach, exhaustive solution of the problem could not be obtained.

This problem was first studied consistently by Khokhlov and Semenov[80] by using the continuum approach. They expressed the interaction potential as the sum of steric and electrostatic contributions

$$U = U_{ster} + U_{att} \tag{8.17}$$

Adopting the approach of Parsons, the respective free-energy was expressed as

$$f_{ster} = \frac{4NT\ell}{\pi d} \left|\ell n(1-\varphi)\right| \int f(\hat{e}_1) f(\hat{e}_2) \sin\gamma \, d\Omega_1 \, d\Omega_2 \tag{8.18}$$

and

$$f_{att} = -\frac{\ell N \varphi}{2d} \left(u_0 + u_a \overline{P_2}^2\right) \tag{8.19}$$

Here u_0 and u_a are constants describing, respectively, the isotropic and anisotropic components of the attractive forces. Equation (8.19) can be rewritten as

$$f_{att} = -\frac{\ell N \varphi \Theta}{d} \left(1 + \frac{u_a}{u_0} \overline{P_2}^2\right) \tag{8.20}$$

Here Θ $(= u_0/2)$ is the theta temperature of the polymer solution which is defined as the temperature at which the osmotic second virial coefficient vanishes. In the actual calculation it was assumed that $u_a/u_0 = 0$.

For a solution of persistent macromolecules the phase diagrams were calculated[80] for the liquid crystalline transition in terms of φ and Θ/T for several values of L/d. It was found that in the region of relatively high temperatures, a narrow corridor of phase separation into isotropic and anisotropic phases lying in the dilute solution region exists. Contrary to it, at low temperatures the region of phase separation is very broad such that an isotropic, practically fully dilute phase, and a concentrated, strongly anisotropic phase, coexist. These two regions are separated by the interval between the triple-point temperature T_t and the critical temperature T_c $(T_c > T > T_t)$, in which there are two regions of phase separation between isotropic and anisotropic phases, and between two anisotropic phases having differing degrees of anisotropy. The temperature T_t and T_c substantially exceed the Θ–temperature. The interval between T_c and T_t becomes narrower as the ratio L/d decreases and drops out when $(L/d)_{c_1} = 125$. When $L/d < 125$ there are no critical or triple points on the diagram, and one can refer only to the cross over temperature T_{cr} between the narrow high-temperature corridor of phase separation and the very broad low-temperature region of separation. With decreasing L/d, the temperature T_{cr} decreases. For $(L/d)_{c_2} \approx 50$, one obtains triple and critical points corresponding to an additional phase transition between two isotropic phases. The concentration of one of these phases is extremely low. The influence of attractive interaction on the order parameter of a nematic solution of persistent macromolecules at the transition point was also analysed. It was found that the attractive forces affect the value of $\overline{P_2}$ very substantially when $T/\Theta \leq 2$.

8.6.2 Nematic Order in Polymer Melts

The liquid crystalline polymer melts are dense system and often called "thermotropic polymeric liquid crystals". The applications of methods discussed above cannot be applied in a straightforward manner because the large parameter ℓ / d can not be used.

8.6.2.1. Melts of linear homopolymers

In thermotropic solutions neither the concentration c nor the volume fraction φ can be used as the external parameter. Khokhlov and Semenov[81] studied the role of pressure in detail with the results that the influence of normal atmospheric pressure on the mesophase ordering is negligibly small. A substantial increase in the region of stability of the nematic phase can be expected only at high pressure of the order of 10^3 atm. When the pressure becomes infinitely large in melt of any particle that are anisodiametric to any degree and have a rigid steric core of interaction, a liquid crystalline phase can be observed.

The phase diagram for a melt of long persistent chain was calculated[81] as a function of T/Θ and L/d for $u_a/u_0 = 0.1$ at the atmospheric pressure. From the results the following features are evident; the existence of three phases – isotropic and anisotropic melts, and also a gas like phase (at high temperatures) was observed. The nematic order can occur only if the asymmetry parameter L/d is smaller than the critical value $(L/d)_c \approx 50$. When $(L/d) > (L/d)_c$, the melt in the equilibrium state is always a nematic (at any temperatures). However, the critical value $(L/d)_c$ depends on the mechanism of the flexibility of the chain. For example, for freely linked flexibility it was found[81] that $(L/d)_c \approx 7$, while for a melt of rigid rods $(L/d)_c \approx 3.5$. In the formal limit $L/d \to 0$, it is seen that the steric interactions are inessential and the mesophase ordering occurs only due to the anisotropy of the attractive forces. The results obtained for L/d = 0 are close to those obtained by Rusakov and Shliomis[82] in the limit $L/d \to 0$. It is worth mentioning that these results are exact to any extent only at small values of L/d, while in the most interesting region L/d >> 1 the steric interactions always dominate and the anisotropy of attractive forces is always a secondary one.

Some additional features were observed[82,83] which do not depend qualitatively on whether steric forces are accounted or not. These works show that the order parameter at the transition point depends on the length of the persistent macromolecules. When $\ell \ll L$, $\overline{P}_{2NI} = 0.43$. As ℓ increases the value of $\overline{P}_{2NI}$ decreases and reaches to a value 0.36. Regarding the conformations of the macromolecules it was found that in the nematic phase the macromolecules are stretched along the axis of ordering (the Z–axis). The degree of extension can be characterized by the parameter

$$y_0 = \langle \mathbf{R}_z^2 \rangle / \langle \mathbf{R}_x^2 \rangle$$

Here $\mathbf{R}_x$ and $\mathbf{R}_z$ are the projections of the vectors joining the ends of the polymer chain. The magnitude of this parameter at the transition point depends on the length of the macromolecules. When $\ell << L$, $y_0 = 3.25$. As ℓ increases, y_0 decreases reaching the minimum value $y_0 = 2.77$, and then increases substantially to a value $y_0 = 14.4$ in the limit of very long persistent chains ($\ell >> L$). With decreasing temperature of the nematic melt (for $\ell >> L$) the polymer chains unfold further; the value of y_0 increases exponentially.

8.6.2.2. Melts of linear copolymers

The nematic ordering in melts of linear copolymers was studied[84–87] by using a generalization of the lattice method[85] for an athermal melt as a function of L/d and φ_0 (φ_0 is the volume fraction of the flexible component) for the case $\ell_0 / d = 1.5$ (ℓ_0 is the length of the effective segment of the flexible region). Two sets of transitions, isotropic-nematics and weakly anisotropic-strongly anisotropic, were observed. The second transition between two nematic ends at a critical point with coordinates $(L/d)_c = 10$, $\propto_{0c} = 0.2$. Upon decreasing the flexibility of the flexible fragments of the chain (i.e. decreasing the ratio ℓ_0 / d), the region of the stability of the weakly anisotropic phase narrows and is shifted towards larger $\propto_0$.

8.6.2.3. Melts of Comb-like Polymers

In the mean-field approximation of Maier-Saupe type, the nematic ordering in the melts of comb-like polymers with rigid fragments in the side chain was studied[87] by incorporating only the interaction of the mesogenic fragments. It was found that the orientational properties of the system substantially depend on the length ℓ' of the fragment of the main chain between adjacent branches. The length of the spacer linking the mesogenic group with the main chain is assumed to be very small. For $\ell' > L$, the mesogenic ordering occurs in a similar manner as the low molecular-weight system. The order parameter at the transition is $\overline{P}_{2NI} = 0.43$. When the density of the side branches increases (i.e. ℓ' / L increases), the order parameter at the transition decreases monotonically. As the length of the flexible fragments ℓ' decreases $\overline{P}_{2NI}$ either increases or decreases very slightly. Thus, the order parameter for the comb-like polymer, always stays smaller than for their linear analogs. This conclusion is in agreement with the experimental data[88-90].

When the density of the side branches is sufficiently large ($L/\ell' > (L/\ell')_c$) the $\overline{P}_{2NI}$ becomes negative. In this case the main chains are oriented along the director, while conversely the mesogenic groups lie preferentially in the easy plane. The critical value of density for a persistent chain is found to be[87]

$$\left(\frac{L}{\ell'}\right)_c = 18$$

Figure 8.6 shows the phase diagram of a melt of comb-like macromolecules as a function of L/ℓ' and T. N_+ and N_- correspond to the different conformations of the macromolecules and N_b is a biaxial nematic phase.

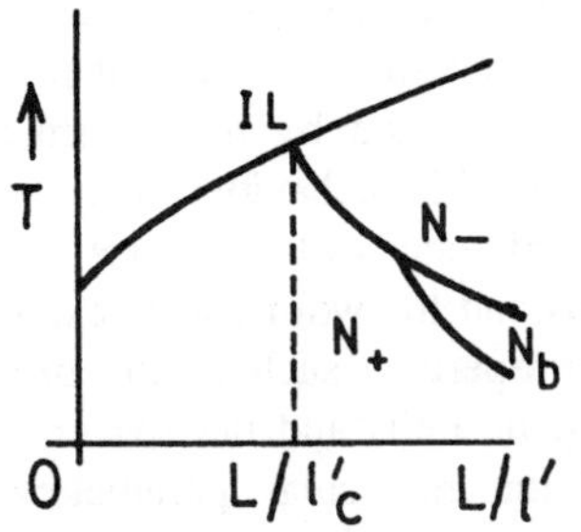

Fig. 8.6. Phase diagram of a melt of comb-like macromolecules in the variable L/ℓ' (effective density of side branches), and temperature (Ref. 87).

8.6.3 Nematic Polymers with Varying Degree of Flexibility

Within the lattice model, Flory[58,91] initiated the study of transitions in semiflexible polymers and presented a simple heuristic treatment of the properties of concentrated, self avoiding polymers as a function of their stiffness. He found a first-order transition from a disordered state to an entirely ordered state. Kim and Pincus[92] gave the mean-field treatments of such transitions. Assuming polymers to be nonchiral, a simple model for the interacting semiflexible polymer was developed by Petschek[93]. He used Ursell–Mayer perturbation expansion technique to systemize the calculation of the properties of the system. Using group theoretical arguments, the details of this expansion were analysed and a relationship with field theory could be established. It was found that the isotropic–nematic transition in systems of long, semiflexible polymers can be controlled either by the usual tensor order parameter or by a hidden, vector-like order parameter. This vector like order parameter is expected to control the behaviour only if the fluctuations are important in determining the behaviour of the material near the transition. Which of the two order parameters is the controlling order parameter depends on the details of the configuration and interaction energies of the polymer system. However, the results of the calculations on simple models have no immediate consequences for the behaviour of the realistic polymer system and it is not evident what physical systems, if any, would be expected to have transitions which are controlled by vector-like order parameters.

Wang and Warner[94] modelled main chain liquid crystal polymers as either worms or jointed rods. These authors presented a model, which accounts for the molecular parameters such as the lengths of the mesogenic group and the spacer units and the interactions between them, to describe the non-homogeneous nematic polymers. The mesogenic groups have been modelled as rods in a quadrupolar potential whereas the spacer were treated exploiting the spheroidal approach[95,96]. The spacers have been found to have an order differing from the mesogenic units. If the spacer is not very long and thus in effect is inflexible, one end of the spacer can retain to some extent the orientation of the other end, allowing orientational correlation between spacers mediated by the intermediate mesogenic unit. This provides the chain a global rods like behaviour as the nematic field becomes strong or the temperature low. The influence of the physical linkage and the van der Waals interactions between the rods and the worms were examined. The nematic-isotropic transition and some other properties, such as the orientational order of the two components, i.e., the mesogenic units and the spacer, and the latent entropy were calculated as function of the molecular parameters. In accordance with the experimental observations it was found that the reduced transition temperature (the temperature reduced by T_{NI} of MS model for pure rods) decreases significantly when the length of the flexible spacer increases, while the latent entropy increases. The effect of interactions between the mesogenic units u_{aa}, the spacer units u_{bb} and the mesogenic-spacer units u_{ab}, on the nematic-isotropic transition temperature was studied. The ration u_{bb}/u_{aa} does not visibly affect the properties of the polymer. The order parameter of the mesogenic unit at the transition does not vary, remaining at about 0.434, while there is a significant variation in that of the flexible spacer.

8.7 SYNTHETIC ROUTES AND STRUCTURE-PROPERTY RELATIONS

8.7.1. Synthetic Routes

The properties of polymers are dependent upon the manner of synthesis. There are two steps to the synthesis of liquid crystal polymers. First, the conventional synthetic route to generate the mesogenic units and second, the polymerization reaction to obtain the desired liquid crystal polymers. The synthetic routes to the monomer units are, in general, analogous to those employed [20] in case of low molar mass mesogens with the additional requirement of inserting a polymerisable functional unit at some site in the monomer material. Several polymerisation methods have been employed in the synthesis of main chain and side chain liquid crystalline polymers.

The main chain liquid crystal polymers and copolymers are synthesized by step polymerisation[20,21] which are based on reversible and irreversible reactions. Several polymerisation methods have been used in case of side chain liquid crystal polymers with well defined molecular weights and narrow molecular weight distributions. These methods include cationic polymerisation of mesogenic vinyl ethers, cationic ring opening polymerisation of mesogenic cyclic imino ethers, group transfer polymerisation of mesogenic methacrylates, polymerisation of methacrylates with methylaluminum porphyrin catalysts, etc. Among all these methods the cationic polymerisation has been proved to be the most successful method because the initiators are very simple and this method can be employed to polymerize under living conditions mesogenic vinyl ethers containing a large variety of functional groups.

We give some representative examples of the strategy adopted for the preparation of liquid crystal polymers.

Kelvar (compound T8.1c, see Table 8.1) is a well know polymeric material which exhibits a nematic phase upon dissolution in sulfuric acid. It is a simple nylon–type polymer that can be synthesized (see Table 8.1) by heating the dicarboxylic acid (T8.1a) and diamine (T8.1b).

The method of transesterification has been found to be suitable for the synthesis of many polymeric materials. Table 8.2 shows the preparation of a main chain liquid crystal polymer (compound T8.2f). When dimethyl terephthalate (T8.2a) is heated to 200^{0}C with ethylene glycol (T8.2b), the methyl easter is converted to a new ester (T8.2c) containing the glycol unit. At 280^{0}C another transesterification takes place to generate the poly (ethylene terephthalate) polymer (T8.2d). When this material is further heated with 4-hydroxybenzoic acid (T8.2e), another transesterification places the 4-hydroxybenzoic acid units randomly within the new polymer chain to generate a main chain liquid crystal polymer (T8.2f).

Table 8.1. Synthesis of Kelvar.

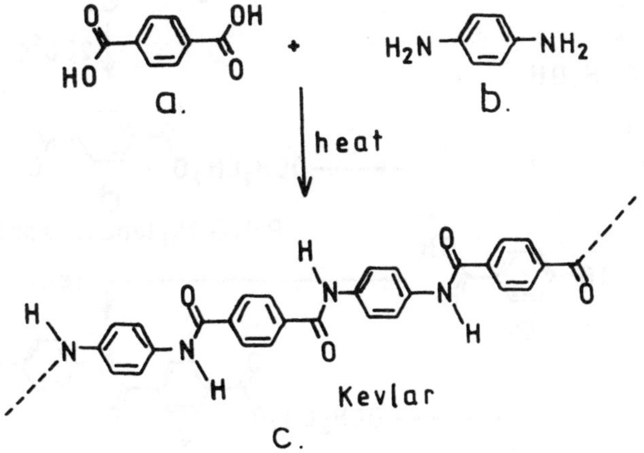

The soluble oligo(p-phenylene)s are prepared by placing methyl substituents on the phenyl ring[97]. The thermal behaviour and phase transitions of these oligomers have been studied[21,98]; after substitution, their solubility increases but their melting points decrease. The high molecular weight polymers displaying liquid crystalline phase[99,100] have been synthesized by the catalyzed coupling, for example, soluble poly (p-2, 5-di-n-alkylphenylene)s were prepared by Ni catalyzed coupling of 1,4-dibromo-2,5-di-n-alkylbenzene, and Pd(0)-catalysed coupling of 4-dibromo-2,5-di-n-alkylbenzeneboronic acid. The methods of Ni(II) catalyzed polymerisation, and anionic polymerisation have been used[101] to synthesize regioregular phenylated poly (p-phenylene)s.

Metal containing poly(yne)s were prepared[102,103] by a copper chloride triethylamine catalyzed coupling of appropriate metal helide with a dialkyne. This class of rigid rod-like polymers exhibiting lyotropic liquid crystal phases[104] have been prepared by employing two novel methods : by the reaction of bis-trimethylstannyl (acetylide)s with trans–[$M(PBu_3)_2Cl_2$][104], and by the reaction of [$Rh(PMe_3)_4Me$] with diacetylenes[105,106]. Reviews are available[107–110] which give the detail of the synthesis of metal containing liquid crystal polymers. The synthetic procedures for the preparation of many main-chain polymeric materials such as spinal columnar liquid crystal polymers, soluble aromatic polyamides and polyesters, rod-like soluble polyimides, thermotropic poly (1,4-arylenevinylene)s, etc., have been developed and are available[21] in the literature.

Table 8.2. Synthesis of a main-chain liquid crystal polymer.

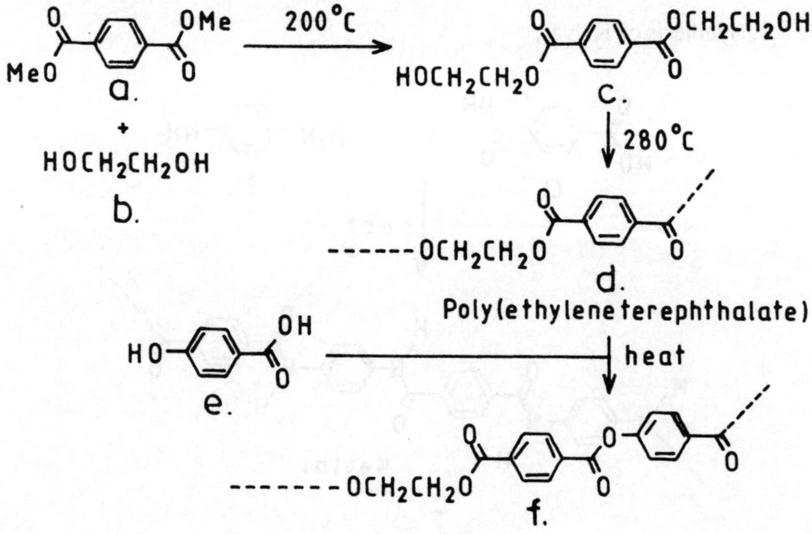

The subject of side-chain liquid crystal polymers has been reviewed by McArdle[9], and Percec[21]. As mentioned above several polymerisation methods have been developed for the preparation of SCLCPs. Using a living cationic mechanism, vinyl ethers containing nucleophilic groups[111] (e.g. methoxybiphenyl), electron-withdrawing groups[111–115] (e.g. cyanobiphenyl), nitrobiphenyl and cyanophenylbenzoate[111], double bonds such as in 4-alkoxy-∝-methylstilbene[114], double bonds and cyano groups like those in 4-cyano-4'-∝-cyanostilbene[113], aliphatic aromatic esters[116], acidic protons and perfluorinated groups[111,116], crown ethers and triple bonds[117,118], etc., all can be polymerized. Perfectly aligned single crystal liquid crystalline polymer films have been obtained by the polymerisation in the liquid crystal phase with aligned films of liquid crystalline monomers[119].

8.7.2. Structure-Property Relations

We shall examine and discuss here how the molecular structure and combination of structural units affect the generation and physical properties of liquid crystalline polymers; in order to make the points clear some representative examples shall be considered. A more detail discriptions are given in Refs. 20 and 21.

8.7.2.1. Main-chain liquid crystal polymers

Figure 8.7 shows the effect of structure on the phase transitions in oligophenylenes[21]. It is seen that due to methyl substituents on the phenyl ring, their solubility increases and their melting points decrease. In accordance with the theoretical predictions, hexamethylsexiphenyl, with axial ratio $x_0 \approx 4.8$, and lower oligomers do not exhibit an enantiotropic mesophase; octamethyloctaphenyl ($x_0 \approx 6.4$), and longer oligomers ($x_0 > 6.2$, critical value) show an enantiotropic nematic phase.

Synthetic routes for the preparation of soluble aromatic polyamides and polyesters were developed by Gaudiana and coworkers[120–122]. A summary of these procedures and how the various molecular factors such as the position, polarizability and size of the substituents influences the solubility of aromatic polyamides are available[123]. These work suggest that the noncoplanar conformation of the biphenyl moiety, the presence of structurally dissimilar units, and the statistical distribution of the biphenyl enantiomers, all increase the solubility. A combination of these features diminishes or even eliminates the interchain correlations which is essential for the close packing in the crystal. Aromatic polyamides with alkyl side chains were also prepared[124–126]. When each structural units have four alkoxy groups, based on their length, these polymers display a board like, "Sanidic" mesophase[21]. However, when there are only two alkoxy groups per one of the two monomeric units, these polyamides exhibit a new layered liquid crystal phase. Similar results were obtained in case of corresponding polyesters. A large group of soluble aromatic polyesters has been synthesized[21] by connecting various large substituents such as phenylalkyl[127,128], phenoxy and ter-butyl[129] perfluoroalkyl[130], phenyl and biphenyl[131], and twisted biphenyl[132–134] units. The dependence of the solubility and thermal properties of polyesters based on substituted hydroquinone and substituted terephthalic acid has been examined and reviewed. The presence of dissimilar substituents on both monomers help in generating the highly soluble polymers.

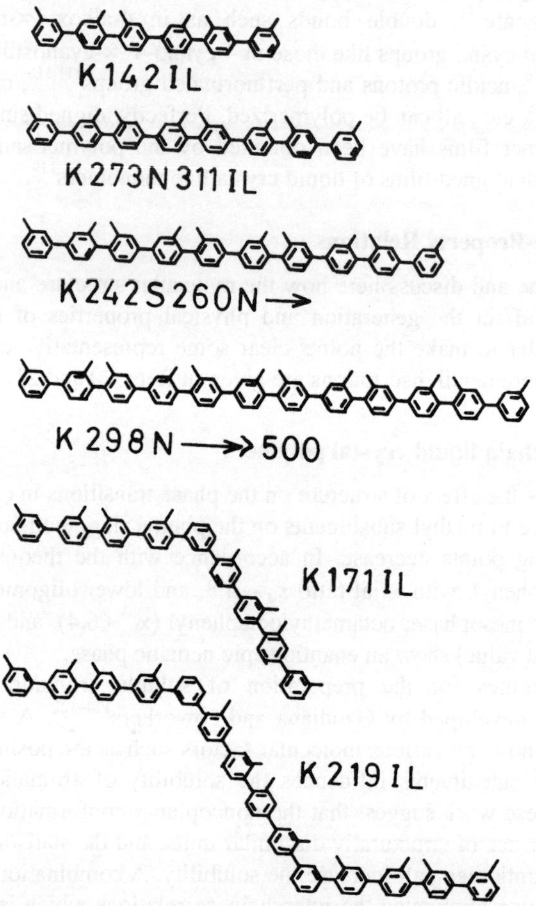

Fig. 8.7. Effect of structure on phase transitions in oligophenylenes.

Table 8.3 shows the relation between the properties of low molar mass liquid crystals based on benzyl ether and methyleneoxy flexible units and their structures[135,136]. Some representative examples of low molar mass liquid crystals based on diphenylethane, cyclohexylphenylethane, dicyclohexylethane and dicyclooctylphenylethane are listed in Table 8.4. It can be seen from these tables that the flexible mesogenic units in both follows the same trend. For example, the phase behaviour of compounds T8.4a and T8.4b is almost identical to that of compounds T8.3a and T8.3b. Further, hydrogenating a phenyl ring from compound T8.4a increases the thermal stability of the nematic phase of compound T8.4b by about 75^{0}C. Replacing a phenyl ring from T8.4a with a bicyclooctane ring (compound T8.4c) increases the thermodynamic stability of the nematic phase by 137^{0}C.

8.7.2.2. Side-chain liquid crystal polymers

The field of side-chain liquid crystal polymers is very rich. We shall discuss here very briefly how various parameters (for example, the nature of the flexible spacer and its length, mesogenic units, and nature and flexibility of polymer backbone and its degree of polymerization) affect the mesomorphic phase behaviour of SCLCPs.

Table 8.3. Effect of structure on phase transitions of some low molar mass liquid crystals based on benzyl ether and methyleneoxy flexible units (see Ref. 21).

No.	Structure	Phase transitions (°C)
a.	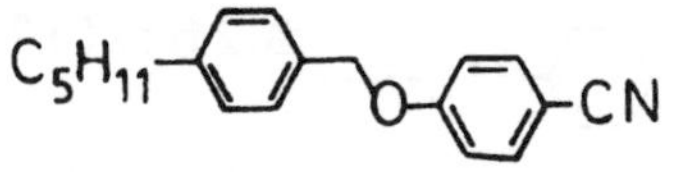	K 49 N [-20] IL
b	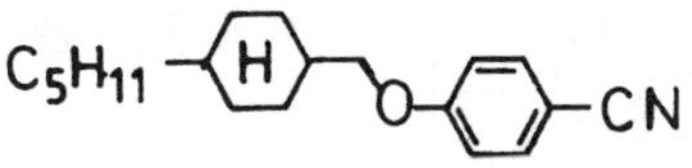	K 74.3 N (28.6) IL
c.	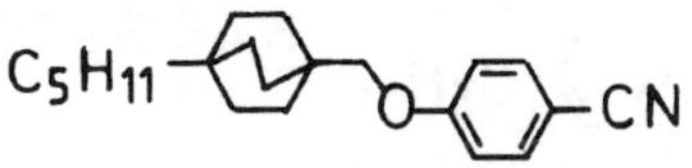	K 72 N 73 IL
d.	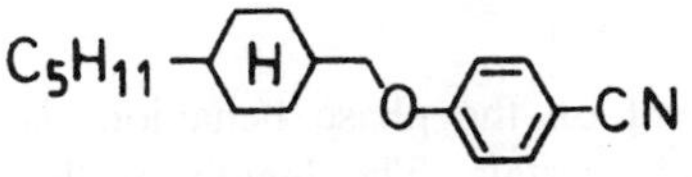	K 35 S_B(31) N [20] IL
e.	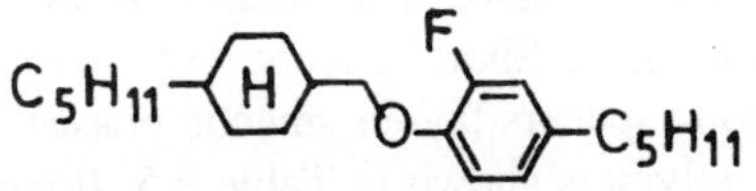	K 43 N [-10] IL
f.	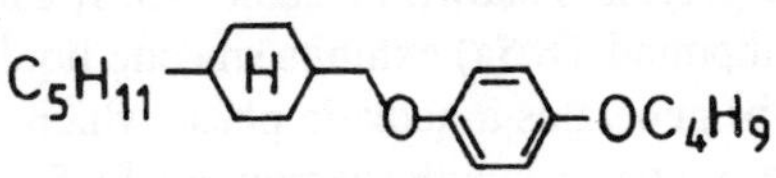	K 55 S_B(47) N (53) IL
g.	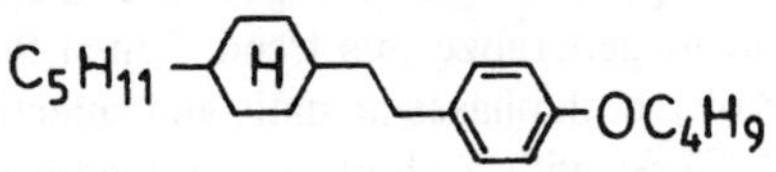	K 22 S_B 44.5 N 45.5 IL

[ ] Virtual, ( ) Monotropic

Table 8.4. Effect of structure on the phase transitions of some low molar mass liquid crystals based on disubstituted ethane derivatives.

No	Structure	Phase transitions (°C)
a.	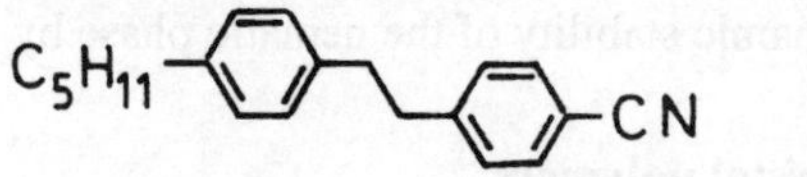	K 62 N [-24] IL
b.	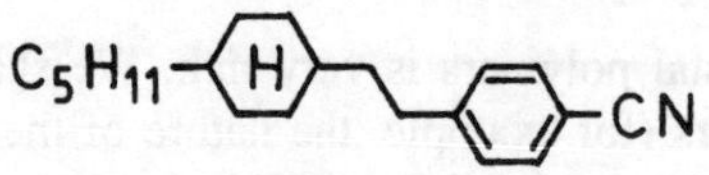	K 30 N 51 IL
c.	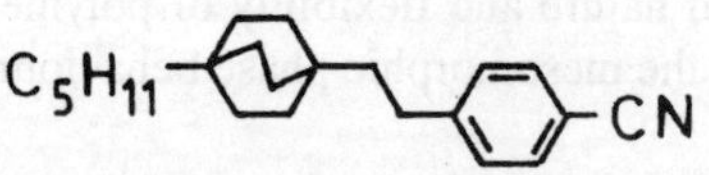	K 76 N 113 IL
d.	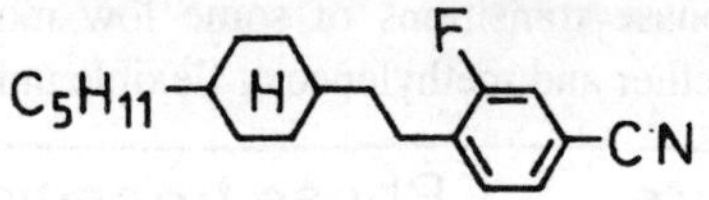	K 45 N 54 IL
e.	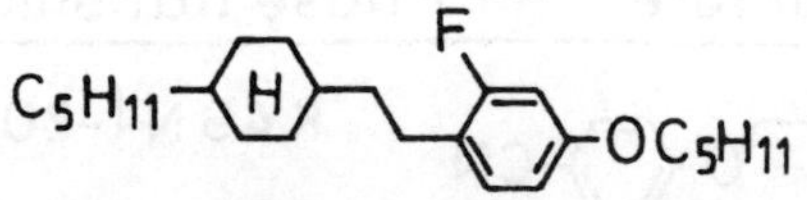	K 26 N (12) IL
f.	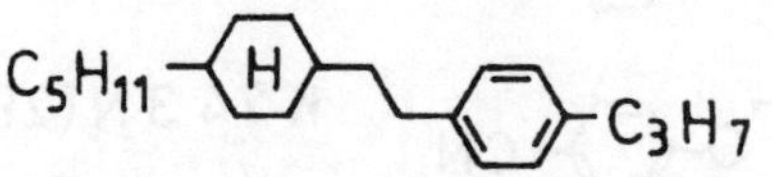	K 4 S_B 18 IL
g.	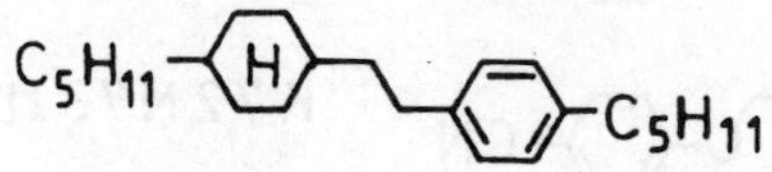	K 46 S_B 109 IL

The influence of the spacer unit on the phase behaviour is similar to that observed in low molar mass liquid crystals. The length of the flexible spacer determines the nature of the mesophase. Short spacers with short terminal chains favour nematic phases, whereas long spacers favour smectic phases[137,138]. This trend is exhibited by the methacrylate polymers shown in Table 8.5. It can be seen that a polymer without spacer unit (compound T8.5a) exhibits smectic liquid crystal phase. A short spacer (compound T8.5b) generates a nematic phase which provides way to smectic phases with the increase of spacer length (compound T8.5c,d). However, as shown in Table 8.6, it is difficult to generalize this trend. When the spacer length increases to n = 6 (compound T8.6h), both phases nematic and smectic are exhibited by the polymer. The compound T8.6e with a short spacer length n = 2 is purely smectic and shows a higher phase stability. Most importantly, it is observed that the glass transition temperature decreases with increasing spacer length. The change in the spacer length affects both the freedom of the mesogenic unit from the polymer backbone and the overall length of the side-chain units. Consequently, a combination of these factors determines the phase behaviour with the spacer length. Thus, the spacer does influence the phase behaviour of SCLCPs significantly. However, it is often difficult to generalize the pattern of this influence.

Table 8.5. Effect of spacer length on the liquid crystalline properties of methacrylate polymers.

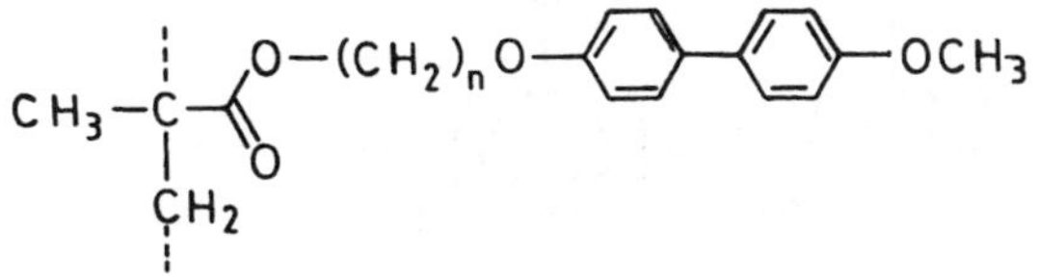

Compound	n	Transition temperatures (°C)
a	0	S 255IL
b	2	g 120N 152IL
c	6	K119 S136 IL
d	11	g 54 S_C 87 S_A 142 IL

Table 8.6. Effect of spacer length and terminal chain on mesomorphic properties of acrylate polymers.

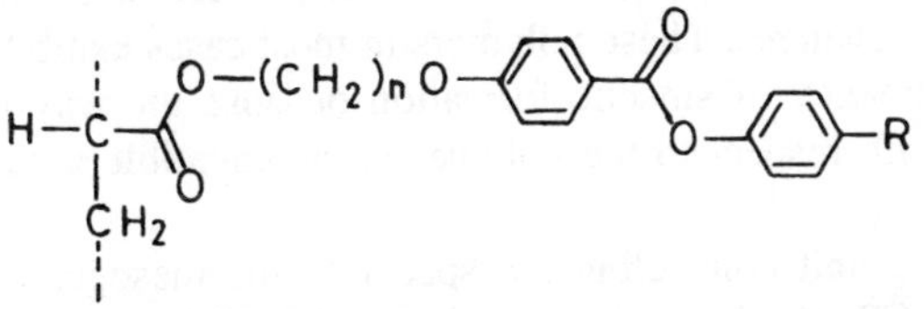

Compound	n	R	Transition temperature (°C)
a.	0	CH_3O	g110K180S296IL
b.	0	C_4H_9O	g120K180S321IL
c.	2	CH_3O	g25K55S116IL
d.	2	CH_3O	g62N116IL
e.	2	C_4H_9O	g30K64S119S154IL
f.	6	CH_3O	g5K20S86S104-118IL
g.	6	CH_3O	g35S97N123IL
h.	6	C_4H_9O	g5K30S103N114 IL
i.	6	$C_6H_{13}O$	g28S130IL

A large number of mesogenic units are available which can easily be used as a side-chain units for generating SCLCPs. These units influence the phase behaviour of polymers quite significantly. A typical template for some possible mesogenic side chain units is shown in Fig. 8.8. These units in SCLCPs can also be connected to the spacer at a lateral position of the calamitic unit.

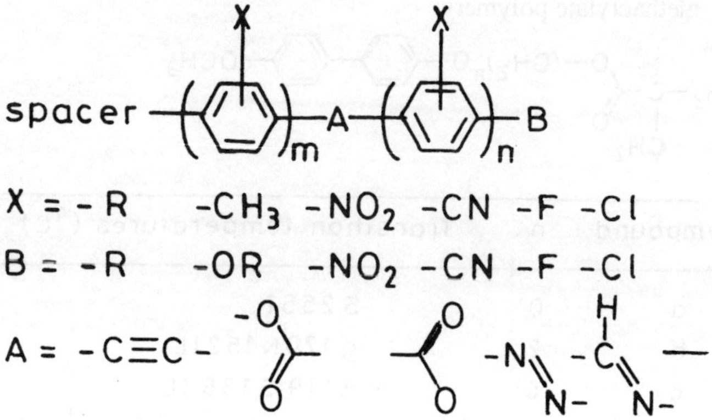

Fig. 8.8. A template structure for possible mesogenic side-chain units (m and n are usually one or two).

In general, the mesomorphic properties of laterally attached SCLCPs are less affected by the structural changes. These polymers in most cases exhibit the nematic phase and show low tendency of smectic formation because the way in which the mesogenic units align with respect to the polymer is incompatible with the smectic phase generation.

The influence of the unit connecting the spacer to the mesogenic unit on the phase behaviour of SCLCPs has been shown in Table 8.7. The first two compounds (T8.7a and T8.7b) differ only in the unit linking the spacer and the mesogenic group. In case of polymer T8.7a higher values of transition temperatures are observed as compared to the compound T8.7b in which the ether oxygen has been removed. The use of an ester linkage (polymer T8.7c) raises both the transition temperatures but the increase in the clearing point is not to the level of polymer T8.7a. The effect of the length of the mesogenic unit on the mesomorphic properties can be seen in Table 8.8. As the length of the mesogenic unit increases, substantial increase in both the transition temperatures are observed. The polymers listed (Table 8.8) exhibit nematic phase in preference to the smectic phase because the spacer and terminal chain lengths are short. As shown in the Table 8.6, a longer terminal chain shows the tendency of the smectic formation. While the polar terminal units (for example, CN, NO_2; compound T8.9a) tend to generate the smectic phases, the nonpolar terminal groups (e.g. CH_3O, CH_3; compound T8.9b) favour the formation of nematic phase.

Table 8.7. Effect of the unit connecting the spacer to the cyanobiphenyl mesogenic unit on the mesomorphic properties.

a. CH$_3$-C(-CH$_2$-)-O-(CH$_2$)$_{11}$O-⟨⟩-⟨⟩-CN

  g 40 S$_A$ 121 IL

b. CH$_3$-C(-CH$_2$-)-O-(CH$_2$)$_{11}$-⟨⟩-⟨⟩-CN

  g 30 S$_A$ 81 IL

c. CH$_3$-C(-CH$_2$-)-O-(CH$_2$)$_{10}$-C(=O)-O-⟨⟩-⟨⟩-CN

  g 45 S$_A$ 93 IL

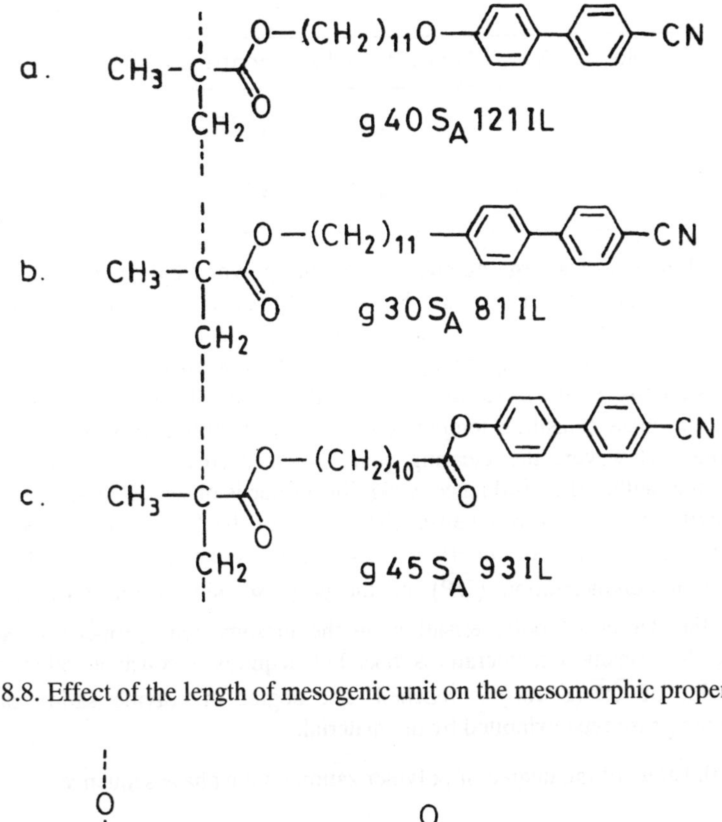

Table 8.8. Effect of the length of mesogenic unit on the mesomorphic properties.

CH$_3$-Si-(CH$_2$)$_3$O-⟨⟩-C(=O)-O-(⟨⟩)$_n$-OCH$_3$

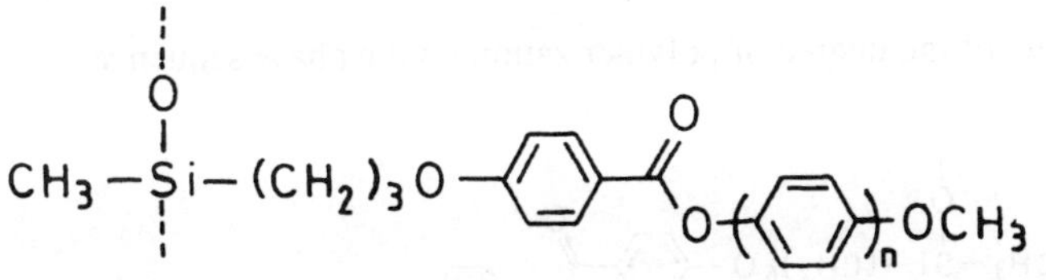

S.No.	n	Transition temperatures (°C)
a	1	g 15 N 61 IL
b	2	K 139 N 319 IL
c	3	K 200 N 360 IL

Table 8.9. Effect of the terminal units on the mesomorphic properties.

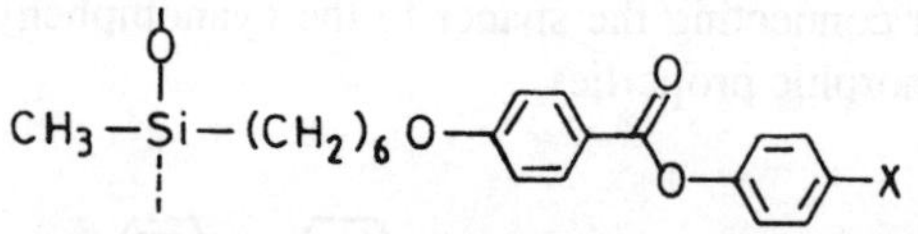

S No.	X	Transition temperatures (°C)
a.	CN	g 14 K 37 S_A 165 IL
b	CH_3O	g 0 K 23 N 97 IL

A combination of a large number of possible polymer backbones are available which can be employed to generate SCLCPs. However, only a few of these (for example, poly(siloxanes), poly(acrylates)) have been widely used[20] in their synthesis and evaluation. The most important aspect of a polymer backbone with regard to liquid crystallinity is its flexibility. As the flexibility of the polymer backbone increases, the glass transition temperature is reduced, giving a wider liqid crystal phase range. However, the clearing point often decreases with the increased flexibility but neither their influence is significant nor is observed in all the cases. It is extremely difficult to rationalise the effect of backbone flexibility on the mesomorphic properties. An important aspect regarding the backbone is related with the degree of polymerisation ($\overline{DP}$) and the polydispersity. Table 8.10 shows the effect of the degree of polymerisation on the mesomorphic properties. As $\overline{DP}$ increases, the transition temperatures rise, but acquires a constant value when a certain value of $\overline{DP}$ ($\geq$ 100) is reached. The degree of polymerisation can also influence the phase type exhibited by the material.

Table 8.10. Effect of the degree of polymerization on the phase sequence.

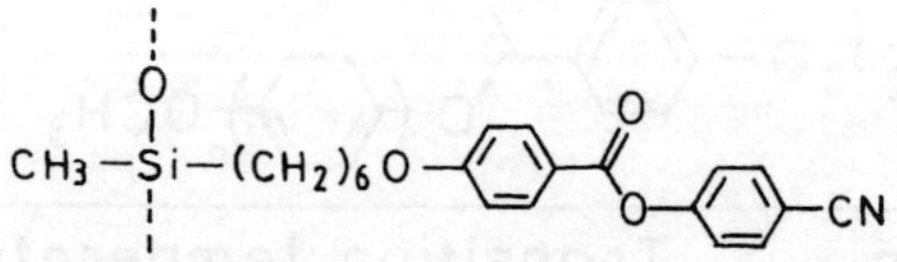

S No.	$\overline{DP}$	Transition temperatures (°C)
a.	40	g 2 S_A 128 IL
b	55	g 9 S_A 137 IL
c	84	g 13 S_A 142 IL
d.	107	g 14 S_A 145 IL

8.8 ELASTIC CONSTANTS OF POLYMER LIQUID CRYSTALS

As compared to the study of elastic behaviour of thermotropic, short (~ 30 Å), hard rod liquid crystals, relatively little attention has been focused[139] on the polymer liquid crystals of long (~ 1000 Å) nearly rigid or semiflexible molecules. The molecular flexibility allows the additional range of fluctuations which can lead to qualitatively new behaviour of the system.

Adopting the Onsager method as generalized to the case of an inhomogeneous systems the elastic constants of an athermal solution of long rigid rods have been calculated by a number of workers[140–145]. The final result [142] always satisfies the relationship.

$$K_2 = \frac{1}{3}K_1 \qquad \text{and} \qquad K_3 = K_1/(1-\overline{P}_2) \tag{8.21}$$

Here the order parameter $\overline{P}_2$ is relatively large. Thus the bend constant is found to be substantially larger than the rest of the two. Vroege and Odijk[146] studied the influence of the electrostatic interactions of the rods on the elastic constants of a nematic solution of strongly charged macromolecules and found a small decrease in the values of K_1 and K_2 due to electrostatic repulsion but the mutual relationship between them still remains in force. It was also observed that the relative change in the elastic constants is approximately proportional to a parameter $h \sim 1/\ell n\, \omega$, where $\omega = \rho^2 r_D/\epsilon_0 T$ with ϵ_0 being the permittivity of the solvent and r_D the Debye radius. While at the low concentrations of the solution K_3 decreases slightly with increasing h, it increases appreciably with h at the higher concentrations.
de Gennes[147] studied the elastic moduli of the melt of very long persistent macromolecules. The case of a solution of persistent (and freely linked) chains of arbitrary lengths was studied by Grosberg and Zhestkov[142]. The elastic moduli of a solution of long ($\ell >> L$) persistent macromolecules of low orientational order were found to obey the relationship

$$K_1 \approx K_3 = \frac{50}{14\pi}TL\varphi\frac{\overline{P}_2^2}{d^2}; \qquad K_2 = \frac{1}{10}K_1; \qquad \overline{P}_2 << 1. \tag{8.22}$$

Here d be the molecular diameter. Thus, similar to the case of the solution of rigid rods, the twist constant K_2 has the smallest value. On the other hand, for the limiting case $\overline{P}_2 \to 1$, the values of the elastic moduli are given[142] by

$$K_1 \approx \frac{T}{4d}\Phi^{1/3}\exp(3.82\Phi^{1/2} - 7.1\Phi^{1/6}) \tag{8.23}$$

$$K_2 = \frac{(1-\overline{P}_2)TL\varphi}{3\pi d^2} \tag{8.24}$$

$$K_3 = \frac{2\overline{P}_2 TL\varphi}{\pi d^2} \tag{8.25}$$

where $\Phi = \varphi L/d >> 1$. Thus the value of K_2 is the smallest and as $\overline{P}_2 \rightarrow 1$, K_3 approaches a constant value. The ratio K_3/K_2 for a solution of persistent macromolecules is approximately twofold larger than for a solution of rigid rods. The splay constant K_1 increases exponentially rapidly with increasing concentration of the solution.

The above studies[140–143,145–147] consider long-wavelength elastic distortion in which the elastic energy is proportional to q^2. As a result the elastic constants become wave-vector independent leading to a terminology of Frank elastic constants. However, in case of polymer liquid crystals the rod lengths (~ 1000 Å) can be comparable in size to the wavelength of distortions and therefore, the q^2-dependent Frank elastic energy is no longer sufficient and one ought to consider a generalization of wave-vector dependent elastic constants, i.e., a nonlocal theory of elasticity. Such a theory has been constructed by Lo and Pelcovits[144] for a nematic polymer composed of rigid or semiflexible molecules. In this approach the wave-vector dependent elastic constants were defined as

$$- k_B T \,\ell n\, Z/Z_0 = \frac{1}{2}\,(\Delta n)^2\, K_i(\mathbf{q}_i) q^2 + 0(\Delta n)^4 \tag{8.26}$$

Here Z_0 and Z are the partition functions of the nematic polymer without and with distortions, respectively. Since a general evaluation of the left hand side of eq. (8.26) is extremely difficult, a mean field approach[83,148], developed to study the nematic-isotropic transition of a warm-like chain, was used. The partition function of a single warm like chain can be written as a functional integral over the local tangent vector $\hat{u}(s)$:

$$Z = \int d\mathbf{R} \int D\hat{u}(s) \exp\left[-\frac{1}{2}\beta \in_p \int_0^{\ell} ds \left| \frac{d\hat{u}(s)}{ds} \right|^2 - \beta \int_0^{\ell} ds\; u[\hat{u}(s)\cdot \hat{n}(\mathbf{r}(s))] \right] \tag{8.27}$$

where a point on the chain is represented by $\mathbf{r}(s)$ and the parameter s measures the chemical distance along the chain from one end (s = 0). The chain end s = 0 is located at the point $\mathbf{R}$ in the BF frame. $\in_p$ is the elastic modulus associated with the chain flexibility and u represents the local interaction between the chain and the nematic mean field $\hat{n}(\mathbf{r}(s))$. For the purpose of analytic calculation a Maier-Saupe form for u was chosen and the expression for the elastic moduli were derived[144] which are quite general and free-from any assumption about the flexibility of chains or strength of nematic order. For a chain of arbitrary length and infinite rigidity, i.e., rod-like molecules it was found that the bend elastic modulus shows pronounced dispersion relatively independent of the strength of nematic order, while the splay and twist constants show less dispersion as the nematic order saturates. For the flexible chain analytic results could not be obtained for the nonlocal elastic constants. However, the results were applicable to the flexible molecules of short length as compared to the global persistent length.

A microscopic molecular statistical theory was developed by Petschek and Terentjev[149] for the thermotropic nematic polymers composed of semiflexible chains of bonded nematogenic monomers. In this approach, based on the basic ideas of the flexible polymer chain statistics, the recurrence equation for the partition function was written as

$$M_k(\hat{e}_i) = \frac{1}{4\pi} \int P^B(\hat{e}_i,\hat{e}_j)\, M_{k-1}(\hat{e}_j) e^{-u(\hat{e}_j,\hat{n}_j)/T}\, d\hat{e}_j \tag{8.28}$$

This equation accounts for the conditional probability of the transfer from the (k-1)th link to the next, the kth one, and for the action of the external orienting field u on this link. Here $P^B(\hat{e}_i,\hat{e}_j)$ represents the probability that a monomer on a free-polymer chain with position $\mathbf{r}_i$ and the orientation $\hat{e}_i$ is followed by another monomer having position $\mathbf{r}_j$ and orientation $\hat{e}_j$. It is assumed that P^B does not depend on the parameters of monomers other than the two directly connected by a bond.

For a long polymer chain in which the majority of monomers are located in positions with large k the total partition function of the chain with N monomers was estimated as

$$Z = \frac{1}{4\pi}\int M_N(\hat{e}_i)\, d\hat{e}_i$$

$$\approx \frac{1}{4\pi} C_0 e^{-\Lambda_0 N/\omega} \int \left[W_0(\hat{e}_i) + \frac{C_n}{C_0} e^{-(\Lambda_n-\Lambda_0)N/\omega}\, W_n + \ldots\right] e^{u(\hat{e}_i)/T} d\hat{e}_i \tag{8.29}$$

where $\Lambda_n[\approx 2(n+1)\sqrt{\omega J/T}]$ is the eigenvalue of the first-excited state and Λ_0 (the ground-state energy) $\approx 2(\omega J/T)^{1/2}$. ω is the chain stiffness parameter, and C_0 and C_n are the expansion coefficients appearing in the expansions of ψ. W_0 and W_n are the eigenfunctions of the first two levels which are chosen to be normalized.

Within the framework of mean-field approximation an expression for the curvature elastic energy was derived[149]

$$\Delta A_e \approx \rho\frac{T}{\omega}(A_{00} - A_{01}^2/\Delta) + \frac{128}{9}\rho J \frac{A_{01}^2}{\Delta^2}\left(\frac{\omega J}{T}\right)^{3/2} \exp\left[-2\sqrt{\frac{\omega J}{T}}\right]$$

$$+\frac{1}{2}\rho^2\sqrt{\frac{\omega J}{T}} < \left(\frac{G}{D}(\delta_{\mu\nu} - e_\mu e_\nu) + \frac{2}{3}TdL^4 e_\mu e_\nu(\hat{e}_i \times \hat{e}_j)\right)$$

$$\times\, (e_\alpha e_\beta(\nabla_\mu\, n_\alpha \nabla_\nu\, n_\beta + n_\alpha\nabla_\mu\nabla_\nu\, n_\beta)) + 2\sqrt{\frac{\omega J}{T}}(e_\alpha e_\beta e_\nu e_\delta n_\alpha n_\beta \nabla_\mu n_\gamma \nabla_\nu n_\delta) > \tag{8.30}$$

where

$$A_{00} \approx -\frac{3J}{2\omega T} L^2(\nabla\cdot\hat{n})^2 - \frac{J}{2\omega T} L^2(\hat{n}\cdot\nabla\times\hat{n})^2 - \frac{J}{T}\sqrt{\frac{J}{\omega T}} L^2[\hat{n}\times(\nabla\times\hat{n})]^2 \tag{8.31}$$

$$A_{01} \approx -(2J/T)\, L(\nabla\cdot\hat{n}) \tag{8.32}$$

and G represents the isotropic attractive coupling constant and J the mean-field coupling constant.

The expression (8.30) shows the influence of hairpin regions on the chain where $W_0(\hat{e}_i)$ differs considerably from $W_1(\hat{e}_i)$. It is quite important to note that the integration over these regions contributes only to the splay elastic constant K_1. So K_1 is much greater than the other two elastic constants K_2 and K_3;

$$K_1 \approx \frac{16}{9}\rho\,\frac{JL^2}{\omega^2}\left(\frac{\omega J}{T}\right)^{5/2} \exp[2(\omega J/T)^{1/2}] + \cdots \tag{8.33}$$

$$K_2 \approx \frac{1}{4}\rho^2\frac{G}{d}\left(\frac{\omega J}{T}\right)^{1/2} + \frac{1}{8}\rho^2\left(\frac{3G}{d} + \frac{1}{3}TdL^4\right) + \cdots \tag{8.34}$$

and

$$K_3 \approx \frac{1}{12}\rho^2 TdL^4\left(\frac{\omega J}{T}\right)^{1/2} + \frac{1}{2}\rho^2\left(\frac{G}{d} - \frac{1}{4}TdL^4\right) + \cdots$$

As obvious the mean-field coupling constant J plays a significant role in all the elastic constants. Thus, in a main-chain thermotropic liquid crystal polymers, as in low molecular weight nematic polymers, the influence of long-range attractive forces on K_i is quite considerable. The anisotropic attraction between monomers if accounted the three elastic constants will differ more in magnitude but the qualitative features may remain the same.

REFERENCES

1. A. Blumstein, ed., (a) Mesomorphic Order in Polymers and Polymerization in Liquid Crystalline Media, ACS Symp. Series No. **74** (American Chemical Society, Washington, D.C., 1978), (b) Liquid Crystalline Order in Polymers (Academic Press, New York, 1978).
2. A. Ciferri, W.R. Krigbaum, and R.B. Meyer, eds, Polymer Liquid Crystals (Academic Press, New York, 1982).
3. L.L. Chapoy, ed., Recent Advances in Liquid Crystalline Polymers (Elsevier Applied Science Publishers, 1985).

4. A. Blumstein, ed., Polymer Liquid Crystals (Plenum, New York, 1985).

5. H. Finkelmann, Angew Chem. Int. Ed. Engl. **26**, 816 (1987).

6. G. Vertogen, and W.H. de Jeu, Thermotropic Liquid Crystals : Fundamentals (Springer-Verlag, 1988).

7. C.B. McArdle, ed., Side chain Liquid Crystal Polymers (Blackie, Glasgow, 1989).

8. A.N. Semenov, and A.R. Kokhlov, Sov. Phys. Usp. **31**, 988 (1988).

9. C.B. McArdle, Side Chain Liquid Crystal Polymers (Chapman and Hall, New York, 1989).

10. R.A. Weiss, and C.K. Ober, eds, Liquid Crystalline Polymers, ACS Symp. Series No. 435 (American Chemical Society, Washington D.C., 1990).

11. A. Ciferri, ed., Liquid Crystallinity in Polymers : Principles and Fundamental Properties (VCH, New York, 1991).

12. H. Finkelmann, W. Meier, and H. Scheuermann, Liquid Crystal Polymers, in "Liquid Crystals Applications and Uses, ed, B. Bahadur (World Scientific, 1992) Chapt. 22, p. 345.

13. G.J. Vroege, and H.N.W. Lekkerkerker, Rep. Prog. Phys., **55**, 1241 (1992).

14. G. Sigaud, Phase Transitions and Phase Diagrams in Liquid Crystalline Polymers, in, Phase transitions in Liquid Crystals, eds., S. Martellucci, and A.N. Chester (Plenum Press, New York, 1992) Chapt. 24, p. 375.

15. X.J. Wang, and M. Warner, Molecularly Non-homogeneous Nematic Polymers, in, Phase Transitions in Liquid Crystals, eds., S. Martellucci, and A.N. Chester (Plenum Press, New York, 1992) Chapt. 25, p. 399.

16. A.M. White, and A.H. Windle, Liquid Crystal Polymers (Cambridge University Press, Cambridge, 1992).

17. A.A. Collyer, ed., Liquid Crystal Polymers : From Structures to Applications (Elsevier, Oxford, 1993).

18. N.A. Plate, ed., Liquid Crystal Polymers (Plenum, New York, 1993).

19. C. Carfagna, ed., Liquid Crystal Polymers (Pergamon, Oxford, 1994).

20. P.J. Collings, and M. Hird, Introduction to Liquid Crystals : Chemistry and Physics (Taylor and Francis Ltd., London, 1997) Chapts. 5 (p. 93) and 8 (p. 147).

21. V. Percec, From Molecular to Macromolecular Liquid Crystals, in, Hand book of Liquid Crystal Research, eds, P.J. Collings, and J.S. Patel (Oxford Univ. Press, 1997) Chapt. 8, p. 259.

22. D. Demus, Mol. Cryst. Liquid Cryst. **165**, 45 (1988); Liquid Crystals **5**, 75 (1989).
23. H. Ringsdorf, B. Schlarb, and J. Venzmer, Angew. Chem. Int. Ed. Engl. **27**, 113 (1988).
24. R.D.C. Richards, W.D. Hawthorne, J.S. Hill, M.S. White, D. Lacey, J.A. Semlyen, G.W. Gray, and T.C. Kendrick, J. Chem. Soc. Chem. Commun. **1990**, 95 (1990).
25. V. Percec, and B. Hahn, J. Polym. Sci. Polym. Chem. Ed. **27**, 2367 (1989).
26. V. Percec, C.G. Cho, C. Pugh, and D. Tomazos, Macromolecules **25**, 1164 (1992).
27. V. Percec, P. Chu, and M. Kawasumi, Macromolecules **27**, 4441 (1994).
28. H.Y. Kim, J. Am. Chem. Soc. **114**, 4947 (1992); Adv. Mater. **4**, 764 (1992).
29. V. Percec, Pure Appl. Chem. **67**, 2031 (1995).
30. V. Percec, P. Chu, G. Ungar, and J. Zhou, J. Am. Chem. Soc. **117**, 11441 (1995).
31. W. Gleim, and H. Finkelmann, in, Side Chain Liquid Crystal Polymers, ed, C.B. McArdle (Chapman and Hall, New York, 1989) p. 287.
32. R. Zentel, Angew. Chem. Int. Ed. Engl. **28**, 1407 (1989).
33. V. Percec, and M. Zuber, Polym. Bull. **25**, 695 (1991).
34. A. Roviello, and A. Sirigu, J. Polym. Sci. Polym. Lett. Ed, **13**, 455 (1975).
35. P.G. de Gennes, C.R. Acad. Sci. Paris **B281**, 101 (1975).
36. M. Gordon, and N.A. Platé, eds, Liquid Crystal Polymers I. Adv. Polym. Sci. Vol. **59** (Springer –Verlag, 1984).
37. H. Finkelmann, Angew. Chem. Int. Ed. Engl. **26**, 816 (1987).
38. R.B. Blumstein, and A. Blumstein, Mol. Cryst. Liquid Cryst. **165**, 361 (1988).
39. M. Ballauff, Angew. Chem. Int. Ed. Engl. **28**, 253 (1989).
40. J. Economy, Mol. Cryst. Liquid Cryst. **169**, 1 (1989); Angew. Chem. Int. Ed. Engl. **29**, 1256 (1990).
41. M. Warner, Phil. Trans. R. Soc. London **A344**, 403 (1993); in, Side Chain Liquid Crystal Polymers, ed., C.B. McArdle (Chapman and Hall, New York, 1989) p. 7.
42. P. Davidson, and A.M. Levelut, Liquid Crystals **11**, 469 (1992).
43. F. Kuschel, A. Madicke, S. Diele, H. Utschik, B. Hisgen, and H. Ringsdorf, Polym. Bull. **23**, 373 (1990).

44. H. Mattoussi, R. Ober, M. Veyssie, and H. Finkelmann, Europhys. Lett. **2**, 233 (1986).
45. V. Percec, B. Hahn, M. Ebert, and J.H. Wendorff, Macromolecules **23**, 2092 (1990).
46. P Davidson, L. Noirez, J.P. Cotton, and P. Keller, Liquid Crystals **10**, 111 (1991).
47. F. Hardouin, S. Mery, M.F. Achard, L. Noirez, and P. Keller, J. Physique II **1**, 511 (1991).
48. C. Noël, Makromol. Chem. Macromol. Symp. **22**, 95 (1988); in, Side Chain Liquid Crystal Polymers, ed., C.B. McArdle (Chapman and Hall, New York, 1989) p. 159.
49. L. Noirez, P. Keller, and J.P. Cotton, Liquid Crystals **18**, 129 (1995).
50. T.I. Gubina, S.G. Kostromin, R.V. Tal'rose, V.P. Shibaev, and N.A. Platé, Vysokomol. Soed. **B28**, 394 (1986).
51. P. Le Barny, J.C. Dubois, C. Friedrich, and C. Noël, Polym. Bull. **15**, 341 (1986).
52. S.G. Kostromin, V.P. Shibaev, and S. Diele, Makromol. Chem. **191**, 2521 (1990).
53. C. Legrand, A. Le Borgne, C. Bunel, N. Lacoudre, P. Le Barny, N. Spassky, and J.P. Vairon, Makromol. Chem. **191**, 2979 (1990).
54. V. Percec, and M. Lee, J. Mater. Chem. **2**, 617 (1992).
55. L. Onsager, Ann. N.Y. Acad. Sci. **51**, 627 (1949).
56. T. Odijk, Macromolecules **19**, 2313 (1986).
57. H.N.W. Lekkerkerker, P. Coulon, R. van der Haegen, and R. Deblieck, J. Chem. Phys. **80**, 3427 (1984).
58. P.J. Flory, Proc. Roy. Soc. **A234**, 73 (1956).
59. P.J. Flory, and G. Ronca, Mol. Cryst. Liquid Cryst. **54**, 289 (1979).
60. (a) P.J. Flory, and G. Ronca, Mol. Cryst. Liquid Cryst. **54**, 311 (1979); (b) F. Dowell, Phys. Rev. **A28**, 1003 (1983), ibid **A31**, 3214 (1985).
61. R. Deblieck, and H.N.W. Lekkerkerker, J. Phys. Lett. (Paris) **41**, 351 (1980).
62. J.K. Moscicki, and G. Williams, Polymer **24**, 85 (1983).
63. T. Odijk, and H.N.W. Lekkerkerker, J. Phys. Chem. **89**, 2090 (1985).
64. W.E. McMullen, W.M. Gelbart, and A. Ben-Shaul, J. Chem. Phys. **82**, 5616 (1985).
65. P.J. Flory, and A. Abe, Macromolecules **11**, 1119 (1978).

66. A. Abe, and P.J. Flory, Macromolecules **11**, 1122 (1978).
67. P.J. Flory, and R.S. Frost, Macromolecules **11**, 1126 (1978).
68. R.S. Frost, and P.J. Flory, Macromolecules **11**, 1134 (1978).
69. T.M. Birshtein, B.I. Kolegov, and V.A. Pryamitsyn, Polym. Sci. USSR **30**, 316 (1988).
70. S. Singh, Phys. Rep. **324**, 107 (2000).
71. A.R. Khokhlov, Phys. Lett. **A68**, 135 (1978).
72. A.R. Khokhlov, and A.N. Semenov, Physica **A108**, 526 (1981); Macromolecules **17**, 2678 (1984).
73. A.R. Khokhlov, and A.N. Semenov, Physica **A112**, 605 (1982).
74. S.K. Nechaev, A.N. Semenov, and A.R. Khokhlov, Polym. Sci. USSR **25**, 1063 (1983).
75. G.J. Vroege, and T. Odijk, Macromolecules **21**, 2848 (1988).
76. R.R. Matheson, and P.J. Flory, Macromolecules **14**, 954 (1981).
77. T.M. Birshtein, and A.R. Merkuŕeva, Polym. Sci. USSR **27**, 1208 (1985).
78. A.R. Khokhlov, Polym. Sci. USSR **21**, 2185 (1979).
79. M. Warner, and P.J. Flory, J. Chem. Phys. **73**, 6327 (1980).
80. A.R. Khokhlov, and A.N. Semenov, J. Statistical Physics **38**, 161 (1985).
81. A.R. Khokhlov, and A.N. Semenov, Macromolecules **19**, 373 (1986).
82. V.V. Rusakov, and M.I. Shliomis, J. Phys. Lett. (Paris) **46**, 935 (1985).
83. X.J. Wang, and M. Warner, J. Phys. **A19**, 2215 (1986).
84. S.V. Vasilenko, V.P. Shibaev, and A.R. Khokhlov, Makromol. Chem. – Rap. Commun. **3**, 917 (1982).
85. S.V. Vasilenko, A.R. Khokhlov, and V.P. Shibaev, Polym. Sci. USSR **26**, 606 (1984); Macromolecules **17**, 2270 (1984).
86. P. Corradini, and M. Vacatello, Mol. Cryst. Liquid Cryst. **97**, 119 (1983).
87. V.V. Rusakov, and M.I. Shliomis, Polym. Sci. USSR **29** (1987).
88. V.P. Shibaev, and N.A. Plate, Adv. Polym. Sci. **60/61**, 173 (1984).
89. H. Finkelmann, and G. Rehage, Adv. Polym. Sci. **60/61**, 99 (1984).
90. N.A. Platé, and V.P. Shibaev, Comblike Polymers and Liquid Crystals, Khimiya, Moscow (1980, in Russian).
91. P.J. Flory, Proc. Natl. Acad. Sci. (USA) **79**, 5410 (1982).
92. Y.H. Kim, and P. Pincus, Biopolymers **18**, 2315 (1979).

93. R.G. Petschek, Phys. Rev. **A34**, 1338 (1986).

94. X.J. Wang, and M. Warner, Liquid Crystals **12**, 385 (1992), Molecularly Nonhomogeneous Nematic Polymers, in, Phase Transitions in Liquid Crystals, eds., S. Martellucci, and A.N. Chester (Plenum Press, New York and London, 1992) Chapt. 25, p. 399.

95. M. Warner, J.M.F. Gunn, and A. Baumgärter, J. Phys. **A18**, 3007 (1985).

96. X.J. Wang, and M. Warner, J. Phys **A19**, 2215 (1986).

97. W. Kern, W. Gruber, and H.O. Wirth, Makromol. Chem. **37**, 198 (1960).

98. W. Heitz (a) Chemiker Zeitung **110**, 385 (1986); (b) Makromol. Chem. Macromol. Symp. **26**, 1 (1989); (c) ibid **47**, 111 (1991).

99. T. Wahlenkamp, and G. Wegner, Macromol. Chem. Phys. **195**, 1933 (1994).

100. H. Wilteler, G. Lieser, G. Wegner, and M. Schulze, Makromol. Chem. Rapid Commun. **14**, 471 (1993).

101. A. Noll, N. Siegfried, and W. Heitz; Makromol. Chem. Rapid Commun. **11**, 485 (1990).

102. S. Takahashi, H. Morimoto, E. Murata, S. Kataoka, K. Sonogashira, and N. Higara, J. Polym. Sci. Polym. Chem. Ed. **20**, 565 (1982).

103. S. Takahashi, Y. Takai, H. Morimoto, and K. Sonogashira, J. Chem. Soc. Chem. Commun. **1984**, 3 (1984).

104. S.J. Davis, B.F.G. Johnson, M.S. Khan, and J. Lewis, J. Chem. Soc. Chem. Commun. **1991**, 187 (1991).

105. M.H. Chisholm, Angew. Chem. Int. Ed. Engl. **30**, 673 (1991).

106. H.B. Fyfe, M. Miekuz, D. Zargarian, N.J. Taylor, and T.B. Marder, J. Chem. Soc. Chem. Commun. 1991, 188 (1991).

107. P. Espinet, M.A. Esteruelas, L. A. Oro, J.L. Serrano, and E. Sola, Coord. Chem. Rev. **117**, 215 (1992).

108. A.M. Giroud-Godquin, and P.M. Maitlis, Angew, Chem. Int. Ed. Engl. **30**, 375 (1991).

109. S.A. Hudson, and P.M. Maitlis, Chem. Rev. **93**, 861 (1993).

110. A.P. Polishchuk, and T.V. Timofeeva, Russ. Chem. Rev. **64**, 291 (1993).

111. H. Jonsson, V. Percec, and A. Hult, Polym. Bull. **25**, 115 (1991).

112. V. Percec, M. Lee, and H. Jonsson, J. Polym. Sci. Polym. Chem. Ed. **29**, 327 (1991).

113. V. Percec, A.D.S. Gomes, and M. Lee, J. Polym. Sci. Polym. Chem. Ed., **29**, 1615 (1991).

114. V. Percec, C.S. Wang, and M. Lee, Polym. Bull. **26**, 15 (1991).

115. V. Percec, M. Lee, and C. Ackerman, Polymer **33**, 703 (1992).

116. V. Percec, Q. Zheng, and M. Lee, J. Mater. Chem. **1**, 611 (1991).

117. V. Percec, and G. Johansson, J. Mater. Chem. **3**, 83 (1993).

118. R. Rodenhouse, and V. Percec, Polym. Bull. **25**, 47 (1991).

119. H. Jonsson, V. Percec, V.W. Gedde, and A. Hult, Makromol. Chem. Macromol. Symp. **54-55**, 83 (1992).

120. R.A. Gaudiana, R.A. Minns, H.G. Rogers, R. Sinta, L.D. Taylor, P.S. Kalyaranaman, and C. McGowan, J. Polym. Sci. Polym. Chem. Ed. **25**, 1249 (1987).

121. H.G. Rogers, R.A. Gaudiana, W.C. Hollinsed, P.S. Kalyaranaman, J.S. Manello, C. McGowan, R.A. Minns, and R. Sahatjian, Macromolecules **18**, 1058 (1985);

H.C. Rogers, and R.A. Gaudiana, J. Polym. Sci. Polym. Chem. Ed. **23**, 2669 (1985).

122. R. Sinta, R.A. Minns, R.A. Gaudiana, and H.G. Rogers, J. Polym. Sci. Polym. Lett. Ed. **25**, 11 (1987).

123. R.A. Gaudiana, R.A. Minns, R. Sinta, N. Weeks, and H.G. Rogers, Prog. Polym. Sci, Japan **14**, 47 (1989).

124. M. Ballauff, Ber. Bunsen-Ges. Phys. Chem. **90**, 1053 (1986).

125. H. Ringsdorf, P. Tschirner, O.H. Schoenherr, and J.H. Wendorff., Makromol. Chem. **188**, 1431 (1987).

126. O.H. Schoenherr., J.H. Wendorff., H. Ringsdorf, and P. Tschirner, Makromol. Chem. Rapid. Commun **7**, 791 (1986).

127. W. Brugging, U Kampschulte, H.-W. Schmidt, and W. Heitz, Makromol. Chem. **189**, 2755 (1988).

128. W. Vogel, and W. Heitz, Makromol. Chem. **191**, 829 (1990).

129. W. Heitz, and N. Niessner, Makromol. Chem. **191**, 225 (1990).

130. L. Freund, H. Jung, N. Niessner, H.-W. Schmidt, and W. Heitz, Makromol. Chem. **190**, 1561 (1989).

131. W. Heitz, Makromol. Chem. Macromol. Symp. **47**, 111 (1991); ibid **48-49**, 15 (1991).

132. M.S. Classen, H.-W. Schmidt, and J.H. Wendorff, Polym. Adv. Technol. **1**, 143 (1990).

133. H.-W. Schmidt, and D. Guo, Makromol. Chem. **189**, 2029 (1988).

134. M. Hohlweg, and H.-W. Schmidt, Makromol. Chem. **190**, 1587 (1989).

135. N. Carr, and G.W. Gray, Mol. Cryst. Liquid Cryst. **124**, 27 (1985).

136. M.A. Osman, Mol. Cryst. Liquid Cryst. Lett. **82**, 47, 295 (1982).

137. A.A. Craig, and C.T. Imrie, (a) Macromolecules **28**, 3617 (1995); (b) J. Mater. Chem. **4**, 1705 (1994).

138. V. Percec, and D. Tomazos, Adv. Mater. **4**, 548 (1992).

139. S. Singh, Phys. Rep. **277**, 283 (1996).

140. R.G. Priest, Phys. Rev. **A7**, 720 (1973).

141. J.P. Straley, Phys. Rev. **A8**, 2181 (1973).

142. A.Y. Grosberg, and A.Z. Zhestkov, Polym. Sci. USSR **28**, 97 (1986).

143. G. Vroege, and T. Odijk, Macromolecules **21**, 2848 (1988).

144. W.S. Lo, and R.A Pelcovits, Phys. Rev. **A42**, 4756 (1990).

145. T. Odijk, Liquid Crystals **1**, 553 (1986).

146. G.J. Vroege, and T. Odijk, J. Chem. Phys. **87**, 4223 (1987).

147. P.G. de Gennes, Mol. Cryst. Liquid Cryst. **34**, 177 (1977).

148. M. Warner, J.M.F. Gunn, and A. Baumgärter, J. Phys. **A18**, 3007 (1985); See also A.R. Khokhlov, and A.N. Semenov, Physica **A108**, 546; **A112**, 605 (1982).

149. R.G. Petschek, and E.M. Terentjcv, Phys. Rev. **A45**, 930 (1992).

9

Chiral Liquid Crystals

Chirality in liquid crystals has been the subject of intense research in recent years, and is directly responsible for their important technological applications. An object (or molecule) which cannot be transformed into their mirror image by rotations or translations is called chiral. The term chiral implies that a molecular structure is asymmetric and handed. Chiral liquid crystalline materials as well as chiral molecules forming liquid crystal phases are very special and unusual. Table 1.3 summarizes the phase types occurring in chiral compounds. Because of the symmetry properties, the chiral nematic (cholesteric) phase is nonferroelectric, whereas the chiral smectic phases exhibit ferroelectric properties. The aim of the present chapter is to discuss the special nature and some important properties of chiral liquid crystal phases[1-4].

The presence of chiral molecules in a liquid crystal can have various consequences. The chirality may cause an intrinsic helical structure of the director field. Instead of the uniform alignment of the director field occurring in the achiral phases, the respective chiral phases exhibit a helical structure; examples of some important chiral phases are chiral nematic, blue phases, chiral smectic phases, twist grain boundary (TGB) phases. Apart from inducing a helical director field, the second fundamental type of consequence of chirality is the appearance of polar physical properties which is described by a vector or a third-rank tensor. A polar property described by a vector can never occur perpendicular to a rotational axis or mirror plane of the considered structure. A very important property of this kind is the spontaneous polarization $\mathbf{P}_S$ which gives to the ferroelectric properties of certain liquid crystals and their electrooptic applications. For achiral liquid crystals, such as nematic phase (point group $D_{\infty h}$), smectic A (point group $D_{\infty h}$), smectic C phase (point group C_{2h}), spontaneous polarization never occurs. The presence of chiral molecules reduces the local symmetry of the respective phases, even when the helical director field is unwound, to the space groups D_∞ (chiral nematic and chiral smectic A phases) and C_2 (chiral smectic C phase) because of lack of mirror symmetry. The chirality can also give certain nonlinear optical effects[5,6] (for example, nonlinear susceptibilities found in chiral smectic C phase).

9.1. CHIRAL NEMATIC (CHOLESTERIC) PHASE

The chiral nematic (N^*) or cholesteric phase is thermodynamically equivalent to a uniaxial nematic except for the chiral-induced twist in the director (see Fig. 1.4a). Historically, the chiral nematic phase is called the cholesteric phase because the first thermotropic liquid crystalline materials exhibiting this phase were cholesterol derivatives. However, today many different types of chiral materials are available that generate the chiral nematic phase and these have no resemblance to the cholesterol derivatives. Locally, the phase structure of a chiral nematic is very similar to a nematic material, i.e., the centers of gravity have no long-range order and the molecular orientation shows a preferred axis indicated by a director $\hat{\mathbf{n}}$. However, the asymmetry of the constituent molecules causes a small gradual rotation of the director $\hat{\mathbf{n}}$. This gradual rotation of $\hat{\mathbf{n}}$ describes a helix which has a specific temperature-dependent pitch $p_0 = 2\pi/q_0$. The pitch of the helix is defined as the longitudinal distance in which the angle of twist has made one complete 360° revolution. Thus the chiral nematic has twist along one axis. If the helical axis coincides with the Z-axis, $\hat{\mathbf{n}}$ has the following structure,

$$
\begin{aligned}
n_x &= \cos(q_0 z + \phi) \\
n_y &= \sin(q_0 z + \phi) \qquad (9.1) \\
n_z &= 0
\end{aligned}
$$

Both the helical axis (Z) and the value of ϕ are arbitrary. The states $\hat{\mathbf{n}}$ and $-\hat{\mathbf{n}}$ are equivalent. The structure is periodic along Z with spatial period L equal to one-half of the pitch $L = \pi/|q_0|$. Both the magnitude and sign of helical wave number q_0 are important. The sign signifies the handedness of the helices. A particular material at a given temperature forms helices of the same sign.

Compounds that generate the chiral nematic are very common and in many respects similar to those that exhibit nonchiral nematic phase except that a chiral unit is present. As can be seen in Fig. 9.1, the first thermotropic cholesterogens discovered by Reinitzer has 8 chiral centers giving a total of 256 stereoisomers; however, only one occurs in nature. Accordingly, it is the chiral unit that is of interest. The chirality can be introduced into a liquid crystal system in a number of ways[1–4]. Nonchiral nematic phases can be induced to be chiral by the addition of small amount of chiral dopant molecules which may not be mesogenic materials at all. Many empirical rules have been derived[8] which suggest relationships between the spatial configuration of the chiral center of the molecules and the twist direction of the helical structure. These rules work well for simple molecules exhibiting a chiral nematic, but for more complex systems sometimes are less appropriate.

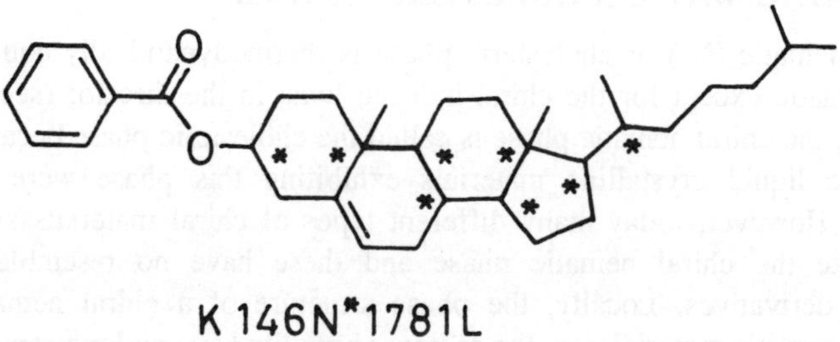

K 146N*178 I L

Fig. 9.1. Cholesteryl benzoate.

9.1.1. Optical Properties

The optical properties of a mesophase characterize their response to high frequency electromagnetic radiation and encompass the properties of reflection, refraction, polarization, optical absorption, optical activity, harmonic generation, optical waveguiding, light scattering, etc. Many applications[5-6] of liquid crystalline phases are based on their optical properties and response to changes of external field and physicochemical factors. At the simplest level the optical properties of materials can be interpreted in terms of refractive indices[7]. Based on the theoretical work of Mauguin[9a], Oseen[9b] and de Vries[10], the unique optical phenomena occurring in the chiral nematic phase can be explained[1–4,11–13] quite rigorously in terms of the helical structure as shown in Fig. 1.4a.

When white light is incident on a planar sample, with optic axis perpendicular to the glass surfaces, selective reflection occurs. The wavelengths of the reflected maxima vary with the angle of incidence in accordance with Bragg's law. In case of normal incidence, the reflected wave is strongly circularly polarized. Along its optic axis, the medium possesses a very high rotatory power. The polarization state of the reflected and transmitted waves depends on the pitch length of the material. If the helical pitch is much larger than the incident light wavelength, both the reflected and transmitted waves are plane polarized and periodically modulated by the pitch of the helical structure. In case the pitch is much smaller than the incident wavelength, there are two circularly polarized waves in the medium. When the incident wavelength is comparable to the helical pitch, the reflected light is strongly circularly polarized. One circular component is almost totally reflected, while the other passes through practically unchanged. The reflected wave preserves its sense of circular polarization as that of the incident wave.

9.1.2. Field – Induced Nematic – Cholesteric Phase Change

When a magnetic-field is applied at right angles to the helical axis of unbounded cholesteric phase composed of molecules of positive diamagnetic anisotropy the cholesteric structure gets distorted. At a critical value of field H_c, the pitch increases logarithmically. As the applied field H becomes greater than H_c ($H > H_c$) the helix is destroyed completely and the cholesteric phase transforms to a nematic phase.

The electric field effects are more complicated due to conduction[14]. The electric field–induced nematic-cholesteric transition has been used for displays[15]. The cholesteric phase is initially aligned with helical axis parallel to the glass substrates. The incident light is scattered and the cell appears milky white. When the applied field exceeds 10^5 V/cm, it unwinds the helix and gives an aligned nematic phase. The cell then becomes transparent. With the decreasing value of the voltage, an intermediate metastable nematic phase appears[16] in which the directors near the surfaces remain homeotropic but in the bulk are slightly tilted. As a result a hysteresis loop is observed which can be used for the optical storage. When the voltage is removed completely, the directors return to their initial scattering states within a few milliseconds.

9.1.3. Elastic Continuum Theory for Chiral Nematics

While deriving the elastic free-energy relation for a nematic system (chapt. 5, sec. 5.1.1) we ignored all terms linear in gradients of $\hat{\mathbf{n}}$. In case of a chiral nematic phase this argument does not hold because the equilibrium configuration is twisted. There are two such terms : $\nabla\cdot\hat{\mathbf{n}}$ and $\hat{\mathbf{n}}\cdot\nabla\times\hat{\mathbf{n}}$. Since the states $\hat{\mathbf{n}}$ and $-\hat{\mathbf{n}}$ are equivalent, $\nabla\cdot\hat{\mathbf{n}}$ term does not survive. Considering the term $\hat{\mathbf{n}}\cdot\nabla\times\hat{\mathbf{n}}$ in eq. (5.14), the distortion free-energy density for the chiral nematics reads

$$f_e = \frac{1}{V}\Delta A_e = \frac{1}{2}K_1(\nabla\cdot\hat{\mathbf{n}})^2 + \frac{1}{2}K_2(\hat{\mathbf{n}}\cdot\nabla\times\hat{\mathbf{n}} + q_0)^2 + \frac{1}{2}K_3(\hat{\mathbf{n}}\times\nabla\times\hat{\mathbf{n}})^2 \tag{9.2}$$

In eq. (9.2) there exists a term linear in the gradients

$$K_2\, q_0\, \hat{\mathbf{n}}\cdot\nabla\times\hat{\mathbf{n}}$$

and a constant term

$$\frac{1}{2}K_2 q_0^2 .$$

In case of pure twist

$$n_x = \cos\theta(z)$$
$$n_y = \sin\theta(z) \tag{9.3}$$
$$n_z = 0$$

eq. (9.2) reduces to

$$f_e = \frac{1}{2}K_2\left(\frac{\partial\theta}{\partial z} - q_0\right)^2 \tag{9.4}$$

This shows that the equilibrium distortion corresponds to a helix of wave vector $\partial\theta/\partial z = q_0$.

In eq. (9.2) both the $\nabla\hat{\mathbf{n}}$ and q_0 must be small on the molecular scale. When q_0 is large ($q_0 a \sim 1$) the structure of f_e becomes more complicated and Jenkins[17] corrections have to be included. However, in practice, $q_0 a \sim 10^{-3}$ and so the Jenkins corrections are not important. The minimization of eq. (9.2) describes the static distortion.

When external fields (**E** or **H**) are present, additional terms involving these fields have to be added to the eq. (9.2). The contribution of **E** or **H** to the elastic free-energy density can be expressed as

$$f_{e,el} = -\frac{1}{8\pi}\epsilon_a(\mathbf{E}\cdot\hat{\mathbf{n}})^2 \tag{9.5}$$

and

$$f_{e,mag} = -\frac{1}{2}\chi_a(\mathbf{H}\cdot\hat{\mathbf{n}})^2 \tag{9.6}$$

where $\chi_a = \chi_\parallel - \chi_\perp$ and $\epsilon_a = \epsilon_\parallel - \epsilon_\perp$. In these equations only the $\hat{\mathbf{n}}$ dependent terms have been taken into account.

9.1.4. Factors Influencing the Pitch

The unique optical properties of a chiral nematic material often depend critically on the value of the spiral pitch p_0. There are some typical agents that can be used to alter the pitch. This property has been exploited for the various interesting applications of the chiral nematic materials.

In most cholesteric materials, the pitch is a decreasing function of the temperature

$$\frac{dp_0(T)}{dT} < 0.$$

The order of magnitude of dp_0/dT is often surprisingly large; in a typical example[18], for cholesteryl nonanoate, around 74.6°C,

$$\frac{1}{p_0}\left|\frac{dp_0}{dT}\right| \sim 100\,\text{deg}^{-1}.$$

The temperature dependence of the pitch has been explained by an analogy with the theory of thermal expansion in crystals[19]. Also the increase in p_0 with the decrease in T may be attributed due to the onset of a short-range order of the smectic type.

The strong temperature dependence of the pitch has a number of practical applications. The material is chosen such that the pitch is of the order of the wavelength of visible light in the temperature range of interest. This is achieved with suitable multicomponent mixtures. The changes of p_0 are reflected in colour changes of the scattered light in the visible spectrum. This technique has been useful for visual display of surface temperatures, infrared lasers and microwave patterns, etc.

A chiral nematic phase of large pitch can be derived by dissolving an optically active material of small concentration c in a nematic material. For low concentration of the cholesteric, pitch is inversely proportional to the concentration[20]

$$p_0 c = \text{constant}.$$

However, at the higher concentrations this law is not obeyed[21]. The pitch versus composition curve attains a maximum at a certain concentration and beyond that it decreases. When certain gases are absorbed in a cholesteric film, a significant change in pitch is observed. Based on this finding certain devices have been invented which change an ultraviolet image into a visible image[22].

Pollman and Stegemeyer[23] studied the effect of pressure on the pitch of a mixture of cholesteryl chloride and cholesteryl oleyl carbonate (COC) and found that the pitch increases rapidly with pressure. As the concentration of COC increases the effect becomes more pronounced.

The structure of a chiral nematic can be distorted by an electric or a magnetic field. Thus a pronounced influence on the pitch is observed due to the presence of external fields[11,12,24,25]. We discuss below the case of magnetic field.

9.1.5. Distortion of Structure by Magnetic Field

Many experiments have shown that when a planar sample of chiral nematic is subjected to magnetic field, either the effects of untwisting of helicoidal structure or the effects of instability under field is observed[12,26,27]. We describe these effects here.

In a chiral nematic sample with negative diamagnetic anisotropy ($\chi_a < 0$) the free-energy is minimum when the helical axis is along the field. In this configuration the director $\hat{\mathbf{n}}$ is normal to the field at all points. No effect of distortion occurs and the helicoidal structure remains independent of the field. On the other hand, for $\chi_a > 0$, the story is different (see below).

9.1.5.1. Unwinding of the helicoidal structure under magnetic field : The cholesteric-nematic transition

When a magnetic field is applied at right angles to the helical axis of a bulk cholesteric sample with $\chi_a > 0$, above a critical field $\mathbf{H}_c$, the cholesteric is completely untwisted and cholesteric-nematic transition is observed. Sackmann et al.[26] were first to observe this effect. Many experiments have been performed to study the distortion of helicoidal structure with the magnetic field. At low fields the structure (pitch) remains essentially unperturbed but at the higher fields structure is distorted; pitch increases with the field and finally diverges at the critical field. The critical field $\mathbf{H}_c$

is found to be inversely proportional to the unperturbed pitch p_0. In actual experiment with the usually available magnetic field strength, it is desirable to work with a thick cholesteric of relatively large pitch; wall effect is eliminated by choosing a thick sample.

Figure 9.2 shows the unwinding of a cholesteric spiral by a magnetic field. The unperturbed (and low fields) situation is shown in Fig. 9.2(a). The influence of intermediate fields $H \leq H_c$ is shown in Fig. 9.2(b). In regions such as "A, A′, ..." the molecules are favourably aligned along the field. On the other hand, in regions such as "B, B′,..." they find an unfavourable orientation with respect to the field. When the field becomes strong enough region "A" will expand whereas region "B" cannot contract very much. As a result an overall increase in pitch p(H) will be observed with the field. At somewhat higher fields, this leads to a succession of 180° walls separating large "A" regions. Each wall has a finite thickness, of order 2ξ(H), and thus the twist is essentially limited to a region of magnitude

$$\xi = \left(\frac{K_2}{\chi_a}\right) H^{-1} \tag{9.7}$$

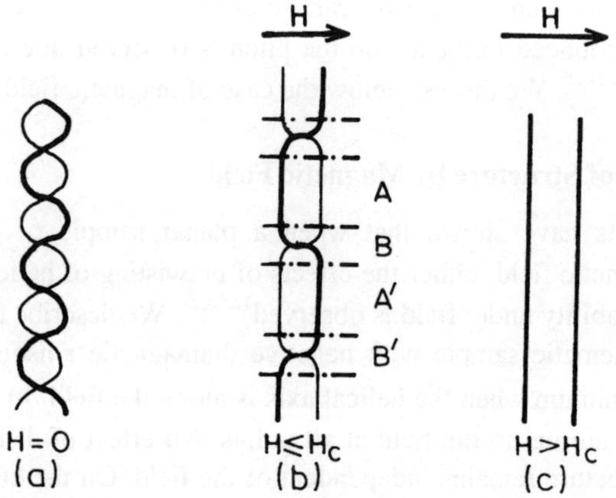

Fig. 9.2. Schematic representation of unwinding of a cholesteric under magnetic field (Ref. 12).

The distance between walls now becomes much larger than ξ. At H_C, the wall becomes infinitely separated; pitch diverges $p \rightarrow \infty$, and a nematic structure results.

From a theoretical point of view, the Cl-N_U transition can be studied from the continuum free-energy relations (eqs. (9.4) and (9.6)),

$$f_e = \frac{1}{2} K_2 \left(\frac{\partial \theta}{\partial z} - q_0\right)^2 - \frac{1}{2} \chi_a (\mathbf{H} \cdot \hat{\mathbf{n}})^2 \tag{9.8}$$

For $H \rightarrow H_c$, well separated walls are observed (Fig. 9.2(b)), and the interaction between walls become negligible. We assume a one-dimensional situation of pure twist ($n_x = \cos\theta(z)$, $n_y = \sin\theta(z)$). Minimization of eq. (9.8) gives

$$\xi^2 \frac{d^2\theta}{dz^2} = \sin\theta \cos\theta \qquad (9.9)$$

θ is the angle of the director with the direction of the field **H** in the cholesteric plane. Equation (9.9) has a first integral

$$\xi^2 \left(\frac{d\theta}{dz}\right)^2 = \sin^2\theta \qquad (9.10)$$

ensuring that $\frac{d\theta}{dz} \rightarrow 0$ for $\theta = \pi$.

The free-energy (per unit area) of one wall, compared to the energy of the nematic state, reads

$$f_w = \int \left[\frac{1}{2} K_2 \left\{\left(\frac{d\theta}{dz} - q_0\right)^2 - q_0^2\right\} + \frac{1}{2}\chi_a H^2 \sin^2\theta\right] dz \qquad (9.11)$$

In view of eqs. (9.7) and (9.10), one obtains

$$\frac{f_w}{\chi_a H^2} = 2 - \pi q_0 \xi \qquad (9.12)$$

Conditions become unfavourable for the wall formation when $\xi(H) < 2/\pi q_0$. This corresponds to a critical field (eq. (9.7))

$$H_c = \int \frac{1}{2}\pi \left(\frac{K_2}{\chi_a}\right)^{\frac{1}{2}} q_0 = \pi^2 \left(\frac{K_2}{\chi_a}\right)^{\frac{1}{2}} \frac{1}{p_0} \qquad (9.13)$$

Equation (9.13) was derived independently by Meyer[28] and by de Gennes[29]; it shows that, for fixed K_2 and χ_a, H_c is inversely proportional to the unperturbed pitch p_0.

For the situation $H < H_c$, the energy of single wall f_w becomes negative; thus walls tend to pile up in the sample until the interactions (proportional to $\exp[-p/2\xi]$) between neighbouring walls lead to an equilibrium. The pitch p(**H**) is only logarithmically divergent for $H = H_C$.

9.1.5.2. Instability under magnetic field : The square grid pattern

Our discussion in sec. 9.1.5.1 has been restricted to the situation of pure twist with the helical axis normal to the field. For the bulk sample this is the usual situation.

However, some other possibilities have also been pointed out which we shall discuss here.

Let us consider that the field is acting along the helical axis of a planar cholesteric film. For positive χ_a, if the boundary effects are absent, a 90° rotation of the helical axis is possible because $\frac{1}{2}(\chi_{\|} + \chi_{\perp}) > \chi_{\perp}$. On the other hand, if $K_3 << K_2$, with increasing field a transition from a helical conformation (**q** normal to **H**) to a conical conformation[30,31] described by

$$\begin{aligned} n_x &= \sin\psi \\ n_y &= \cos(qx)\cos\psi \\ n_z &= \sin(qx)\cos\psi \end{aligned} \tag{9.14}$$

should occur. ψ is a function of **H** only, the conical axis is parallel to the field. However, in practice, $K_3 > K_2$ and so this conical phase has not been observed.

The possibility of an another kind of distortion, a corrugation of the layers (Fig. 9.3) was pointed out first by Helfrich[32] mainly with the electric field. A more elaborate discussion with both magnetic and electric fields is due to Hurault[14]. This distortion has been observed[33–35] experimentally with both the fields at a much lower threshold. It results in the so called square grid pattern. We consider here the magnetic field case and discuss a coarse-grained version[12] of the continuum theory for cholesteric.

Let us consider the cholesteric to be a quasilayered structure. In the unperturbed state the cholesteric planes are equidistant (interval p_0) and parallel; unperturbed helical axis be the Z-axis. In a slightly deformed state, each plane is displaced by u(**r**) along Z. For small gradients $\nabla u << 1$, the coarse-grained free-energy density can be written as

$$f_{cg} = \frac{1}{2}B\left(\frac{\partial u}{\partial z}\right)^2 + \frac{1}{2}\tilde{K}\left(\frac{\partial^2 u}{\partial x^2} + \frac{\partial^2 u}{\partial y^2}\right)^2 \tag{9.15}$$

There are no terms involving $(\partial u/\partial x)^2$ and $(\partial u/\partial y)^2$ because they correspond to a uniform rotation of the layers and do not contribute to the free-energy. Term proportional to $\frac{\partial u}{\partial z}\left(\frac{\partial^2 u}{\partial x^2} + \frac{\partial^2 u}{\partial y^2}\right)$ is not included due to incompatibility with the existence of a two-fold axis of symmetry in the unperturbed state. Equation (9.15) may be written in a slightly more general form in term of a unit vector $\hat{\mathbf{d}}(\mathbf{r})$ normal to the cholesteric plane

$$f_{cg} = \frac{1}{2}B\left(\frac{p}{p_0} - 1\right)^2 + \frac{1}{2}\tilde{K}(\nabla\cdot\hat{\mathbf{d}})^2 \tag{9.16}$$

where p is the pitch in the distorted state.

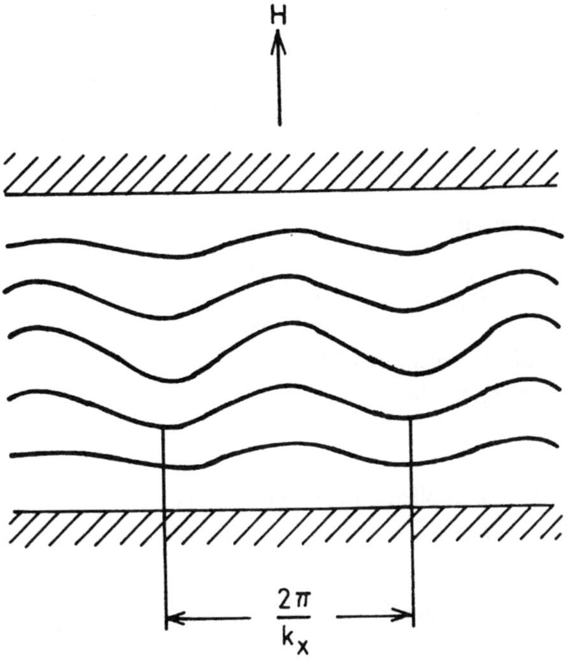

Fig. 9.3. Distortion of a planar texture due to a magnetic field acting along the helical axis ($\chi_a > 0$).

The energy associated with the twist distortion can be expressed in terms of the pitch as $\frac{1}{2}K_2(q-q_0)^2$; a comparison with the first term of eq. (9.16) gives

$$B \approx K_2 q_0^2$$

To obtain $\tilde{K}$ we assume that the film is rolled up into a cylinder. The director components in the cylindrical coordinates (r,ϕ,z) are

$$n_r = 0$$

$$n_\phi = \cos\theta(r)$$

$$n_z = \sin\theta(r)$$

Now the local free-energy density is given by

$$f_{eg} = \frac{1}{2}K_2\left(\frac{d\theta}{dr} - q_0 - \frac{\sin\theta\,\cos\theta}{r}\right)^2 + \frac{1}{2}K_3\frac{\cos^4\theta}{r^2} \qquad (9.17)$$

Imposing the periodicity $\theta(r+p_0) \equiv \theta(r)$ (no change in pitch), the optimum value of $\theta(r)$ corresponds to

$$\frac{d\theta}{dr} = q_0 + \frac{1}{r} \sin\theta \cos\theta \tag{9.18}$$

Averaging over $\cos^4\theta$, we obtain

$$\tilde{K} = \frac{3}{8} K_3 \tag{9.19}$$

and the free-energy density reads

$$f_{eg} = \frac{1}{2} K_2 q_0^2 \left(\frac{\partial u}{\partial z}\right)^2 + \frac{3}{16} K_3 \left(\frac{\partial^2 u}{\partial x^2} + \frac{\partial^2 u}{\partial y^2}\right)^2 \tag{9.20}$$

Let us express the displacement variable u as

$$u = u_0 \cos k_x x \ \sin k_z z \tag{9.21}$$

with $k_z = \pi/d$.

When a magnetic field is applied normal to the layers

$$f_e = f_{eg} + f_{mag} \tag{9.22}$$

with

$$f_{mag} = -\frac{1}{4} \chi_a \left(\frac{\partial u}{\partial x}\right)^2 H^2 \tag{9.23}$$

Equation (9.22) is a coarse-grained version of the continuum theory and is only valid when the pitch of the layers is large enough as compared to the associated distortion. The K_2 term is analogous to an energy of compression of the layers, and the K_3 term to an energy of curvature of the layers.

9.1.6. Convective Instabilities under Electric Field : The Square-Lattice Distortion

Let us consider[12] that a thin sample of nematic-cholesteric mixture with a planar texture is subjected to a static electric field E_0 parallel to the helical axis. The distortion occurs in the texture with the field. When field exceeds a certain threshold E_C, the texture breaks into small domains which have different orientations (different $\mathbf{q}_0$) and give rise to a strong scattering of light. When the field E_0 is switched off, the domains persist and so the process has been described[36] as the "storage mode" or the "memory effect" in cholesteric liquid crystals.

A separation of instability was worked out by Helfrich[37a]. The material is characterized by two conductances $\sigma_\parallel$, along the local director $\hat{\mathbf{n}}$, and $\sigma_\perp$, normal to $\hat{\mathbf{n}}$. Let $\sigma_\parallel > \sigma_\perp$. Similarly the two dielectric constants are $\epsilon_\parallel$ and $\epsilon_\perp$. Let us discuss the influence of small distortion (Fig. 9.4) on a cholesteric planar texture. A superposition of two distortions, as shown in Fig. 9.4, oriented at right angles (along X and Y-axes) gives a square-lattice distortion. When the spatial wavelengths involved are much larger than the pitch, a coarse-grained description can be used. The displacement u(**r**) of the cholesteric plane can be considered as the variable; along the Z-axis u is measured as well as the field is applied. The Helfrich distortion corresponds to

$$u = u_0 \exp(i k_x x) \sin(k_z z) \tag{9.24}$$

where k = π/D, D is the sample thickness. u = 0 for both the cases z = 0 and z = D. Let $\hat{\mathbf{d}}$ be a unit vector normal to the layers

$$d_z \approx 1$$

$$d_x = -\frac{\partial u}{\partial x} \tag{9.25}$$

At the coarse-grained level, we can write

$$\boldsymbol{\sigma} = \sigma_{\perp\hat{d}} + (\sigma_{\parallel\hat{d}} - \sigma_{\perp\hat{d}})\hat{d}:\hat{d} \tag{9.26}$$

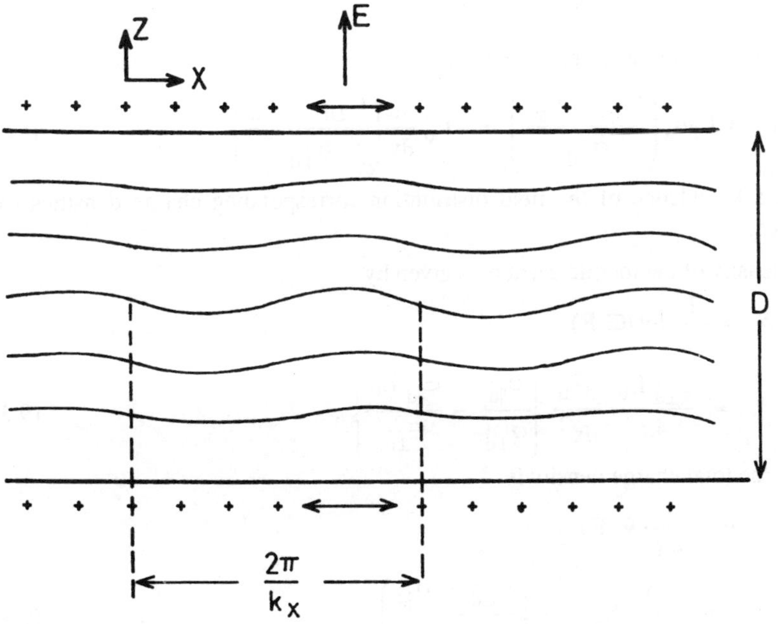

Fig. 9.4. The Helfrich distortion for a cholesteric planar texture under an electric field ($\in_a > 0$) (Ref. 12).

where

$$\sigma_{\|\hat{d}} = \sigma_{\perp} \qquad \text{along } (\hat{d})$$

$$\sigma_{\perp\hat{d}} = \frac{1}{2}(\sigma_{\|} + \sigma_{\perp}) \qquad \text{(normal to } \hat{d}) \tag{9.27}$$

The current present in the distorted structure is given by

$$\mathbf{J} = \boldsymbol{\sigma}\cdot\mathbf{E} \tag{9.28}$$

where the total current **E** is the sum of E_0 and the fields caused by the Carr-Helfrich charges.

Treating the components E_x and d_x to be small, one finds that J_z is a constant, and

$$J_x = \sigma_{\perp\hat{d}}\, E_x + (\sigma_{\|\hat{d}} - \sigma_{\perp\hat{d}})\, E_0\, d_x \tag{9.29}$$

The condition of charge conservation

$$\boldsymbol{\nabla}\cdot\mathbf{J} = 0$$

gives

$$\frac{\partial J_x}{\partial x} = 0 \qquad \text{or} \qquad J_x = 0$$

Now the lateral field component is given by

$$E_x = E_0 d_x \left(\frac{\sigma_{\perp\hat{d}} - \sigma_{\|\hat{d}}}{\sigma_{\perp\hat{d}}}\right) = -E_0 \frac{\partial u}{\partial x}\left(\frac{\sigma_{\perp\hat{d}} - \sigma_{\|\hat{d}}}{\sigma_{\perp\hat{d}}}\right) \tag{9.30}$$

From a knowledge of the field distribution corresponding charge densities can be derived.

The density of the mobile carriers is given by

$$p_c = \frac{1}{4\pi}\,\mathrm{div}\,(\in\cdot\mathbf{E})$$

$$= \frac{\in_{\perp\hat{d}} E_0}{4\pi}\,\frac{\partial^2 u}{dx^2}\left\{\frac{\sigma_{\|\hat{d}}}{\sigma_{\perp\hat{d}}} - \frac{\in_{\|\hat{d}} E_0}{\in_{\perp\hat{d}}}\right\}, \tag{9.31}$$

while the total charge density is

$$\rho = \frac{1}{4\pi}(\boldsymbol{\nabla}\cdot\mathbf{E})$$

$$= -\frac{1}{4\pi}\, E_c\, \frac{\partial^2 u}{dx^2}\left(\frac{\sigma_{\perp\hat{d}} - \sigma_{\|\hat{d}}}{\sigma_{\perp\hat{d}}}\right) \tag{9.32}$$

The total electric force acting on the layers is

$$\phi_z = \phi_c + \phi_d \tag{9.33}$$

where ϕ_c is the vertical electric force due to the mobile carriers

$$\phi_c = \rho_c E_0$$

and ϕ_d is the contribution from the dielectric torque, directed along the Y-axis, acting on the molecules

$$\phi_d = E_0^2 \left(\frac{\epsilon_{\parallel d} - \epsilon_{\perp d}}{4\pi} \right) \frac{\sigma_{\parallel \hat{d}}}{\sigma_{\perp \hat{d}}} \frac{\partial^2 u}{dx^2}$$

Equation (9.32) reads

$$\phi_z = E_0^2 \frac{\epsilon_\parallel}{4\pi} \left(\frac{\sigma_\parallel - \sigma_\perp}{\sigma_\perp} \right) \frac{\partial^2 u}{dx^2} \qquad (9.34)$$

Against ϕ_z an elastic restoring force ϕ_{el} will act. A relation for ϕ_{el} can be derived using the coarse-grained version of the continuum theory. The elastic free-energy density is given by

$$f_{cg} = \frac{1}{2} K_2 q_0^2 k_z^2 u^2 + \frac{3}{16} K_3 k_x^4 u^2 \qquad (9.35)$$

Thus

$$\phi_{el,z} = -\left(K_2 q_0^2 k_z^2 + \frac{3}{8} K_3 k_x^4\right) u \qquad (9.36)$$

The threshold condition is

$$\phi_z = -\phi_{el,z} \qquad (9.37)$$

The ratio

$$-\frac{\phi_{el,z}}{\phi_z} = \left(K_2 q_0^2 k_z^2/k_x^2 + \frac{3}{8} K_3 k_x^2\right) \left(\frac{\sigma_{\perp \hat{d}}}{\sigma_{\perp \hat{d}} - \sigma_{\parallel \hat{d}}} \right) \frac{4\pi}{\epsilon_{\parallel \hat{d}} E_0^2} \qquad (9.38)$$

will be minimum if

$$k_x^2 = q_0 k_z \left(\frac{K_2}{\frac{3}{8} K_3} \right)^{\frac{1}{2}} \qquad (9.39)$$

Thus the wavelength of the perturbation is proportional to the geometric mean of the pitch $p_0 = 2\pi/q_0$ and the sample thickness $D = \pi/k_z$. This analysis requires that $k_z << q_0$ which means $D >> p_0$. Now the Helfrich relation for the threshold field is

$$E_c^2 = \frac{8\pi}{\epsilon_{\parallel \hat{d}}} \left(\frac{\sigma_{\perp \hat{d}}}{\sigma_{\perp \hat{d}} - \sigma_{\parallel \hat{d}}} \right) \left(\frac{3}{8} K_2 K_3 \right)^{\frac{1}{2}} q_0 \frac{\pi}{D} \qquad (9.40)$$

Thus the threshold field is proportional to $-(p_0 D)^{1/2}$.

The instability has also been studied under a.c. field[37]. However, the behaviour at the higher frequency is not fully understood.

9.1.7. Light Propagation in the Chiral Nematic Phase

It is well known that the theory of light propagation in the chiral nematic phase is based on the Maxwell equations[38]

$$\nabla \cdot \mathbf{D} = 0 \tag{9.41a}$$

$$\nabla \cdot \mathbf{B} = 0 \tag{9.41b}$$

$$\nabla \times \mathbf{E} = -\frac{\partial \mathbf{B}}{\partial t} \tag{9.41c}$$

$$\nabla \times \mathbf{B} = \frac{1}{\epsilon_0 c^2} \frac{\partial \mathbf{D}}{\partial t} \tag{9.41d}$$

Here the **r** dependence of the dielectric tensor ϵ_{ij} poses difficulty in obtaining the solutions of these equations; ϵ_{ij} connects the fields **D** and **E** as

$$D_i = \epsilon_0 \epsilon_{ij} E_j \tag{9.42}$$

Explicit expression for ϵ_{ij} can be obtained in the uniaxial approximation of the chiral nematic phase. In this approximation, the cholesteric phase is assumed to be locally uniaxial like the nematic phase. With Z-axis parallel to the helix axis, the dielectric tensor can be expressed as

$$\epsilon_{ij} = \epsilon_\perp \delta_{ij} + (\epsilon_\parallel - \epsilon_\perp)\, n_i(z)\, n_j(z) \tag{9.43}$$

Assuming that the director $\hat{\mathbf{n}}(z)$ forms a right-hand helix, we write

$$n_x = \cos(q_0 z), \qquad n_y = \sin q_0(z) \qquad \text{and} \qquad n_z = 0$$

Now the explicit form of the dielectric tensor is given by

$$\hat{\epsilon}(z) = \begin{pmatrix} \frac{\epsilon_\parallel + \epsilon_\perp}{2} & 0 & 0 \\ 0 & \frac{\epsilon_\parallel + \epsilon_\perp}{0} & 0 \\ 0 & 0 & \epsilon_\perp \end{pmatrix} + \frac{\epsilon_\parallel - \epsilon_\perp}{2} \begin{pmatrix} \cos(2q_0 z) & \sin(2q_0 z) & 0 \\ \sin(2q_0 z) & -\cos(2q_0 z) & 0 \\ 0 & 0 & 0 \end{pmatrix} \tag{9.44}$$

This expression (9.44) has two parts. The first one representing the ϵ_{ij} averaged on a volume with dimensions much larger than the pitch p_0 is position independent. This averaged tensor is uniaxial with the symmetry axis parallel to the helix. The second part of eq. (9.44) represents a purely biaxial (traceless) tensor rotating along the Z-axis. Here the period of rotation is $p_0/2$ because the wavevector is $2q_0$.

For simplicity, we consider here the propagation of light in the direction of cholesteric helix. In such a case, the electric field is orthogonal to Z and only its X and Y components are nonzero,

$$E_x = E_x(z)\, e^{-i\omega t} \tag{9.45a}$$

$$E_y = E_y(z)\, e^{-i\omega t} \tag{9.45b}$$

and the Maxwell's curl equations read

$$\hat{e}_z \times \frac{\partial \mathbf{E}}{\partial z} = i\omega \mathbf{B} \tag{9.46a}$$

and

$$\hat{e}_z \times \frac{\partial \mathbf{B}}{\partial z} = -\frac{i\omega}{\epsilon_0 c^2}\mathbf{D} \tag{9.46b}$$

where $\hat{e}_z$ is a unitary vector along Z–axis.

Using eqs. (9.42) and (9.46), one obtains

$$\frac{d^2E_i}{dz^2} = -\left(\frac{\omega}{c}\right)^2 \epsilon_{ij} E_j \qquad ; \; ij = x,y \tag{9.47}$$

With the eq. (9.44), one gets the following coupled equations for the amplitudes of electric field

$$-\frac{d^2E_x}{dz^2} = k_0^2 E_x + k_a^2 [\cos(2q_0 z) E_x + \sin(2q_0 z) E_y] \tag{9.48a}$$

$$-\frac{d^2E_y}{dz^2} = k_0^2 E_y + k_a^2 [\sin(2q_0 z) E_x - \cos(2q_0 z) E_y] \tag{9.48b}$$

where

$$k_0^2 = \left(\frac{\omega}{c}\right)^2 \frac{\epsilon_\parallel + \epsilon_\perp}{2} \tag{9.49a}$$

and

$$k_a^2 = \left(\frac{\omega}{c}\right)^2 \frac{\epsilon_\parallel - \epsilon_\perp}{2} = \frac{1}{2}k^2 \epsilon_a \tag{9.49b}$$

In terms of new variables

$$E_\pm = E_x \pm i E_y \tag{9.50}$$

eq. (9.48) becomes

$$-\frac{d^2E_+}{dz^2} = k_0^2 E_+ + k_a^2 e^{-i2q_0 z} E_- \tag{9.51a}$$

and

$$-\frac{d^2E_-}{dz^2} = k_0^2 E_- + k_a^2 e^{-i2q_0 z} E_+ \tag{9.51b}$$

Let us assume that the solutions of eq. (9.51) are of the type

$$E_+(z) = A\, e^{i(\ell + q_0)z} \tag{9.52a}$$

$$E_-(z) = B\, e^{i(\ell - q_0)z} \tag{9.52b}$$

Now the following equations result for the amplitudes A and B

$$[(\ell + q_0)^2 - k_0^2]A - k_a^2 B = 0 \tag{9.53a}$$

$$[(\ell - q_0)^2 - k_0^2]B - k_a^2 A = 0 \tag{9.53b}$$

In view of eq. (9.50), the amplitudes of the electric field can be written as

$$E_x(z) = \frac{E_+(z) + E_-(z)}{2} \tag{9.54a}$$

$$E_y(z) = \frac{E_+(z) - E_-(z)}{2} \tag{9.54b}$$

Now

$$E_x = \mathrm{Re}[E_x(z)\, e^{-i\omega t}] = \frac{1}{2}A\cos[(\ell + q_0)z - \omega t] + \frac{1}{2}B\cos[(\ell - q_0)z - \omega t] \tag{9.55a}$$

and

$$E_y = \mathrm{Re}[E_y(z)\, e^{-i\omega t}] = \frac{1}{2}A\sin[(\ell + q_0)z - \omega t] - \frac{1}{2}B\sin[(\ell - q_0)z - \omega t] \tag{9.55b}$$

Figure 9.5 shows the polarization of light wave for z = 0 and different values of A and B. For $A \neq 0$ and $B = 0$ (wave E_+), this polarization is circular clockwise, while for $A = 0$ and $B \neq 0$ (wave E_-), it is circular anticlockwise. For $A = \pm B$, it is linear and elliptic everywhere else. In general, the polarization of light wave can be represented as a superposition of two waves with opposite circular polarizations; E_+ with amplitude A and E_- with amplitude B.

For $z \neq 0$, the analysis of polarization can be carried out in exactly the same manner except for the fact that the phase shift of $\Delta\varphi = 2q_0 z$, between the clockwise and anticlockwise circularly polarized wave components, has to be accounted. This phase difference arises due to the difference of $2q_0$ between the corresponding wavevectors $\ell + q_0$ and $\ell - q_0$. However, this phase difference can be eliminated in the reference frame whose X-axis rotates with the cholesteric helix. Consequently, if the polarization is elliptical for z = 0, it remains elliptical for any z but the axes of the ellipse rotate in space with the cholesteric helix.

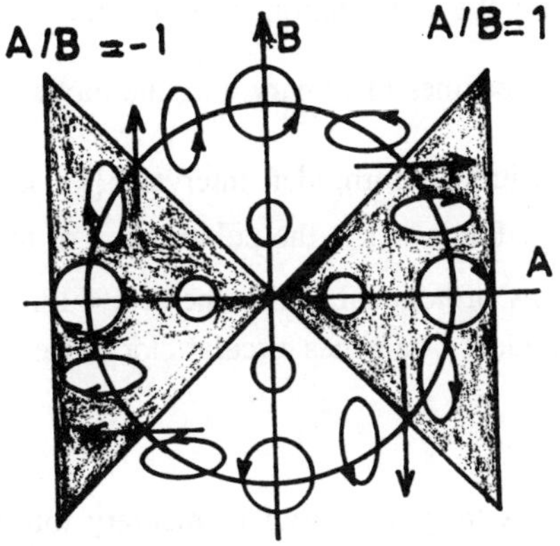

Fig. 9.5. Polarization of light wave for z = 0 and different values of the amplitudes A and B.

Equations (9.53) leads to the dispersion relation which connects the frequency ω, through K_a and K_0, to the wavevector ℓ

$$[(\ell+q_0)^2 - k_0^2]\,[(\ell-q_0)^2 - k_0^2] - k_a^4 = 0 \qquad (9.56)$$

For a given ω, this equation yields four solutions that form two pairs with opposite signs

$$\pm\ell_2, \pm\ell_1 = \pm\sqrt{(k_0^2+q_0^2)\pm(4k_0^2 q_0^2+k_a^4} \qquad (9.57)$$

When k >> q_0, i.e. the pitch p_0 is much larger than the wavelength λ, the two roots of eq. (9.56) are

$$\ell_2^2, \ell_1^2 \approx k_0^2 \pm k_a^4 \qquad (9.58a)$$

or

$$\ell_1 \approx \sqrt{\epsilon_\perp}\,k = n_0\,k \qquad \text{and} \qquad \ell_2 \approx \sqrt{\epsilon_\parallel}\,k = n_e\,k \qquad (9.58b)$$

Thus one finds that the modes ℓ_1 and ℓ_2 have orthogonal linear polarizations. When the ratio k/q_0 becomes smaller, the polarizations of modes ℓ_1 and ℓ_2 become elliptical clockwise and elliptical anticlockwise, respectively. The ratio A/B of mode ℓ_2 decreases monotonically to zero with k in the limit $k \to 0$ and the polarization becomes perfectly circular anticlockwise. The behaviour of mode ℓ_1 is much more complex. When k is decreasing, first its ellipticity increases. However, when k approaches a typical value k_u given by

$$k_u = \frac{q_0}{\sqrt{\epsilon_\perp}}$$

the ratio A/B → −1 and the polarization becomes linear again. Between k_u and $k_\ell \left(= \frac{q_0}{\sqrt{\epsilon_\parallel}}\right)$ the root ℓ_1 becomes imaginary, i.e., the mode can not propagate in the cholesteric phase. The width of forbidden interval $[k_\ell, k_u]$ is a function of the anisotropy $\epsilon_a = \epsilon_\parallel - \epsilon_\perp$. For $k = k_\ell$, the polarization of the mode is linear, as for k_u, but its direction is now orthogonal to the one at $k = k_u$. Finally, when $k/q_0 \to 0$, the polarization of the mode ℓ_1 becomes circular clockwise.

9.2. BLUE PHASES

Our aim in this section is to give a brief elementary introduction to our current understanding of blue phases. There are several reviews[39–50] which deal with the various aspects of the theory and experiment of blue phases.

Blue phases are distinct thermodynamic mesophases appearing in a very small temperature range (~ 1^0C) at the helical (chiral nematic)-isotropic boundary of highly chiral nematics of sufficiently short pitches, less than about 700 nm. In the absence of electric fields, in certain systems upto three blue phases are thermodynamically stable : BP I and BP II, both of which have cubic symmetry, and BP III, which possesses the same symmetry as the isotropic liquid phase. These phases got their name "blue phases" due to their blue appearance in early investigations. However, they are not always blue. It is known now that they can reflect the light of other colours as well.

Evidence for the existence of two lower temperature phases BP I and BP II was found by Bergmann and Stegemeyer[51] in 1979 by DSC, thermographs, selective reflection and optical rotation measurements, while the possibility of a third higher temperature phase BP III was pointed out[52] in 1980 by polarizing microscopy. Convincing evidence in support for the stability of a BP III was found by collings[53] from the optical rotatory dispersion measurement. The BP II – BP III and the BP III – IL transitions were resolved by Kleiman et. al.[54] by high precision specific heat measurements. From these observations the thermodynamic stability of BP III became quite obvious. There are several observations[55–58] and other studies[45,49] which confirm that BP I is a body-centred cubic lattice (crystallographic space group $I4_132$ or O^8), BP II a simple cubic lattice ($P4_232$ or O^2) and BP III probably amorphous. Several authors[59-61] have pointed out that the BP III may be quasi-crystalline.

The most striking feature of the blue phases is the mosaic of bright colours they can display, in contrast to the relatively featureless helical (chiral nematic) phase and isotropic liquid. Secondly, like the helical phase, the blue phases are optically active and therefore rotate the direction of polarization of linearly polarized light, although with rotatory powers that are orders of magnitude less than those of the helical phase.

However, unlike the helical phase, they are optically isotropic and are not birefringent. Another important feature of the blue phases is that of frustration[46]. In a frustrated system, the conditions responsible for a local energetic minimum cannot be extended globally.

9.2.1. General Experimental Overview

The very first observation of a blue phase was described by Reinitzer himself in his historic letter to Lehmann. He reported a bright blue-violet reflection just below the clearing point of cholesteryl benzoate which is actually caused by BP I. Similar phenomena was observed in many other chiral nematic compounds[62] and it was suggested that these materials exhibit a new thermodynamic stable phase at temperatures between the isotropic liquid and the chiral nematic phase. However, the stability of the blue phases has been verified only in 1970s. The selective Bragg reflection of particular wavelengths indicates that the structure is periodic. In reflection, platelets scattering a well defined colour under particular Bragg conditions are observed; the scattered waves are circularly polarized. In transmission between crossed polarizers, in the absence of Bragg reflection, one observes that the field of view is black.

The existence of all three distinct thermodynamically stable blue phases is now firmly established by thermal, optical and viscoelastic measurements. The initial experiments show density discontinuities[63] and new peaks in DSC traces[40,51,64,65]. Additional evidence is provided by heat capacity measurements[54,66,67]. In cholesteryl nonanoate three peaks riding on the edge of the helical-isotropic peak are observed[66]; these peaks separate the helical, BP I, BP II, BP III and the isotropic phases. Figure 9.6 shows the phase diagram of chiral CE2 in a chiral racemic mixture. From the similar diagrams[68–71] it is inferred that the blue phases always appear in the sequence BP I, BP II and BP III as the chirality is increased. Further, while BP I and BP III appear to be increasingly stable at high chirality, BP II only exists over a limited region.

A number of experimental studies are available that elucidate the important influence of pitch on the stability of the blue phases. These phases are found to occur only in chiral nematic systems with short pitches. As the pitch is decreased, the temperature range over which the blue phases are available generally increases. When the pitch exceeds a critical value p_c, blue phases no longer exist. The critical pitch does not adopt a universal value but depends on both the molecular structure and the composition of the system. The nonuniversality of p_c has been explained using Landau free-energy expansion[72]. Within a homologous series the homologues (see Ref. 44) with the smallest pitch exhibit the largest blue phase range.

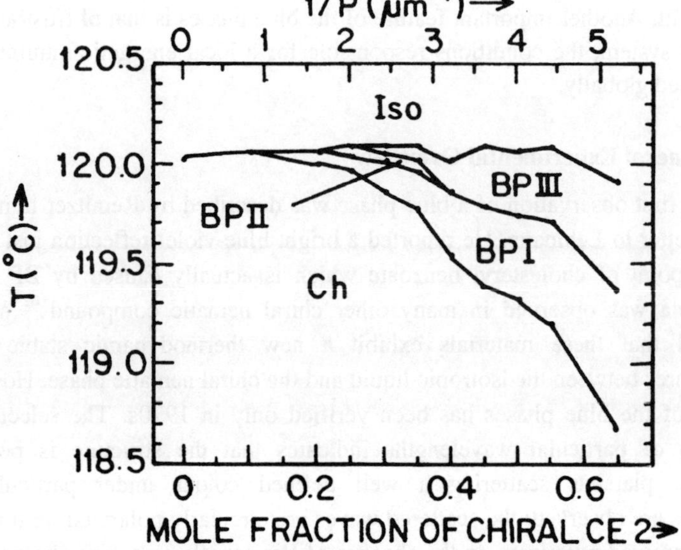

Fig. 9.6. Phase diagram of chiral CE2 in a chiral recemic mixture.

It has been found[73] that in chiral mixtures exhibiting blue phases, the viscosity is anomalously large in the blue phase region; sharply rises to a factor of upto 10^6 times its value in the chiral nematic phase as the temperature is raised, and decreases rapidly in the isotropic phase.

In polymorphic mesogenic systems the phase sequence

$$S_A \rightarrow N^* \rightarrow BP \rightarrow IL$$

occurs, on increasing temperature in systems with small pitches. However, exception to this sequence has been observed[44,74] in a mixture of cholesteryl myristate (CM) and 4,4′-di-n-decylazoxybenzene (C_{10}). It has been found that on increasing the C_{10} concentration the cholesteric range decreases finally to zero at 15 mol percent whereas the BP I/BP II ranges remain nearly constant. In a mixture of 15 to 18 mol percent C_{10} a direct $S_A \rightarrow$ BP I transition occurs. These results show that the occurrence of a blue phase does not require the simultaneous existence of a chiral nematic phase.

9.2.2. BP I and BP II – Experiment

Several experimental techniques, such as Bragg scattering, Kossel diagrams, platelet growth morphology, rotatory power measurements, viscoelastic measurements, etc., have been employed to determine the structure of BP I and BP II.

It is well known that N* phase with a planar texture will selectively rflect light. Selective reflection requires light with a circularly polarized component of same handedness as that of helix and a wavelength satisfying λ = np where n is an average refractive index and p is the pitch. However, in cubic blue phases, corresponding to various crystallographic planes, there may be several selective reflection wavelengths such that none of these coincide with the N* selective reflection wavelength. Meiboom et. al.[75] measured the transmitted light on an unaligned blue phase sample. Such a measurement, at each selective reflection wavelength, gives a step in the transmitted intensity. Figure 9.7 shows the selective reflection wavelengths λ_{SR} of a chiral CB15/E9 (nematic) mixture[76]. It can be seen that BP I supercools into the N^* phase, whereas the BP I – BP II transition is reversible. In both the phases the wavelengths have the ratios λ_0, $\lambda_0/\sqrt{2}$, $\lambda_0/\sqrt{3}$, ..., where λ_0 is the longest selective reflection wavelength. These ratios give the signature of Bragg scattering from either a bcc (body-centred cubic) or sc (simple cubic) lattice. Thus it can be inferred that BP I and BP II are cubic; the lattice parameters are of the order of visible light wavelengths and are different for both phases. In order to distinguish bcc from sc, the polarization selection rules were worked out[61a,77,78] for each possible bcc and sc space group. It has been seen that each set of Bragg planes contains five order parameter coefficients - ϵ_0, $\pm\,\epsilon_1$, and $\pm\,\epsilon_2$, each of which reflects circularly polarized light in a unique way. However, the results obtained using this technique[79] violate these rules. Despite efforts to resolve the selection rule contradiction, better techniques, such as Kossel diagrams, and platelet growth morphology, can be used to identify the structure reliably.

The Kossel diagrams technique is a special case of Bragg scattering. For the blue phase analysis this technique was first used by Pieranski and coworkers[57,80–82] and later by Miller and Glesson[83,84]. From an analysis of the Kossel diagrams, it is clear that the symmetry of the Kossel diagrams is just the symmetry of the crystal itself. Thus the longest wavelength bcc line [bcc(110)] and the longest wavelength sc line [sc(100)] can be distinguished because only the sc(100) Kossel line has four-fold symmetry. In addition, the angles and crystal-plane spacing can be determined from the quantitative measurements of the Kossel diagrams. This method has been successful in observing some new electric-field-induced blue phase and in confirming that the BP II symmetry is $O^2(P4_232)$.

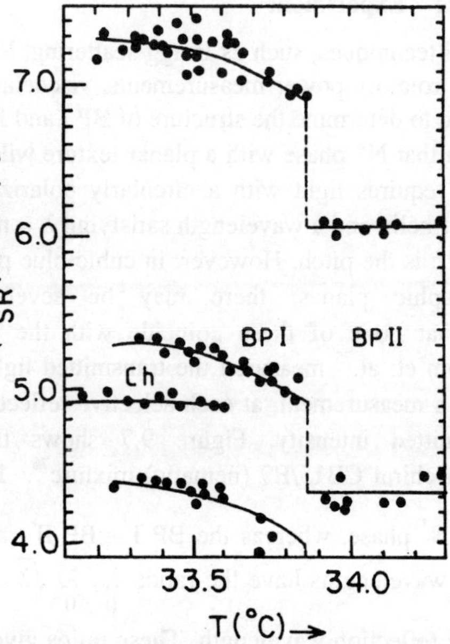

Fig. 9.7. Selective reflection wavelengths λ_{SR} versus temperature for a chiral CB15/E9 (nematic mixtures) (Ref. 76).

Microscopic examination of the textures[85] has provided useful initial information on the blue phases. A more reliable and convenient method to determine a crystal's space group is to carefully grow the crystal and observe its shape which has the unit cell symmetry. This can be done in the two-phase region of a multicomponent mixture, and/or in a temperature gradient. The work done in this direction is reviewed by Stegemeyer et. al.[44]. From an analysis of grown crystallite in BP I, together with the Kossel diagrams information, it has been concluded that BP I is body-centred cubic with space group O^8 ($I4_132$), and BP II is simple cubic with space group O^2 ($P4_232$). Once the cubic nature of these blue phases was established, efforts were made[54,56,86] to measure the elastic moduli using very sensitive techniques. For cholestery nonanoate, using the torsional oscillators configured as cup viscometers, the shear elasticity was measured by Kleiman et. al.[54]. The data obtained at various frequencies were extrapolated to O HZ to obtain the static properties. These authors claimed that the shear elasticity is nearly zero in the chiral nematic phase and about > 10 dyn cm^{-2} in BP I. However, BP I also has viscosity, it behaves like that of a viscoelastic solid.

9.2.2.1. BP I and BP II in an electric field

The electric field effects on the blue phases have been investigated by several groups[25,44,48–50]. An external electric field interacts with the local dielectric anisotropy of a blue phase and contributes[87] to the energy with a term $\Delta\epsilon E^2/4\pi$. The cubic lattice is distorted by the field and thus a change in the angular (spectral) positions of Bragg reflection results. In addition, important phenomena such as field-induced phase transitions[44,88,89] and field-induced optical biaxiality[92] have been discovered.

If the field is weak, no phase transitions occur and a cubic lattice is distorted according to the sign of the local dielectric anisotropy of the medium[93]. The effect can be described by the strain tensor

$$u_{ij} = \sum_{k=1} \sum_{\ell=1} \gamma_{ijk\ell} E_k E_\ell \qquad (9.59)$$

where $\gamma_{ijk\ell}$ is the fourth-rank electrostriction tensor.

The distortion of the lattice causes a change in the Fourier harmonics of dielectric tensor[94] and the corresponding changes in the optical properties. Figure 9.8 shows the changes in the wavelengths of the Bragg peaks with the applied field strength[93a] for a CB15/E9 mixture (49.2-50.8%) with $\Delta\epsilon > 0$.

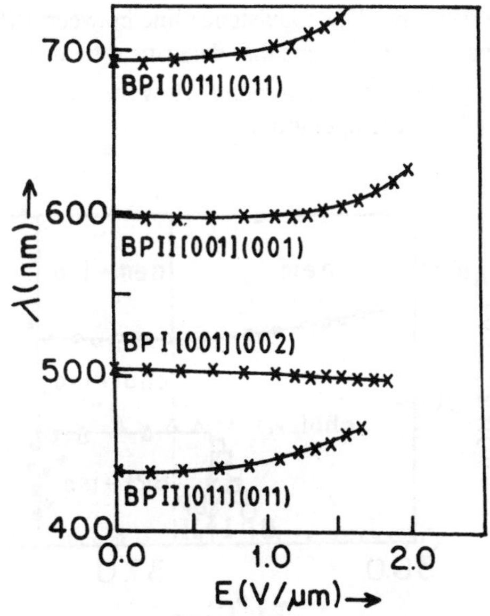

Fig. 9.8. Wavelengths of the Bragg peaks corresponding to planes (hkm) with applied field strength for BPI (34.1°C) and BP II (34.3°C) orientated with either a fourfold [001] or a two-fold [011] axis parallel to the field (CB15/E9 mixture) (Ref. 93a).

It is obvious that the changes in the wavelengths of the Bragg peaks, with the increasing field, depend on the field direction with respect to the crystallographic axes. When $\Delta\in < 0$, the signs of the components of $\gamma_{ijk\ell}$ are inverted and the field–induced red shifts of the Bragg maxima are replaced by blue shifts and vice-versa[93b].

It is well known that the pitch can be increased by an electric field upto infinity and thus transforming the N^* helical structure into a nematic. Efforts were made to increase the helical pitch by an electric field upto values $p > p_c$ so that the field-induced BP-N^* transitions can result. With the increasing field, a series of field-induced transitions BP I – N^*, BP II – N^*, and then N^* - N are observed. This is shown in a voltage-temperature phase diagram[44] (Fig. 9.9) for a mixture (47-53 mol%) of chiral 2-methylbutyl derivative (CB15) with polar compound 4-n-hexyloxy-4′-cyanobiphenyl (6OCB). A BP II - N^* transition as reported by Finn and Cladis[95] was not observed in this mixture. The observed BP I/BP II – N^* transitions occur at field strength at which the helical pitch is only increased by about 2 percent with respect to zero-field value. Thus the magnitude of the pitch does not play an important role during the field-induced phase transitions. The inclination of the coexistence line for BP I – N^* transition is towards the ordinate; this shows that the transition temperature is field-dependent. This is not the case for the BP I – BP II and the BP II – N^* transitions. The coexistence line between BP II and the N^* phase runs parallel to the abscissa; this means that the critical field strength E_{BPN^*} for BP II – N^* transition is independent of temperature. However, it is not clearly understood why E_{BPN^*} is independent of temperature.

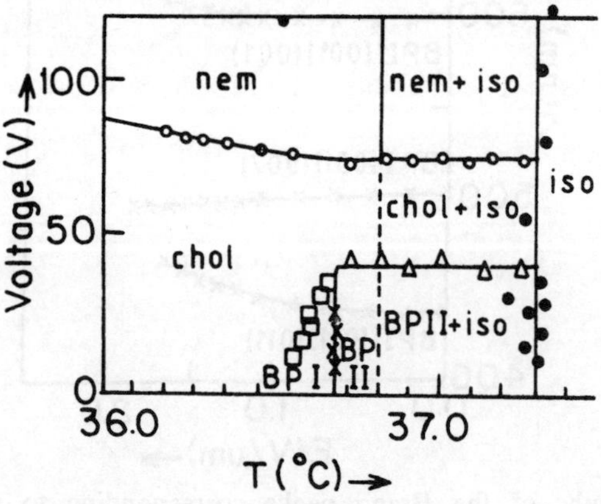

Fig. 9.9. Voltage–temperature phase diagram for 47-53 mol.% mixture of CB15 and 6OCB (Ref. 44).

Many field-induced novel blue phases have been observed in several mixtures. Among them, a tetragonal (BPX)[91], and two hexagonal[89] (a three-dimensional BPH[3d], and a two-dimensional BPH[2d]) were distinguished in CB15/E9 mixture. In a voltage-temperature phase diagram[90] (Fig. 9.10) of the same mixture a new phase BPE has been detected. However, the BPE structure is not yet determined.

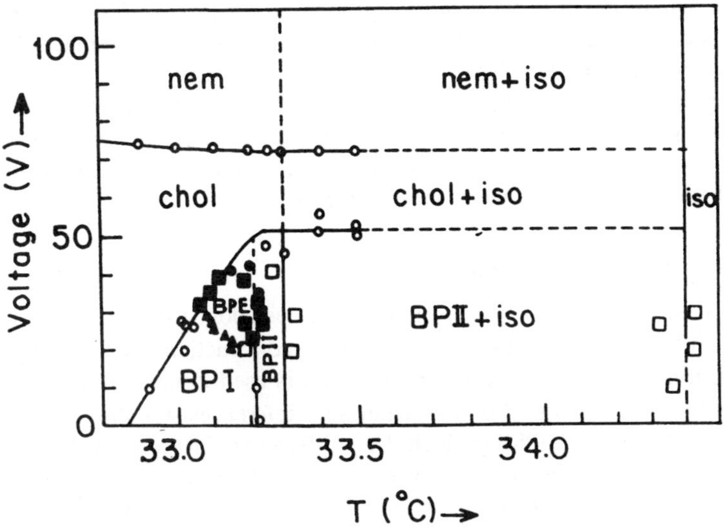

Fig. 9.10. Voltage–temperature phase diagram for a 49.6% mixture of chiral CB15/E9 (nematic) Mixture (Ref. 90).

9.2.3 BP III - Experiment

As mentioned above, the blue phase III, designated variously as grey phase[40], the fog phase[85a], the blue fog[95], and BP III[75c], is amorphous and not cubic. Several reviews[41b,48,50,97] have appeared covering the various aspects of this phase; we present here some of the salient features :

As shown in Figure 9.6, BP III is the highest temperature phase occurring either between BP II and IL or at higher chirality, between BP I and IL. Its visual appearance is foggy, and often it is bluish or grayish in colour. At very high chirality, to the naked eye, it may be invisible or indistinguishable from the isotropic liquid phase. Similar to BP I, the temperature range of its appearance increases monotonically with the increasing chirality. Like the other two blue phases, BP III selectively reflects circularly polarized light[75(b,c)]. However, unlike the cubic blue phases, the spectrum is very broad (~ 100 nm)[98]. Corresponding to the various crystal planes, BP I and BP II exhibit several Bragg peaks, while BP III exhibits only one peak. With temperature and chirality the magnitude of BP III rotatory power decreases[53a]. At the BP I – BP II, BP II – BP III and BP III – IL transitions, discontinuous jump in the rotatory power is observed.

From the visual appearance, BP III structurally appears to be closer to the isotropic phase than to BP I or BP II. In contrast to it, from the heat capacity data[66], BP III appears to be closer in structure to BP II.

Koistinen and Keyes[100] have suggested that both phases, BP III and IL, possess the same symmetry. In view of this suggestion, it is expected that the BP III – IL coexistence line (Fig. 9.6) should end to a critical point at some higher chirality in the temperature-chirality plane. Collings and coworkers[101-103], using the light scattering and rotatory power techniques, have shown that the discontinuities found in rotatory power at low-chirality BP III – IL transition do not exist at very high chirality. Additional evidence for a critical point is provided by the adiabatic and nonadiabatic scanning calorimetry measurements[102,103].

9.2.4 Theory for the Blue Phases

Here we describe the theoretical work that have been done to describe the cubic blue phases – BP I and BP II. These work involve the construction of a free-energy whose minimization (approximately done in this case) is used to predict the structure and properties of these phases (for the detail see References 48–50). Two groups of workers have approached this problem from different points of view. However, both the approaches share the common theme of a phenomenological free-energy. The first-approach, initiated by Brazovskii and Dmitriev[106] and developed more fully by Hornreich and Shtrikman[47], the Landau theory, uses most general tensorial order parameter with cubic symmetry which is relaxed over a cubic unit cell for minimizing the free-energies of various structures. The second approach due to Meiboom and coworkers[75b,104], the defect theory, uses a Frank free-energy for a chiral nematic phase and constructs blue phases from regular arrays of line defects (disclinations). However, these two approaches are complementary rather than competing[48]. Wright and Mermin[48] have argued that the Landau theory is appropriate when the molecular chirality is strong, i.e., the gradient term in the free-energy (see below) dominate. The second approach is appropriate in the other limit where the chirality is low, and the bulk term in the free-energy is dominant.
Despite differences, these two theories, have many similarities. Both theories result in double-twist regions, which are exploited to form 3D structures. Both approaches involve defects; in the defect theory they are explicit disclination lines where order parameters fall abruptly to zero, while in the Landau theory defects appear implicitly as regions where the order parameter becomes zero gradually. If these defects are described in Fourier space in terms of spatial Fourier components, the two approaches differ only in number of Fourier components included in the order parameter; the defect theory involves many higher order components, while the Landau theory includes very few Fourier components.

9.2.4.1 Application of Landau Theory

Detailed reviews of Landau theory have been given by several workers[43,47–50,97]. We give here the general picture of the approach without mathematical detail.

Let us define a symmetric, traceless tensor order parameter **Q** in terms of the local dielectric constant $\in$

$$Q_{ij} = \in_{ij} - \frac{1}{3}(\mathrm{Tr}\in)\delta_{ij} \tag{9.60}$$

The Landau free-energy density is written as

$$f = f_{grad} + f_{bulk} \tag{9.61}$$

where the gradient free-energy density is expressed as a linear combination of three independent terms of the form

$$f_{grad} = \frac{1}{4}K_1[(\nabla\times\mathbf{Q})_{ij} + 2q_0Q_{ij}]^2 + \frac{1}{4}K_0[(\nabla\cdot\mathbf{Q})_i]^2 \tag{9.62}$$

and through fourth order the general form of the bulk free-energy density is

$$f_{bulk} = c\,\mathrm{Tr}(\mathbf{Q}^2) - \sqrt{6}\,b\,\mathrm{Tr}(\mathbf{Q}^3) + a[\mathrm{Tr}(\mathbf{Q}^2)]^2 \tag{9.63}$$

The coefficient of cubic term is defined with the explicit factor of $\sqrt{6}$ to simplify subsequent expressions. The coefficient a ought to be positive for the stability requirement. We take c to drop linearly with decreasing temperature through a range that includes the value 0; while we consider b and a as temperature independent. In eq. (9.63) the tensor $\nabla\times\mathbf{Q}$ and vector $\nabla\cdot\mathbf{Q}$ are defined as

$$(\nabla\times\mathbf{Q})_{ij} = \in_{ist}\nabla_s Q_{tj}$$

$$(\nabla\cdot\mathbf{Q})_i = \nabla_j Q_{ji}$$

The corresponding three independent parameters are taken to be two elastic constants K_1 and K_0, and an inverse length q_0 which characterizes the helical pitch of the chiral nematics.

Wright and Mermin[48] have shown that the stability of the full free-energy f requires K_1 to be positive and K_0 to be greater than $-\frac{1}{2}K_1$. Further, throughout this range of stability the gradient free-energy remains non-negative and minimized by order parameters that satisfy

$$\nabla\times\mathbf{Q} = -2q_0\,\mathbf{Q} \tag{9.64}$$

The equilibrium structure of a chiral nematic can be determined by minimizing the full free-energy, which is the volume integral of the free-energy density. However, since the gradient and bulk energies favour different structures, this minimization problem has not been solved. So the limiting cases have to be considered. We discuss two solvable limiting cases below.

a. The nematic limit : infinite pitch

When $q_0 = 0$ the gradient free-energy attains its minimum value of 0 for any **Q**, irrespective of its form. Then minimizing the free-energy we recover a homogeneous uniaxial structure which characterizes a nonchiral (ordinary) nematic.

b. The second order term : vanishing cubic term

When b = 0, the bulk free-energy depends only on $\mathrm{Tr}(\mathbf{Q}^2)$, and it is minimized by any **Q** with the appropriate value of $\mathrm{Tr}(\mathbf{Q}^2)$. If f_{bulk} is to be minimized everywhere, $\mathrm{Tr}(\mathbf{Q}^2)$ must be independent of position. The condition (9.64) that **Q** minimize f_{grad} is uniquely satisfied by a structure proportional to

$$\begin{aligned} Q_{ij}^{bh} &= \mathrm{Re}[(\mathbf{x}-i\mathbf{y})_i\,(\mathbf{x}-i\mathbf{y})_j\, e^{2iq_0 z}] \\ &= (\hat{x}_i\hat{x}_j - \hat{y}_i\hat{y}_j)\cos(2q_0 z) + (\hat{x}_i\hat{y}_j + \hat{y}_i\hat{x}_j)\sin(2q_0 z) \end{aligned} \tag{9.65}$$

or by linear combinations of such structures. The order parameter Q_{ij}^{bh} is referred[48] as the "biaxial helix"; it is constant in directions perpendicular to the pitch axis Z and rotates uniformly along Z. The eigenvalues of $\mathbf{Q}^{bh}$ are 1, 0 and –1. It is thus, in a sense, maximally biaxial. While uniaxial structures have two degenerate eigenvalues and maximize $|\mathrm{Tr}(\mathbf{Q}^3)|$ for given $\mathrm{Tr}(\mathbf{Q}^2)$, the structure $\mathbf{Q}^{bh}$ minimizes it. Although $\mathbf{Q}^{bh}$ depends on position, its square does not,

$$[(\mathbf{Q}^{bh})^2]_{ij} = \lambda^2(\hat{x}_i\hat{x}_j + \hat{y}_i\hat{y}_j) \tag{9.66}$$

Thus $\mathrm{Tr}(\mathbf{Q}^{bh})^2 = 2\lambda^2$. For b = 0, $\mathbf{Q}^{bh}$ can minimize f_{bulk} as well as $f_{grad.}$ with a suitable choice of the constant amplitude λ.

c. The coherence length

It is useful to define a coherence length ξ which characterizes the nematic ($q_0 = 0$) phase at the T_{NI}. Let us assume that the amplitude of the order parameter deviates slightly from its value λ_0 which minimizes the free-energy,

$$\mathbf{Q} = \frac{\lambda}{\sqrt{6}}\begin{bmatrix} -1 & 0 & 0 \\ 0 & -1 & 0 \\ 0 & 0 & 2 \end{bmatrix}; \qquad \lambda = \lambda_0 + \delta\lambda \tag{9.67}$$

If δλ is constrained to be nonzero in a region, its relaxation back to zero will be governed by the condition for minimizing δf,

$$\nabla^2\delta\lambda = \delta\lambda/\xi^2 \tag{9.68}$$

where the coherence length ξ is given by

$$\xi = (aK_1/b^2)^{1/2} \tag{9.69}$$

The length ξ provides an important scale in chiral as well as nonchiral nematics. Usually this parameter is referred to as the "coherence length" even when $q_0 \neq 0$ or K_0 differs from K_1.

d. Free-energy density in terms of rescale variables

It is convenient to define dimensionless parameters. Let us define a dimensionless free-energy φ, effective temperature τ, order parameter χ, and length scale r′ as follows

$$\varphi = (a^3/b^4)\,f \quad , \quad \tau = (a/b^2)\,c$$
$$\boldsymbol{\chi} = (a/b)\,\mathbf{Q} \quad , \quad r' = 2q_0\,r \tag{9.70}$$

This rescaling is singular in the limits $b \to 0$ and $q_0 \to 0$. In terms of these variables the dimensionless free-energy density can be expressed as

$$\varphi = \varphi_{grad} + \varphi_{bulk} \tag{9.71}$$

with

$$\varphi_{grad} = \kappa^2\{[(\nabla\times\boldsymbol{\chi})_{ij} + \chi_{ij}]^2 + \eta[(\nabla\cdot\boldsymbol{\chi})_i]^2\} \tag{9.71a}$$

$$\varphi_{bulk} = \tau\,\mathrm{Tr}(\boldsymbol{\chi}^2) - \sqrt{6}\,\mathrm{Tr}(\boldsymbol{\chi}^3) + [\mathrm{Tr}(\boldsymbol{\chi}^2)]^2 \tag{9.71b}$$

where κ is the "chirality parameter"[72a]

$$\kappa = (aK_1q_0^2/b^2)^{1/2} = q_0\,\xi = 2\pi\xi/p_0 \tag{9.72}$$

and $\eta = K_0/K_1$.

Thus the rescaling indicates that the strength of the chirality parameter κ determines whether it is the bulk (small κ) or the gradient (large κ) terms that dominate the free-energy.

d.1. General features of high chirality limit

Based on the original approach of Brazovskii and Dmitriev[106], the high chirality behaviour of blue phases has been discussed in detail by Hornreich and Shtrikman[47,107]. In the limit of large κ, the free-energy (9.71) is dominated by the gradient free-energy. The φ_{grad} is non-negative and its high degree of degeneracy is obtained from the linearity of the minimization condition (9.64) which is terms of rescaled quantities assumes the form

$$\nabla\times\boldsymbol{\chi} = -\boldsymbol{\chi} \tag{9.73}$$

The role of φ_{bulk} term is simply to reduce this degeneracy by selecting the most favourable linear combination of the biaxial helical order parameters. As $\kappa \to \infty$ the helical order parameter becomes a single biaxial helix and the nonhelical phases existing in the neighbourhood of the transition will have an order parameter obtained from a linear combination of biaxial helices (eq. (9.65)) of the form

$$\chi_{ij}^{bh} = \mathrm{Re}[(\mathbf{a}-i\mathbf{b})_i\,(\mathbf{a}-i\mathbf{b})_j\,e^{i\mathbf{c}\cdot\mathbf{r}}] \tag{9.74}$$

where **a**, **b** and **c** are any triad of orthonormal vectors obtained as a result of suitable rotation of the Cartesian unit vectors. This solution has been termed by Wright and Mermin[48] as the biaxial helix. It corresponds to a pitch axis defined by the wavevector **c** and is constant in the **a** – **b** plane. However, it is a biaxial tensor with eigenvalues 1, 0 and –1.

The transitions between blue phases, blue phase-isotropic liquid and the blue phase – chiral nematic are first-order in nature. The first-order transitions in Landau theory require a nonzero cubic term in the free-energy. A superposition χ of terms of the form χ^{bh} for different choices of the orthonormal triads will give a nonzero value of the term $\mathrm{Tr}(\boldsymbol{\chi}^3)$ only if there is a triad of wavevectors **c**, **c′** and **c″** in the superposition that add up to zero. Eight such distinct trios satisfying this condition are possible; which structure is most stable depends on detailed calculations. The real lattice structure is body-centered cubic. A simple cubic structure can also result if nonunit wavevectors obtained by the linear combinations of the unit wavevectors **c** are allowed. Figure 9.11 shows the phase diagram thus obtained. It can be seen that the free-energy difference between the different cubic structures is small and so the inclusion of higher order terms in Landau free-energy can change this picture. The cubic phases will appear only when chirality exceeds a critical value which in the Hornreich-Shtrikman treatment can be given an upper bound of 0.47.

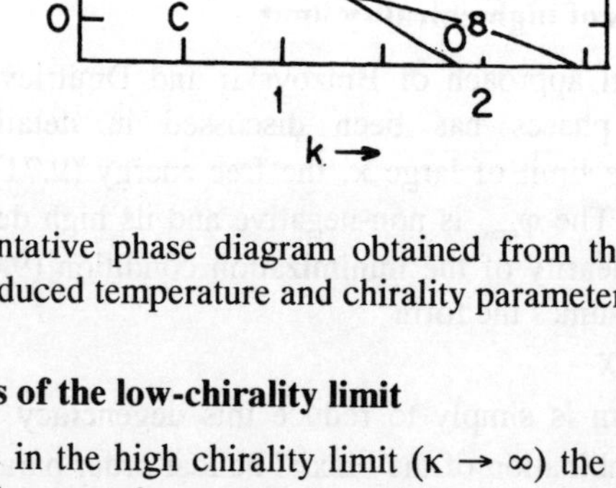

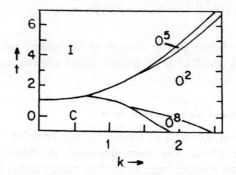

Fig. 9.11. A representative phase diagram obtained from the Landau theory[47] in terms of reduced temperature and chirality parameter.

d.2 General features of the low-chirality limit

As discussed above in the high chirality limit ($\kappa \to \infty$) the role of φ_{bulk} term has been only to select the most favourable form from the large degenerate family of order parameters that minimize the dominant φ_{grad} term. In the limit of low-chirality ($\kappa \to 0$) the situation is reversed, φ_{bulk} term dominates. Thus the starting point of the approach, adopted by Meiboom et. al.[75b,104,105], is to minimize the nongradient terms alone. It has been shown that the minimum of these terms corresponds to a uniaxial order parameter **Q** (due to the presence of the cubic term). Taking a uniaxial order parameter, the free-energy density can be reexpressed in terms of a unit length director field $\hat{\mathbf{n}}$ and an amplitude λ. Within the one elastic constant approximation, the free-energy density terms have the form

$$\varphi_{\text{grad}} = \kappa^2\left[(\Delta\lambda)^2 - \frac{1}{2}\lambda^2 + 3\lambda^2(\nabla_i n_j + \frac{1}{2}\epsilon_{ijk}\, n_k)^2\right] \tag{9.75}$$

and

$$\varphi_{\text{bulk}} = \tau\lambda^2 - \lambda^3 + \lambda^4 \tag{9.76}$$

Here ϵ_{ijk} is the antisymmetric tensor.

As $\kappa \to 0$, the bulk free-energy density (9.76) dominates the total free-energy. The amplitude λ is thus required everywhere to have the fixed amplitude λ_0 that minimizes eq. (9.76).

With the assumption λ = constant, it has been shown[48] that the free-energy is minimized in the helical state. For $\kappa = 0$ this refers to ground state. For $\kappa \neq 0$, the assumption of constant λ can be relaxed and nonhelical phases can have lower energy. If the chiral molecules are visualized as screw-like objects these phases and their corresponding lower energies can be understood. However, it has been seen that a double twist in the director field of a mesophase is more favourable in a local scale. This can be understood from the gradient free-energy. In the helical phase given by

$$\hat{\mathbf{n}}(\mathbf{r}) = \hat{x}\cos q_0 z + \hat{y}\sin q_0 z$$

the gradient term gives $\frac{1}{4}k^2\lambda_0^2$ for a constant λ_0. However the gradient term is bounded from below by the value $-\frac{1}{2}k^2\lambda_0^2$. This value can be obtained locally if $\nabla_i n_j + \frac{1}{2}\epsilon_{ijk}\, n_k = 0$. This equation can be satisfied along lines, and by continuity the local bulk free-energy density will then be lower than that of helical phase in some neighbourhood of those lines as well. Using cylindrical coordinates centred on a line, the solution

$$\hat{\mathbf{n}} = \hat{z}\cos\frac{1}{2}r - \hat{\phi}\sin\frac{1}{2}r$$

does indeed give $\nabla_i n_j + \frac{1}{2}\epsilon_{ijk}\, n_k = 0$ when $r \to 0$. This situation precisely refers to double twist[48]. In fact, it is a delicate and very difficult problem to build up the blue phase structures from the double twist cylinders. Detailed computer calculations have been performed[75b,104c,108] to determine the free-energies of these structures. In these calculations the director field $\hat{\mathbf{n}}$ is assumed to have a fixed magnitude, falling abruptly to zero at the cores of the defects. The core energy of the defects is inferred from the experimental data. The energies of these structures have been calculated for two specific cases; one elastic constant approximation and $K_1 = K_2 = 2K_2$. In the former case the O^2 structure appears to be most favourable, while in the latter O^{8-} is preferred. These results are in accordance with the experiments; several materials exhibit two crystalline blue phases (BP I and BP II) with space groups O^8 and O^2, respectively.

9.2.4.2 Defect Theory

The macroscopic chirality in liquid crystals arises because the molecules themselves are chiral. Consequently, the interactions between adjacent molecules are also chiral. In such systems twist arises because the interaction energy of two neighbouring molecules is minimum when they are at slight angle to each other. This condition is partially met in a usual helical phase; twist occurs along the single axis $\hat{\mathbf{k}}$ but not along axes perpendicular to the twist axis. This gives the possibility of lowering energy further by allowing twist in all directions perpendicular to the local director. Such a configuration is referred to "double twist".

One can envision the blue phase as a lattice of double-twist tube or a lattice of disclinations[48,50]. Evaluation of free-energy for a lattice of defects is a difficult task. For a single disclination line, the free energy per unit length F_{disc} can be written as

$$F_{disc} = F_{el} + F_{surf} + F_{core} + F_{int} \tag{9.77}$$

where the first term represents the elastic free-energy associated with the defect,

$$F_{el} \sim K \ln (R_o/R_c) \tag{9.78}$$

This term is obtained from the calculation of usual Frank free-energy, with one elastic constant approximation, outside a disclination core of radius R_c but inside a cutoff radius size R_o.

The second term is an additional elastic term which is usually ignored. However, when the integration is performed over the inner surface of the defect core, it cannot be ignored. This term pulls the energy of the defect below zero

$$\begin{aligned} F_{surf} &= \frac{1}{2}\int K\,\boldsymbol{\nabla}\cdot[(\hat{\mathbf{n}}\cdot\boldsymbol{\nabla})\,\hat{\mathbf{n}} - \hat{\mathbf{n}}\,(\boldsymbol{\nabla}\cdot\hat{\mathbf{n}})]\,d^3r \\ &\approx -K \end{aligned} \tag{9.79}$$

The third term

$$F_{core} = \alpha\,(T_{N^*I} - T)\,\pi R_0^2 \tag{9.80}$$

arises because the energy cost required to maintain a highly strained, nonzero order parameter at the core is higher as compared to the energy required to drive the core itself isotropic; T_{N^*I} is the usual helical-isotropic transition temperature. The quantity $\alpha\,(T_{N^*I} - T)$ is the difference between the free-energies of the isotropic and helical phases.

The interfacial energy between the core and the chiral material outside is given by

$$F_{int} = 2\pi \sigma R_c$$

where σ is the surface tension.

Considering all the above terms it has been found that the free-energy is positive for $T << T_{N^*I}$, while for the temperatures close to (but still less that T_{N^*I}) T_{N^*I}, it becomes negative. This result opens the possibility for a transition to a blue phase lattice just below the T_{N^*I}. The idea is to assemble a lattice of defects with a particular space group symmetry and fill the interstices with nematic material. The director is then allowed to relax everywhere, except at the defects, and the free-energy is evaluated. This gives the existence of an interpenetrating lattice of double-twist tubes and it is seen that the structure is stable between the helical (cholesteric)- and isotropic phases.

9.2.4.3 The bond-orientational order and fluctuation models

Despite theoretical successes of the above described approaches (Landau- and defect theories), the theoretical phase diagrams (Fig. 9.11) are missing many important features of the experimental phase diagrams (e.g. Fig. 9.6). The space groups of the chiral nematic, BP I and BP II, and their order with the increasing temperature are correctly predicted theoretically. However, the predicted bcc O^5 phase has not been detected experimentally. In the theoretical phase diagrams, with the increasing chirality the BP II region becomes broader, whereas BP II vanishes at high chirality in the experimental observations. The theoretical phase diagrams neither show the BP III phase nor reproduce the experimentally observed critical points which terminates the BP III – IL coexistence line at high chirality.

Trebin and coworkers[109] considered the application of cubic bond-orientational order (BOO) model, originally developed for crystals, to rectify these theoretical problems. The idea is that while retaining the orientational order of the bonds, the positional order of the atoms can be lost due to generation of defect pairs or fluctuations. In the blue phases, the atoms and bonds are replaced by the unit cell's corners and edges, respectively. The aim is to convert the higher temperature O^5 phase to a cubic phase with only bond-orientational order, which might then be the amorphous-appearing BP III phase. This was accomplished by adding one additional fluctuation term to the usual Landau free energy. Trebin and coworkers[109b] have also proposed a fluctuation dominated model of the blue phases. In this approach the free-energy is separated into a mean-field and a fluctuating part. In the resulting phase diagram the O^5 phase does not appear for large transition temperatures and BP II disappears at high chirality. However, theory provides only a signature of the occurrence of a second isotropic phase, BP III, and of a critical point at high chiralities.

9.3. CHIRAL SMECTIC PHASES

As we have already mentioned in Chapter 1 (Table 1.3), there exist many different kinds of chiral smectic (S_C^*, S_I^* and S_F^*) phases and chiral crystal smectic (S_J^*, S_G^* and S_K^*, S_H^*) phases which exhibit ferroelectric properties. Among all of these mesophases the S_C^* phase is by and far the most important and investigated one; it is least ordered and least viscous, and is employed in the ferroelectric display devices. In this section the major part of the discussion will be confined to the S_C^* phase[3,4,110]. However, the same arguments could equally be extended and applied to other tilted chiral smectic phases.

9.3.1. The Origin of Ferroelectricity

The ferroelectric effect was observed by Valasek[111] in 1921 in crystals of Rochelle salt (potassium sodium tartrate). When a sample of Rochelle salt is placed between two capacitor plates and subjected to an electric field, a marked departure from the normal dielectric behaviour is observed; a residual polarization is observed after the electric field is removed. Valasek suggested that this behaviour could be due to a hysteresis in the crystal which is analogous to ferromagnetic hysteresis, and his experiments confirmed this theory. Therefore, the term ferroelectric was derived from the analogous phenomenon of ferromagnetism. However, the ferroelectricity is not related to iron. The ferroelectric effect can be understood in terms of the physical changes which occur in a material under mechanical stress. Rochelle salt is unusual in that it exhibits a piezoelectric effect, i.e., an induced polarization is caused due to the structural change in the crystal when subjected to mechanical stress. In this case the ferroelectricity arises as a result of the lowering of the symmetry elements; the crystal changes from an orthorhombic structure to a monoclinic structure under mechanical stress. The sample then becomes polarized as a result of dipole alignment when subjected to an applied electric field.

In 1975, Meyer[112] predicted from the symmetry arguments that the liquid crystals in a tilted smectic phase (S_C^* or S_H^*) could also exhibit ferroelectricity. According to the symmetry conditions proposed by Meyer, spontaneous polarization can arise not only in thermotropic liquid crystal phases (S_C^*, S_H^*) but also in any other system having a layered structure, tilt and chirality of the constituent molecules. Systems of this kind could include, for example, solid layered structures, lyotropic liquid crystals, many biological objects, etc.

9.3.2. Structure, Symmetry and Ferroelectric Ordering in Chiral Smectic C Phase

The structure of achiral smectic C (S_C) phase is lamellar and the molecules within the layers are tilted at a temperature-dependent angle θ(T) from the layer normal (Fig. 1.3b). Within each layer there is no positional order of the molecules with the correlation length extending only over a few molecular centers, but the tilt direction remains relatively uniform over large domains. The chiral smectic C (S_C^*) phase is composed of optically active molecules. The optical activity arises due to molecular asymmetry. The molecules in S_C^* phase are periodically disposed along one axis Z which produces the layered structure; each molecule in the layer is tilted with an angle $\theta(T)$. As a result of precession of the molecular tilt about an axis perpendicular to the layer planes, a macroscopically helical structure is formed. The tilt direction is rotated through an azimuthal angle (ϕ) on moving from one layer to the next. The rotation occurs in a constant direction leading to a helix which is either left handed or right-handed. However, the helix manifests itself in a different way from the helix in the chiral (nonferroelectric) nematic phase.

In S_C phase the local symmetry elements consists of a center of inversion, a mirror plane normal to the layers and a C_2–axis parallel to the layers and normal to the tilt direction (see Fig. 9.12). Due to combination of these elements the symmetry group of S_C phase is C_{2h}. When some chiral molecules are present in the mesophase, the center of symmetry and mirror planes are lost and only the C_2–axis remains (see Fig. 9.13). Thus the symmetry group of S_C^* phase is C_2 which permits a permanent electric dipole along the C_2–axis and this makes the material ferroelectric.

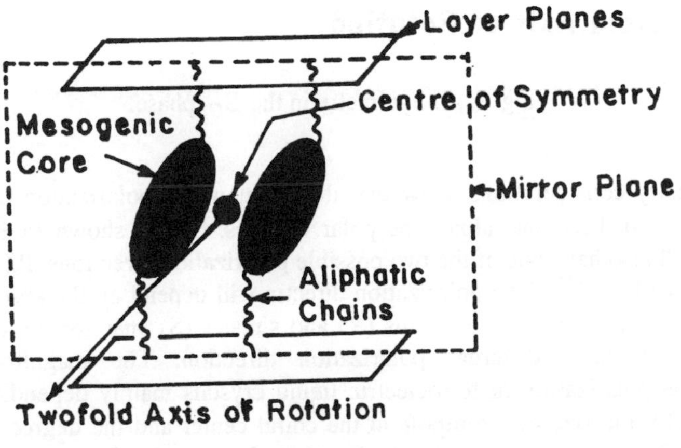

Fig. 9.12. Symmetry in the S_C phase.

In case of polar molecules an imbalance arises with respect to the molecular dipoles along the C_2–axis[113]. The dipoles tend to align along the C_2–axis. This time dependent alignment of the dipoles causes the spontaneous polarization (P_s) to develop along this direction and parallel to the layer planes. Each individual layer, therefore, essentially has a spontaneous polarization associated with it. As the layers are stacked one on top of the other generating a helical structure (Fig. 9.14), a macroscopic sample of S_C^* phase is not ferroelectric because the layer polarizations are consequently averaged out to zero along one pitch of the sample, and the phase is described as helielectric[114]. However, if the helix is unwound, the layer polarizations point in the same direction and then the phase becomes ferroelectric.

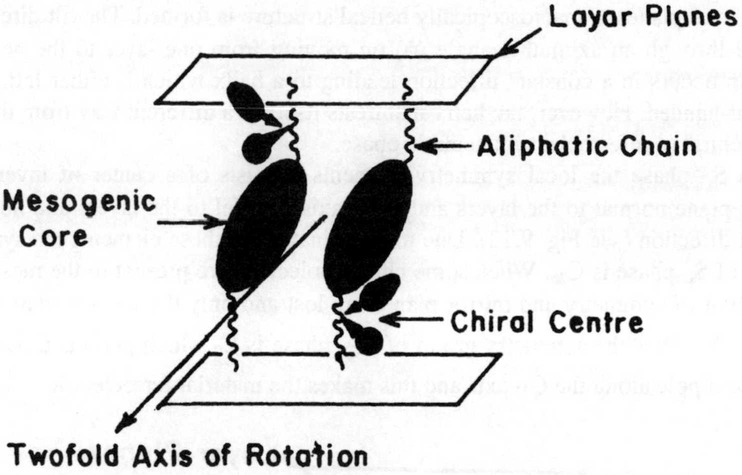

Fig. 9.13. Symmetry in the S_C^* phase.

Symmetry considerations show that the spontaneous polarization can point in one of the two directions along the polar C_2–axis. So, as shown in Fig. 9.15, a material will also have one of the two possible polarization directions, $P_s(+)$ and $P_s(-)$ associated with it[115,116]. The polarization directin will depend on the absolute spatial configuration of the molecule; rectus (R) and sinister (S) enantiomers of the same molecule will have different polarization direction. The magnitude of the spontaneous polarization in ferroelectric liquid crystals mainly depends on the tilt angle[117] θ(T), the size of the dipole at the chiral center and the degree of freedom that the chiral center has to rotate about the long molecular axis. Its value is relatively low as compared to its value in inorganic ferroelectric materials. Several reasons may be assigned for its low values in ferroelectric liquid crystals. Since the chiral smectic molecules are always in a state of rapid reorientational motion about their long molecular axes, only the time-dependent alignment of the lateral dipoles will contribute to the polarization. The molecules of ferroelectric smectogens are considerably less polar than inorganic ferroelectric materials. As the mesophase is fluid, the molecules relax rapidly when subjected to an applied electric field.

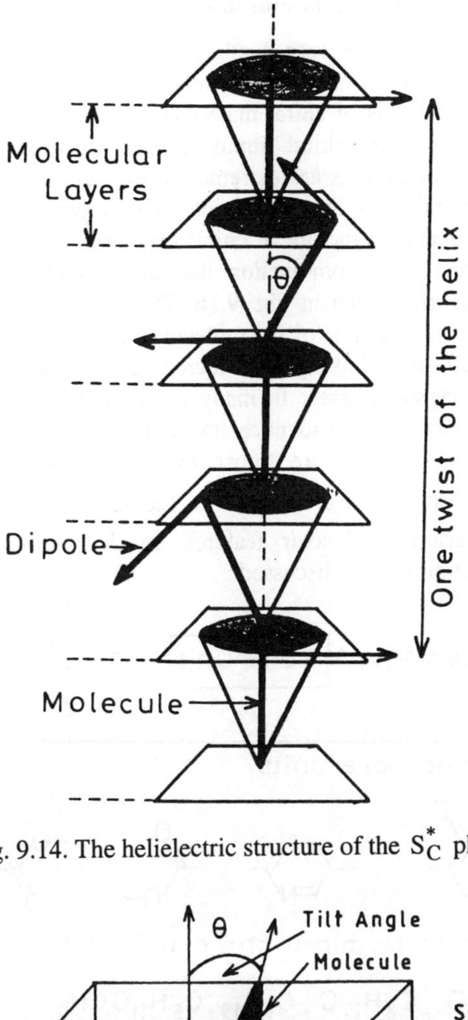

Fig. 9.14. The helielectric structure of the S_C^* phase.

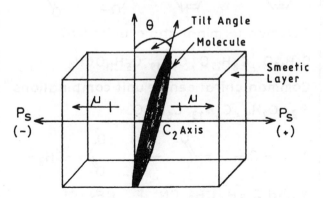

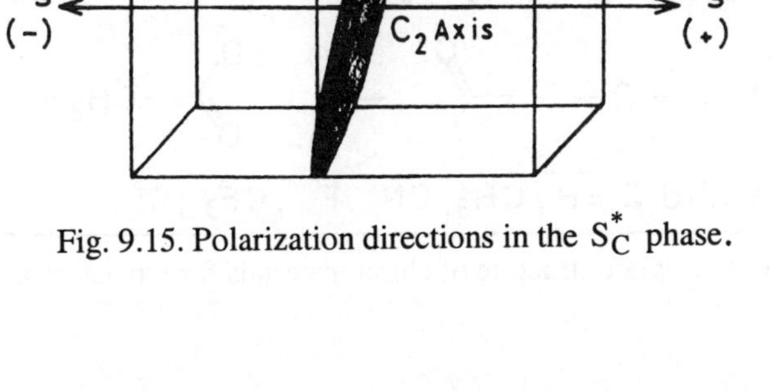

Fig. 9.15. Polarization directions in the S_C^* phase.

9.3.3 Structure-Property Correlations in Chiral Smectic Phases

Several chiral compounds have been synthesized[1–4] and some understanding about the structure-property relationships with respect to the spontaneous polarization has been achieved. The synthesis of chiral materials is special in many respects; chiral materials and not necessarily chiral liquid crystals are involved because several nonmesogenic chiral compounds are extremely useful as additives to liquid crystal mixtures. Usually chirality is introduced into a mesogenic compound through a terminal chain. The general structure of chiral materials, mesogenic or not, and the types of moieties and their combination that are suitable in the generation of ferroelectric mixtures are shown in Fig. 9.16. The use of ether and ester moieties gives the required lateral dipoles for the molecular tilting and the use of conjugated cinnamate and Schiff's base linking units increases the polarizability to ensure the formation of liquid crystal phases. In many respects the synthetic routes for the generation of chiral smectics are identical to their achiral analogues; but are often difficult and expensive. Therefore, most of the structural considerations for generating S_C^* phase are identical to those for the achiral analogues. In this sectin, the influence of various molecular features on the mesophase sequence and spontaneous polarization will be discussed.

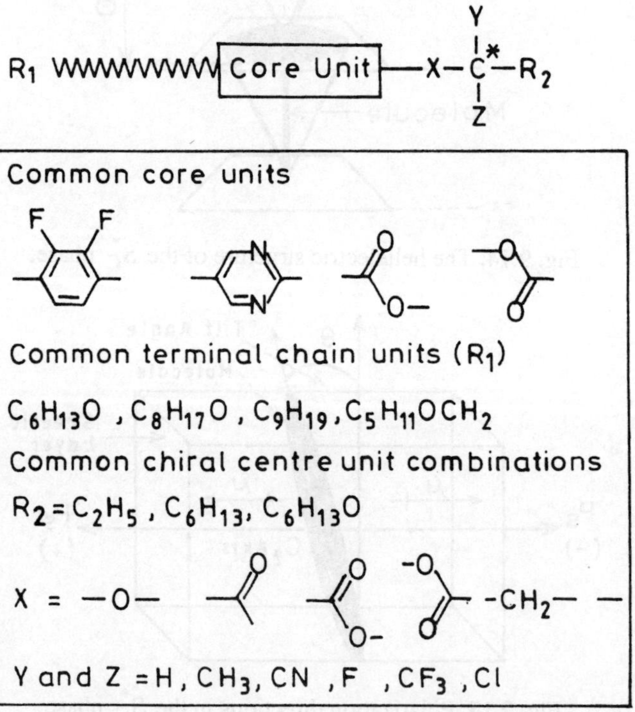

Fig. 9.16. A template structure of chiral materials for ferroelectric mixtures.

Some typical examples of ferroelectric liquid crystals are given in Table 9.1. The first S_C^* phase was discovered[112] in (S)-4-n-decyloxybenzylideneamino-2-methylbutylcinnamate (DOBAMBC) in 1974 (see compound T 9.1a). However, it is photochemically unstable, and has a very low P_S (4nc cm^{-2}) and relatively high viscosity. So the compound T 9.1a is not suitable for use in devices. Its analogues have been prepared (see Table 9.2) which are very useful in understanding the role of molecular structure on the phase morphology, transition temperatures and the magnitude of spontaneous polarization.

Table 9.1. Some examples of ferroelectric liquid crystals; phase sequence (transition temperatures ^{0}C) is also given.

a. 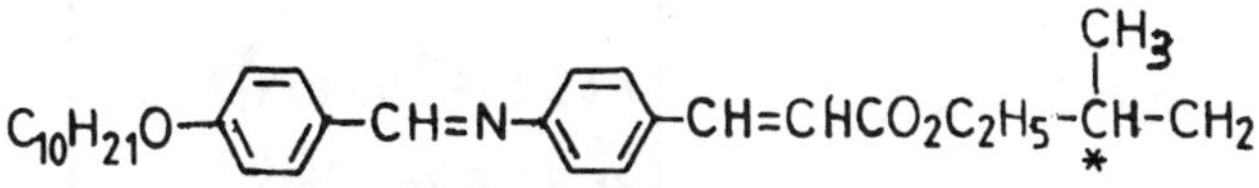

K 76.0 (S_I^* 63.0) S_C^* 95.0 S_A 117.0 IL

b. 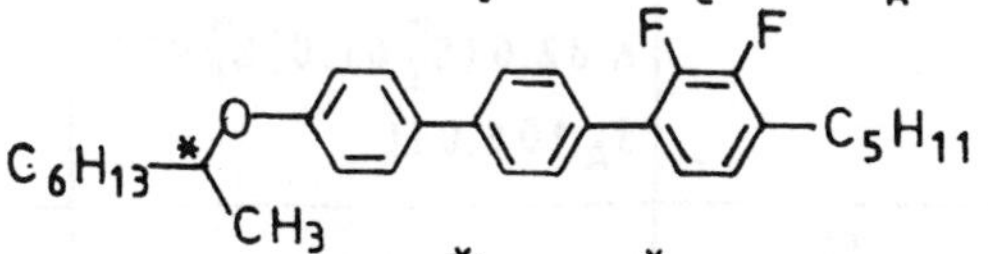

K 38.5 S_C^* 51.5 N^* 59.5 BPII 61.5 IL

c. 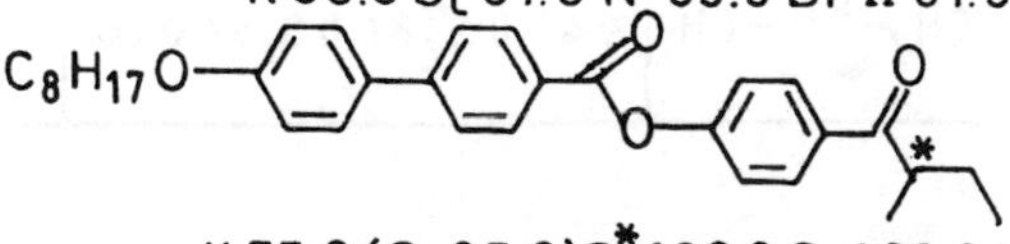

K 75.0 (S_X 65.0) S_C^* 130.0 S_A 186.0 IL

d. 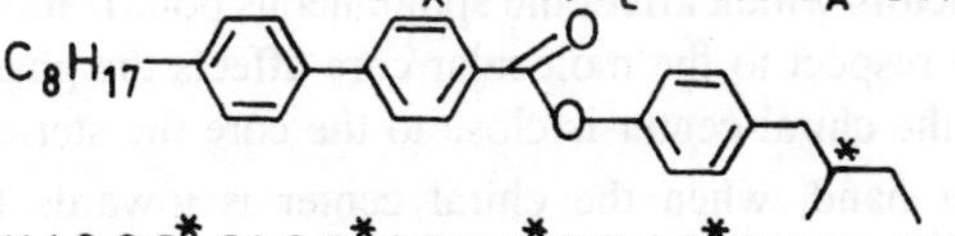

K 48.0 S_G^* 61.0 S_J^* 67.0 S_I^* 70.0 S_C^* 87.0 S_A 135.0 N^* 140.0 IL

e. 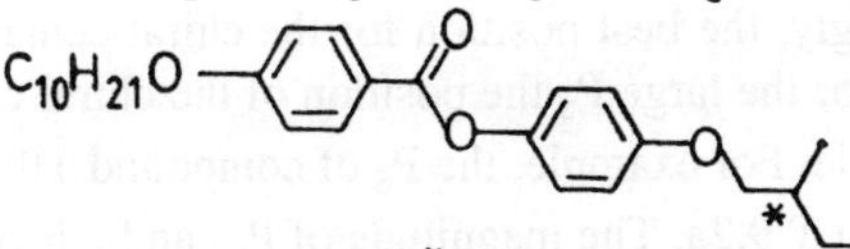

K 45.5 S_C^* 50.0 S_A 63.0 IL

f. 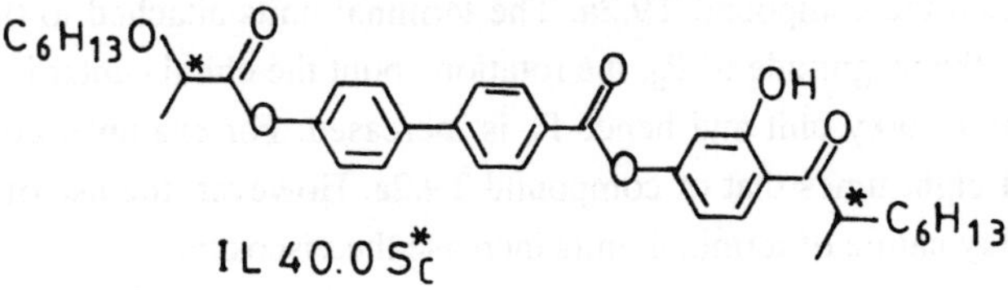

IL 40.0 S_C^*

Table 9.2. Effects of chiral center on the transition temperatures and P_S.

$$C_{10}H_{21}O-\phenyl-CH=N-\phenyl-CH=CHCO_2R^*$$

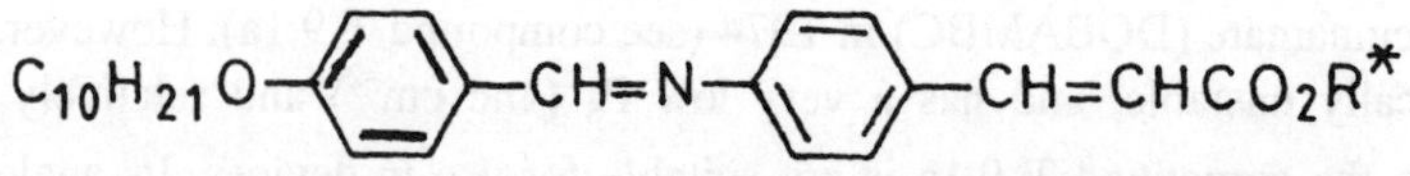

R^*	Transition temperatures (°C)	P_S (nc cm^{-2})
a. $C_2H_5-\overset{CH_3}{\underset{*}{CH}}-CH_2$	K 76.0 (S_I^* 63.0) S_C^* 95.0 S_A 117.0 IL	4
b. $C_2H_5-\overset{Cl}{\underset{*}{CH}}-CH_2$	K 65.0 (S_J^* 63.0) S_I^* 74.5 S_E^* 81.0 S_A 136.0 IL	15
c $C_2H_5-\overset{CH_3}{\underset{*}{CH}}$	K 82.0 (S_I^* 61.0) S_C^* 91.0 S_A 106.0 IL	18
d. $CH_3-\overset{CH_3}{\underset{*}{CH}}-CH_2-\overset{Cl}{CH}-CH_2$	K 49.0 S_C^* 80.0 S_A 94.0 IL	34

There are several factors which affect the spontaneous polarization. The position of the chiral center with respect to the molecular core affects the phase stability and the magnitude of P_S. If the chiral center is close to the core the steric effect is more excessive. On the other hand, when the chiral center is towards the end of the terminal chain, the steric effect is somewhat diluted and so liquid crystal phase stability is upheld. Accordingly, the best position for the chiral center may be at the end of the chain. However, for the large P_S the position of the chiral center should be as close to the core as possible. For example, the P_S of compound T9.2c is four times higher than that of compound T 9.2a. The magnitude of P_S can be increased by using a polar group at the chiral center; the chloro-substituted compound T9.2b has a much larger P_S than the compound T9.2a. The terminal units attached to the chiral center also affects the magnitude of P_S; the rotation about the chiral center is reduced due to a long and/or bulky unit and hence P_S is increased. For example, compound T9.2d has a P_S of eight times that of compound T9.2a. However, the use of a polar group, and the bulky nature of terminal units increase the viscosity.

Table 9.3 shows another example of how the P_S is affected by the intermolecular rotational restriction. It can be seen that there is a considerable increase in the value of P_S as the alkyl chain is lengthened; P_S is maximum for n = 5 and then decreases after this point as the effect of dipole is diluted with the increase of the overall size of the molecules. This is because the spontaneous polarization is proportional to the dipole density which in turn is affected by the molecular size. Similar effects have been observed[3] in other systems also.

Table 9.3. Effect of alkyl chain length on the P_S in a series of (R)-and (S)-1-methylalkyl-4-(n-octanoyloxy) biphenyl-4-carboxylates.

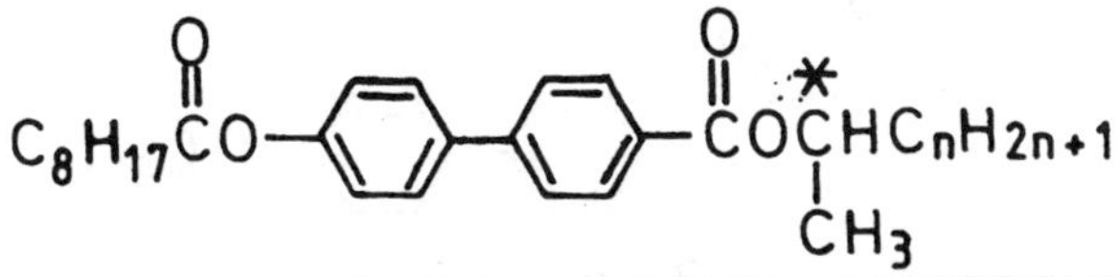

n	Ps (nc Cm^{-2})
2	7.7
3	32.0
4	30.7
5	48.8
6	42.2

The rotational freedom of the chiral center can be restricted in a number of ways. In order to achieve this more efficiently compounds possessing two chiral centers have been synthesized (see Table 9.4). These compounds show much higher P_S values in comparison with most of the one chiral center compounds. Compound T9.4a, synthesized by Kawada et. al.[118] possess two chiral centers linked together by an alkyl chain which forms part of an alicyclic ring, and P_S = 260 nC cm^{-2}. Ikemoto et. al.[119] have synthesized compounds which include a chiral valerolactone ring and displaying a large polarization value. In case of compound T9.4c it can be seen that not only a dipole, but also the chiral center itself is introduced into the central core which restricts the rotation of both the dipole moment and the chiral center. Compound T9.4b is a chiral lactone with two chiral centers as part of a ring unit. This structural features minimizes molecular broadening and provides a high P_S. In compound T9.4d a polar cyano group has been used at a chiral center which is part of a ring structure and very close to the aromatic core; combination of polarity and restricted rotation ensure a high P_S. The rotational freedom of the chiral center can also be restricted by intramolecular hydrogen bonding[120].

Table 9.4. Compounds with two chiral centers.

a. 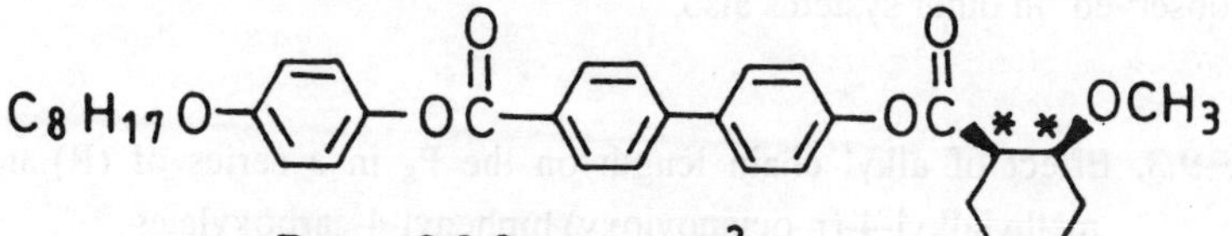

P_S = 260 nC cm^{-2}

b. 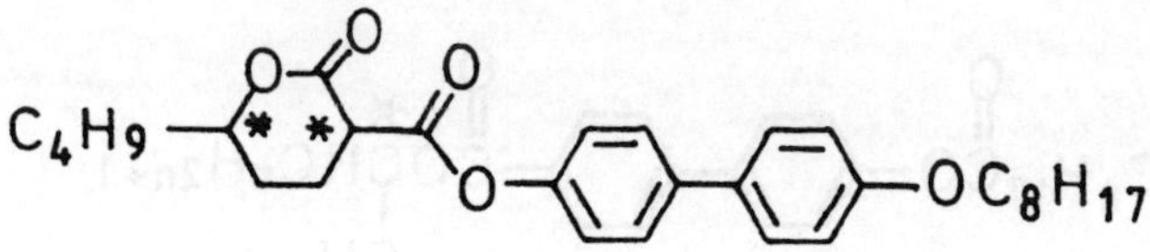

P_S = 384 nC cm^{-2}

c. 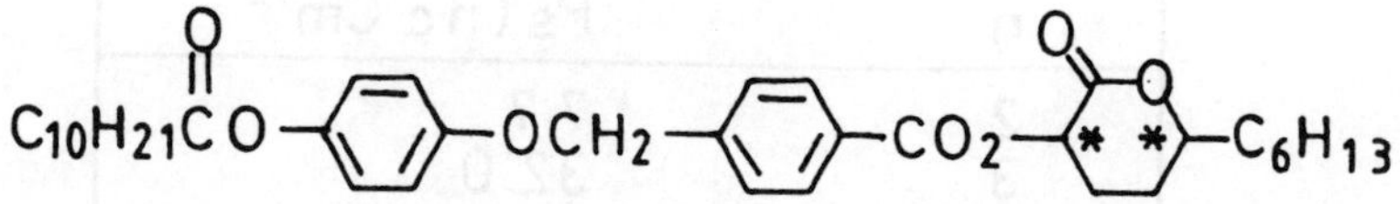

d. 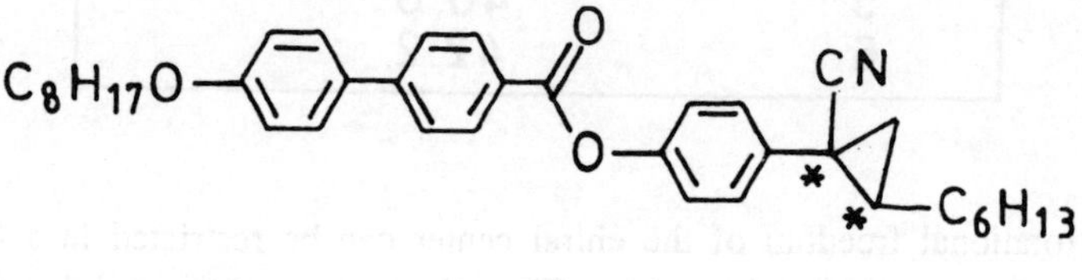

Another factors influencing P_S are conformational structure, magnitude of the lateral dipole moment located near to the chiral center, etc. Compounds have been reported[121–124] in which the sign of P_S inverts with decreasing temperature. The influence of the strength of the dipole at the chiral center on P_S has been investigated[125] via the introduction of fluorine atoms at the asymmetric center (see Table 9.5). The fluorine atoms were positioned around the chiral center to investigate how these affects P_S. Table 9.5 compares the values of P_S of several fluorinated compounds with the nonfluorinated analog. It can be seen that a larger value of P_S is observed in fluorinated compounds than the nonfluorinated analog.

Table 9.5. The influence of the position of fluorine atoms at the chiral center on the values of P_S.

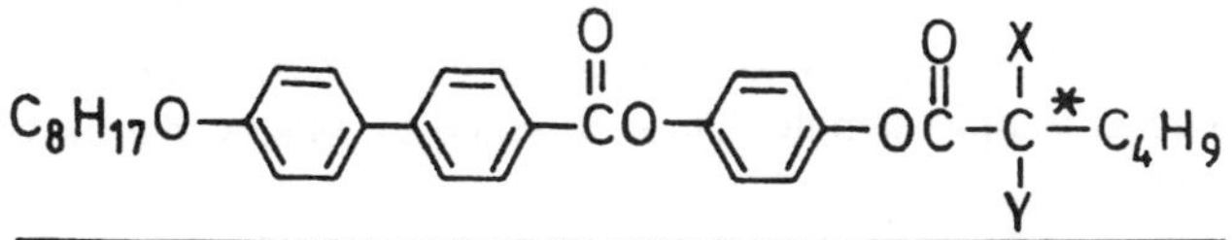

X	Y	P_S (nc cm^{-2})
CH_3	F	135
CF_3	H	95
F	H	85
CF_3	F	25
CH_3	H	15

9.3.4. Continuum Description of Chiral S_C^* (S_I^*, S_F^* and S_K^*)

The symmetry element characterizing these systems is C_2 only. It is both perpendicular to the normal to the layers $(\hat{\mathbf{k}})$ and to the tilt direction $(\hat{\mathbf{c}})$. The elastic free-energy for these systems can be constructed[12] in terms of $\hat{\mathbf{k}}$, $\hat{\mathbf{c}}$ and $\hat{\mathbf{p}}$ $(=\hat{\mathbf{n}}\times\hat{\mathbf{c}})$ vectors. In case of a nonchiral S_C phase all terms linear in $\nabla\hat{\mathbf{k}}(\hat{\mathbf{c}},\hat{\mathbf{p}})$ were prohibited in f_e because they are incompatible with the inversion symmetry. In the chiral smectic phases some of these terms become allowed which are the analogues of $(\hat{\mathbf{n}}\cdot\nabla\times\hat{\mathbf{n}})$ term of eq. (9.2) for cholesterics. The first-order part of the elastic free-energy density can be written as

$$f_e = D_1(\hat{\mathbf{c}}\cdot\nabla\hat{\mathbf{c}}\cdot\hat{\mathbf{p}}) - D_2(\hat{\mathbf{p}}\cdot\nabla\hat{\mathbf{k}}\cdot\hat{\mathbf{c}}) + D_3(\hat{\mathbf{k}}\cdot\nabla\hat{\mathbf{c}}\cdot\hat{\mathbf{p}}) + D_4(\hat{\mathbf{p}}\cdot\nabla\gamma) \tag{9.81}$$

In the compressible limit, the D_4 term ($\gamma = du/dz$) may be omitted. In terms of $\mathbf{\Omega}$ vector field

$$\Omega_x = \frac{\partial u}{\partial y} \quad ; \quad \Omega_y = -\frac{\partial u}{\partial x} \quad ; \quad \Omega_z$$

f_e can be expressed as

$$f_e = D_1\frac{\partial\Omega_z}{\partial x} + D_2\frac{\partial\Omega_x}{\partial x} + D_3\frac{\partial\Omega_z}{\partial z} + D_4\frac{\partial\Omega_x}{\partial z} \tag{9.82}$$

Equations (9.81) and (9.82) suggest complementary interpretation. While D_1 term tend to bend the $\hat{\mathbf{c}}$ director (i.e. splay the electric polarization), the D_2 term tends to transform a flat layer into a twisted ribbon.

In many real situations D_1 has no observable effect. Secondly, in such cases where $\hat{\mathbf{c}}$ is not continuous (no singular lines) the volume integral of curl $\hat{\mathbf{c}}$ may be transformed into a surface integral. As a result D_1 does not contribute to the volume energy. If the molecules are strongly anchored at the limiting surface, the surface term D_1 is unobservable. The only possible effect of D_1 then would be to generate a finite density of defects. A D_2 type structure is entirely acceptable for a single layer. In case of many layer system it is not compatible with the role of constant inter-layer spacing for a macroscopic sample. Thus, it would be observable only in case any defect is present.

The D_3 term, which is coupled to a simple twist of the $\hat{c}$ director, gives a strong as well as observable effect. The D_4 term is also not compatible with the role of a constant inter-layer distance and requires the existence of a dislocation density.

9.3.5. Antiferroelectric and Ferrielectric Chiral Smectic C Phases

Historically, the antiferroelectric liquid crystals (AFLCs) were discovered accidently; the related compounds (see Table 9.6) were synthesized[126-128] several years before without realizing that they exhibit antiferroelectric phase. Certain properties of these compounds were measured[129-131]. The most important observation was made in MHPOBC (compound T9.6a) in 1988 before the identification of antiferroelectric phase; from the electro-optic effect and the switching current measurements, tristable switching with a sharp threshold and a double hysteresis was observed[132] – two switching current peaks appear when sharp transmittance changes occur. It is easily realized that a switching device could be made by using the threshold and the hysteresis.

Table 9.6. Chemical structures of initially synthesized three AFLCs.

a. $n-C_8H_{17}O-\langle\underline{}\rangle-\langle\underline{}\rangle-COO-\langle\underline{}\rangle-COO-R$

$R = \overset{*}{C}H(CH_3)-n-C_6H_{13}$

b. $R_1 = CH-\langle\underline{}\rangle-N=CH-\langle\underline{}\rangle-CH=N-\langle\underline{}\rangle-CH=R_2$

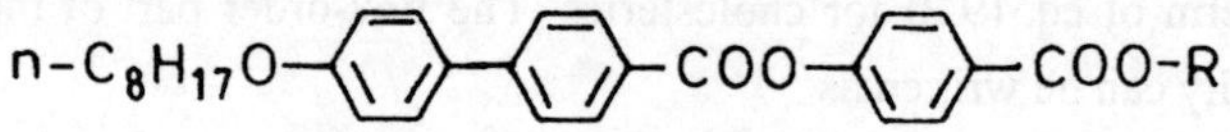

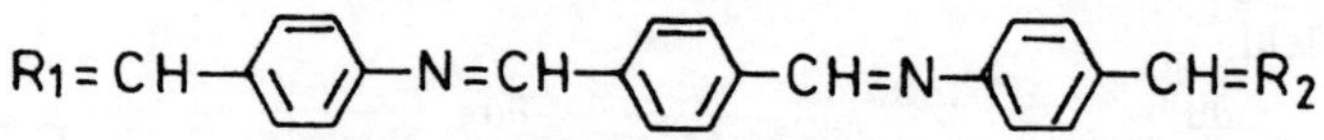

$R_1 = CHOCO(CH_3)\,H\overset{*}{C}C_6H_{13}$
$R_2 = CHCOO\overset{*}{C}H(CH_3)C_6H_{13}$

c. $n-C_{10}H_{21}O-\langle\underline{}\rangle-COO-\langle\underline{}\rangle-\langle\underline{}\rangle-COO-R$

$R = \overset{*}{C}H(CH_3)-n-C_4H_9$

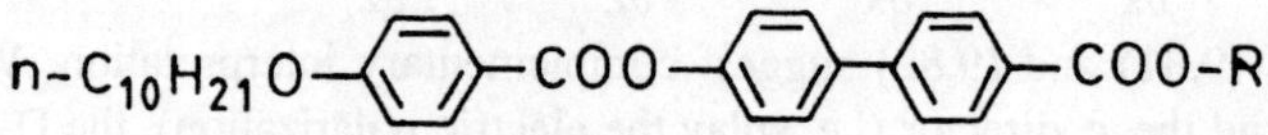

The helicoidal structures of ferroelectric (S_C^*) and antiferroelectric (S_{CA}^*) liquid crystals are shown in Fig. 9.17. These structures have been identified by a set of experimental observations – tristable switching[132,133], transmittance spectra in oblique incidence[134], interferential microscopy of a droplet[135–138], ellipsometry for free-standing films[139], microscope observation of defects[140,141], conoscope observations[142,143], dielectric measurements[144,145], etc.

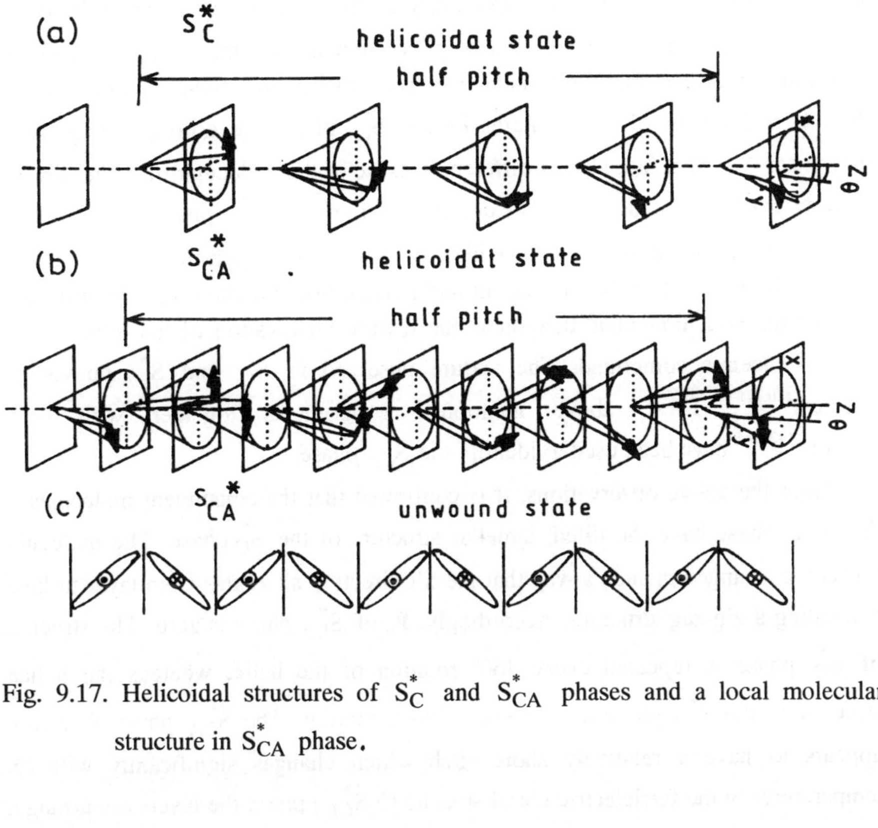

Fig. 9.17. Helicoidal structures of S_C^* and S_{CA}^* phases and a local molecular structure in S_{CA}^* phase.

When light propagates along the helicoidal axis the S_C^* and S_{CA}^* phase structures look apparently the same because in both the structures half the pitch is optically one period. Contrary to this, for the oblique incidence in the ferroelectric structure the optical period is a full pitch, while in the antiferroelectric structure, it is still half the structure pitch. Consequently, in case of obliquely incident light the molecular orientations in the layers separated by half the pitch are different. The two different structures can be easily differentiated by the selective reflection at oblique incidence[134]. In the S_C^* phase full pitch band is observable at the wavelength of $2\lambda_{SR}$, while the full pitch band must be absent in the S_{CA}^* phase. From an interferential microscope observation[135] of a droplet of racemic MHTAC (compound T9.6b) on a glass surface it was confirmed[136–138] that the molecular tilt sense changes from layers to layers. The local molecular arrangement, i.e. herring bone structure, in S_{CA}^* was also confirmed by an ellipsometry using free-standing films[139]; the phase difference $\Delta(=\Delta_p-\Delta_s)$ between s and p components of the obliquely incident transmitted light was measured under positive and negative fields in free-standing films as a function of temperature. In the S_C^* phase the corresponding phase differences Δ_+ and Δ_- are always different because of the molecular tilt either toward or away from the incident laser beam. However, in the S_{CA}^* phase, for two layer films $\Delta_+ = \Delta_-$, while for three layer films $\Delta_+ \neq \Delta_-$. This experiment clearly reveals the layer-by-layer alternation of the tilt and polarization direction. Another effective tool for the structural identification is the texture observation of the homeotropic cells of racemic compound. The texture observed in S_C^* and S_{CA}^* phases are different[140,141]. These above mentioned experiments together with other methods[142–145] have been used to identify the S_{CA}^* phase.

From the above observations, it is confirmed that the constituent molecules of the S_{CA}^* phase have the tilted, lamellar structure of the S_C^* phase. The molecular layers are arranged in such a way that the tilt direction alternates from layer to layer generating a zig-zag structure. Accordingly, $\mathbf{P_S}$ of S_{CA}^* phase is zero. The structure of S_C^* phase is repeated every 360° rotation of the helix, whereas the helical structure of the S_{CA}^* phase repeats every 180°C rotation. The S_{CA}^* phase, therefore, appears to have a relatively short pitch which changes significantly with the temperature. In the ferrielectric chiral smectic C(S_{CF}^*) phase the layers are arranged in such a way that a net overall spontaneous polarization is observed. In this phase, the alternation in the layer structure is not symmetrical and more layers are tilted in one direction than the other. Accordingly, the phase generates a $\mathbf{P_S}$ which depends upon the degree of alternation of the tilt directions. Several antiferroelectric and ferrielectric subphases, namely $S_{C\alpha}^*$, $S_{C\gamma}^*$, AF, etc., have also been observed to occur in antiferroelectric compounds. AF phase is a reentrant antiferroelectric phase between $S^*_{C\alpha}$ and $S^*_{C\gamma}$. A more detail description about all these phases is available elsewhere[146].

The materials exhibiting the S^*_{CA} phase are required to have a branched terminal chain with chiral center located near to the rigid aromatic core. The terminal chain facilitates the self-assembly of layers with alternating tilt direction. The usual molecular core structures for antiferroelectric phases consist of three phenyl rings (see Fig. 9.18(a,b)) or two phenyl rings and one heterocyclic ring (Fig. 9.18c). Most of the moieties that promote the generation of S^*_C phase have similar structure based around the chiral center. In compounds displaying antiferroelectric behaviour, the degree of molecular chirality is normally relatively high. In these systems the motion of the chiral center about the long molecular axis is damped which increases the degree of the molecular chirality.

a.

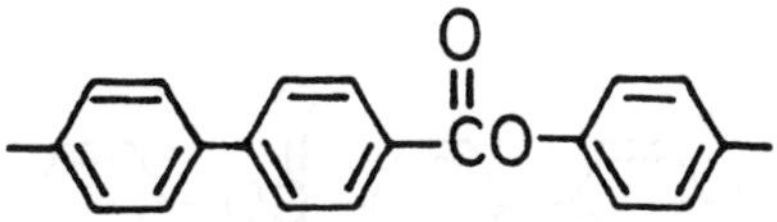

b.

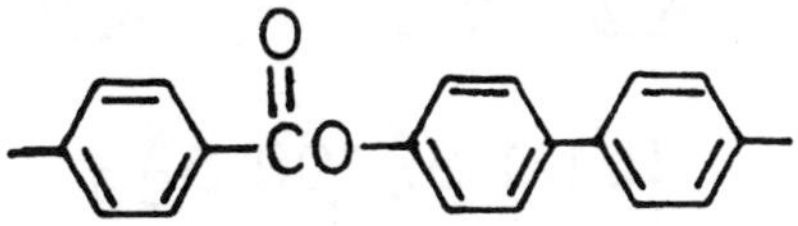

c.

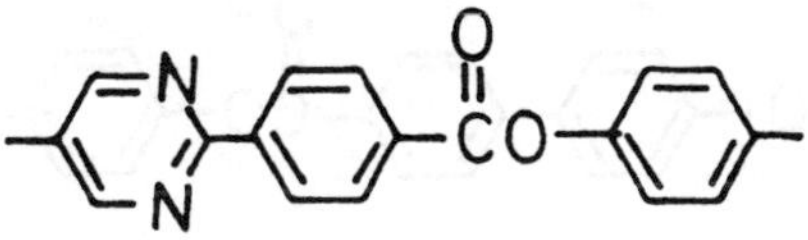

Fig. 9.18. Molecular core structures of compounds generating S^*_{CA} phase.

There are several factors that affect the generation of S^*_C and S^*_{CA} phases (see Tables 9.7 and 9.8). It can be seen that (a) the replacement of the lateral methyl group with an ethyl group at the chiral center increases the stability of S^*_C phase[147] (Table 9.7), and (b) fluorine atoms at the chiral center (compound T9.8a) appear to increase the stability of S^*_{CA} phase[148], but when hydrogen atoms are also present in the lateral chiral branch (compound T9.8b) the S^*_{CA} phase does not occur. The orientation of ester groups in the molecule also influences the formation of S^*_{CA} phase.

Table 9.7. The effect of the replacement of a lateral methyl group with an ethyl group at the chiral center on the phase stability.

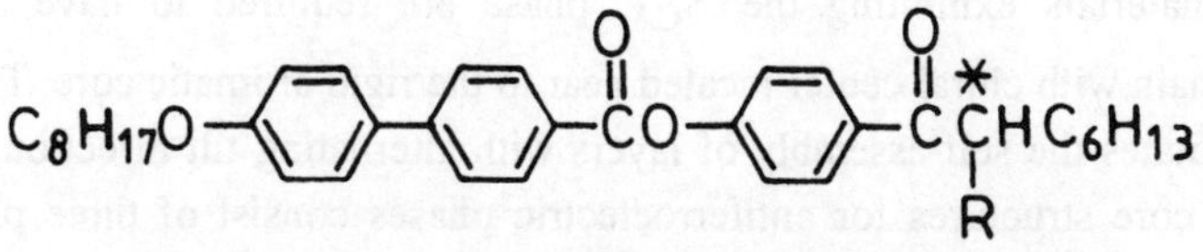

a. $R=CH_3$, IL 150 S_A^* 132 S_C^* 65 S_{CA}^*

b. $R=C_2H_5$, IL 180 S_A^* 100 S_C^* 85 S_{CA}^*

Table 9.8. The effect of introducing fluorine atoms at the chiral center.

a. 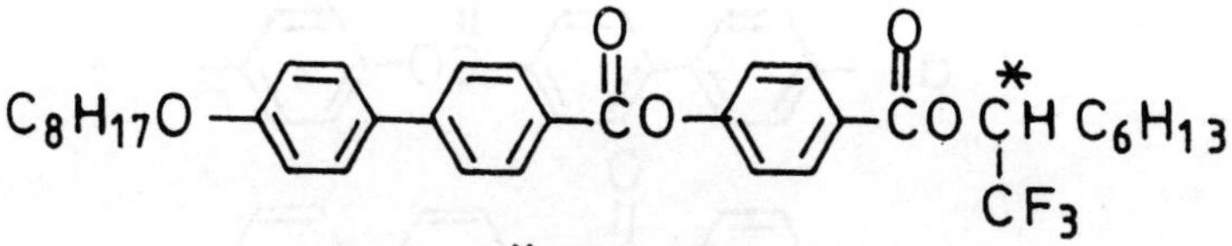

IL 121 S_A^* 109 S_{CA}^*

b. 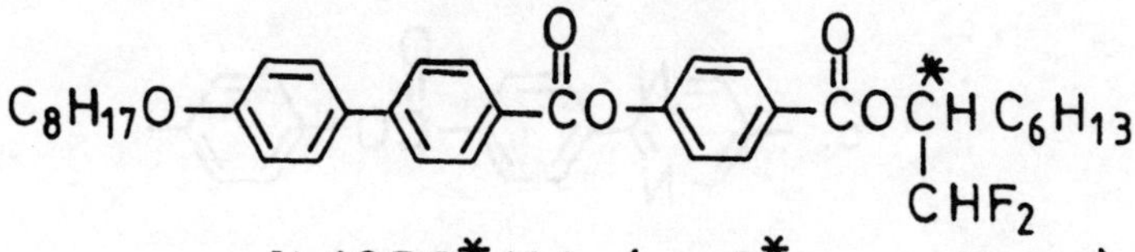

IL 125 S_A^* 114 (no S_{CA}^* phase)

The S_{CA}^* phase has a great potential in display devices. The application of a strong electric field to an antiferroelectric material induces a ferroelectric ordering. When the field is removed, the antiferroelectric ordering regenerates. The switching from antiferroelectric to the ferroelectric state can be investigated by measuring the tilt angle as a function of applied voltage. At low voltages no change in the tilt angle occurs, but the phase changes from antiferroelectric to ferroelectric at a critical field. At this field the tilt angle increases substantially and reaches maximum of S_C^* phase. The presence of a sharp switching threshold in antiferroelectrics makes these materials very useful in display applications that require multiplexing with grey scales.

9.4. CHIRAL DISCOTIC PHASES

Chirality in a discotic system can be introduced by incorporating a chiral unit into one or more of the peripheral units surrounding the discotic core (see Table 9.9). As can be seen in compound T9.9a all of the peripheral units are chiral and only one of the peripheral acetylene units is chiral in compound T9.9b. Both the compounds exhibit solely a chiral nematic discotic phase because the steric effect of the branched chains at the chiral center disrupts the ability of the molecules to pack in columns. The liquid crystalline behaviour depends critically on the type of chiral peripheral chains. The large size of the planar aromatic core (compound T9.9a) gives a high clearing point.

Table 9.9. Examples of chiral discotic liquid crystals.

a.

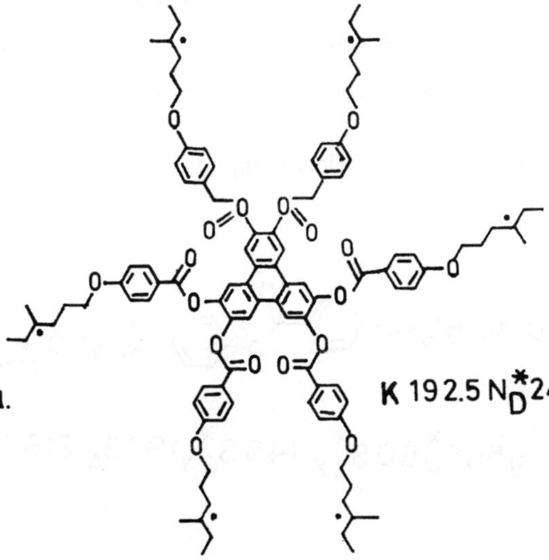

K 192.5 N_D^* 246.5 IL

b.

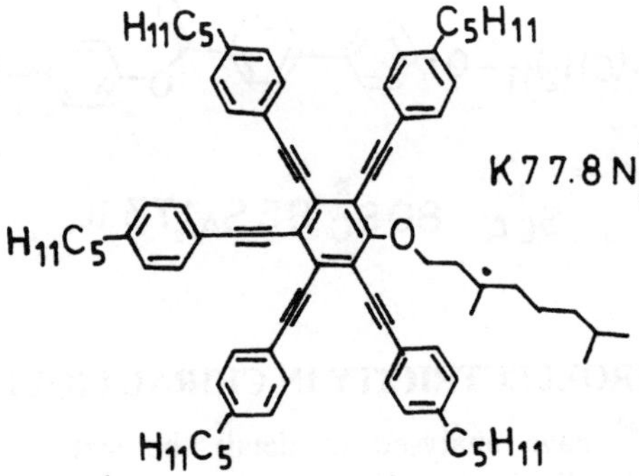

K 77.8 N_D^* 98.4 IL

Only few examples of chiral discotic nematic mesogens exist. The N_D^* phase has an analogous structure to the calamitic chiral nematic phase with a gradual rotation of the molecular director through the phase which describes a helix.

9.5. CHIRAL POLYMERIC LIQUID CRYSTALS

Chirality in the liquid crystal polymers are usually introduced by incorporating a chiral moiety, as a part of the terminal chain, within the structure. These polymers have potential use as non-linear optical materials or as pyroelectric detectors. The ferroelectric polymers have a spontaneous polarization and can be switched in the same manner as the analogous low molar mass materials. However, these polymers are extremely viscous and switching times are quite large. Table 9.10 shows typical examples of ferroelectric liquid crystal polymers. Compound T9.10a exhibits a range of chiral liquid crystal phases with two S_C^* phases. Polymer T9.10b exhibits an antiferroelectric S_C^* phase below a ferroelectric S_C^* phase.

Table 9.10. Typical examples of ferroelectric liquid crystal polymers.

a.

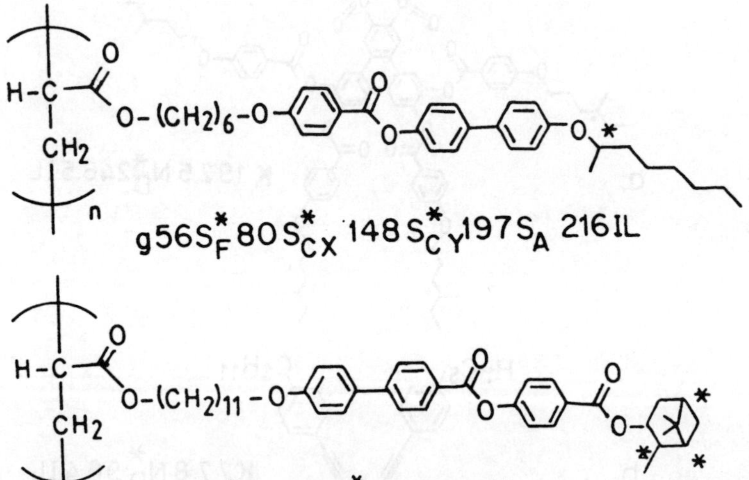

b.

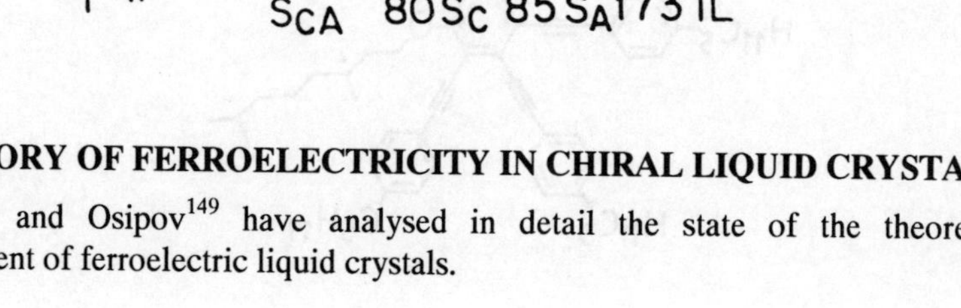

9.6. THEORY OF FERROELECTRICITY IN CHIRAL LIQUID CRYSTALS

Pikin and Osipov[149] have analysed in detail the state of the theoretical development of ferroelectric liquid crystals.

9.6.1. Application of Landau Theory

The first complete description, within the context of Landau theory, of the thermodynamic properties of S_C^* phase has been given by Pikin and Indenbom[150]. Blinc and Žekš[151] described the dynamics of the helical structure. First we shall identify the parameters for the description of the phase transitions in ferroelectric liquid crystals and then shall write the Landau expansion in terms of free-energy invariants with the components of these parameters. Finally the analysis of the generalized Landau model will be presented.

For the description of the state of director $\hat{\mathbf{n}}$, the symmetry properties and one dimensional structure of smectic liquid crystals, an order parameter of orientational transition has to be identified. It can be defined as a combination of certain physical quantities and related magnitudes. The density wave $\psi = \psi_0 \exp[i(\mathbf{kr} + \alpha)]$, where ψ_0 is the amplitude, α is the wave phase, and $\mathbf{k}$ is the wave-vector, can be used to describe a one-dimensional crystal structure. The magnitude $\mathbf{kr}$ = kz is invariant to symmetry operations in the groups $D_{\infty h}$, D_∞, C_{2h} and C_2.

In the chiral smectic phases (e.g. S_C^*) the director tilts with respect to normals by an angle $\theta \neq 0$ and the rotation around the long-molecular axes becomes biased. The in-plane component of the spontaneous polarization $\mathbf{P}_S$ becomes nonzero in a direction perpendicular to the tilt. Assuming ψ_0 = constant and $\mathbf{k}$ to be parallel to the Z–axis, the order parameter can be expressed as a combination of two components[150–153]

$$\xi = (\xi_1, \xi_2) \tag{9.83}$$

$$\xi_1 = n_x n_z = \theta \cos \varphi(z) \tag{9.84a}$$

$$\xi_2 = n_y n_z = \theta \sin \varphi(z) \tag{9.84b}$$

If the molecules have no constant dipole moments, even in the absence of spontaneous polarization, a similar order parameter can serve to describe the orientational $S_A^* - S_C^*$ phase transition. In such cases the components (ξ_1,ξ_2) are transformed according to a two dimensional representation E_1 of the group D_∞.

As one moves from one layer to the other the tilt vector and the direction of in-plane spontaneous polarization $\mathbf{P}_S = (P_x,P_y,0)$ precess around the normal. Accordingly a helical structure is generated. (P_x,P_y) transform according to the same representation E_1. So a linear relation[149] between (ξ_1,ξ_2) and (P_x,P_y) must exist

$$(P_x, P_y) \sim (\xi_2, -\xi_1) \tag{9.85}$$

Equation (9.85) is common to the point group symmetry D_∞ and can describe the appearance of the spontaneous polarization both at the $S_A^* - S_C^*$ phase transition and at other phase transitions (e.g. first order $N^* - S_C^*$ transition). In view of this equation the components P_x and P_y can transform like ξ_2 and $-\xi_1$ under the application of the symmetry operations of the group D_∞ (i.e. tilt through an angle around the C_∞ axis and tilt through an angle π around the axes perpendicular to the C_∞ axis). It follows from eq. (9.85) that in the S_C^* phase of molecules having at least some lateral electric dipole moments a spontaneous polarization appears as a result of the final tilt of the molecules through an angle θ. This parameter can serve to describe the $S_A^* - S_C^*$ transition. At low magnitudes of θ the proportionality $P \sim \theta$ exists[149].

The polarization effects of various kinds in ferroelectric liquid crystals can be studied by writing the Landau free-energy density in terms of the variants formed from the components of the various parameters of the order. Two kinds of invariants appear in this expansion; certain invariants are permitted by the symmetry group whereas some are connected with the chiral aspect of the system.

For small tilt angle θ and the absence of spontaneous deformations $\left(\frac{\partial \xi_1}{\partial x} + \frac{\partial \xi_2}{\partial y}\right)$, the generalized Landau expansion of the free-energy density can be written as

$$f = a(\xi_1^2+\xi_2^2) + b(\xi_1^2+\xi_2^2)^2 + c(\xi_1^2+\xi_2^2)^3 + \frac{1}{2\chi_0}(P_x^2+P_y^2) + \frac{1}{4}\eta(P_x^2+P_y^2)^2$$

$$-\mu_p(P_x\xi_2 - P_y\xi_1) + \Omega_p(P_x\xi_2 - P_y\xi_1)^2 + \frac{1}{2}K^0\left[\left(\frac{\partial \xi_1}{\partial z}\right)^2 + \left(\frac{\partial \xi_2}{\partial z}\right)^2\right]$$

$$+\lambda_0\left(\xi_1\frac{\partial \xi_2}{\partial z} - \xi_2\frac{\partial \xi_1}{\partial z}\right) - \mu_f^0\left(P_x\frac{\partial \xi_1}{\partial z} + P_y\frac{\partial \xi_2}{\partial z}\right) + \lambda_0^1(\xi_1^2+\xi_2^2)\left(\xi_1\frac{\partial \xi_2}{\partial z} - \xi_2\frac{\partial \xi_1}{\partial z}\right)$$

$$-\mu_f^1(\xi_1^2+\xi_2^2)\left(P_x\frac{\partial \xi_1}{\partial z} + P_y\frac{\partial \xi_2}{\partial z}\right) + \delta\mu_f^1(P_x\xi_2 - P_y\xi_1)\left(\xi_1\frac{\partial \xi_2}{\partial z} - \xi_2\frac{\partial \xi_1}{\partial z}\right) \qquad (9.86)$$

where $a = a_0(T-T_0)$, $a_0 > 0$, $b > 0$, $c > 0$, $\chi_0 > 0$, $K^0 > 0$.

The first few terms of eq. (9.86) are the usual free-energy invariants containing powers with respect to the components (ξ_1,ξ_2) and (P_x,P_y). Next two invariants involving piezoelectric coefficients μ_p and Ω_p take account of the piezoeffect. Minimization of the sum of these invariants with respect to the polarization components (P_x,P_y) gives

$$P \sim P_p = \chi_0\,\mu_p\,\theta \qquad (9.87)$$

where the magnitude of θ can be obtained by minimizing the functional $f(\theta)$. The invariant containing coefficient K^0 arises due to the elastic free-energy density.

The invariants containing the coefficients $\underset{0}{\lambda}$ and μ_f^0, permitted by the symmetry group D_∞, arise, respectively, due to the effect of space modulation of the orientational order parameter and the flexoelectric effect. The later one with μ_f^0 is also permitted by the symmetry group $D_{\infty h}$ and is no way connected with the chiral aspect of the system. At the thermodynamic equilibrium and at low temperature, the terms involving coefficients μ_f^1 and $\delta\mu_f^1$ are associated with certain effects caused by the invariants of a flexoelectric nature of order $P\theta^3$. It is clear that the invariant involving μ_f^0 and μ_f^1 contribute to f,

$$-\,(\mu_f^0 + \mu_f^1\,\theta^2)\left(\frac{1}{2}P_0 \sin(2\theta)\,\frac{\partial\varphi}{\partial z} - P_1 \cos 2\theta \frac{\partial\theta}{\partial z}\right),$$

where P_0 is the projection of the polarization vector **P** on the C_2 axis (in the xy plane) and P_1 is the projection of **P** on the component of the director $\hat{\mathbf{n}}$ in the plane xy and perpendicular to C_2 axis. The invariant involving $\delta\mu_f^1$ includes only a contribution proportional to the value

$$\delta\,\mu_f^1\,P_0\,\theta^3\left(\frac{\partial\varphi}{\partial z}\right)$$

Thus, the term $\dfrac{\partial\varphi}{\partial z} \neq 0$ and $\dfrac{\partial\theta}{\partial z} \neq 0$ contribute to the polarization components,

$$P_0 \sim \mu_f\,\theta\left(\frac{\partial\varphi}{\partial z}\right) \qquad \text{and} \qquad P_1 \sim (\mu_f - \delta\mu_f)\left(\frac{\partial\theta}{\partial z}\right)$$

which have corresponding flexoelectric coefficients μ_f, $\mu_f - \delta\mu_f$, differing by a value of the order θ^2. In the vicinity of the second order $S_A^* - S_C^*$ transition the difference $\delta\mu_f$ is not significant. But far away from the transition point in the S_C^* phase the coefficient μ_f and $\mu_f - \delta\mu_f$ can differ considerably and the magnitude $\delta\mu_f$ can have either sign.

From the variation of f (eq. (9.86)) with respect to the polarization components P_x (= $P \sin\varphi(z)$), P_y (= $P\cos\varphi(z)$), one obtains (θ = constant << 1),

$$P = \chi_0\left(\mu_p - \mu_f\,\frac{\partial\varphi}{\partial z}\right) \tag{9.88}$$

Ignoring some higher order terms, we can expressed eq. (9.86) as[149]

$$f = A\theta^2 + B\theta^4 + \frac{1}{2}\,K\theta^2\left(\frac{\partial\varphi}{\partial z} - q_0\right)^2 \tag{9.89}$$

with

$$A = a - \frac{1}{2}\chi_0 \mu_p^2 - \frac{(\lambda_0 + \chi_0 \mu_p \mu_f^0)^2}{2K} \tag{9.89a}$$

$$q_0 = -\frac{\lambda + \chi_0 \mu_p \mu_f}{K} \tag{9.89b}$$

$$\lambda = \lambda_0 + \lambda' \theta^2 \tag{9.89c}$$

$$\mu_f = \mu_f^0 + (\mu_f^1 + \delta\mu_f^1)\theta^2 \tag{9.89d}$$

$$K = K^0 - \chi_0 \mu_f^{0^2} \tag{9.89e}$$

It follows from eq. (9.86) that the minimal free-energy is characterized by

$$\varphi(z) = \varphi_0(z) = q_0 z \tag{9.90}$$

and the helical distribution of the polarization with the amplitude

$$P_s = \chi_0 \left|\mu_p - \mu_f q_0\right| \theta \tag{9.91}$$

Here

$$q_0 = q_c + q' \theta^2 + ... \tag{9.92}$$

with

$$q_c = \frac{\lambda_0 + \chi_0 \mu_p \mu_f^0}{K} \tag{9.93}$$

The above equations include revalued effective coefficients A, B, K, λ′ and q′ and the phenomenological dependence $q_0' = q_0'(\theta^2)$. The temperature dependence of the spiral pitch

$$h_0(T) = 2\pi/q_0(\theta^2) \tag{9.94}$$

is determined by the value

$$q' = -[\lambda' + \chi_0 \mu_p(\mu_f' - \delta\mu_f')]/K.$$

The value of q_c (= $q_0(T_c)$) at the phase transition temperature can be determined from

$$T_c = T_0 + \frac{1}{2a'}(\chi_0 \mu_p^2 + K q_c^2) \tag{9.95}$$

In the S_C^* phase the finite tilt angle of the director for $T < T_c$ is given by

$$\theta = \theta_0 = \left(-\frac{A}{2B}\right)^{1/2} \tag{9.96}$$

It is obvious from eq. (9.91) that, to a first approximation, $P_s \sim \theta$. However, including terms of higher order in (9.86) with respect to θ and/or considering the orientation of the short molecular axis, one obtains $P_S \sim \theta^3$. Small corrections to the free-energy induced by the piezoelectric effect are of not great interest because they can be characterized by the same relaxation times as the main contribution. Below the $S_A^* - S_C^*$ phase transition point the ordering of the short molecular axes has to occur, as a secondary phenomenon, with its own relaxation time. The components (η_1, η_2),

$$\eta_1 = Q_t(n_x'^2 - n_y'^2), \qquad \eta_2 = 2Q_t n_x' \, n_y' \tag{9.97}$$

can serve as the parameter of this ordering. Here Q_t ($\neq 0$) is the order parameter for the short-axis and $\eta_x' = -\sin\varphi$, $\eta_y' = \cos\varphi$. The components (η_1, η_2) are transformed with respect to irreducible vector representation E_2 of the symmetry group D_∞. Now the expression for f has an invariant of the form

$$\mu_t[(P_y\xi_1 + P_x\xi_2)\eta_1 + (P_y\xi_2 - P_x\xi_1)\eta_2] \approx \mu_t \, P\,\theta\,Q_t \tag{9.98}$$

where μ_t is the phenomenological coefficient.

The qualitative dependence $Q_t(\theta)$ can be determined from an additional term in eq. (9.86) of the type

$$\delta f(Q_t) = \frac{1}{2\chi_t}(\eta_1^2 + \eta_2^2) + g[\eta_1(\xi_1^2 - \xi_2^2) + 2\eta_2\xi_1\xi_2] \approx -\, g\,\theta^2 Q_t + \frac{1}{2\chi_t} Q_t^2 \tag{9.99}$$

where g and $\chi > 0$ are phenomenological coefficients. Minimization of (9.99) with respect to Q_t gives

$$Q_t = g\,\chi_t\,\theta^2 \tag{9.100}$$

and thus the spontaneous polarization of the S_C^* phase is proportional to

$$P_t \sim \mu_t\,\chi_0\,\chi_t\,g\,\theta^3 \tag{9.101}$$

The behaviour of chiral smectics in an external field **X** (**E** or **H**) can be studied by adding a term – **PX** in eq. (9.86). Relations for the functions $\theta(\mathbf{r})$ and $\varphi(\mathbf{r})$ can be obtained from the variation of the functional f – **PX** with respect to **P**, θ and φ. The incommensurate structure of S_C^* phase is distorted under an external field. In weak fields the distortions $\theta(\mathbf{r}) - \theta_0$ and $\varphi(\mathbf{r}) - \varphi_0(z)$ are insignificant. In strong fields these distortions are caused by the spiral unwinding of the dipole moments and the increase of the tilt angle θ.

The most important aspect of the model (9.86) is that in addition to the chiral bilinear coupling between the polarization and the tilt, $P\theta$, a large nonchiral biquadratic coupling, $P^2\theta^2$, is also present. For the small tilt, close to T_C, only the bilinear coupling term is important, whereas for large θ, well below T_C, the biquadratic term dominates.

The model (9.86) has been applied to study the behaviour of chiral smectics and a detail analysis is available in the literature[149,152]

9.6.2. Molecular Statistical Theory For The Ferroelectric Order in Smectics

In the molecular theory of ferroelectric liquid crystals the most important problem is the origin of the intermolecular interactions which can cause the spontaneous polarization. It is obvious that the ferroelectric ordering in the S_C^* phase can be explained only by considering intermolecular interactions which are sensitive to the molecular chirality. The corresponding potentials, discussed in the following section, must reflect the chirality of interacting molecules.

9.6.2.1. Microscopic models for the ferroelectric ordering in the S_C^* phase

In order to define the interaction potential which can be used to understand the origin of ferroelectric ordering (spontaneous polarization) let us consider the orientation of a given molecule in S_C^* phase as shown in Fig. 9.19. The unit vectors $\hat{\mathbf{a}}_i$ and $\hat{\mathbf{b}}_i$ specify, respectively, the orientation of the long and short axes of the molecule i, $(\hat{\mathbf{a}}_i \, \hat{\mathbf{b}}_i) = 0$.

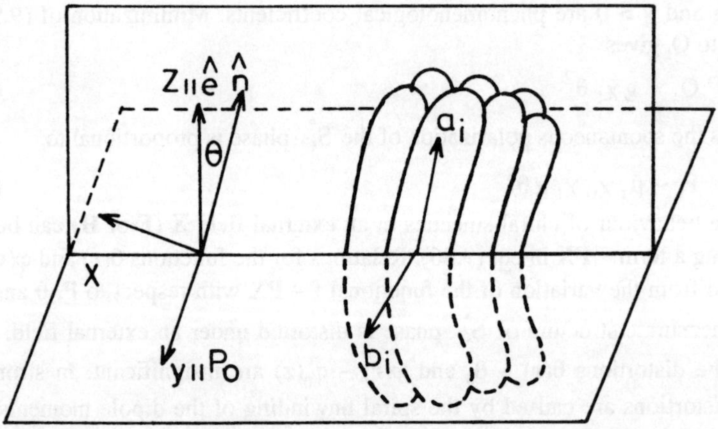

Fig. 9.19. Relative orientation of the director $\hat{\mathbf{n}}$, spontaneous polarization $\mathbf{P}_S$, and the smectic plane normal $\hat{\mathbf{e}}$ in the S_C^* phase. The unit vectors $\hat{\mathbf{a}}_i$ and $\hat{\mathbf{b}}_i$ specify the orientation of molecule i (Ref. 149).

In the theory of liquid crystals, the expression for the chiral interaction energies was first proposed by Goossens[154]. The most simple chiral interaction between molecules i and j can be written as

$$U^*(i,j) = - J^*(r_{ij})\,(\hat{\mathbf{a}}_i\,\hat{\mathbf{a}}_j)\,([\hat{\mathbf{a}}_i\,\hat{\mathbf{a}}_j]\hat{\mathbf{r}}_{ij}) \tag{9.102}$$

The interaction energy U^*, being a scalar, remains invariant under inversion. The pseudoscalar $([\hat{\mathbf{a}}_i\,\hat{\mathbf{a}}_j]\hat{\mathbf{r}}_{ij})$ changes sign under an inversion operation. The coupling constant J* also changes sign when the handedness of the molecules are changed. The interaction potential (9.102) has been widely used in the description of helicoidal twisting in cholesterics and S_C^* phases[153–156]. However, it is obvious that this potential, being cylindrically symmetric, can not cause the ferroelectric ordering.

For describing the ferroelectric ordering a model potential must depend on the direction of the short molecular axis $\hat{\mathbf{b}}_i$, i.e., the direction of the transverse dipole. The most simple interaction potential between chiral and polar molecules can be written as

$$U^*(i,j) = U_{ij} + U_{ji}$$

with

$$U_{ij} = - U_{ij}^*(r_{ij})\,(\hat{\mathbf{a}}_i\,\hat{\mathbf{a}}_j)\,(\hat{\mathbf{a}}_i\,\hat{\mathbf{r}}_{ij})\,([\hat{\mathbf{b}}_i\,\hat{\mathbf{a}}_j]\hat{\mathbf{r}}_{ij}) \tag{9.103}$$

The potential (9.103) is sensitive to the direction of the molecular transverse dipole. Osipov and Pikin[158] used this potential in the description of ferroelectric ordering. Considering, for simplicity, a S_C^* phase with a perfect orientational and translational order the internal energy is given by

$$\Delta U = \frac{1}{2}\sum_{i,j} \left\langle U^*(i,j)\right\rangle_{r_{ij},\hat{\mathbf{b}}_i} \tag{9.104}$$

Assuming that $U^*(i,j)$ is short-range and considering only interactions between nearest neighbours, the average interaction (9.104) can be expressed as

$$\Delta U = U_0^*\,\rho^2\left(\frac{3}{2}\sigma - 1\right)(\hat{\mathbf{n}}\,\hat{\mathbf{e}})\,([\hat{\mathbf{n}}\,\hat{\mathbf{e}}]\hat{\mathbf{b}}_i) \tag{9.105}$$

Here ρ is the number density, σ is a fraction of the nearest neighbours located in the same plane as the central molecule, and

$$U_0^* = \int [U_{ij}^*(r_{ij}) + U_{ji}^*(r_{ij})]\,r_{ij}^2\,dr_{ij} \tag{9.106}$$

If the short axes $\hat{\mathbf{b}}_i$ is parallel to the transverse molecular dipole $\mathbf{d}_\perp$, the spontaneous polarization can be expressed as

$$\mathbf{P}_s = \rho \mathbf{d}_\perp < \hat{\mathbf{b}}_i > \qquad (9.107)$$

With (1.107), we obtain

$$\Delta U = -\mu(\hat{\mathbf{n}}\,\hat{\mathbf{e}})\ ([\hat{\mathbf{n}}\,\hat{\mathbf{e}}]\,\mathbf{P})$$

where

$$\mu = U_0^* \rho\, d_\perp^{-1} \qquad (9.108)$$

Expression (9.105) has the same form as that of the piezoelectric term in the Landau free-energy expansion (9.86). This piezoelectric term is responsible for the spontaneous polarization

$$\mathbf{P}_s = -\mu\chi[\hat{\mathbf{n}}\,\hat{\mathbf{e}}] \qquad (9.109)$$

where χ be the dielectric susceptibility.
Using a single particle potential

$$U(\phi) = -a_1\,\theta\cos\varphi - a_2\,\theta^2\cos 2\varphi \qquad (9.110)$$

for the rotation of the molecule around its long axis. $\tilde{Z}$ek$\tilde{s}$ [153] studied the influence of molecular tilt on the ordering of the transverse dipoles. Here φ determines the orientation of the transverse molecular dipole. The first term (eq. (9.110)) which is linear in the tilt induces polar ordering, $<\cos\varphi> = 0$, in the direction perpendicular to the tilt where $\varphi = 0$. The second quadratic term in θ is not of chiral character; for $a_2 < 0$ it gives bipolar (quadrupolar) ordering, $<\cos 2\varphi> \neq 0$, in the direction perpendicular to the tilt.

The potential (9.110) leads to a free-energy expansion in powers of polar, $<\cos\varphi>$, and quadrupolar, $<\cos 2\varphi>$, order parameter[152]

$$f_1 = \frac{1}{2}C_1P^2 + \frac{1}{4}C_2P^4 + \frac{1}{2}\tilde{C}_2Q^2 + \frac{1}{4}\tilde{C}_2Q^4 - a_1\theta P - a_2\theta^2 Q - a_3P^2Q \qquad (9.111)$$

where

$$Q = K_1 < \cos 2\varphi >$$

$$P = K_2 < \cos\varphi >$$

$$<\cos\varphi> = \frac{\int_0^{2\pi} \cos\varphi \exp(-U(\varphi)/k_BT)\,d\varphi}{\int_0^{2\pi} \exp(-U(\varphi)/k_BT)\,d\varphi}$$

Assuming $\tilde{C}_2 = 0$ and minimizing f_1 with respect to Q, one obtains

$$Q = K_1 < \cos 2\varphi > = \frac{1}{C_1} [a_2\theta^2 + a_3 < \cos \varphi >^2] \tag{9.112}$$

Thus the quadrupolar ordering is proportional to the square of the tilt and the square of the polar ordering. With eq. (9.112) one obtains from eq. (9.111) a biquadratic coupling between the tilt and the polarization $P^2\theta^2$. A θ^4 term is also obtained. Further, the quadrupolar ordering is found to be always much larger than the polar one

$$|< \cos 2\varphi >| \quad >> \quad < \cos \varphi >$$

Equation (9.111), with the additional elastic and Lifshitz terms, has the same form as the generalized Landau model (eq. (9.86)). However, the origin of the anomalously large biquadratic coupling term between the polarization and the tilt does not become clear. For the homogeneous case, the polarization is given by the solution of the equation

$$\left(\frac{1}{\in} - \Omega\theta^2\right) P + \eta P^3 - C\theta = 0 \tag{9.113}$$

Close to T_C, $\mathbf{P}/\theta = \in/C$, and far below T_C, $P/\theta = \left(\frac{\Omega}{\eta}\right)^{1/2}$.

The above discussion is in no way rigorous one and only demonstrates the main ideas. In order to describe a chiral interaction responsible for the ferroelectric ordering let us assume a schematic model of a chiral molecule as shown in Fig. 9.20(a). The chirality of molecule i is determined by the orientation of the substitution group with the long axis $\hat{\mathbf{o}}_i$ with respect to the long axis of the whole molecule $\hat{\mathbf{a}}_i$. When the vectors $\hat{\mathbf{a}}_{ij}$, $\hat{\mathbf{m}}_i$ and $\hat{\mathbf{o}}_i$ are coplanar, $([\hat{\mathbf{a}}_i\, \hat{\mathbf{m}}_i\,]\,\hat{\mathbf{o}}_i) \neq 0$, the molecule becomes chiral. The effective bend angle $\in << 1$. The molecule i possesses[159] a so called steric dipole $|\mathbf{s}_i| = \in D$ with D as molecular diameter.

For molecules possessing large transverse dipoles in the chiral part, the chiral interaction is represented by the induction interaction between this dipole and the polarizable core of the neighbouring molecule. The induction interaction energy averaged over the rotations of the molecule i about its long axis is given by

$$U^*_{ind}(\hat{\mathbf{o}}_i, \hat{\mathbf{a}}_i, \hat{\mathbf{r}}_{ij}) = J_1(i,j)[(\hat{\mathbf{o}}_i\hat{\mathbf{a}}_j - 3(\hat{\mathbf{o}}_i\hat{\mathbf{r}}_{ij})] \; ([\hat{\mathbf{o}}_i\hat{\mathbf{a}}_j]\,\hat{\mathbf{r}}_{ij}) \tag{9.114}$$

where

$$J_1(i,j) = 3r_{ij}^{-7}\, \Delta\alpha_j\, d^2_{\perp i}(\hat{\mathbf{o}}_i\hat{\mathbf{a}}_i) \; ([\hat{\mathbf{o}}_i\hat{\mathbf{a}}_i\,]\,\hat{\mathbf{m}}_i) \tag{9.114a}$$

Here $\Delta\alpha_j$ is the anisotropy of the polarizability of molecule j. However, this potential (9.114) cannot be directly used in the theory of ferroelectricity because it does not depend on the direction of the short molecular axis. So the polar asymmetry must be accounted. In Fig. 9.20(b) the anisotropic polarizability of molecule j is located at two effective centers o_{i1} and o_{j2} with point polarizabilities $\boldsymbol{\alpha}_{j1}$ and $\boldsymbol{\alpha}_{j2}$. The two induced dipoles make a small angle $\in << 1$. Now the chiral induction interaction between the molecules i and j can be expressed as

$$U^*(i,j) = U^*_{ij} + U^*_{ji} \tag{9.115}$$

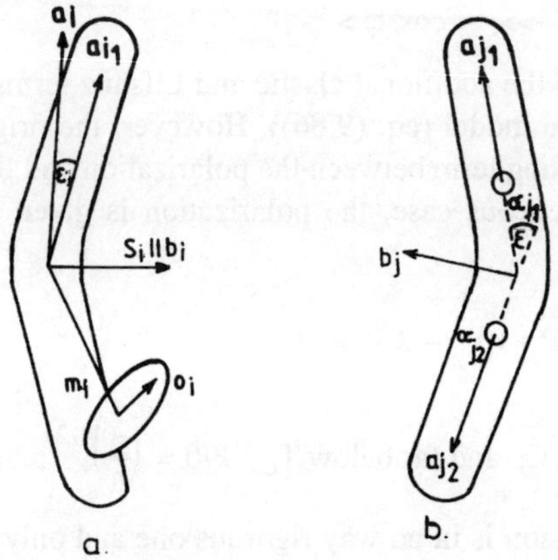

Fig. 9.20. (a) Model for a chiral molecule with a transverse steric dipole $\mathbf{s}_i$ and the substitution group with the long axis $\hat{\mathbf{o}}_i$, (b) model for an asymmetric molecule with two polariable centers (Ref. 149).

Here

$$U^*_{ij} = U^*_{ind}(\hat{\mathbf{o}}_i, \hat{\mathbf{a}}_{j1}, \mathbf{r}_{ij1}) + U^*_{ind}(\hat{\mathbf{o}}_i, \hat{\mathbf{a}}_{i1}, \mathbf{r}_{ij1})$$

with

$$\mathbf{r}_{ij\alpha} = \left|\mathbf{r}_{ij\alpha}\right| \hat{\mathbf{r}}_{ij\alpha} = \mathbf{r}_{ij} + \ell \hat{\mathbf{a}}_{j\alpha} \quad , \quad \alpha = 1,2$$

$$\boldsymbol{\alpha}_{j1,2} = \pm\, \hat{\mathbf{a}}_j + \in \hat{\mathbf{b}}_j$$

ℓ is the distance between two polarizable centers.

9.6.2.2. Free energy and ferroelectric ordering of S^*_C phase

In S^*_C phase the dipole ordering is very weak, and so the associated energy is small as compared to orientational interaction energy. Consequently, the perturbation expansion method can be employed to describe the ferroelectric ordering. The orientational singlet distribution function can be written as

$$f_1(i) = f_0(i)\,[1+g_1(i)] \qquad (9.116)$$

where $f_0(i)$ corresponds to the nonchiral S_C phase and the function $g_1(i)$ depending on the direction of the short molecular axis $\hat{\mathbf{b}}_i$ is small. Now the spontaneous polarization of the S_C^* phase can be expressed as

$$\mathbf{P} = \rho\,\mathbf{d}_\perp \langle \hat{\mathbf{b}}_i \rangle = \rho\,\mathbf{d}_\perp \int \hat{\mathbf{b}}_i\, f_0(i)\, g_1(i)\, \delta(\hat{\mathbf{a}}_i\, \hat{\mathbf{b}}_i)\, d^2\hat{\mathbf{a}}_i\, d\hat{\mathbf{b}}_i \qquad (9.117)$$

Here the short axis $\hat{\mathbf{b}}_i$ is taken in the direction of the transverse molecular dipole $\mathbf{d}_\perp$.

Adopting the usual perturbation expansion method the free-energy of the system can now be written in the form

$$A(r_i) = -\rho\, k_B T \int f_1(i) \ln f_1(i)\, \mathbf{d}(i) + \frac{1}{2}\rho^2 k_B T \int \delta(\xi_{ij} - r_{ij})\, f_1(i)\, f_1(j)\, d^3\mathbf{r}_{ij}\, \mathbf{d}(i)\, \mathbf{d}(j)$$
$$+\ \rho^2 \int \delta(\xi_{ij} - r_{ij})\, u_{att}(i,j)\, f_1(i)\, f_1(j)\, d^3\mathbf{r}_{ij}\, \mathbf{d}(i)\, \mathbf{d}(j) \qquad (9.118)$$

where $\mathbf{d}(i) = \delta(\hat{\mathbf{a}}_i\, \hat{\mathbf{b}}_i)\, d^2\hat{\mathbf{a}}_i\, d\hat{\mathbf{b}}_i$. ξ_{ij} is the closest distance of approach between molecules i and j, and $\delta(\xi_{ij} - r_{ij})$ determines the excluded volume for the centers of two molecules,

$$\delta(\xi_{ij} - r_{ij}) = 0 \qquad , \text{if } r_{ij} < \xi_{ij}.$$
$$= I \qquad , \text{if } r_{ij} > \xi_{ij}.$$

The first two terms of eq. (9.118) are, respectively, the mixing entropy and the packing entropy; the last term is the internal energy.

Minimization of free energy, subjected to the normalization conditions for $f_1(i)$ and $f_0(i)$, determines the values of distribution functions. The function $g_1(i)$ is to be determined by the chiral part of the attractive and repulsive interactions. From symmetry considerations, it is obvious that $\mathbf{P}_S$ vanishes if u_{att} and $\delta(\xi_{ij} - r_{ij})$ are odd in $\hat{\mathbf{b}}_i$, i.e., in case of nonpolar molecules in transverse direction. Thus in S_C^* phase the dipolar ordering is determined by the chiral and polar terms of interaction potential. Further, the experimental evidences show that the polarization in S_C^* phase is more sensitive to the dipole than the molecular shape. Therefore, it can be assumed that the attractive interaction is spherically symmetric $u_{att}(i,j) = u_{att}(r_{ij})$, the chirality of the molecular shape is negligible and only the chiral part of the dispersion interaction u^*_{att} contributes to the free-energy

$$\mathbf{P}_s = -\beta\rho^2 \mathbf{d}_\perp \int \hat{\mathbf{b}}_i\, \delta(\xi_{ij} - r_{ij})\, u^*_{att}(i,j)\, d^3\mathbf{r}_{ij}\, \mathbf{d}(i)\, \mathbf{d}(j) \qquad (9.119)$$

This equation shows that the spontaneous polarization can be determined from the chiral interaction u^*_{att} modulated by the anisotropic repulsion. The existence of $\mathbf{P}_s$

requires that the product $\delta(\xi_{ij}-r_{ij})\,u^*_{att}$ must be odd in $\hat{\mathbf{b}}_i$. Now ξ_{ij} can be approximated as

$$\xi_{ij} = \xi^0_{ij} + [(\mathbf{S}_i\,\hat{\mathbf{r}}_{ij}) + (\mathbf{S}_j\,\hat{\mathbf{r}}_{ij})]\left(\frac{D}{4L}\right) \tag{9.120}$$

where $\mathbf{S}_i\ (= D \in \hat{\mathbf{S}}_i)$ is the transverse steric dipole of molecule i and $\in << 1$ is the effective bend angle. ξ^0_{ij} corresponds to the molecules without steric dipoles, i.e., $\xi_{ij} \to \xi^0_{ij}$ if $\in \to 0$.

Considering a simplified case of an ideal orientational and transitional order and thus assuming that the influence of the variation in orientational and translational order on the spontaneous polarization can be neglected, one obtains

$$\mathbf{P}_s = -\beta\rho^2 \mathbf{d}_\perp\,[(I-\sigma)\int W(\hat{\mathbf{n}},\hat{\mathbf{r}}_{ij})\ \delta(\hat{\mathbf{r}}_{ij}\,\hat{\mathbf{e}})]\ d^2\,\hat{\mathbf{r}}_{ij} + \sigma\,W(\hat{\mathbf{n}},\hat{\mathbf{e}}) \tag{9.121}$$

where the effective potential is given by

$$W(\hat{\mathbf{n}},\hat{\mathbf{r}}_{ij}) = \int \hat{\mathbf{b}}_i \Big\{\int \delta(\xi^0_{ij} - r_{ij})\ u^*_{att}(i,j)\ r^2_{ij}\ d\mathbf{r}_{ij} + \left(\frac{D}{4L}\right) u^*_{att}(i,j)\Big|_{r_{ij}=} (\xi^0_{ij})^2\,[(\mathbf{S}_i\,\hat{\mathbf{r}}_{ij}) + (\mathbf{S}_j\,\hat{\mathbf{r}}_{ij})]\Big\} d\,\hat{\mathbf{b}}_i\ d\,\hat{\mathbf{b}}_j \tag{9.122}$$

Here σ is the fraction of nearest neighbour of a given molecule located in the same place. In eq. (9.122) the first term is the contribution from the interaction of molecules located in the same smectic plane, whereas the second one refers to the molecules located in the adjacent smectic layers. The effective potential $W(\hat{\mathbf{n}},\hat{\mathbf{r}}_{ij})$ which determines the spontaneous polarization has two terms. The first term gives the contribution from the asymmetric polar part of the u^*_{att} modulated by the uniaxial molecular shape, while the second term is a contribution from the symmetric part of the u^*_{att} modulated by the asymmetric molecular shape.

The influence of chiral interactions on the spontaneous polarization in S^*_C phase has been explained using two different molecular models[149,159]. Osipov[159] considered a model of a chiral molecule with a large dipole in the chiral part. In this case predominantly the chiral attraction is represented by the induction interaction between the dipole $\mathbf{d}_i$ of the molecule i and the polarizability of the neighbouring molecule j. The simplest chiral induction interaction contributing to the spontaneous polarization is the dipole-dipole, dipole-quadrupole induction interaction,

$$U_{ddq}(i,j) = U_{ij} + U_{ji} \tag{9.123}$$

with

$$U_{ij} = r_{ij}^{-7} \sum_{nj} (E_{oj} - E_{nj})^{-1} \langle o_j | u_{dq}(i,j) | o'_j \rangle \langle o'_j | u_{dd}(i,j) | o_j \rangle \tag{9.124}$$

Here $\langle o_j |$ and $\langle o'_j |$ represent the ground state and the excited state of molecule j, respectively, and $(E_{oj} - E_{nj})$ is the excitation energy of the molecule. The dipole-dipole, and dipole-quadrupole interaction potentials are given by

$$u_{dd}(i,j) = \int e(\boldsymbol{\rho}_j)[3(\boldsymbol{\rho}_j\, \hat{\mathbf{r}}_{ij})(\mathbf{d}_i\, \hat{\mathbf{r}}_{ij}) - (\boldsymbol{\rho}_j\, \mathbf{d}_i)]\, d^3\boldsymbol{\rho}_j \tag{9.125}$$

$$u_{dq}(i,j) = \int e(\boldsymbol{\rho}_j)\left[\frac{3}{2}\, d_i^2(\boldsymbol{\rho}_j\, \hat{\mathbf{r}}_{ij}) + 3(\mathbf{d}_i\, \boldsymbol{\rho}_j)(\mathbf{d}_i\, \hat{\mathbf{r}}_{ij}) - \frac{15}{2}(\mathbf{d}_i\, \hat{\mathbf{r}}_{ij})^2 (\boldsymbol{\rho}_j\, \hat{\mathbf{r}}_{ij}) - \frac{3}{2}(\boldsymbol{\rho}_j)^2 (\mathbf{d}_i\, \hat{\mathbf{r}}_{ij})^2 - 3(\mathbf{d}_i\, \boldsymbol{\rho}_j)(\boldsymbol{\rho}_j\, \hat{\mathbf{r}}_{ij}) + \frac{15}{2}(\boldsymbol{\rho}_j\, \hat{\mathbf{r}}_{ij})^2 (\mathbf{d}_i\, \hat{\mathbf{r}}_{ij})\right] d^3\boldsymbol{\rho}_j \tag{9.126}$$

where $e(\boldsymbol{\rho}_j)$ is the charge density of molecule j, and $\boldsymbol{\rho}_j$ is the vector pointing from the center of mass of molecule j to the given point.

Substituting eq. (9.123) into (9.121), the spontaneous polarization of S_C^* phase can be written as

$$\mathbf{P}_s = \mu(\hat{\mathbf{n}}\, \hat{\mathbf{e}})\, [\hat{\mathbf{n}}\, \hat{\mathbf{e}}] \tag{9.127}$$

where

$$\mu = -\frac{1}{4}\beta\rho^2(\mathbf{Sd})\, \Delta(D^6/L)\;\; \sigma(2\chi_\perp + 15\Delta\chi + \Delta\chi_\perp) \tag{9.128}$$

with

$$\Delta = (\mathbf{d}_i\, \hat{\mathbf{a}}_i)\, ([\mathbf{d}_i\, \hat{\mathbf{m}}_i]\, \hat{\mathbf{a}}_i) \tag{9.129}$$

Here $\chi_\perp (= \chi_{xx} + \chi_{yy})$ is the transverse molecular polarizability, and $\Delta\chi_\perp (= \chi_{xx} - \chi_{yy})$ is anisotropy of the transverse polarizability.

It is interesting to note that the spontaneous polarization (9.127) is proportional to σ, the fraction of nearest neighbours located in the same smectic plane. Also $\mathbf{P}_s$ depends on the angle between the steric dipole **S** and the electric dipole **d** of the same molecule. Thus this model[159] indicates that the ferroelectric ordering in S_C^* phase of molecules with large steric dipoles is mainly due to the induction interaction between the chiral group and the polarizable core of the neighbouring molecules modulated by short-range steric forces.

In case of molecules with no steric dipoles, the ferroelectric ordering can be caused owing to the polar part of the chiral dispersive induction interaction. The spontaneous polarization has been explained considering the chiral attraction which changes sign under a 180^0 rotation of the molecular coordinate frame about the long axis $\hat{\mathbf{a}}_i$. The simplest chiral and polar dispersion interaction potential is given by

$$u_{dq}(ij) = u_{dqj} + u_{dqi} \tag{9.130}$$

with

$$u_{dqi} = r_{ij}^{-8} \sum_{ki} (E_{oi} - E_{ki})^{-1} \left\langle o_i \middle| u_{dq}(i,j) \middle| k_i \right\rangle \left\langle k_i \middle| u_{dq}(i,j) \middle| o_i \right\rangle \tag{9.131}$$

where $u_{dq}(i,j)$ is given by eq. (9.126). Now with eq. (9.124) the spontaneous polarization is given by

$$\mathbf{P}_s = \mu\,(\hat{\mathbf{n}}\,\hat{\mathbf{e}})\,[\hat{\mathbf{n}}\,\hat{\mathbf{e}}] \tag{9.132}$$

with

$$\mu = -0.45(\beta\rho^2)\,\Delta D^{-5}\sigma(Q_{xx} + Q_{yy})\,\chi_\perp \tag{9.133}$$

where Q_{xx} and Q_{yy} are the components of the molecular quadrupole. Here μ is determined by the interaction between the permanent quadrupole of a given molecule and the induced dipoles of the nearest neighbour.

Thus in both the models the spontaneous polarization is proportional to the pseudoscalar parameter Δ which characterizes the chirality of a molecule with a substitution group. Δ measures the variation of $\mathbf{P}_S$ with a change in position and orientation of the chiral group with respect to the molecular hard core. The polarization also depends on the molecular dipole **d**; $\mathbf{P}_S \sim d^3$ within the model of molecules with steric dipoles.

9.7. THEORY FOR THE ANTIFERROELECTRIC SUBPHASES

Based on the Ising model and XY model theoretical methods have been developed for investigating the structure of antiferroelectric subphases. In the Ising model, molecules locally lay on a plane and are allowed to tilt left or right. The XY model considers that molecules are distributed at any azimuthal angles on a cone. The key idea is that the dipole-dipole interaction or steric interaction stabilizes an antiferroelectric structure, while the excluded volume effect stabilizes a ferroelectric structure.

One-dimensional Ising model with long-rang repulsive interaction has been considered by Bak and Bruinsma[160,161]. This work suggests that the ferroelectric ordering is excited in the antiferroelectric ordering due to the competing interaction of ferroelectricity and antiferroelectricity, and that a long-range repulsive interaction between the ferroelectric orderings is responsible for the uniform distribution of ferroelectrically ordered positions[162]. The model has been successful in describing the structures of major subphases $S^*_{C\gamma}$ and AF (reentrant antiferroelectric phase).

A modified ANNNI + J_3 model, i.e., axial next-nearest neighbour Ising model with the third neighbouring interaction, has been adopted by Yamashita and Miyazima[163] and Yamashita[164]. This model writes the Hamiltonian as

$$H = -J \sum_{(i,j)} s_i s_j - J_1 \sum_i s_i s_{i+1} - J_2 \sum_i s_i s_{i+2} - J_3 \sum_i s_i s_{i+3} \tag{9.134}$$

where the Ising spin assumes values of ± 1 referring to the molecular tilting senses of the ith smectic layer. The first term considers the summation over nearest-neighbouring pairs (i,j) in the same smectic layer , while the other summations correspond to the first, second, and third neighbouring pairs in the axial direction parallel to the layer normal. It is required that J_2 should be negative to ensure competition, and J_3 may be positive or negative. Yashamita[165] also analysed the role of the directional sense of the molecular long axis and gave justification for the use of these long-range interactions. He has showed that four ground state appear (see Fig. 9.21) – S^*_{CA} $(q = \frac{1}{2})$, $S^*_{C\gamma}$ $(q = \frac{1}{3})$, AF$(q = \frac{1}{4})$ and S^*_C $(q = 0)$.

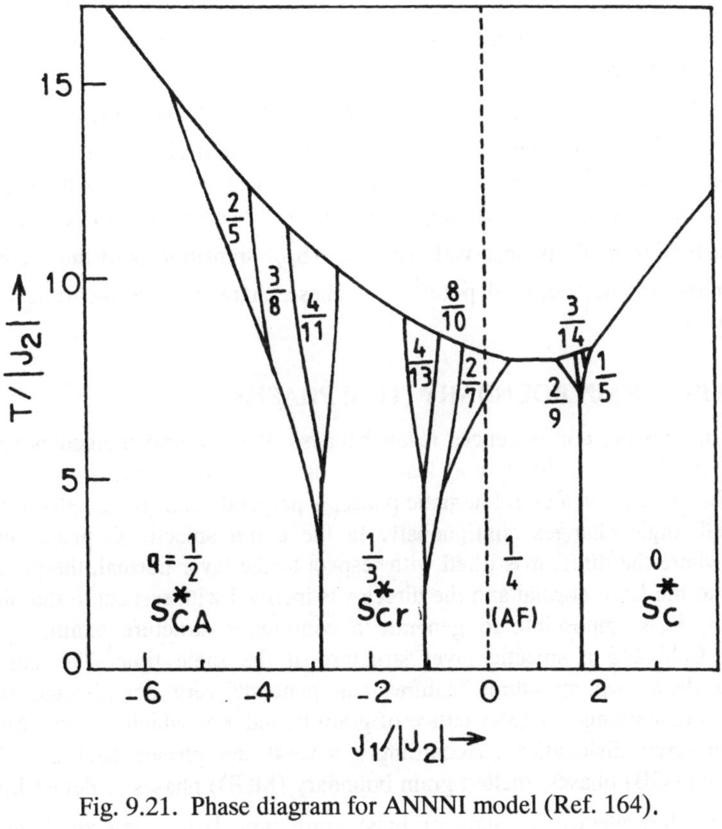

Fig. 9.21. Phase diagram for ANNNI model (Ref. 164).

The application of phenomenological Landau model has been considered by several authors by incorporating the coupling of ferroelectric and antiferroelectric order parameters in bilayer[166-170] and ANNN interaction[171,172]. In the bilayer models no layer spacing change along the layer normal has been observed and hence they are not suitable for the $S^*_{C\gamma}$ phase. Using the ANNNXY model (XY character introduced in the ANNN interaction), Roy and Modhusudana[172a] calculated the phase sequence and observed the sequence $S_A - S^*_{C\alpha} - S^*_C - FI_H - FI_I\ (S^*_{C\gamma}) - FI_L - S^*_{CA}$. These authors also calculated[172b] the molecular orientation under external electric field. From a resonant X-ray scattering using the free-standing films of sulfur containing antiferroelectric mesogens possessing subphases, Mach et. al.[174] have concluded that the clock model[171–173] is successful in explaining the experimental result. The helical structures with a very short pitch consisting of several layers result from the XY model, while the experimental observation[175] gives macroscopic helical in the S^*_{CA} and other subphases. So the XY model, particularly the clock model, is not a realistic model for various subphases characterized by macroscopic helices.

The $S^*_{C\alpha}$ phase has been characterized by several experiments[176–183]; it is a tilted phase[176,177] and within the phase several structural changes with temperature and applied electric field occur[178,179]. The helical structure with either very short[177,180] or very long[181] pitch may occur. In the dynamic light scattering[182], a very slow fluctuation mode is observed. The $S^*_A - S^*_{C\alpha}$ transition is of the second order and is close to the tricritical point[177,183]. These characteristics theoretically is not understood.

9.8. TWIST GRAIN BOUNDARY (TGB) PHASES

Several interesting consequences result because of the chiral molecules forming a liquid crystal phase. Chirality may lead to an intrinsic helical structure of the director field. The director, in a chiral nematic phase, is perpendicular to the helix axis and its azimuthal angle changes continuously. In the chiral smectic C phase and those phases where the director is tilted with respect to the layer normal, the pitch axis is parallel to the layer normal and the director is inclined with respect to the pitch axis. However, it is impossible to generate a continuous structure exhibiting both a director field and a smectic layer structure at the same time. The competition between these two structural features can generate very complicated frustrated structures containing a regular lattice of grain boundaries which in turn consist of a lattice of screw dislocations. Accordingly, several new phases, such as twist grain boundary (TGB) phases, melted grain boundary (MGB) phases, a defect line liquid (N^*_L), antiferroelectric crystals of twist grain boundaries, smectic blue phases, smectic-Q phase, ..., have been predicted and/or observed. We give here a very brief account of these phases. More details can be found in some excellent reviews[184–190] which are focused on different aspects.

The twist grain boundary (TGB) phases have been known[191] since 1988; their structures are being investigated extensively at present. The structure of a TGB phase is characterized by the distance ℓ_d between dislocations in a layer, the distance ℓ_b between grain boundary, and the separation d between smectic layers. Usually the TGB phases[191–195] occur in the temperature range between a N^* phase with short pitch and a smectic phase, such as S_A or S_C^*. These phases are expected to appear close to a $N^* - S_A - S_C^*$ triple point[195]. They exhibit unique properties – the selective reflection of circularly polarized light, and the formation of layered structures as evident from the X-ray investigations. The former property shows that the director field possesses a helical structure similar to the N^* phase, while the layered structure indicates a smectic – like behaviour[196]. The combination of the N* and smectic properties can be clearly seen from the textures observation of TGB phase in a polarizing microscope. A N^*-like Grandjean texture appears for the parallel alignment of both the substrates, while a dark area reminiscent of a S_A or S_C phase is seen at the same temperature at locations of homeotropic alignment. However, a TGB phase can be distinguished from the other phases by the filaments appearing in the latter structure. In a material, if the smectic phases occur at lower temperatures, the pitch of the TGB phases usually increases with decreasing temperature[196–198]. Contrary to it, in materials exhibiting reentrant N* phase[199–202], the TGB phases occur in the temperature range below a smectic phase and a reversal of pitch dependence is observed, i.e., the pitch increases with increasing temperature.

The structure of TGB_A phase[191,196,202] consists of smectic slabs separated by the defect walls; neighbouring slabs are tilted with respect to each other leading to a helical structure. A very convincing evidence for the existence of smectic slabs is given by the electron microscopy of freeze fractures[203]. The order within each slab is equivalent to the S_A phase. The molecules are uniaxially aligned along the local director $\hat{\mathbf{n}}$ and arranged in layers. The layer normal $\mathbf{Q}$ is parallel to $\hat{\mathbf{n}}$. The layer spacing d $(= 2\pi/|\mathbf{Q}|)$ is in the range of a few nm. The slabs have a typical thickness ℓ_b of some 10 nm. The grain boundaries between the slabs are defect walls consisting of parallel defect lines (twist dislocations). The distance ℓ_d between the defect lines is similar to the slab thickness ℓ_b in TGB_A. The director $\hat{\mathbf{n}}$ forms a helical structure due to the tilt between the neighbouring slabs. The helix axis is perpendicular to the director and the pitch is in the range from a few hundred nm to a few μm. The director field of TGB_A phase is very similar to the director field of a N^* phase. Thus, when the wavelength of light is of the same order of magnitude as the pitch, a TGB_A phase with parallel surface alignment of the director shows selective reflection of circularly polarized light. Similar to that of N^* phase, the Bragg wavelength is given by

$$\lambda_0 = np\cos\theta$$

where θ is the angle between the pitch axis and the direction of light propagation. In spite of the similarities, there is a marked difference between the director fields of the N^* and TGB_A phases; a continuous twist is seen in the director field of N^* phase, whereas the TGB structure behaves optically like a stack of discrete birefringent blocks of finite thickness having a uniform orientation of the optic axis. The high resolution X-ray diffraction studies on well aligned samples show that the structure factor of the TGB phase has a cylindrical shape[191,196]. If the helical axis $\hat{\mathbf{h}}$ is oriented along the X-axis, the diffraction pattern in the (y,z)-plane is a ring. The intensity profile of this ring along the radial direction is a δ-function which is an indication of a long-range smectic order along the director. However, in principle, the long-range translational order can be destroyed due to positional fluctuations, resulting in a algebraic singularities, if α is irrational. The available X-ray data seem to indicate that both commensurate and incommensurate TGB_A phases can occur; an incommensurate phase has been found in 14P1M7[196], while 10BTF2O1M7[202] exhibits a commensurate phase. The latter compound shows a diffraction ring consisting of 40 to 60 distinct spots; the number of spots decrease with increasing temperature.

Several systems exhibit different behaviour under high pressure[204–206]. In 14P1M7 the transitions ($S_C^* - TGB_A$, TGB_A – IL) were found to be shifted to lower temprature with increasing pressure. The same effect was observed for the S_C^* – TGB_A transition is 12FBTFMO1M7. Above certain critical values of the pressure the temperature range of the existence of the TGB_A phase becomes narrower. At a very high pressure the TGB_A phase disappears. A binary mixture of CE8 and CI2 exhibits TGB phases close to a virtual $N^* - S_A - S_C^*$ triple point. It was observed[205] that the appearance of the TGB phase between S_A and N^* can be induced by pressure.

The TGB_C phases[207–210] consist of a helical arrangement of smectic slabs like the TGB_A phase, but the director $\hat{\mathbf{n}}$ is tilted with respect to the layer normal. The tilt direction becomes an additional parameter suggesting the possibility of different structures, such as TGB_{C_p}, TGB_{C_t}, TGB_{C^*}, etc. In the TGB_{C_p} phase both the director $\hat{\mathbf{n}}$ and the layer normal **Q** are perpendicular to the pitch axis; this phase corresponds to a true ferroelectric phase. The director of the TGB_{C_t} phase is perpendicular to the helix axis $\hat{\mathbf{h}}$, but the smectic layers are tilted within the plane defined by $\hat{\mathbf{n}}$ and $\hat{\mathbf{h}}$; $\mathbf{P}_S$ is perpendicular to $\hat{\mathbf{n}}$, and shows a helical arrangement so that the structure exhibits helielectric properties. The TGB_{C_t} structure was first proposed by Dozov and Durand[211] and called the Melted Grain Boundary (MGB) phase. TGB_{2q} slabs, first predicted by Luk'yanchuk[212], is a superposition of two equivalent S_C^* populations with different tilt directions of the layer normal with respect to the pitch axis. The TGB_{C^*} phase[213–215], initially predicted by Renn[213], shows a twisted director field within each smectic slab. Consequently, there is a superposition of N^*-like and S_C^*-like helical structures with the two local pitch axes perpendicular to each other; the orientation of the smectic pitch axis changes from the slab to slab.

Theoretical predictions suggest that the existence of TGB phases requires a high optical purity and the molecular abilities to generate a small N^* pitch as well as a smectic layer structure. It is evident from the phase diagrams of chiral-racemic mixtures (Fig. 9.22) that the TGB phases occur only for sufficiently high enantiomeric excess[216–218]. However, a number of studies on the structure-property correlations show that the balance of different properties is rather difficult. Goodby[219] has pointed out that the formation of a smectic layer structure is facilitated by an extended molecular core of three or more benzene rings, whereas the occurrence of a twisted structure with small pitch is supported by a zig-zag molecular structure in combination with an internal molecular twist.

The existence of a TGB_A phase was proved first[193] in the homologous series of (R) or (S)-1-methylheptyl-4′-(4″-n-alkoxyphenyl-propioloyloxy) biphenyl-4-carboxylates (1M7nOPPBC). An S_A – IL transition was observed in the homologues with short chain length $n \leq 12$, while for $n \geq 16$ an S_C^* – IL transition was found. The TGB phase occurs for the compounds with n = 13, 14 or 15 just below the clearing point. The TGB_A phase appears close to virtual S_A – S_C^* – IL meeting point in an (n,T) phase diagram of this homologous series. A low value of the enthalpy (~ 1 kJ/mol) at the clearling point in the homologues shows the occurrence of a TGB phase, whereas a broad nonsingular DSC peaks indicate the existence of a further new phase in the isotropic region, called as "isotropic flux phase"[220]. This additional phase is possibly identical with the twisted line liquid N_L^* [221]. The racemic mixtures of these chiral compounds exhibit an S_A phase and an S_C phase; neither TGB nor N_L^* phase occurs[216]. The predicted phase sequence S_A – TGB_A – N* has been observed in several mixtures and pure compounds, such as mixture of cholesteryl nonanoate and nonyloxybenzoic acid[222], chiral derivatives of 4-biphenylbenzoate[216]. A chloro substitution in the chiral chain of phenylpropiolate derivatives[120] generates a compound with the sequence S_A – TGB_A – N* – IL. It is speculated that the carboxyl group in the core which is present in many TGB compounds plays a significant role by bending the overall structure of the molecule.

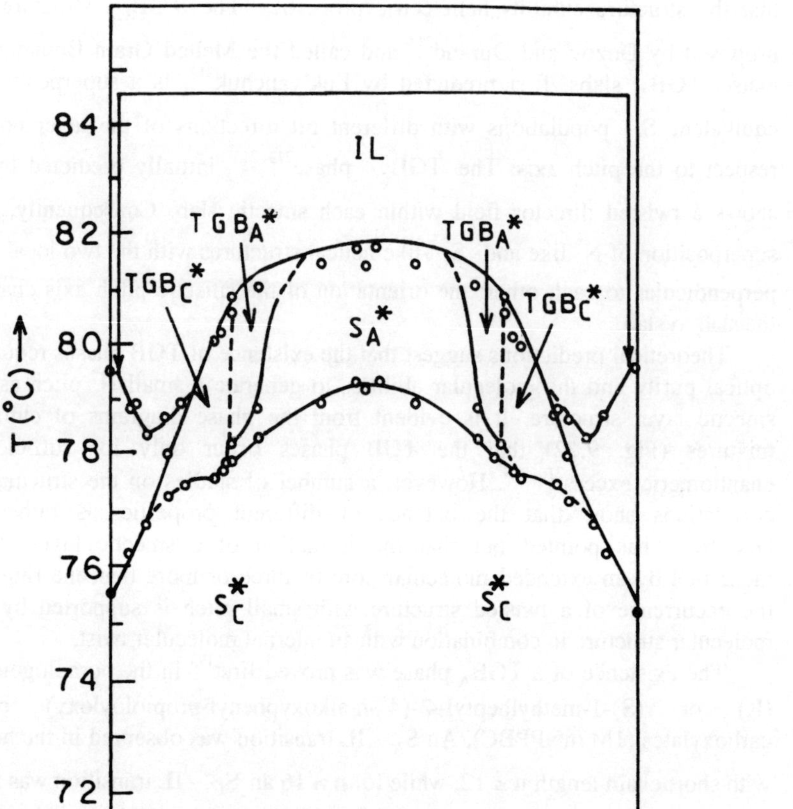

Fig. 9.22. Chiral racemic mixture phase diagram between transition temperature and the concentration ratio of two enantiomers (Ref. 218).

The TGB_C phase was observed for the first time in the homologous series of 3-fluoro-4-[(R) or (S)-1-methylheptyloxy]-4′-(4″-alkoxy-2″,3″-difluorobenzoyloxy) tolanes ($nF_2BTFO_1M_7$)[208]. Further, calorimetric studies[223] have shown that for some homologs of this series even two different TGB_C phases occur. Variation of the number of the lateral fluorine atoms and their positions generates a series of compounds[184,197,224] which exhibits both TGB_C and TGB_A phases. It has been suggested that a transverse dipole moment at the linking group is needed to stabilize the S_A phase with respect to the N^* phase, and an additional electron attracting effect of the chiral group favours the generation of TGB phases[208,197]. The steric hindrance due to lateral fluorine atoms supports, in general, the formation of tilted phases and thus the occurrence of TGB_C phases.

The occurrence of TGB phases is very sensitive to the molecular structure; even small variations in the molecular structure influence the generation of these phases. In spite of this sensitive dependence these mesophases have been observed in several liquid crystals (see Ref. 184). These phases are found in pure compounds as well as in a large variety of mixtures, including systems of a nonchiral liquid crystal and a non-mesogenic chiral dopants[225–228]. However, TGB phases occur only in a narrow temperature interval of a few degree centrigrade.

REFERENCES

1. P.J. Collings, and M. Hird, Introduction to Liquid Crystals : Chemistry and Physics (Taylor and Francis Ltd., 1997) Chapt. 6 , p. 111 and Chapt. 8, p. 147.
2. A.W. Hall, J. Hollingshurst, and J.W. Goodby, Chiral and Achiral Calamitic Liquid Crystals for Display Applications, in, Handbook of Liquid Crystal Research, eds., P.J. Collings, and J.S. Patel (Oxford Univ. Press, New York, Oxford, 1997) Chapt. 2, p. 17.
3. J.W. Goodby, R. Blinc, N.A. Clark, S.T. Lagerwall, M.A. Osipov, S.A. Pikin, T. Sakurai, K. Yoshino, and B. Žekš, eds., Ferroelectric Liquid Crystals : Principles, Properties and Applications (Gordon and Breach Science Publishers, 1999).
4. H.S. Kitzerow, and Ch. Bahr, eds., Chirality in Liquid Crystals (Springer-Verlag, New York, 2001).
5. B. Bahadur, ed., Liquid Crystals : Applications and Uses (World Scientific, Vol. 1, 1990; Vol. 2, 1991; Vol. 3, 1992).
6. I.C. Khoo, and S.T. Wu, Optics and Nonlinear Optics of Liquid Crystals (World Scientific, 1993).
7. D.A. Dunmur, Physical Origin of Liquid Crystal Optical Properties, in, The Optics of Thermotropic Liquid Crystals, eds., S. Elston, and R. Sambles (Taylor and Francis Ltd., 1998) Chapt. 1, p. 5.
8. (a) G.W. Gray, and D.G. McDonnell, Mol. Cryst. Liquid Cryst. **34**, 211 (1977);

   (b) D.G. McDonnell, in, Thermotropic Liquid Crystals, ed., G.W. Gray (Chichester, Wiley, 1987) p. 120.

9. (a) M.C. Mauguin, Bull. Soc. Franc. Miner. Crist. **34**, 71 (1911); (b) C.W. Oseen, Trans. Faraday Soc. **29**, 883 (1933).
10. H. de Vries, Acta. Cryst. **4**, 219 (1951).
11. S. Chandrasekhar, Liquid Crystals (Cambridge Univ. Press, 1992) Chapt. 4, p. 213.
12. P.G. de Gennes, and J. Prost, The Physics of Liquid Crystals (Oxford Univ. Press, 1993) Chapt. 6, p. 263.
13. H.L. Ong, Phys. Rev. **A37**, 3520 (1988).
14. J.P. Hurault, J. Chem. Phys. **59**, 2068 (1973).
15. A. Mochizuki, H. Gondo, T. Watanula, K. Saito, K. Ikegami, and H. Okuyama, SID Technical Digest **16**, 135 (1985).
16. C.G. Lin-Hendel, Appl. Phys. Lett. **38**, 615 (1981); J. Appl. Phys. **53**, 916 (1982).
17. J.T. Jenkins, Unpublished D. Phil. Thesis, Johns Hopkins University (1969).
18. J. Fergason, in, Liquid Crystals, Proceedings of The Second Kent Conference, ed., G. Brown, G. Dienes, and M. Labes, (Gordon and Breach Science Publishers, New York, 1966) p. 89.
19. P.N. Keating, Mol. Cryst. Liquid Cryst. **8**, 315 (1969).
20. J. Adams, W. Haas, and J. Wysocki, Liquid Crystals and Ordered Fluids, ed., R.S. Porter, and J.F. Johnson (Plenum Press, New York, 1970) p. 463.
21. T. Nakagiri, H. Kodama, and K.K. Kobayachi, Phys. Rev. Lett. **27**, 564 (1971).
22. J. Adams, W. Haas, and J. Wysocki, Mol. Cryst. Liquid Cryst. **7**, 371 (1969).
23. P. Pollmann, and H. Stegemeyer, Chem. Phys. Lett. **20**, 87 (1973).
24. Ref. 6, Chapt. 2, p. 100.
25. L.M. Blinov, Electric Field Effects in Liquid Crystals, in, Handbook of Liquid Crystal Research, eds., P.J. Collings, and J.S. Patel (Oxford Univ. Press, New York, Oxford, 1997) Chapt. 5, p. 125.
26. E. Sackmann, S. Meiboom, and L.C. Snyder, J. Am. Chem. Soc. **89**, 5892 (1967).
27. M. Kleman, Points, Lines and Walls (John Wiley and Sons, 1983) Chapter 4, p. 77.
28. R.B. Meyer, Appl. Phys. Lett. **14**, 208 (1969).
29. P.G. de Gennes, Solid State Commun. **6**, 163 (1968).
30. R.B. Meyer, Appl. Phys. Lett. **12**, 281 (1968).
31. F.M. Leslie, Mol. Cryst. Liquid Cryst. **12**, 57 (1970).

32. W. Helfrich, J. Chem. Phys. **55**, 839 (1971).

33. C. Gerritsma, and P. Van Zanten, Phys. Lett. **A37**, 47 (1971).

34. F. Rondelez, and J.P. Hulin, Solid State Commun. **10**, 1009 (1972).

35. F. Rondelez, and H. Arnould, C.R. Acad. Sci. **B273**, 549 (1971); T.J. Scheffer, Phys. Rev. Lett. **28**, 593 (1972).

36. G.H. Heilmeier, and J.E. Goldmacher, Appl. Phys. Lett., **13**, 132 (1968).

37. (a) See, L.D. Landau, and E.M. Lifshitz, Fluid Mechanics, (Pergamon, London, 1952) Chapt. 2, p. 17.

(b) H. Arnould, and F. Rondelez, Mol. Cryst. Liquid Cryst. **26**, 11 (1974).

38. P. Pieranski, Classroom Experiments with Chiral Liquid Crystals, in, Chirality in Liquid Crystals, eds., H.S. Kitzerow, and Ch. Bahr (Springer–Verlag, 2001) Chapt. 2, p. 28.

39. K. Bergmann, and H. Stegemeyer, Bur. Bunsenges Phys. Chem. **82**, 1309 (1978).

40. H. Stegemeyer, and K. Bergmann, in, Liquid Crystals of One and Two Dimensional Order, eds. W. Helfrich, and G. Heppke (Springer–Verlag, 1980) p. 161.

41. P.P. Crooker, (a) Mol. Cryst. Liquid Cryst. **98**, 31 (1983); (b) Liquid Crystals **5**, 751 (1989).

42. R. Barbet-Massin, P.E. Cladis, and P. Pieranski, Recherche **15**, 548 (1984).

43. V.A. Belyakov, and V.E. Dmitrienko, Sov. Phys. Usp. **28**, 535 (1985).

44. H. Stegemeyer, Th. Blümel, K. Hiltrop, H. Onusseit, and F. Porsch, Liquid Crystals 1, 3 (1986).

45. P.E. Cladis, in, Theory and Applications of Liquid Crystals, eds., J.L. Ericksen, and D. Kinderlehrer (Springer, New York, 1987) p. 73.

46. J.P. Sethna, in, Theory and Applications of Liquid Crystals, eds., J.L. Ericksen, and D. Kinderlehrer (Springer–Verlag, 1987) p. 305.

47. R.M. Hornreich, and S. Shtrikman, Mol. Cryst. Liquid Cryst. **165**, 183 (1988).

48. D.C. Wright, and N.D. Mermin, Rev. Mod. Phys. **61**, 385 (1989).

49. R.A. Pelcovits, Theory and Computation, in, Handbook of Liquid Crystal Research (Oxford Univ. Press, 1997) Chapt. 3, p. 71.

50. P.P. Crooker, Blue Phases, in, Chirality in Liquid Crystals, eds., H.S. Kitzerow, and Ch. Bahr (Springer–Verlag, 2001) Chapt. 7, p. 186.

51. K. Bergmann, and H. Stegemeyer, Z. Naturforsch **34a**, 251, 1031 (1979).

52. H. Stegemeyer, and K. Bergmann, Springer Ser. Phys. Chem. **11**, 161 (1980).

53. P.J. Collings, (a) Phys. Rev. **A30**, 1990 (1984); (b) Mol. Cryst. Liquid Cryst. **113**, 277 (1984).
54. R.N. Kleiman, D.J. Bishop, R. Pindak, and P. Taborek, Phys. Rev. Lett. **53**, 2137 (1984).
55. H. Onusseit, and H. Stegemeyer, Z. Naturforsch **36a**, 1083 (1981).
56. P.E. Cladis, P. Pieranski, and M. Joanicot, Phys. Rev. Lett. **52**, 542 (1984).
57. P.E. Cladis, T. Garel, and P. Pieranski, Phys. Rev. Lett. **57**, 2841 (1986).
58. J.H. Flack, P.P. Crooker, D.L. Johnson, and S. Long, in, Liquid Crystals and Ordered Fluids, Vol. 4, eds., A.C. Griffin, and J.F. Johnson (Plenum, New York, 1984) p. 901.
59. D.S. Rokhsar, and J.P. Sethna, Phys. Rev. Lett. **56**, 1727 (1986).
60. V.M. Filev, JETP Lett. **43**, 677 (1986).
61. (a) R.M. Hornreich, and S. Shtrikman, Phys. Rev. **A24**, 635 (1981); (b) Phys. Rev. Lett. **59**, 68 (1987).
62. O. Lehmann, Z. Phys. Chem. **56**, 750 (1906).
63. D. Armitage, and F.P. Price, J. Chem. Phys. **66**, 3414 (1977).
64. D. Armitage, and F.P. Price, J. Phys. (Paris) Colloq. **36**, C1-133 (1975).
65. D. Armitage, and R.J. Cox, Mol. Cryst. Liquid Cryst. Lett. **64**, 41 (1980).
66. J. Thoen, Phys. Rev. **A37**, 1754 (1987).
67. G. Voets, H. Martin, and W. van Dael, Liquid Crystals **5**, 871 (1989).
68. K. Tanimoto, P.P. Crooker, and G.C. Koch, Phys. Rev. **A32**, 1893 (1985).
69. J.D. Miller, P.R. Battle, P.J. Collings, D.K. Yang, and J.P. Crooker, Phys. Rev. **A35**, 3959 (1987).
70. M.B. Atkinson, and P.J. Collings, Mol. Cryst. Liquid Cryst. **136**, 141 (1986).
71. D.K. Yang, and P.P. Crooker, Phys. Rev. **A35**, 4419 (1987).
72. H. Grebel, R.M. Hornreich, and S. Shtrikman, (a) Phys. Rev. **A28**, 1114 (1983); (b) Phys. Rev. **A30**, 3264 (1984).
73. H. Stegemeyer, and P. Pollmann, Mol. Cryst. Liquid Cryst. **82**, 123 (1982).
74. H. Onusseit, and H. Stegemeyer, Z. Naturforsch **39a**, 658 (1984).
75. (a) S. Meiboom, M. Sammon, and D. Berreman, Phys. Rev. **A28**, 3553 (1983); (b) S. Meiboom, and M. Sammon, Phys. Rev. Lett. **44**, 882 (1980); (c) S. Meiboom, and M. Sammon, Phys. Rev. **A24**, 468 (1981).
76. D.L. Johnson, J.H. Flack, and P.P. Crooker, Phys. Rev. Lett. **45**, 641 (1980).
77. R.M. Hornreich, and S. Shtrikman, Phys. Rev. **A28**, 1791 (1983).

78. V.A. Belyakov, V.E. Dmitrenko, and S.M. Osadchii, Sov. Phys. JETP **56**, 322 (1982).

79. K. Tanimoto, and P.P. Crooker, Phys. Rev. **A29**, 1566 (1984).

80. P. Pieranski, E. Dubois-Violette, F. Rohten, and L. Strelecki, J. Phys. **42**, 53 (1981).

81. P. Pieranski, and P.E. Cladis, Phys. Rev. **A35**, 355 (1987).

82. B. Je'rôme, P. Pieranski, V. Godec, G. Haran, and C. Germain, J. Phys. France **49**, 837 (1988).

83. R.J. Miller, and H.F. Gleeson, Phys. Rev. **E52**, 5011 (1995).

84. R.J. Miller, and H.F. Gleeson, J. Phys. II France **6**, 909 (1996).

85. M. Marcus, (a) J. Phys. (Paris) **42**, 61 (1981); (b) Phys. Rev. **A25**, 2272 (1982).

86. N. Clark, S.T. Vohra, and M.A. Handschy, Phys. Rev. Lett. **52**, 57 (1984).

87. V.E. Dmitrienko, Liquid Crystals **5**, 847 (1989).

88. B. Jérôme, and P. Pieranski, Liquid Crystals **5**, 799 (1989).

89. G. Heppke, B. Jérôme, H.S. Kitzerow, and P. Pieranski, Liquid Crystals **5**, 813 (1989).

90. F. Porsch, and H. Stegemeyer, Liquid Crystals **2**, 395 (1987).

91. P. Pieranski, P.E. Cladis, and R. Barbet-Massin, J. Phys. (Paris) Lett. **46**, 973 (1985).

92. F. Porsch, H. Stegemeyer, and K. Hiltrop, Z. Naturforsch **39a**, 475 (1984).

93. G. Heppke, B. Jérôme, H.S. Kitzerow, and P. Pieranski, (a) J. Phys. (Paris) **52**, 2991 (1991); (b) ibid **50**, 549 (1989).

94. V.A. Belyakov, Difraktsionnaya Optika Periodicheskikh Sred Slozhnoi Struktury (Diffraction Optics of Periodical Media with Complex Structure) (Moscow : Nauka, 1988).

95. P.L. Finn, and P.E. Cladis, Mol. Cryst. Liquid Cryst. **84**, 159 (1982).

96. H.S. Kitzerow, Mol. Cryst. Liquid Cryst. **207**, 981 (1991).

97. T. Seideman, Rep. Prog. Phys. **53**, 659 (1990).

98. E.I. Demikhov, V.K. Dolganov, and S.P. Krylova, JETP Lett. **42**, 15 (1985).

99. E.I. Demikhov, and V.K. Dolganov, Sov. Phys. JETP Lett. **38**, 445 (1983).

100. E.P. Koistenen, and P.H. Keyes, Phys. Rev. Lett. **74**, 4460 (1995).

101. J.B. Becker, and P.J. Collings, Mol. Cryst. Liquid Cryst. **265**, 163 (1995).

102. Z. Kutnjak, C.W. Garland, J.L. Passmore, and P.J. Collings, Phys. Rev. Lett. **74**, 4859 (1995).

103. Z. Kutnjak, C.W. Garland, C.G. Schatz, P.J. Collings, C.J. Booth, and J.W. Goodby, Phys. Rev. **E53**, 4955 (1996).

104. (a) S. Meiboom, J.P. Sethna, P.W. Anderson, and W.F. Brinkman, Phys. Rev. Lett. **46**, 1216 (1981); (b) S. Meiboom, M. Sammon, and W.F. Brinkman, Phys. Rev. **A27**, 438 (1983).

105. J.P. Sethna, Phys. Rev. **B31**, 6278 (1985).

106. S.A. Brazovskii, and S.G. Dmitriev, Sov. Phys. JETP **42**, 497 (1975).

107. R.M. Hornreich, and S. Shtrikman, Bull. Isr. Phys. Soc. 25, 46 (1979), J. Phys. (Paris) **41**, 335 (1980); **42**, 367(E) (1980); Phys. Lett. **A82**, 345 (1981); **A84**, 20 (1981).

108. M. Sammon, Mol. Cryst. Liquid Cryst. **89**, 305 (1982).

109. (a) L. Longa, and H.R. Trebin, Phys. Rev. Lett. **71**, 2757 (1993); Liquid Crystals **21**, 243 (1996);

(b) J. Englert, L. Longa, H. Stark, and H.R. Trebin, Phys. Rev. Lett. **81**, 1457 (1998).

110. (a) L.M. Blinov, and L.A. Beresnev, Sov. Phys. Usp. **27**, 7 (1984).

(b) D.C. Ulrich, and S.J. Elston, Optical Properties of Ferroelectric and Antiferroelectric Liquid Crystals, in, The Optics of Thermotropic Liquid Crystals, eds., S. Elston, and R. Sambles (Taylor and Francis Ltd., 1998) Chapt. 9, p. 195.

111. J. Valasek, Phys. Rev. **17**, 475 (1921).

112. R.B. Meyer, 5th Int. Liq. Cryst. Conf. 17-21 (06/1974) Stockholm;

R.B. Meyer, L. Liebert, L. Strzelecki, and P. Keller, J. Phys. (Paris) Lett. **36**, L69 (1975).

113. R.B. Meyer, Mol. Cryst. Liquid Cryst. **40**, 74 (1976).

114. H.R. Brand, P.E. Cladis, and P.L. Finn, Phys. Rev. **A31**, 361 (1985).

115. N.A. Clark, and S.T. Lagerwall, Ferroelectrics **59**, 25 (1984).

116. J.W. Goodby, J. Mater. Chem. **1**, 307 (1991).

117. J.S. Patel, and J.W. Goodby, Mol. Cryst. Liquid Cryst. **144**, 117 (1987).

118. M. Kawada, Y. Uesugi, K. Matsumura, Y. Sudo, K. Kondo, and T. Kitamura, Proceedings of XVIth Japanese Liquid Crystal Conference (Hiroshima, 1990) Abstract, p. 10.

119. T. Ikemoto, K. Sakeshita, S. Hayashi, Y. Kageyama, M. Uematsu, J. Nakauchi, and K. Mori, Proceedings of the XVth Japanese Liquid Crystal Conference (Osaka, 1989) Abstract, p. 18.

120. H. Taniguchi, M. Ozaki, K. Nakano, K. Yoshino, N. Tamasaki, and K. Satoh, Jpn. J. Appl. Phys. **27**, 452 (1988).

121. N. Mikami, R. Higuchi, T. Sakurai, M. Ozaki, and K. Yoshimo, Jpn. J. Appl. Phys. **25**, 1833 (1986).

122. J.W. Goodby, E. Chin, J.M. Geary, J.S. Patel, and P.L. Finn, J. Chem. Soc. Faraday Trans. **I83**, 3429 (1987).

123. G. Scherowsky, Polymers for Advanced Technologies, Vol. **3** (Chichester, Wiley, 1992) p. 219;

G. Scherowsky, B. Brauer, K. Grunegerg, U. Muller, L. Komitov, S.T. Lagerwall, K. Skarp, and B. Stebler, Mol. Cryst. Liquid Cryst. **215**, 257 (1992).

124. N. Koide, T. Kaneko, and Y. Aoyama, Proceedings of the XVth Japanese Liquid Crystal Conference (Osaka, 1989) Abstract, p. 88.

125. N. Shiratori, A. Yoshizawa, I. Nishiyama, M. Fukumasa, T. Hirae, and M. Yamane, Mol. Cryst. Liquid Cryst. **199**, 129 (1991).

126. T. Inukai, K. Furukawa, K. Terashima, S. Saito, M. Isogai, T. Kitamura, and A. Mukoh, Abstract Book of Japan Domestic Liquid Crystal Meeting (Kanazawa, 1985) p. 172.

127. A.M. Levelut, C. Germain, P. Keller, and L. Liebert, J. Phys. (Paris) **44**, 623 (1983).

128. J.W. Goodby, and E. Chin, Liquid Crystals. **3**, 1245 (1988).

129. N. Hiji, A.D.L. Chandani, S. Nishiyama, Y. Ouchi, H. Takezoe, and A. Fukuda, Ferroelectrics **85**, 99 (1988).

130. K. Furukawa, T. Terashima, M. Ichihashi, S. Saitoh, K. Miyazawa, and T. Inukai, Ferroelectrics **85**, 63 (1988).

131. Y. Galerne, and L. Liebert, Abstract book of 2nd International Conference Ferroelectric Liquid Crystals 027 (1989).

132. A.D.L. Chandani, T. Hagiwara, Y. Suzuki, Y. Ouchi, H. Takezoe, and A. Fukuda, Jpn. J. Appl. Phys. **27**, L729 (1988).

133. K. Itoh, Y. Takanishi, J. Yokoyama, K. Ishikawa, H. Takezoe, and A. Fukuda, Jpn. J. Appl. Phys. **36**, L784 (1997).

134. A.D.L. Chandani, E. Gorecka, Y. Ouchi, H. Takezoe, and A. Fukuda, Jpn. J. Appl. Phys. **28**, L1265 (1989).

135. Y. Galerne, and L. Liebert, Phys. Rev. Lett. **64**, 906 (1990).

136. G. Heppke, P. Kleineberg, D. Lötzsch, S. Mery, and R. Shashidhar, Mol. Cryst. Liquid Cryst. **231**, 257 (1993).

137. Y. Takanishi, H. Takezoe, M. Johno, T. Yui, and A. Fukuda, Jpn. J. Appl. Phys. **32**, 4605 (1993).

138. P. Cladis, and H.R. Brand, Liquid Crystals **14**, 1327 (1993).

139. Ch. Bahr, and D. Fliegner, Phys. Rev. Lett. **70**, 1842 (1993).

140. Y. Takanishi, H. Takezoe, A. Fukuda, and J. Watanabe, Phys. Rev. **B45**, 7684 (1992).

141. Y. Takanishi, H. Takezoe, A. Fukuda, H. Komura, and J. Watanabe, J. Mater. Chem. **2**, 71 (1992).

142. T. Fujikawa, K. Hiraoka, T. Isozaki, K. Kajikawa, H. Takezoe, and A. Fukuda, Jpn. J. Appl. Phys. **32**, 985 (1993).

143. E. Gorecka, A.D.L. Chandani, Y. Ouchi, H. Takezoe, and A. Fukuda, Jpn. J. Appl. Phys. **29**, 131 (1990).

144. K. Hiraoka, Y. Ouchi, H. Takezoe, and A. Fukuda, Mol. Cryst. Liquid Cryst. **199**, 197 (1991).

145. K. Hiraoka, A. Taguchi, Y. Ouchi, H. Takezoe, and A. Fukuda, Jpn. J. Appl. Phys. **29**, L103 (1990).

146. H. Takezoe, and Y. Takanishi, Smectic Liquid Crystals : Antiferroelectric and Ferrielectric Phases, in, Chirality in Liquid Crystals, eds., H.S. Kitzerow, and Ch. Bahr (Springer–Verlag, 2001) Chapt. 9, p. 252.

147. I. Nishiyama, A. Yoshizawa, M. Fukumasa, and T. Hirai, Jpn. J. Appl. Phys. Lett. **28**, 2288 (1989).

148. (a) Y. Suzuki, T. Hagiwara, Y. Aihara, Y. Sadamure, and I. Kawamura, Proceedings of the XVth Japanese Liquid Crystal Conference (Osaka, 1989) abstract p. 302;

(b) T. Hagiwara, and Y. Suzuki, ibid, p. 304.

149. S.A. Pikin, and M.A. Osipov, Theory of Ferroelectricity in Liquid Crystals, in, Ferroelectric Liquid Crystals : Principles, Properties and Applications, eds., J.W. Goodby, R. Blinc, N.A. Clark, S.T. Lagerwall, M.A. Osipov, S.A. Pikin, T. Sakurai, K. Yoshino, and B. Žekš (Gordon and Breach Science Publishers, 1999) Part III, p. 249.

150. S.A. Pikin, and V.L. Indenbom, Uspekhi Fiz. Nauk **125**, 251 (1978).

151. R. Blinc, and B. Žekš, Phys. Rev. **A18**, 740 (1978).

152. R. Blinc, Models for Phase Transitions in Ferroelectric Liquid Crystals : Theory and Experimental Results, in, Phase Transitions in Liquid Crystals, eds. S. Martellucci, and A.N. Chester (Plenum Press, New York, 1992) Chapt. 22, p. 343.

153. B. Žekš, Mol. Cryst. Liquid Cryst. **114**, 259 (1984);

B. Urbanc, and B. Žekš, Liquid Crystals **5**, 1075 (1989).

154. W.J.A. Goossens, Mol. Cryst. Liquid Cryst. **12**, 237 (1971).

155. B.W. van der Meer, G. Vertogen, A.J. Dekker, and J.G.J. Ypma, J. Chem. Phys. **65**, 3935 (1976).

156. B.W. van der Meer, and G. Vertogen, A Molecular Model for the Cholesteric Mesophase, in, The Molecular Physics of Liquid Crystals, eds., G.R. Luckhurst, and G.W. Gray (Acad. Press, New York, 1979) Chapt. 6, p. 149.

157. H. Kimura, M. Hosino, and H. Nakano, J. Phys. (Paris) Coll. **40**, C3-174 (1979).

158. M.A. Osipov, and S.A. Pikin, Mol. Cryst. Liquid Cryst. **103**, 57 (1983).

159. M.A. Osipov, Ferroelectrics **58**, 305 (1984).

160. P. Bak, and R. Bruinsma, Phys. Rev. Lett. **49**, 249 (1982).

161. R. Bruinsma, and P. Bak, Phys. Rev. **B27**, 5824 (1983).

162. A. Fukuda, Y. Takanishi, T. Isozaki, K. Ishikawa, and H. Takezoe, J. Mater. Chem. **4**, 997 (1994).

163. M. Yamashita, and S. Miyajima, Ferroelectrics **148**, 1 (1993).

164. M. Yamashita, Mol. Cryst. Liquid Cryst. **263**, 93 (1995); Ferroelectrics **181**, 201 (1996).

165. M. Yamashita, J. Phys. Soc. Japan **65**, 2122 (1996); ibid **65**, 2904 (1996).

166. H. Orihara, and Y. Ishibashi, Jpn. J. Appl. Phys. **29**, L115 (1990).

167. H. Sun, H. Orihara, and Y. Ishibashi, J. Phys. Soc. Japan **62**, 2706 (1993).

168. B. Žekš, R. Blinc, and M. Cepic, Ferroelectrics **122**, 221 (1991).

169. B. Žekš, and M. Cepic, Liquid Crystals **14**, 445 (1993).

170. V.L. Lorman, A.A. Bulbitch, and P. Toledano, Phys. Rev. **E49**, 1367 (1994).

171. M. Cepic, and B. Žekš, Mol. Cryst. Liquid Cryst. **263**, 61 (1995).

172. A. Roy, and N.V. Madhusudana, (a) Europhys. Lett. **36**, 221 (1996); (b) ibid **41**, 501 (1998).

173. V.L. Lorman, Mol. Cryst. Liquid Cryst. **262**, 437 (1995).

174. P. Mach, R. Pindak, A.M. Levelut, P. Barois, H.T. Nguyen, C.C. Hunag, and L. Furenlid, Phys. Rev. Lett. **81**, 1015 (1998).

175. J. Li, H. Takezoe, and A. Fukuda, Jpn. J. Appl. Phys. **30**, 532 (1991).

176. T. Isozaki, K. Hiraoka, Y. Takanishi, H. Takezoe, A. Fukuda, Y. Suzuki, and I. Kawamura, Liquid Crystals **12**, 59 (1992).

177. M. Skarabot, M. Cepic, B. Žekš, R. Blinc, G. Heppke, A.V. Kityk, and I. Musevik, Phys. Rev. **E58**, 575 (1998).

178. K. Hiraoka, Y. Takanishi, K. Skarp, H. Takezoe, and A. Fukuda, Jpn. J. Appl. Phys. **30**, L1819 (1991).

179. Ch. Bahr, D. Fliegner, C.J. Booth, and J.W. Goodby, Phys. Rev. **E51**, 3823 (1995).

180. V. Laux, N. Isaert, H.T. Nguyen, P. Cluseau, and C. Destrade, Ferroelectrics **179**, 25 (1996).

181. K. Yamada, Y. Takanishi, K. Ishikawa, H. Takezoe, A. Fukuda, and M.A. Osipov, Phys. Rev. **E56**, R43 (1997).

182. A. Rastegar, M. Ochsenbein, I. Musevic, T. Rasing, and G. Heppke, Ferroelectrics **212**, 249 (1998).

183. K. Ema, M. Kanai, H. Yao, Y. Takanishi, and H. Takezoe, Phys. Rev. **E61**, 1585 (2000).

184. H.S. Kitzerow, Twist Grain Boundary Phases, in, Chirality in Liquid Crystals, eds., H.S. Kitzerow, and Ch. Bahr (Springer–Verlag, New York, 2001), Chapt. 10, p. 296.

185. C.W. Garland, Liquid Crystals **26**, 669 (1999).

186. W. Kuczynski, Selforganization in Chiral Liquid Crystals (Scientific Publishers OWN, Poznan, 1997).

187. T.C. Lubensky, Physica **A220**, 99 (1995).

188. A. Bouchta, H.T. Nguyen, L. Navailles, P. Barois, C. Destrade, F. Bougrioua, and N. Isaert, J. Mater. Chem. **5**, 2079 (1995).

189. J.W. Goodby, A.J. Slaney, C.J. Booth, I. Nishiyama, J.D. Vuijk, P. Styring, and K.J. Toyne, Mol. Cryst. Liquid Cryst. **243**, 231 (1994).

190. J.W. Goodby, J. Mater. Chem. **1**, 307 (1991).

191. S.R. Renn, and T.C. Lubensky, Phys. Rev. **A38**, 2132 (1988).

192. C.C. Huang, D.S. Lin, J.W. Goodby, M.A. Waugh, S.M. Stein, and E. Chin, Phys. Rev. **A40**, 4153 (1989).

193. J.W. Goodby, M.A. Waugh, S.M. Stein, E. Chin, R. Pindak, and J.S. Patel, Nature **337**, 449 (1989).

194. T.C. Lubensky, T. Tokihiro, and S.R. Renn, Phys. Rev. Lett. **67**, 89 (1991).

195. T.C. Lubensky, and S.R. Renn, Phys. Rev. **A41**, 4392 (1990).

196. G. Srajer, R. Pindak, M.A. Waugh, J.W. Goodby, and J.S. Patel, Phys. Rev. Lett. **64**, 1545 (1990).

197. A. Bouchta, H.T. Nguyen, M.F. Achard, F. Hardouin, and C. Destrade, Liquid Crystals **12**, 575 (1992).

198. Y. Sah, Mol. Cryst. Liquid Cryst. **302**, 207 (1997).

199. V. Vill, and H.W. Tunger, J. Chem. Soc. Chem. Commun. **1995**, 1047 (1995).

200. V. Vill, H.W. Tunger, and D. Peters, Liquid Crystals **20**, 547 (1996).

201. G.S. Iannacchione, and C.W. Garland, Phys. Rev. **E58**, 595 (1998).

202. L. Navailles, B. Pansu, L. Gorre-Talini, and H.T. Nguyen, Phys. Rev. Lett. **81**, 4168 (1998).

203. K.J. Ihn, J.A.N. Zasadzinski, R. Pindak, A.J. Slaney, and J. Goodby, Science **258**, 275 (1992).

204. C. Carboni, H.F. Gleeson, J.W. Goodby, and A.J. Slaney, Liquid Crystals **14**, 1991 (1993).

205. S.K. Prasad, G.G. Nair, S. Chandrasekhar, and J.W. Goodby, Mol. Cryst. Liquid Cryst. **260**, 387 (1995).

206. A. Anakkar, A. Daoudi, J.M. Buisine, N. Isaert, F. Bougrioua, and H.T. Nguyen, Liquid Crystals **20**, 411 (1996).

207. S.R. Renn, and T.C. Lubensky, Mol. Cryst. Liquid Cryst. **209**, 349 (1991).

208. H.T. Nguyen, A. Bouchta, L. Navailles, P. Barois, N. Isaert, R.J. Twieg, A. Maaroufi, and C. Destrade, J. Phys. France **II2**, 1889 (1992).

209. L. Navailles, P. Barois, and H.T. Nguyen, Phys. Rev. Lett. **71**, 545 (1993).

210. L. Navailles, R. Pindak, P. Barois, and H.T. Nguyen, Phys. Rev. Lett. **74**, 5224 (1995).

211. (a) I. Dozov, and G. Durand, Europhys. Lett. **28**, 25 (1994); (b) I. Dozov, Phys. Rev. Lett. **74**, 4245 (1995).

212. I. Luk'yanchuk, Phys. Rev. **E57**, 574 (1998).

213. S.R. Renn, Phys. Rev. **A45**, 953 (1992).

214. P.A. Pramod, R. Pratibha, and N.V. Madhusudana, Current Science **73**, 761 (1997).

215. W. Kuczynski, and H. Stegemeyer, SPIE **3318**, 90 (1997).

216. A.J. Slaney, and J.W. Goodby, Liquid Crystals **9**, 849 (1991).

217. C.J. Booth, D.A. Dunmur, J.W. Goodby, J.S. Kang, and K.J. Toyne, J. Mater. Chem. **4**, 747 (1994).

218. C.J. Booth, J.W. Goodby, K.J. Toyne, D.A. Dunmur, and J.S. Kang, Mol. Cryst. Liquid Cryst. **260**, 39 (1995).

219. J.W. Goodby, Mol. Cryst. Liquid Cryst. **292**, 245 (1997).

220. J.W. Goodby, I. Nishiyama, A.J. Slaney, C.J. Booth, and K.J. Toyne, Liquid Crystals **14**, 37 (1993).

221. R.D. Kamien, and T.C. Lubensky, J. Phys. France I**3**, 2131 (1993).

222. O.D. Lavrentovich, Y.A. Nastishin, V.I. Kulishov, Y.S. Narkevich, and A.S. Tolochko, Europhys. Lett. **13**, 313 (1990).

223. L. Navailles, C.W. Garland, and H.T. Nguyen, J. Phys. France II**6**, 1243 (1996).

224. L. Navailles, H.T. Nguyen, P. Barois, N. Isaert, and P. Delord, Liquid Crystals **20**, 653 (1996).

225. N. Kramarenko, V. Kulishov, L. Kutulya, and N. Shkolnikova, SPIE **2795**, 83 (1996).

226. N. Kramarenko, V. Kulishov, L. Kutulya, V.P. Seminozhenko, A.S. Tolochko, and N.I. Shkolnikova, Kristallografiya **41**, 1082 (1996); Crystallog. Rep. **41**, 1029 (1996).

227. N.L. Kramarenko, V.I. Kulishov, L.A. Kutulya, G.P. Semenkova, V.P. Seminozhenko, and N.I. Shkolnikova, Liquid Crystals **22**, 535 (1997).

228. R.A. Lewthwaite, J.W. Goodby, and K.J. Toyne, Liquid Crystals **16**, 299 (1994).

10

Lyotropic Liquid Crystals

Lyotropic liquid crystals are formed on the dissolution of most surfactants in a solvent (usually water). Surfactants are amphiphilic materials whose constituent molecules are formed from a polar head group and a nonpolar chain (often hydrocarbon). There exist several different types of lyotropic liquid crystal phase structures[1-15]; each of these phases has a different arrangement of molecules within the solvent matrix. The concentration of the solute material in the solvent determines the kinds of lyotropic liquid crystal phase that is exhibited. However, at a given concentration, it is also possible to observe transitions between lyotropic mesophases by changing the temperature.

The lyotropic liquid crystals are frequently encountered in everyday life. For example, most surfactants in water form lyotropic liquid crystal phases; they occur during the dissolution of soaps and detergents. These phases are also exhibited during cooking where cake batters often contain a liquid crystal stabilized emulsion. In industry the best known example of the use of lyotropics is the existence of neat phase during soap manufacture. Of far greater importance is the occurrence of lyotropic mesophases in several materials in living organism. This includes a large number of organelles of cells, such as membrane, myelin and mitochondria. They include such noncellular forms as protein filaments of muscles, the lens of the eye, and the head of the spermatozoon. The product of life activity - bile, urine and blood - also contain substances in liquid crystalline phase. Thus the life processes itself critically depends on the properties of lyotropic liquid crystal phases.

This chapter gives a general introduction to the different aspects of lyotropic liquid crystals.

10.1. AMPHIPHILIC MESOGENIC MATERIALS

The types of molecular structure generating lyotropic liquid crystal phases are amphiphilic. Amphiphilic molecules possess two distinct parts (polar as well as nonpolar) with rather different properties in the same molecule. The hydrophilic (polar) head attracts water, while the lipophilic tail (nonpolar) avoids water. Some typical examples of amphiphilic molecules are given in Table 10.1.

Table 10.1. Some typical examples of amphiphilic molecules.

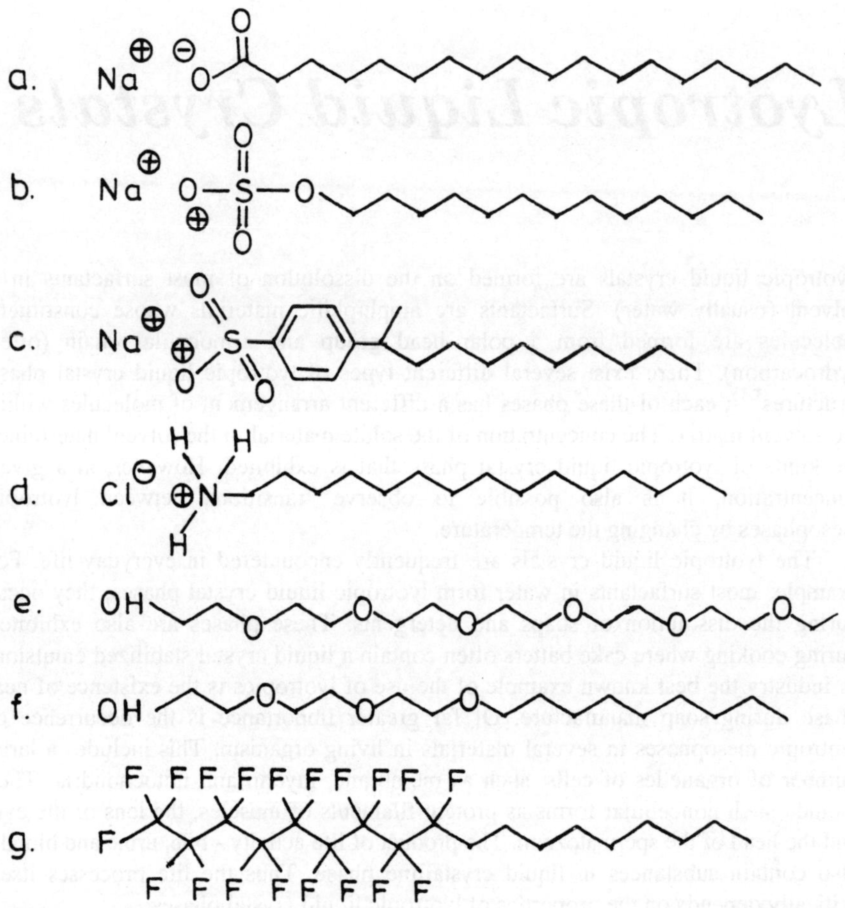

Soaps such as sodium stearate (compound T10.1a) have a polar head group made up of a carboxylate salt and a nonpolar unit of a long hydrocarbon unit. Synthetic detergents, compounds T10.1b (alkyl sulfates) and T10.1c (aromatic sulfonates), are analogous in nature to the compound T10.1a. These compounds T10.1(a-c) are known as anionic surfactants. Cationic surfactant (compound T10.1d) consists of an amine with a long terminal chain. Amphiphilic molecules can also be generated by nonionic species, for example, compound T10.1(e-f). Compound T10.1g is a typical example of an amphiphilic molecule in which the polar hydrophilic head group is made up of a long perfluoroalkyl chain connected directly to a long hydrocarbon chain as the hydrophobic section.

When a solvent (water) is added to solid amphiphilic material following possibilities can occur : The surfactant does not dissolve at all, and remains as a solid crystal plus an aqueous solution of amphiphile monomers. Some of the amphiphiles dissolve to form an aqueous micellar solution. Under certain conditions, larger structures than micelles generate, and these form lyotropic liquid crystals which may dissolve in more water to form an aqueous micellar solution. Amphiphiles that remain almost insoluble in water are nonpolar, and semipolar lipids, and polar surfactants at temperatures below the Krafft point. The Krafft point is defined as the temperature (T_K) below which micelles are insoluble. Above the Krafft point lyotropic mesophases are generated. Usually the temperature must be increased to about 10°C above the Krafft point before lyomesophases are formed. Within the temperature range between the Krafft point and the surfactant melting point mostly the lyotropic liquid crystals exist.

10.2. CLASSIFICATION AND STRUCTURE OF LYOTROPIC LIQUID CRYSTAL PHASES

Amphiphil compounds are highly polar in nature. Their liquid crystalline behaviour can be characterized by a segregation of the polar and the nonpolar parts of the molecules. Based on the type of low-angle X-ray diffraction patterns, three different classes of lyomesophase structures are well recognized. These are the lamellar, the hexagonal and the cubic phases. The first two have a single symmetry axis, while the symmetry of the cubic phases is obvious from the nomenclature. Similar to that of thermotropic liquid crystal phases there exists several different types of lyomesophases (e.g., nematic, smectic).

10.2.1. The Lamellar Lyotropic Liquid Crystal Phases

The lamellar (L_α) lyotropic phase, also known as "neat phase", has the structure as shown in Fig. 10.1. It can be seen that in this phase the amphiphilic molecules are arranged in bilayers separated by water layers. The layers extend over large distances, usually of the order of microns or more. Depending on the water content, the thickness of water layer can vary from approximately 10 A to more than 100 A. The bilayer thickness is generally about 10-30% less than twice the length of an all-trans nonpolar chain. Usually, lamellar mesophases only exist down to 50%. When surfactant is below 50%, transition from lamellar phase to hexagonal phase or an isotropic micellar solution may occur. However, in certain extreme cases the lyotropic lamellar phases is exhibited in extremely dilute solution. In lamellar phases the parallel layers slide over each other easily during shear and hence lamellar phases are less viscous than the hexagonal lyotropic phases.

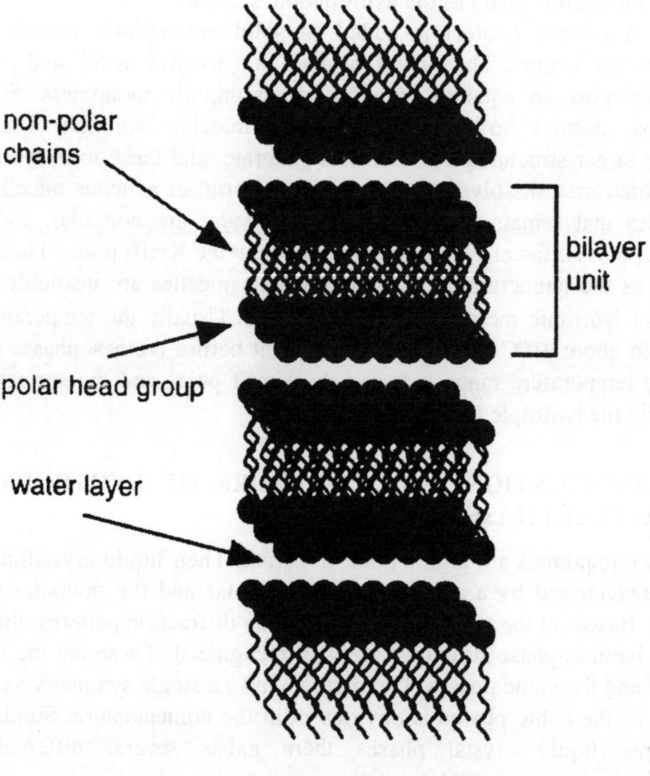

Fig. 10.1. Schematic view of the lamellar lyomesophase.

10.2.2. The Hexagonal Lyotropic Liquid Crystal Phases

In the hexagonal lyotropic liquid crystal phases there exist a hexagonal ordering of molecular aggregate. Two well-established phase structures are identified; the hexagonal phase (H_1) (Fig. 10.2a), and the reversed (or inverse) hexagonal phase (H_2) (Fig. 10.2b). In the H_1 phase the calamitic micelles of indefinite length are packed in an hexagonal array and separated by a continuous region. The diameter of the micellar cylinders is typically 10-30% less than twice the length of an all-trans nonpolar chain. Depending upon the contents of water and surfactants, the spacing between cylinders varies appreciably between 1 and 5 nm. Although hexagonal phase contains high water content, the phase is very viscous.

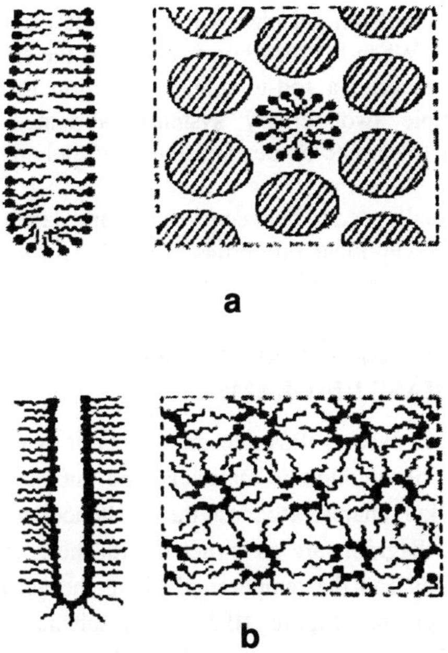

Fig. 10.2. Structure of (a) hexagonal phase and (b) reversed hexagonal phase.

In the reversed hexagonal phase (Fig. 10.2b), the hydrocarbon chains occupy the spaces between the hexagonally packed water cylinders of indefinite length. The water is contained within the cylindrical reversed micelles having a typical diameter of 1 to 2 nm. The nonpolar chains occupying the remaining space overlap to bring the cylinders much closer together than in the hexagonal phase. The H_2 phase occupies much smaller region of the phase diagrams than the H_1 phase.

10.2.3. The Cubic Lyotropic Liquid Crystal Phases

The cubic lyotropic phases, also known as viscous isotropic phases, structurally are not as well characterized as the lamellar or hexagonal phases. In the most well known cubic phase there exists a cubic arrangement of molecular aggregates; these are similar to micelles (I_1 phase) or reversed micelles (I_2 phase). The cubic phases are extremely viscous; even more viscous than hexagonal phases. They are optically isotropic and so are often called the viscous isotropic phases.

10.2.4. Ringing Gels and Sponge Phases

Ringing gels are microemulsion gel phases formed in ternary system of surfactant, hydrocarbon and water. When hydrocarbon is added to the surfactant/ water system the rod-like micelles undergo a rod-to-sphere transition forming the ringing gel phases. As an example two ternary systems fabricated from the nonionic surfactants[16], alkyldimethylamine oxides and the alkyl polyglycol ethers, are worth mentioning.

Sponge phases usually occur between vesicle and lamellar phases. They are isotropic phases with exceptional properties[17] and are formed from the randomly connected distribution of amphiphilic bilayers.

10.3. OCCURRENCE OF MESOPHASES IN AMPHIPHILE/WATER SYSTEMS : PHASE DIAGRAMS

Phase diagrams are often used to illustrate the effects of external variables, such as type and concentration of amphiphilic materials, and number, length and degree of unsaturation of the alkyl chains, etc., on the existence of lyotropic mesophase structures. Such phase diagrams are constructed with the amphiphile concentration along the horizontal axis and the temperature along the vertical axis. Here a few typical examples are given. Figure 10.3 shows schematically a typical phase diagram of a soap (sodium stearate; compound T10.1a) in water. It clearly shows the critical micelle concentration (below which micelles do not exist) and the Krafft point at each temperature. Above the Krafft point, the lyotropic phase structures are generated, while below this point the crystal is insoluble in water. At the relatively low amphiphile concentrations the hexagonal phase exists within a limited temperature range beyond which micellar solution is formed. At relatively high concentrations the lamellar phase is generated. The lamellar phase exists up to a high temperature (wide temperature range) than the hexagonal phase but eventually a micellar solution is obtained at beyond this temperature region. At the extremely high concentrations of the surfactants reversed or inverted phases are formed which on cooling transforms to a crystalline structure.

The phase diagrams have been constructed[1,5,8,18] for the various amphiphile-water systems. In case of a cationic surfactant (compound 10.1d)/water system a strong tendency towards the lamellar phase is observed over a wide temperature range from a 50% (and above) amphiphile concentration; the hexagonal phase occurs over a narrow concentration and temperature ranges. In the phase diagram of non-ionic surfactant (compound T10.1e)/water system the lamellar phase is generated at very high concentrations (70-90%) and up to moderate temperatures (~ 60°C). At the lower temperatures than this and with lower amphiphile concentrations (40-70%) the hexagonal phase is generated. Most importantly at the intermediate concentrations a cubic lyotropic phase is generated. In the phase diagrams of a charged surfactant (AOT)/water system[19] and an non-ionic surfactant (C12E5)/water system[20] the lamellar phase exists over a wide range of concentrations, including the very dilute region. In these systems highly dilute lamellar phases occur due to long-range repulsive interaction.

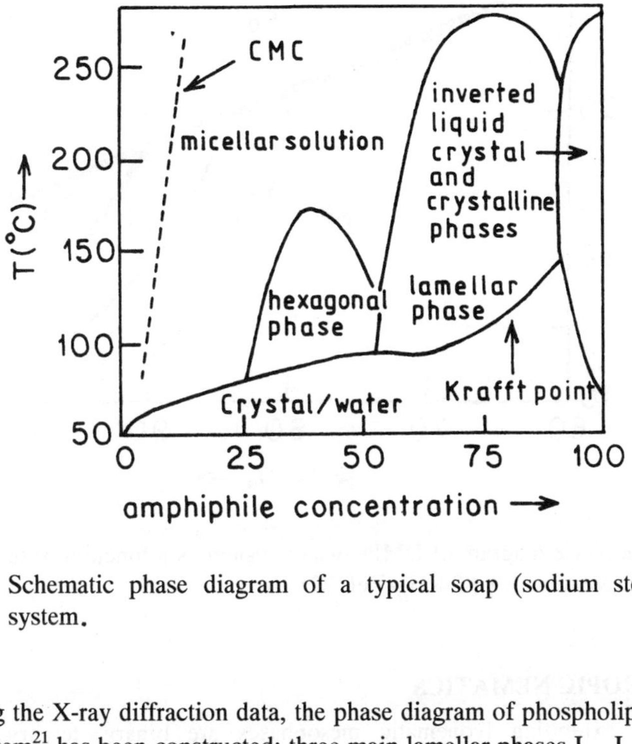

Fig. 10.3. Schematic phase diagram of a typical soap (sodium stearate) water system.

Using the X-ray diffraction data, the phase diagram of phospholipids (DMPC)/ water system[21] has been constructed; three main lamellar phases L_α, $L_{\beta'}$ and P_β were observed. These phases exhibit a stacked water-bilayer structure with a well defined repeat spacing d. The distinguishing feature of these structures is the in-plane order of the hydrocarbon chains. The L_α phase is the highest temperature phase, for any given concentration, in which the chains are liquid like with no regular in-plane structure. In the $L_{\beta'}$ phase, the chains form a closed packed lattice in which the bilayer planes are flat and the chains are tilted with respect to the bilayer normal. In the P_β phase, the chains are still frozen and tilted, but the bilayer planes become rippled with a long wavelength modulation of periodicity ~ 12-20 nm. The temperature-relative humidity (T-RH) phase diagram[22] of this system (DMPC/water, Fig. 10.4) shows L_α, P_β and three newly distinguished phases $L_{\beta F}$, $L_{\beta I}$ and $L_{\beta L}$. Here labels F and L refer to the hexotic liquid crystal phases with the molecules tilting between nearest neighbours and I refers to a hexotic phase where the molecules tilt toward nearest neighbours.

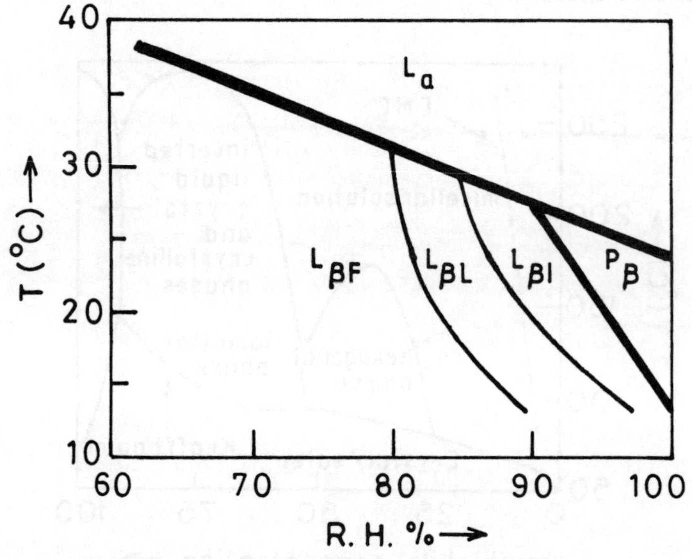

Fig. 10.4. The phase diagram of DMPC/water system as a function of temperature-relative humidity (T-RH) (Ref. 22).

10.4. LYOTROPIC NEMATICS

The systems exhibiting lyonematic mesophases are binary, ternary or more component solutions of amphiphiles. Upto 1980 only two types of lyonematics were known to exist. The first type constructed of cylindrical micelles are known as calamitics (N_C or CM). The micelles of second kind have the form of discs and are called as discotics. They are denoted in various ways : N_D, N_L (lamellar micelle) or DM (discotic micelle). This distinction was based on the idea that the micellar shape was an oblate or a prolate ellipsoids with a long-range orientational order along the director $\hat{\mathbf{n}}$. As discussed in chapter 6 (sec. 6.7), in 1980, Yu and Saupe[23] were first to observe a third lyonematic phase, known as biaxial lyonematic (N_{bx}), in a lyotropic mixture of potassium laurate, 1-decanol and water. This phase was found in the middle between the disc and cylindrical region and it gave rise to a reentrant phenomenon.

From a macroscopic point of view, lyonematics appear as a homogeneous fluid solutions, more viscous than ordering liquids and characterized by very low birefringence ($\Delta n \sim 10^{-3}$) as compared with thermotropics. There are experimental evidences[24-27] which suggest that the intrinsic micellar shape is always biaxial and various nematic phases are the consequence of overall fluctuations.

There are several ternary amphiphilic systems which exhibit all the three lyonematic phases. These lyonematics have been obtained with only a few types of amphiphilic molecules. They include negative amphiphilic ions, such as decylsulfate (DS), Laurate (L), positive ions such as decylammonium (DA), etc. The alcohols, such as decanol (DeOH) and octanol (OeOH), occupy a special place in this lyomesophases formation; they themselves do not form lyomesophases but take part in the generation of practically all the lyonematic phases. Two typical best studied examples are KL/DaCl(DeOH)/H_2O (potassium laurate/ Decylammonium hydrochloride/water)[28] and NaDS/DeOH/H_2O (sodium decylsulfate/decanol/water)[29] systems. Figure 10.5 shows the phase diagram of the NaDS/DeOH/H_2O system constructed from the textures and the small angle X-ray scattering data at a temperature of 22°C. It can be seen that the small nematic region is surrounded by three extensive regions. The lyonematic phases occur in the narrow concentration range : NaDS (37-41%), H_2O (50-56%), and DeOH (5-9%). In this region three nematic phases are distinguished. In between N_D and N_C phases, N_{bx} phase has been found. The N_C and N_D phases occur in the concentration region : N_C – NaDS (39-41%), H_2O (52-55%), DeOH (5-9%); N_D – NaDS (37-39%), H_2O (51-56%), DeOH (5-9%). However, these numbers are only estimates; the exact phase diagram in this region has been constructed only from a few points. One of the main arguments supporting the classification of lyomesophases with asymmetric micelles as nematics is the fact that they generate textures typical of thermotropic nematics. The optical studies on the oriented samples have shown that the N_D phase is uniaxial positive, and the N_C phase uniaxial negative. In an another phase diagram[28b] it has been found that the existence of these lyonematic phases strongly depends upon the concentrations of water and decanol and the purity of the material. In the micellar solution region (Fig. 10.5) two phases (M_1 and M_2) have been found at the different weight concentration of NaDS. The isotropic solution M_1 is characterized by spherical micelles, while in M_2 phase the micelles are no longer spherical. For DeOH/NaDS ratio from 0 to 5% the hexagonal (H) phase exists, and when this ratio is increased up to 8.5% an orthorhombic (R) phase occurs. At the high decanol concentrations a transition takes place to a lamellar (L) phase. The structure and the physical properties (diamagnetic, dielectric, viscoelastic, etc.) of lyonematic phases are described in detail by Sonin[9], and Kuzma and Saupe[2].

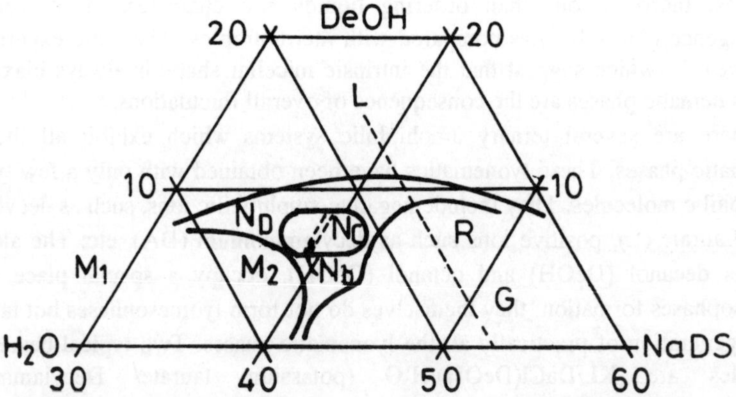

Fig. 10.5. Phase diagram of the NaDS/DeOH/H_2O system at room temperature (Ref. 30).

10.5. LYOTROPIC LIQUID CRYSTAL POLYMERS

If a polymer is to generate lyomesophases, it must be fairly rigid and soluble in a solvent. In order to dissolve fairly rigid structures, drastic solvents such as sulfuric acid are often required. When Kevlar (compound T8.1c, chapter 8) is dissolved in high concentrations in sulfuric acid, polymer lyomesophase is generated. Most of the lyomesophase polymers are not structural materials but are biological molecules of high significance. Biological structures are all water-based and the water is required as the solvent to exhibit lyotropic liquid crystal phases. For example, some derivatives of cellulose form lyomesophases. Linear viruses are an interesting class of lyomesophase polymer. The tobacco mosaic virus with a helical structure is relatively rigid and soluble in sufficiently high quantities to generate liquid crystal phases.

10.6. LYOMESOPHASES IN BIOLOGICAL SYSTEMS

The lipid components of cell biomolecules exhibit[4] a range of lyomesophases and form a periodically ordered long-range organization with a somewhat disordered short-range organization. The mesophase properties of lipid molecules in water are of considerable significance for the various aspects of the organization and function of cell membrane. The structure and properties of lipid molecules, in particular those associated with biomembranes, generating lyomesophases can explain many important aspects of their biological behaviour.

Associated with the biomembranes a whole class of lipid molecules are available[4]. Many of these molecules are based upon two types of alcohols, i.e., glycerol and sphingosine. The variations occurring in the lipid polar groups in the position of substitution and in the various fatty acids linked to these alcohols either by ester or ether linkages give rise to a range of phospholipids and sphingolipids.

10.6.1. Lyomesophases Formed by Lipid Molecules

Phospholipids such as lecithin in water show interesting behaviour. When phospholipids are added in the water a direct transition from the crystalline state to a solution, in general, does not occur; in-between many hydrated phases are exhibited. The lyomesophases thus obtained is a function of both the water content and the temperature. The thermotropic phase transition temperature (T_C) plays a significant role in the melting of hydrocarbon chains. When water diffuses into the lattice it does so into the polar (ionic) region only at a temperature at which the hydrocarbon chains melt. At a higher temperature than this a simultaneous dissociation of the ionic lattice occurs by the penetration of water and the melting of hydrocarbon chain region. The T_C line depends upon the nature of both the hydrocarbon chains and the polar region of the molecule, the water content present and the solutes dissolved in the water. Once the water has penetrated into the amphiphile lattice and then the sample is cooled to below the T_C line, the rearrangements of the hydrocarbon chains into an orderly crystalline lattice occur and the remaining water may not necessarily be expelled from the system. These phases having crystallinc chain region are sometimes called gels.

The X-ray diffraction pattern[31] shows that the lyotropic phase obtained by mixing two parts of water to one part of lecithin gives rise to a lamellar structure, liquid-like hydrocarbon chains and a long spacing of 69 A, about 60% greater than that of solid material. The phase diagram of lecithin (phosphatidylcholine)/water system was studied by Chapman et. al.[32]. Above the T_C line this system exists in the mesomorphic lamellar phase with the hydrocarbon chains in the melted condition. At the maximum hydration the system composition is ~ 40 wt.% water. Addition of more than 50 wt.% water gives rise to a two-phase system – fragments of the lamellar phase at the maximum hydration dispersed in the excess water. When the phosphatidylcholine/water system is cooled to below the T_C line, the hydrocarbon chains adopt an ordered packing. The structure of this gel phase is lamellar where the hydrocarbon chains are packed in a variety of subcells. The phase diagram of different chain length lecithin /water systems are essentially equivalent.

The effects of cholesterol on the lipid phase transitions have been investigated. Due to the presence of cholesterol at high concentrations in the lamellar phase, the lipid endothermic phase transition can be removed. The cholesterol has an ordering effect on the lipid chains above the lipid T_C temperature.

The number and position of double bonds introduced into the lipid hydrocarbon chain appears to be of crucial importance; the behaviour of the system is affected appreciably. The addition of double bonds into the lipid hydrocarbon chain decreases the melting point, the enthalpy, and the entropy of the main chain transition. The introduction of a single double bond in one of the acids decreases the melting temperature, and the addition of two double bonds in a single chain lowers the chain-melting point by about the same as the introduction of a single double bond in both chains. However, the introduction of third or fourth double bond does not show any noticeable effect.

The normal hexagonal (H_1) phase is very widespread in the biological systems. This phase has been found in solutions of DNA, polypeptides and polysaccharides[33], in muscle and in collagen[34]. Any rod-shape biopolymer may under certain conditions form, a hexagonal phase. In the lipid/water systems both the hexagonal phases, normal (H_1) and inverse (H_2), have been observed[35-37]. The H_1 phase structure was recognized by Bernal and Fankuchen[38] in aqueous solutions of tobacco mosaic virus (TMV). The structure of H_1 phase found[35] in hydrated dodecylsulfonic acid (23-70% in water) was elucidated by Luzzati and co-workers[39]. The H_2 phase of phospholipids was discovered[36] in a lipid extract from human brain containing 52% PE, 35% PC, and 13% PI, at 37°C with water contents below 22 wt.%. The same structure was also identified[40] in the ternary surfactant system sodium caprylate/ decanol/water system. In many other systems hexagonal phases have been found and other details about these phases have been given by Chapman[4].

It has been shown[37,41] that in the hydrated lipid systems the lamellar organization is only one of the large variety of lyomesophase structures. Among the non-lamellar arrangements, the cubic phases are the most complex and intriguing[41,42]. Six different structures having cubic symmetry have so far been identified – Q^{212}, Q^{223}, Q^{224}, Q^{227}, Q^{229} and Q^{230}. Here a cubic phase is designated as Q^n where Q stands for cubic and n is the number of the relative space group. The structure of these cubic phases have been determined[42-45] unambiguously. The structure of the Q^{230}, Q^{224} and Q^{229} can each be described in terms of two 3-D networks of joined rods, mutually intertwined and unconnected the rods are, respectively, linked coplanarly three by three, tetrahedrally four by four and cubically six by six. One of the most remarkable property[43] of these cubic structures is that both the water and hydrocarbon media are continuous throughout the structure. These structures can be visualized as a 3-D topological generalizations of the lipid bilayer. In contrast, in the lamellar structure the water and the hydrocarbon media are each subdivided into an infinite number of disjoined planar layers. In the

Q^{212} phase structure one of the two networks of rods is preserved, while the other is replaced by a lattice of closed micelles. In addition, Q^{212} phase structure consists of two continuous disjoined media, one apolar and other polar and can be visualized as a 3-D generalization of the lipid monolayer. The Q^{223} and Q^{227} phases consists of two types of closed and disjoined micelles embedded in a continuous matrix; these structures are nonbicontinuous[44,45].

10.6.2. Lyomesophases Formed by DNA Derivatives

It was reported[46] in 1959 that the high molecular weight DNA might exhibit mesophases; two phases were recognized – one of the hexagonal type[46], and the other cholesteric[47]. It is well known that DNA and RNA solutions can spontaneously undergo transition to a liquid crystalline phase above a certain concentration[48]. A large number of helical biological polymers, such as polynucelotides, polypeptides and polysaccharides, exhibit lyomesophases. The structures of these phases depend on the polymer concentration[48-50]. With the increasing polymer content, there appear cholesteric spherulites within the isotropic phase, and at high concentrations the cholesteric regions expand to occupy the whole preparation. When the concentration is increased further, transition from cholesteric structure to hexagonal phase occurs. In the DNA/water system the structure strongly resembles with those of amphiphiles and discotics; parallel alignments of the elongated molecules forming a hexagonal array. Each molecule can translate longitudinally and can rotate about its long axis[50].

10.7. PHASE CHIRALITY OF MICELLAR LYOTROPICS

The phase chirality of micellar lyomesophases has not yet been investigated fully. The most extensively investigated chiral lyotropics are the chiral nematics (cholesterics). The first report on the existence of this kind of phase is by Radley and Saupe[51] in 1978. Due to high electric conductivity and mostly nonuniform sample orientation it is difficult to obtain the experimental evidence for the ferroelectricity in lyotropics. However, a nonaqueous chiral lamellar phase with ferroelectric properties has been reported[52]. In this work the existence of piezoelectricity has been interpreted as a manifestation of ferroelectricity.

The chiral nematic can be prepared in three possible ways. The micellar lyotropics can be formed by chiral surfactants. The use of chiral solvents can generate lyotropic phase chirality with small twists. Also chiral guest/achiral host systems have been found to form chiral lyomesophases. There exist three types of chiral nematics; the two uniaxial phases N_D^* ($\equiv N_L^*$) and N_C^* and one biaxial phase N_b^*.

A survey over lyotropic mixtures exhibiting intrinsic as well as induced phase chirality has been made first by Boidart et. al.[53]. In the meantime many new materials have been realized. Several chiral surfactants, such as α-Alanine hydrochloride decylester, N-dodecanoyl alaninate, serinate, valinate, etc., exhibiting chiral nematic lyomesophases in water are available. A number of chiral dopants have been obtained[5] which induce phase chirality in the achiral host phases. The chemical structures of some of them are given in Table 10.2.

Table 10.2. Structures of some chiral dopants.

a. Ph–(CH$_2$)$_n$–*CH(OH)–COOH   n=0 Mandelic acid
   n=1 Phenyllactic acid
   n=2 HPBA

b. C$_6$H$_{11}$–*CH(OH)–COOH   Hexahydromandelic acid

c. CH$_3$–*CH(OH)–(CH$_2$)$_n$–COOH   n=0 Lactic acid
   n=1 HBA

d. CH$_3$–CH(CH$_3$)–(CH$_2$)$_n$–*CH(OH)–COOH   n=0 HIVA
   n=1 HMVA

e. H$_3$C–*C(H)(OH)–(CH$_2$)$_5$–CH$_3$ 2-Octanol

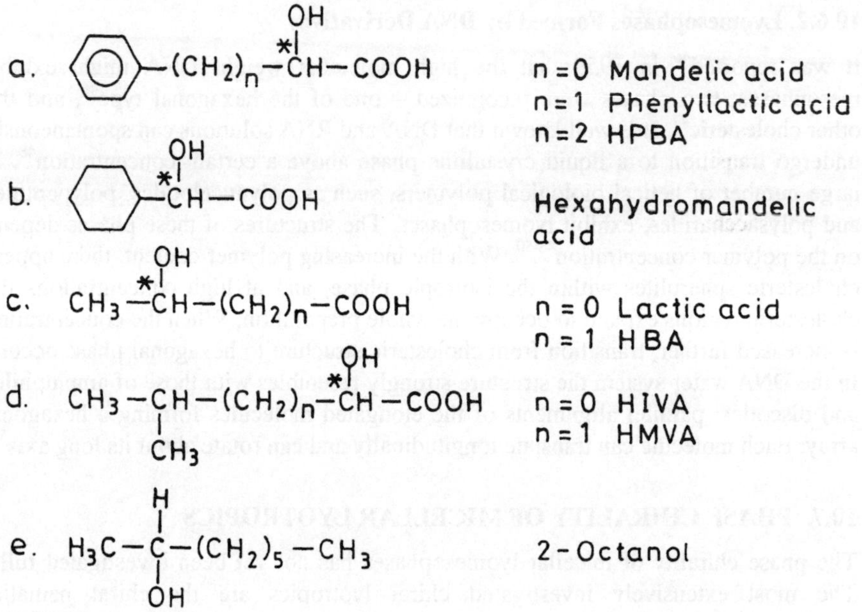

Just as for thermotropic cholesterics, the twist (p^{-1}) of lyotropics depend on the concentration and temperature. The pitches of chiral nematic have been experimentally determined over the full range down to above 1 μm. For the water based systems the twist of lyotropics has been found to be weaker than the thermotropics by an order of magnitude. The twist increases with increasing enantiomeric excess of a chiral surfactant. The same is true for the dopant concentration also. The twist versus dopant concentration curve is linear for the small concentration. However, this curve becomes nonlinear at higher dopant concentration[54]; except few exceptions[55], mostly the absolute value of the slope decreases. In few cases a maximum twist has been found, particularly when a lamellar phase is adjacent to the chiral nematic phase. The helical twisting power $[(\partial p^{-1}/\partial c)_{c \to 0}]$ is influenced by the dopant as well as by the achiral host phase. For the thermotropic cholesteric phase, in most of the cases, the pitch decreases with the increasing temperature. In case of lyotropics both the negative and positive temperature coefficients occur frequently depending on the structure and nature of the mixture[56].

The shape anisotropy of micellar aggregates has been determined[57] using X-ray scattering. There are several parameters, such as the solubilization of dopants[58], the temperature, pH, etc., which can affect the shape and the size of these aggregates. A decrease in pitch with the increasing salt concentration and decreasing pH has been reported by Goozner and Labes[59]. When a salt is added to a solution of ionic surfactants the Debye length of the ions is reduced. Larger micelles decrease the twist of director field. The addition of chiral dopant induces phase chirality as well as modifies several achiral properties of the phase.

From the data on twisting efficiency for tartaric acid, brucine sulfate and cholesterol, it has been found that for a large twisting power of a chiral dopant, it has to be solubilized within the micelles. Only a few studies are available on the influence of certain properties of the chiral surfactants or dopants on the phase chirality. From the investigation on amino acids some rules have been derived[60]. The long and bulky side-chains within the dopant molecule give small values of twisting helical power, while a strong hydrophilic side-chain yields higher helical twisting power. However, these rules found for the chromonic host phase cannot be generalized to other systems.

From the investigation[56] of the influence of host phase for some homologues of the anionic alkyl-ammonium halides and cationic surfactants, it is obvious that the longer the alkyl chain of the surfactant, the smaller twist; and the same is true for the bulkier head group also. Extremely large twists have been found[61] for some sugar/steroid compounds like digitoxine, solanine, etc.

Two models have been proposed[5,51] for the mechanism of chiral interactions. The first one assumes that the micelles adopt a chiral shape and consequently phase chirality, via sterical interaction between adjacent micelles, results. The second model rests on the idea that a chiral dispersion force acts directly between chiral molecules located in different micelles. However, it has been argued[62] that a chiral dispersion interaction between single chiral molecules in adjacent micelles is rather unlikely to be essential for the phase chirality. It is important to mention here that the models for the mechanism of phase chirality in lyotropics has not yet succeeded in providing a conclusive explanation.

REFERENCES

1. P.J. Collings, and M. Hird, Introduction to Liquid Crystals : Chemistry and Physics (Taylor and Francis Ltd., 1997) Chapt. 7, p. 133.

2. M.R. Kuzma, and A. Saupe, Structure and Phase Transitions of Amphiphilic Lyotropic Liquid Crystals, in, Handbook of Liquid Crystal Research, eds., P.J. Collings, and J.S. Patel (Oxford Univ. Press, New York, Oxford, 1997) Chapt. 7, p. 237.

3. S.E. Friberg, Lyotropic Mesophases in Non-Living Systems, in, Liquid Crystals: Applications and Uses, Vol. 2, ed. B. Bahadur (World Scientific, 1991) Chapt. 23, p. 157.
4. D. Chapman, Lyotropic Mesophases in Biological Systems, in Liquid Crystals : Applications and Uses, Vol. 2, ed. B. Bahadur (World Scientific, 1991) Chapt. 24, 185.
5. K. Hiltrop, Phase Chirality of Micellar Lytropic Liquid Crystals, in, Chirality in Liquid Crystals, eds., H.S. Kitzerow, and C. Bahr (Springer-Verlag, 2001) Chapt. 14, p. 447.
6. C. Tanford, The Hydrophobic Effect (Wiley, New York, 1973) .
7. V. Degiorgio, and M. Corti, eds., Physics of Amphiphiles : Micelles, Vesicles and Microemulsions (North-Holland, Amsterdam, 1985).
8. C.J.T. Tiddy, Physics Reports **57**, 1 (1980).
9. A.S. Sonin, Sov. Phys. Usp. **30**, 875 (1987).
10. P.A. Winsor, Solvent Properties of Amphiphilic Compounds (Butterworths, London, 1954).
11. P.A. Winsor, Chem. Rev. **68**, 1 (1968).
12. V. Luzzati, in, Biological Membranes, ed. D. Chapman (Academic Press, London and New York, 1968) Chapt. 3, p. 71.
13. P. Ekwall, L. Mandell, and K. Fontell, Liquid Crystals **2**, ed., G.H. Brown (Gordon and Breach Science Publisher Ltd., 1969) Part II, p. 325.
14. P. Ekwall, in, Advances in Liquid Crystals, ed., G.H. Brown (Academic Press, New York, San Francisco and London, 1971) Vol. 1, Chapt. 1, p. 1.
15. K. Fontell, Prog. Chem. Fats Other Lipids **16**, 145 (1978).
16. H. Hoffmann, and G. Ebert, Angew. Chem. Int. Ed. Engl. **27**, 902 (1988).
17. D. Roux, C. Coulon, and M.E. Cates, J. Phys. Chem. **96**, 4174 (1992).
18. S. Martellucci, and A.N. Chester, eds., Phase Transitions in Liquid Crystals (Plenum Press, New York and London, 1992).
19. K. Fontell, J. Colloid Interface Sci. **44**, 318 (1973).
20. R. Strey, R. Schomaecker, D. Roux, F. Nallet, and U. Olsson, J. Chem. Soc. Faraday Trans. **86**, 2253 (1990).
21. V. Luzzati, Biol. Membr. **1**, 71 (1968); A. Tardieu, V. Luzzati, and F.C. Reman, J. Mol. Biol. **75**, 711 (1973).

22. G.S. Smith, C.R. Safinya, D. Roux, and N.A. Clark, Mol. Cryst. Liquid Cryst. **144**, 235 (1987); G.S. Smith, E.B. Sirota, C.R. Safinya, and N.A. Clark, Phys. Rev. Lett. **60**, 813 (1988); E.B. Sirota, G.S. Smith, C.R. Safinya, R.J. Plano, and N.A. Clark, Science **242**, 1406 (1988).

23. L.J. Yu, and A. Saupe, Phys. Rev. Lett. **45**, 1000 (1980).

24. R. Bartolino, G. Chidichimo, A. Golemme, and F.P. Nicoletta, Ref. 18, Chapt. **27**, p. 427.

25. Y. Hendrikx, J. Charvolin, and M. Rawiso, J. Colloid. Interface Sci. **100**, 597 (1984).

26. A.M. Figueiredo Neto, Y. Galerne, A.M. Levlut, and L. Liebert, J. de Phys. Lett. **46**, L499, L999 (1985); A.M. Figueiredo Neto, L. Liebert, and Y. Galerne, J. Phys. Chem. **89**, 3737 (1985).

27. M.B. Lacerda Santos, Y. Galerne, and G. Durand, Phys. Rev. Lett. **53**, 787 (1984); M.B. Lacerda Santos, and G. Durand, J. Phys. (Paris) **47**, 529 (1986).

28 (a) E.A. Oliveira, L. Liebert, and A.M. Figueiredo Neto, Liquid Crystals **5**, 1669 (1989); F.P. Nicoletta, G. Chidichimo, A. Golemme, and N. Picci, Liquid Crystals **10**, 665 (1991).

(b) Y. Hendrikx, Y. Charvolin, M. Rawiso, L. Liebert, and M.S. Holmes, J. Phys. Chem. **87**, 3991 (1983).

29. B.J. Forrest, and L.W. Reeves, Chem. Rev. **81**, 1 (1981).

30. E. Govers, and G. Vertogen, Physica Ser. A (Utrecht) **133**, 337 (1985).

31. R.S. Bear, K.J. Palmer, and F.O. Schmitt, J. Cell. Comp. Physiol. **17**, 355 (1941); K.J. Palmer, and F.O. Schmitt, J. Cell. Comp. Physiol. **17**, 385 (1994).

32. D. Chapman, R.M. Williams, and B.D. Ladbrooke, Chem. Phys. Lipids **I**, 445 (1967).

33. F. Livolant, and Y. Bouligand, J. Phys. France **47**, 1813 (1986).

34. R.J. Hawkins, and E.W. April, Adv. Liq. Cryst. **6**, 243 (1983).

35. S.S. Marsden, and J.W. McBain, J. Am. Chem. Soc. **70**, 1973 (1948).

36. V. Luzzati, and F. Husson, J. Cell Biol. **12**, 207 (1962); W. Stoekenius, J. Cell. Biol. **12**, 221 (1962).

37. J.M. Seddon, Biochim. Biophys. Acta **1031**, 1 (1990).

38. J.D. Bernal, and I. Fankuchen, J. Gen. Physiol. **25**, 111 (1941).

39. V. Luzzati, H. Mustacchi, and A. Skoulios, Discuss. Faraday Soc. **25**, 43 (1958).

40. K. Fontell, P. Ekwall, L. Mandell, and I. Danielsson, Acta. Chem. Scand. **16**, 2294 (1962).

41. K. Fontell, Colloid Polym. Sci. **268**, 264 (1990).
42. V. Luzatti, P. Mariani, and T. Gulik-Krzywicki, in, Physics of Amphiphilic Layers", eds., J. Meunier, D. Langevin, and N. Boccara (Springer-Verlag, Berlin, 1987) p. 131.
43. P. Mariani, V. Luzzatti, and H. Delacroix, J. Mol. Biol. **204**, 165 (1988).
44. J.F. Sadoc, and J. Charvolin, Acta. Cryst. **45A**, 10 (1989).
45. J. Charvolin, and J.F. Sadoc, Colloid. Polym. Sci. **268**, 190 (1990).
46. V. Luzzati, and A. Nicolaieff, J. Mol. Biol. **1**, 127 (1959).
47. C. Robinson, Tetrahedran **13**, 219 (1961).
48. F. Livolant, J. Physique **47**, 1605 (1986).
49. Y.M. Yevdokimov, S.G. Skuridin, and V.I. Salyanov, Liq. Cryst. **3**, 1443 (1988).
50. F. Livolant, and Y. Bouligand, J. Physique **47**, 1813 (1986).
51. K. Radley, and A. Saupe, Mol. Phys. **35**, 1405 (1978).
52. L. Blinov, S. Davidyan, A. Tetrov, A. Todorov, and S. Yablonskii, JETP Lett. **48**, 285 (1988).
53. M. Boidart, A. Hochapfel, and P. Peretti, Mol. Cryst. Liq. Cryst. **172**, 147 (1989).
54. K. Radley, and H. Catley, Mol. Cryst. Liquid Cryst. **250**, 167 (1994).
55. J. Partyka, and K. Hiltrop, Liquid Crystals **20**, 611 (1996).
56. M. Pape, and K. Hiltrop, Mol. Cryst. Liquid Cryst. **307**, 155 (1997).
57. M. Valente Lopes, and A. Figueiredo Neto, Phys. Rev. **A38**, 1101 (1988).
58. H. Dorfler, and C. Swaboda, Tenside Surf. Det. **35**, 126 (1998).
59. R. Goozner, and M. Labes, Mol. Cryst. Liquid Cryst. **116**, 309 (1985).
60. D. Goldfarb, M. Moseley, M. Labes, and Z. Luz, Mol. Cryst. Liquid Cryst. **89**, 119 (1982).
61. E. Figgerneier, and K. Hiltrop, Progr. Colloid Polymer Sci. **89**, 307 (1992).
62. P. Covello, B. Forrest, M. Marcondes Helene, L. Reeves, and M. Vist, J. Phys. Chem. **87**, 176 (1983).

11

Defects and Textures in Liquid Crystals

A rich variety of defects and optical textures[1–11] are exhibited by the liquid crystals when viewed under the optical polarizing microscopy : the Schlieren texture of nematics and tilted smectics, the focal-conic and mosaic textures and various dislocations of chiral systems, platelet textures of blue phases, filamentary defect textures of TGB_A^* phase, etc. We give here, an overview of the main aspects of defects and textures in liquid crystals.

An understanding of the defects and textures in liquid crystalline phases is vital for the several reasons; it allows for the characterization and classification of mesophase type where the combinations of textures can be used to give definitive identification. It provides insight into the mesophase structure and shows how the structures of devices can be compromised. In some cases, even more important role is played by the defects, for example, they stabilize the formation of blue phases and Twist Grain Boundary (TGB) phases. Without the stabilizing effect caused by the introduction of defects in these phases these unusual mesophase structures probably would not exist. Another important implication of the defect formation is their impact on the device fabrication. Liquid crystal electro-optic devices are typically fabricated with the ultimate aim of being defect free. The defects, discontinuities, inhomogeneties in liquid crystal layer that are found in the devices considerably affect the optics, contrast and performance of the display. Thus, for removing or reducing the defects in liquid crystal cells, an understanding of the structures of defects becomes very important. Several alignment methods have been developed to eradicate these defects. However, in some cases these methods are employed to preferentially form certain types of defects that will stabilise a particular mesophase structure. Topological defects may also play significant role in case of defect mediated phase transitions[12].

The characterization of defects becomes very important in the classification of liquid crystal phases. For the identification of mesophase type using optical polarizing microscopy a thin sample of a mesogenic compound is usually sandwitched between a glass microscope slide and glass coverslip. There are two

For the calamitic mesophases, a homeotropic alignment is obtained when the long-axes of the molecules (more importantly their optic axes), on the average, are normal to the surface of the glass plate. For such a orientation of molecules the polarised light remains unaffected by the material and so the light can not pass through the analyzer. Accordingly, an observer looking through the microscope from this vantage point will see complete blackness. In homogeneous (planar) alignment the molecules have their long axes lying parallel or at an angle to the surface of the glass. With homogeneous alignment, a thin film of the mesophase exhibits birefringence and when viewed between crossed polarisers a coloured texture results. When the long molecular axes, however, are in line with either polariser, light is extinguished.

11.1 CLASSIFICATION OF DEFECTS IN LIQUID CRYSTALS

Mathematically a perfect crystal is an exceedingly useful concept; its structure is usually thought of as a rigid network of ions, atoms or molecules, where each has a fixed position in the array or lattice (this excludes small displacements due to thermal vibrations). The pattern repeats indefinitely within the crystal in three dimensions. Most of the real crystals, however, particularly naturally occurring ones, tend to possess defects or imperfections. These defects can exist as either localized faults and minor misorientations in the structure, or in some cases extensive structural discontinuities. Their nature and effects are often very important in understanding the properties of crystals. This is because the structure-insensitive properties of materials are not affected by the presence of defects in a crystal, while many of the structure-sensitive properties of technical importance are greatly affected by the relatively minor changes in crystal structure caused by the defects. As discussed earlier, liquid crystals typically possess a symmetry which is intermediate between that of a crystal and an isotropic liquid. Hence, like solids, they are also known to possess a vast variety of defects and discontinuities in their macroscopic structure. In addition, due to their fluid like characteristics, they can also show strong changes in the orientational order, which results in the formation of disclinations.

Generally, the order parameter of an ordered medium has spatial dependence, Q(**r**). Distortions of Q(**r**) can be of two types : those containing singularities and those without singularities. At singularities Q(**r**) is not defined. For a 3-dimensional medium, the singular regions might be zero-dimensional (points), 1-dimensional (lines), or 2-dimensional (sheets or surfaces). These are the defects. These defects are energetically unfavourable, but are present due to the influence of the sample container and the thermal history of the sample. In most of the cases they can be removed by the specific experimental procedures. However, in certain cases they are the part of the structure of the liquid crystalline phases. Whenever a homogeneous state cannot be arrived (i.e. the nonhomogeneous state cannot be eliminated) by the continuous variations of the order parameter, it is called a topological defect. If the inhomogeneous state does not contain singularities, but nevertheless is not deformable continuously into a homogeneous state, the system is said to contain a topological configuration[13] (or soliton).

There are two main categories of the defects : (i) those that are in thermal equilibrium with the system; as such are always present, and (ii) those that arise kinetically; their existence particularly is dependent on the nucleation processes from which the medium was formed. Examples of the first kind, which are observed in liquid crystals, are those of Frenkel[14] type (interstitial particles and vacancies). Since the size of this kind of defects is comparable to a molecular length, they do not appreciably affect the microscopic defect textures of the mesophases. The structure and presence of other defects may depend on the external effects such as boundary, surface and field effects. This kind of defect, for example, the schlieren defect, is much more important for the study of liquid crystal textures. Both kinds of defect can be classified in several ways; one of these involves characterizing their dimensionality[15]. Point defects are classified as zero-dimensional, linear defects as 1-dimensional and sheet (or wall) defects as 2-dimensional. Point defects tend to occur in restricted geometries and at surfaces. Line defects are the most important in liquid crystals. A sheet discontinuity always tends to smear out into a continuously distorted region of finite thickness. Such structures are called walls. It is also useful to distinguish between defects that have a core singularity and those that have continuous cores[16] (see Table 11.1).

Table 11.1. Classification of defects based on their dimensionality and core-singularity.

	0-dimensional (Point defects)	1-dimensional (Linear defects)	2-dimensional (Walls)
With a core singularity	Point singularities	Translational dislocations (edge and screw) Rotational dislocations (disclinations)	
No core singularity		Surface inversion lines Coreless defect lines	Inversion walls

The most widely used method for the classification of linear defects is due to Volterra[31,17]. There is a topological relationship between the symmetry elements of an ordered medium and the types of defects found in that medium. This relationship for the linear defects is illustrated by the Volterra process. This process describes the creation of a line singularity (dislocation) in such a medium, where the only invariances are spatial translations and rotations. In this method, as shown pictorially in Figure 11.1, a perfect medium is first cut along an arbitrary surface (S) bounded by a line (L) called the disclination line. The two lips (S_1) and (S_2) of the surface is then displaced rigidly relative to each other by a translation, a rotation, or by a combination of both a rotation and a translation. Next either the excess material from the overlapping regions is removed or the material is added to fill in the voids, and the system is allowed to relax. This leads to the formation of a line discontinuity. The dislocation line produced possesses several important characteristics. A dislocation line is always a closed line, whether it ends at infinity, on the surface of the crystal, or on another dislocation. The displacement has the same value along its whole length, and all displacements can be analysed in terms of either translation, or rotation components. The vector of the displacement **d**(r) may be expressed as

$$\mathbf{d}(r) = \mathbf{b} + \Omega(\nu r) \tag{11.1}$$

where **b**, a measure of the translational part of the displacement, is known as Burgers vector. r is the distance from the rotation axis ν about which the material was rotated by an angle Ω. In pure translational dislocations, important in crystalline solids, **b** ≠ 0 and Ω = 0, whereas in pure rotational defects (disclinations), important in liquid crystals, **b** = 0 and Ω ≠ 0. In terms of Volterra process, disclinations may be classified as wedge disclinations (axis of rotation is parallel to the disclination line), and twist disclinations (axis of rotation is perpendicular to the disclination line). The twist disclinations can only exist if they are coupled with the disclinations (translational) terminating on the line.

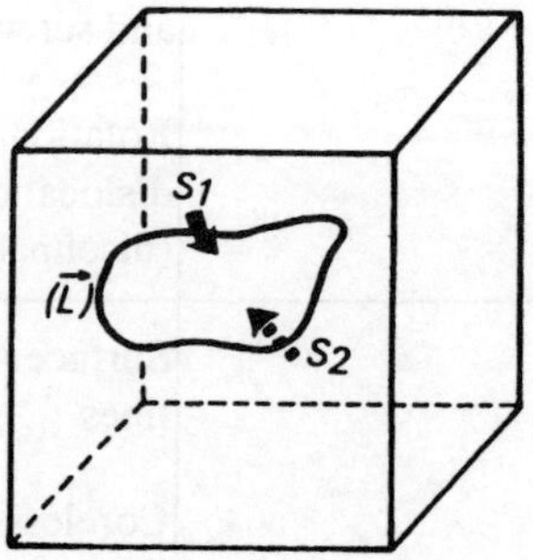

Fig. 11.1 . Classification method for linear defects (Volterra process).

Taking into consideration the energy of the defects, the magnitude of the size of a defect can be estimated by the equation

$$E \sim K[\mathbf{d}(r)]^2$$

where the constant of proportionality K is associated with the elastic deformation of the medium.

11.1.1 Defect Textures Exhibited by Liquid Crystals

The appearance (texture) of a material when viewed between crossed polarisers is a representation of the defects and so is often known as defect textures. Their classification and characterization can be made on the basis of experimental methods adopted for the sample preparation and paramorphosis (phase sequence). Usually the mesophase type identification is done via cooling of the isotropic liquid rather than by heating of the solid. The natural texture of a liquid crystal is the defect texture inherently generated on cooling a mesogenic material from the isotropic liquid (or nematic) into its liquid crystalline phase. Subsequent cooling may give transitions to other mesophases which exhibit paramorphotic textures. The occurrence of these paramorphotic textures is based on the defects of the preceding natural texture[18]. Tables 11.2 and 11.3 list the textures that are commonly found naturally and paramorphotically. The natural textures that are exhibited when the phase is generated directly from the isotropic liquid, or the nematic in case of smectics, are given in the Table 11.2. Table 11.3 lists the paramorphotic textures observed upon cooling after certain smectic – smectic phase transitions. In case of optical microscopic observation mechanical (or other) shearing of the sample may provide additional information about the physical nature of a mesophase. So the results of simple shearing of the sample is also given in the Table 11.2.

Table 11.2. Natural textures and defects displayed by liquid crystals (as viewed between crossed polarisers).

Meso-phase	Homogeneous/planar/ alignment	Homeotropic/ orthogonal alignment	Mechanical shearing	Other
N_u	schlieren homogeneous $s = \pm\frac{1}{2}$ (2-brush) and ± 1 (4 brush) disclinations	extinct black	shear easily	Brownian flashes
N_b	$s = \pm\frac{1}{2}$, ± 1 disclinations hybrid disclinations	extinct black		

N_D	schlieren $s = \pm \frac{1}{2}, \pm 1$ disclinations	extinct black		
S_A	focal-conic homogeneous s = ± 1 disclinations polygonal defects	extinct black	shears to homeotropic	
S_C	focal-conic broken	schlieren s = ± 1 disclinations	shears to schlieren	Brownian motion
S_B	focal-conic	extinct black	shears to homeotropic	mosaics possible
S_I	focal-conic broken	schlieren	shears viscous	schlieren diffuse
$S_{B(cry)}$	mosaic	extinct black	Shears viscous	grain boundaries
S_F	mosaic	schlieren mosaic	Shears viscous	grain boundaries
S_J, S_G, S_E, S_H, S_K	mosaic	mosaic	very viscous	grain boundaries

Table 11.3. Paramorphotic textures associated with liquid crystals.[10]

Mesophases	Paramorphotic textures
S_C	schlieren from S_A homeotropic. Sanded schlieren from isotropic cubic D phases. Broken focal-conic from S_A focal-conic.
S_I	schlieren from S_C schlieren or S_A homeotropic. Broken focal-conic from S_A focal-conic or blurred S_C.
S_B	Extinct homeotropic from S_A homeotropic or S_C schlieren. Focal-conic from S_A focal-conic. Clear focal-conic from S_C broken focal-conic.
$S_{B(cry)}$	Homeotropic from S_A, S_B homeotropic or S_C schlieren. Clear focal-conic domains from S_A, S_B focal-conic or S_C pseudo focal-conic texture squared-off domains from S_A. Transition bars across the focal-conic domains parallel to the layers of the phase textures at the transition to or from S_A.
S_F	schlieren mosaic from S_A, S_B homeotropic or S_C, S_I schlieren. Broken focal-conic from S_A, S_B, S_C, S_I or $S_{B(cry)}$ focal-conic.

S_J	Broken pseudo focal-conic fans, Chunky from S_A, S_B, S_C, S_I, S_F or $S_{B(cry)}$ focal-conic domains. Mosaic from S_A, S_B, $S_{B(cry)}$ homeotropic or S_C, S_I schlieren or S_F mosaic-schlieren.
S_G	As S_J except mosaic from S_J mosaic. Broken pseudo focal-conic fans, Chunky focal-conic domain from S_J. Banded focal-conic from S_B focal-conic.
S_E	Banded focal-conic from S_A, S_B, S_C, S_I, S_F focal-conic domains. Shadowy mosaic domains from S_A, S_B, $S_{B(cry)}$ homeotropic, and S_C schlieren lined mosaic domains from $S_{B(cry)}$, S_J and S_H mosaic.
S_H	Broken pseudo focal-conic domains from S_A, S_B, S_C, S_I, S_F, S_J, S_G focal-conic. Line mosaic domains from $S_{B(cry)}$, S_G, S_J, mosaic and S_C, S_I, S_F schlieren.
S_K	Broken pseudo focal-conic domains from S_A, S_B, S_C, S_I, S_F, S_J, S_G focal-conic. Lines mosaic domains from $S_{B(cry)}$, S_G, S_J mosaic and S_C, S_I, S_F schlieren.

It can be seen from these tables that a variety of textures are formed by the various liquid crystalline phases. The types of defects associated with these textures are influenced by the symmetry (or order) of the mesophases. Certain defects are found to be common to many mesophases, while others are the characteristic to only one mesophase. In many instances combinations of defects may also be observed. The paramorphotic textures (Table 11.3) depend on the thermal and phase history of the material, i.e., which mesophases are exhibited and in what order were they formed on cooling the isotropic liquid.

11.1.2 Dislocations (Edge and Screw)

In liquid crystalline phases, two main types of dislocation exist, the edge and screw dislocations[1,10]. A mixed dislocation also results due to a combination of these two types. In a solid phase, an edge dislocation is formed by inserting an extra half plane of atoms or molecules into a crystal. A negative edge dislocation is observed when the extra half plane is inserted below the slip plane of the crystal (Fig. 11.2), whereas a positive edge dislocation results with the extra half-plane of atoms being inserted below the slip plane. The edge dislocation is exhibited by a number of mesophases. This type of dislocation can also be observed in smectic mesophases by virtue of their layer ordering.

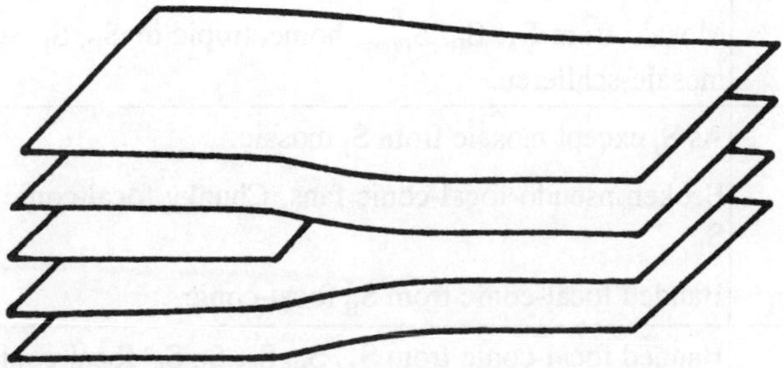

Fig. 11.2. Structure about an edge dislocation.

The structure about the screw dislocation is shown in Fig. 11.3. This defect can occur in several mesophases. In some cases it can appear as a defect which is observable in the microscope and thus is useful in the phase identification. In other cases, for example, TGB phases, screw dislocations provide, on the microscopic scale, the stability to the mesophase. Line singularities can also occur as the screw dislocations, and in certain phases (antiferroelectric and ferrielectric phases), where the molecules are tilted with respect to the layer plane, they are prevalent[19].

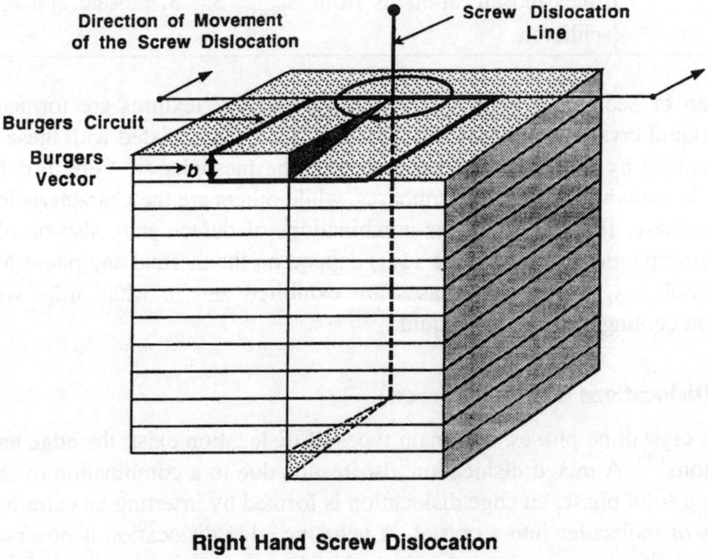

Fig. 11.3. Structure about a screw dislocation.

11.1.3 Schlieren Defects

Schlieren defects are exhibited by the nematic, tilted smectic and discotic nematic phases. When continuously degenerate (tangential or conical, no preferred axis in the plane of the walls) boundary conditions are imposed by the glass walls to the material, often a system of singular points on the surface is observed. These singular points are connected by the black stripes, between the crossed polarisers, showing the regions where the optic axis is parallel to the one of the nicols. The black brushes originating from the points are due to 'line singularities' perpendicular to the layer. The points were called 'noyaux' by G. Friedel. In analogy with dislocations in crystals, Frank[20] coined the term 'disinclinations' which now is termed as 'disclinations'. The general texture resulting from these points is known as 'schlieren texture'.

The black brushes are regions where the director (or the local optic axis) is either parallel or perpendicular to the plane of the polarization of the incident light. The polarization remains unchanged by the sample in these regions, and is therefore extinguished by the crossed analyzer. The observed schlieren defect texture is formed in the polarizing microscope either by a continuous, but sharp, change of molecular orientation within the sample, or by a change in molecular orientation about a point or line singularity. When the molecules are aligned in the direction of either of the two crossed polarizers characteristic dark brushes appear. These brushes appear black because of optical extinction caused by the crossed polarizers and rotate across the area being seen with the rotation of the polarisers, whereas the point source of the brushes remains in a fixed position. At the center of the schlieren a point (Fig. 11.4), or a line singularity is observed.

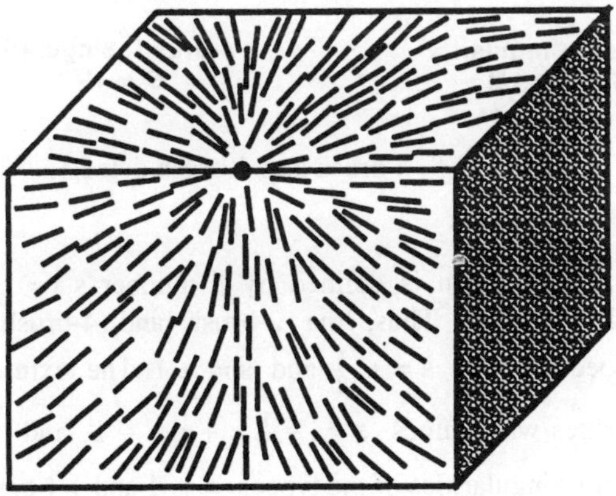

Fig. 11.4. Molecular orientations around a point singularity.

Different types of point singularities are known to occur. Figure 11.5 schematically illustrates 2-brush and 4-brush defects. The schlieren defects can be characterized by observing the defects in the microscope and rotating the polariser and the analyzer. When the brushes rotate in the direction of the rotation of polarisers the singularity is characterized as positive (positive disclination), whereas the singularity is negative (negative disclination) if the two directions of rotations are opposite to each other. The rate of rotation is about equal to that of the polarisers when the disclination has four brushes and is twice as fast when it has only two.

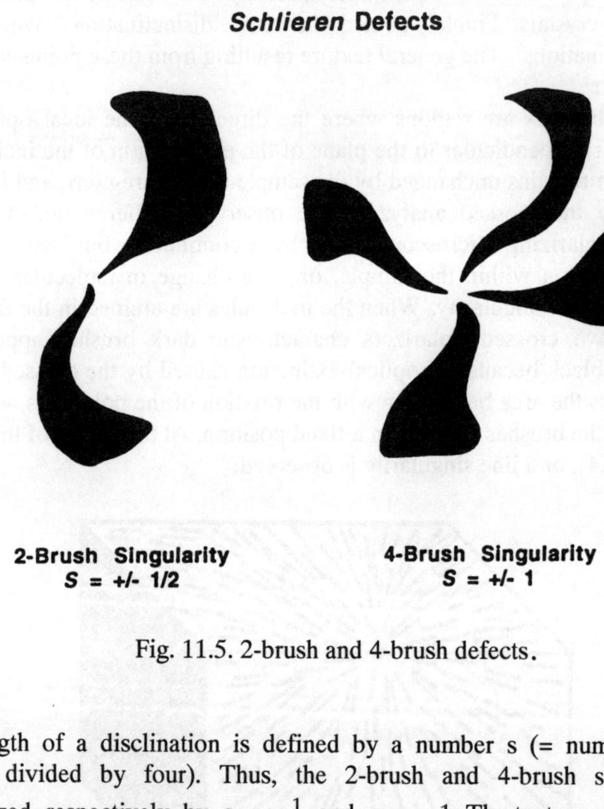

Fig. 11.5. 2-brush and 4-brush defects.

The strength of a disclination is defined by a number s (= number of brushes observed divided by four). Thus, the 2-brush and 4-brush singularities are characterized, respectively, by $s = \pm\frac{1}{2}$ and $s = \pm 1$. The texture of nematic phase possesses singularities with values $s = -\frac{1}{2}, +\frac{1}{2}, -1$ and $+1$, while for the smectic C phase only singularities of the type $s = -1$ and $+1$ are observed.

Some characteristic features become evident by a careful examination of schlieren texture of a phase (see sec. 11.2 and 11.4). The neighbouring singularities connected by the brushes are of the opposite sign. In a sample the sum of all the s-numbers should be equal to zero. Singularities of opposite signs attract each other and eventually merge. Two singularities of same strengths but of opposite signs annihilate one another. However, singularities of different strengths form a new singularity with a s-value equivalent to the addition of the strengths of two original singularities.

11.1.4 Focal-conic Defects

The focal-conic defects are displayed by the smectic phases (layered structure), chiral mesophases because of helicity, and also by some other crystal phases due to paramorphosis[21,22]. The focal-conic or fan-like texture is often produced when a layered structure (smectic) capable of sustaining a bend deformation is supported between two surfaces or nucleation points. These defects usually exist when the molecules develop strong attachments to the surface of the sample preparation or to the nucleation points. Around a nucleation point the molecules adopt a radiating fan-like structure such that the thickness and parallelism of the layers remain invariable. Thus, the layers must curve about the center. However, due to the fluid like character of the mesophase and ease of nucleation, growing nucleation centers are found to coalesce allowing the layers to form Dupin cyclides (Fig. 11.6).

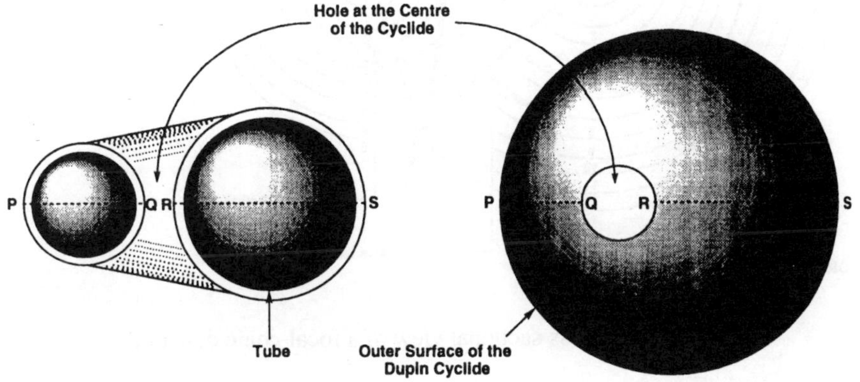

Fig. 11.6 . (a) Cross-sectional view and (b) plan view of a Dupin cyclide.

The Dupin cyclide[23] is composed of a closed hollow ring, with a circular cross-section that varies in length between a maximum (RS) and a minimum (PQ). PQ and RS are known as the principal cross-sections. The geometrical basis for the focal-conic texture consisting of a family of cyclides is explained nicely by Slaney et al.[10], and is shown in Figures 11.7 and 11.8; O and O′ are the centers of the two principal cross-sections. Molecular layers (e.g., 1, 2, 3, 4, …, in Fig. 11.7) are added on about these cross-sections to form Dupin cyclides. The central cyclide (1) has a shape like that of a 'ring doughnut'. More layers are added with a disproportionate rate at O and O′. As a result the doughnut grows more rapidly on one side than the other forming a Dupin cyclide. As the layers are added the hole at the center of the cyclide (shown as point B) becomes closed. After this further addition of any layers leads to the formation of a line intersection which looks like a hyperbola centred about point B (shown by thick line HI in Fig. 11.7). Because of the mismatch of the orientations of the molecules an optical discontinuity occurs along this line.

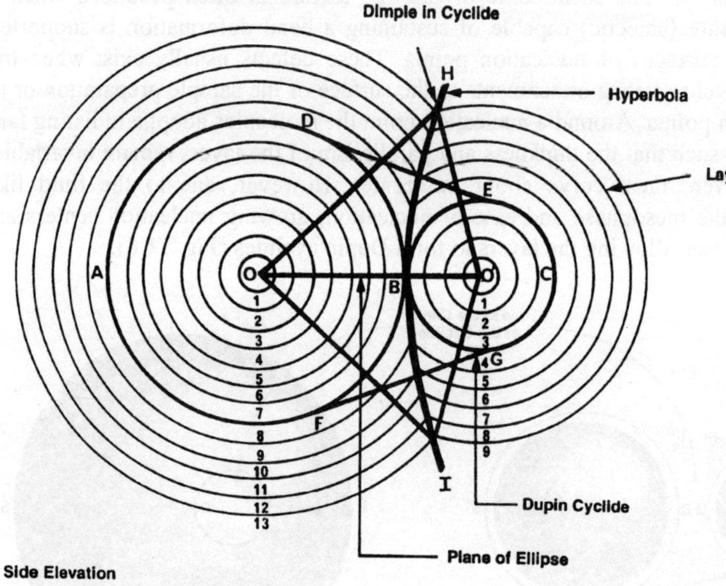

Fig. 11.7. Cross-sectional view of a focal-conic domain.

Let us consider the situation where the center hole at B has just been closed; as described by the ADECGF in Figures 11.7 and 11.8. The surface of the doughnut is completely closed over when the first four cyclides of the smaller principal cross-section have been completed; the 'topside' and 'bottomside' of the cyclide are described, respectively, by DE and FG (Fig. 11.8). However, when fourth cyclide was added the minor principal cross-section has also shrunk to a point O′ (see the plan section of Fig. 11.8). After this if more cyclides are added they remain incomplete with their cross-sections shrinking to points PP′, QQ′ and RR′. As a result, instead of complete rings, hollow crescents are formed. The cross-sectional

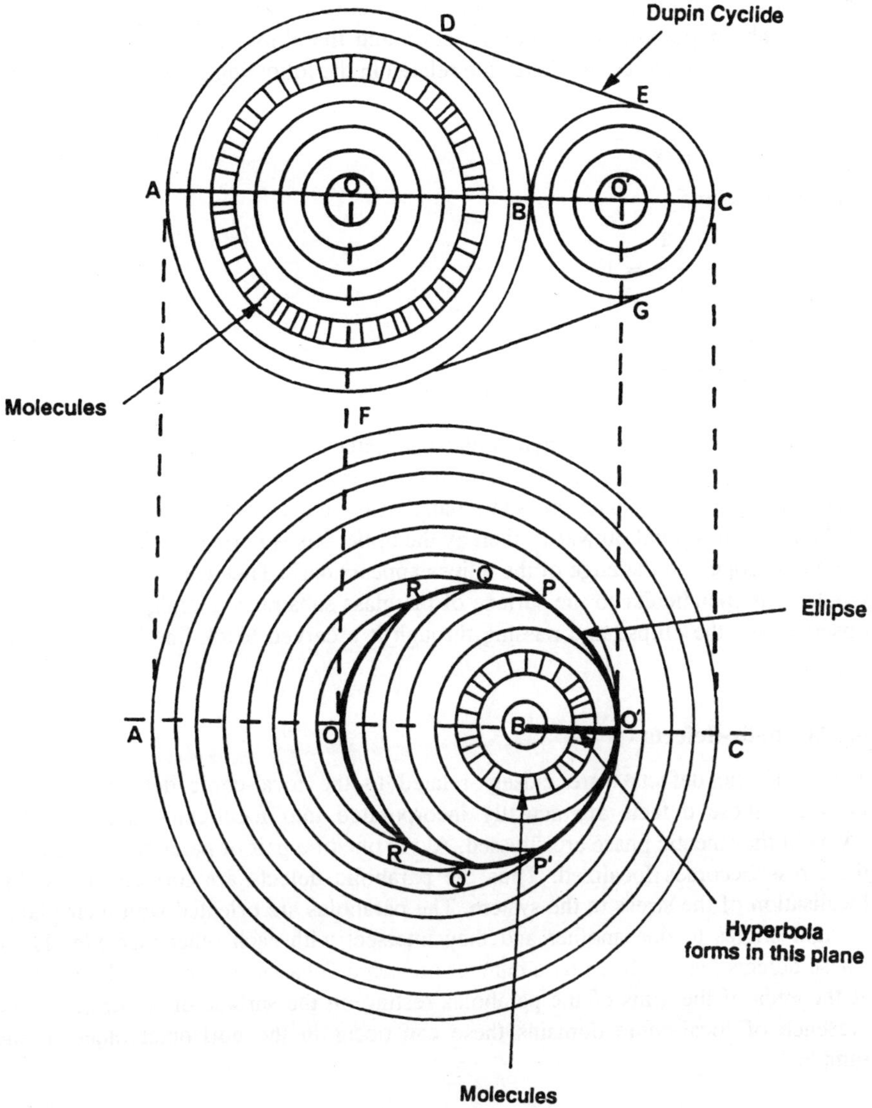

Fig. 11.8. The cross-sectional and plan views of a focal-conic domain; the center hole of the outer cyclide is closed up.

points P, P′, Q, Q′, R, R′ and O and O′ lie on the path of an ellipse with B as one of the foci and O and O′ as vertices. Those cyclides which are situated outside of ADECGF, do not have any central hole, but only a 'dimple' above and below the cyclide. The hyperbola in the cross-section and the ellipse in the plan-section are termed a pair of focal-conics. They are related in the following general way,

$$\frac{x^2}{a^2} + \frac{y^2}{b^2} = 1$$

for the ellipse and

$$\frac{x^2}{a'^2} - \frac{y^2}{b'^2} = 1$$

for the hyperbola, with

$$a' = \sqrt{(a^2 - b^2)} \qquad \text{and} \qquad b' = b.$$

It is seen that along the hyperbolic line defect the long axes of the molecules (i.e. the optic axes of the relative domains) are mismatched, and hence produces a dark line (hyperbola) under crossed polars. At the edge of the ellipse the long axes of the molecules point outwards, thereby the optic axis is normal to the polarisers of the microscope and the edge of the ellipse appears dark. Thus when the plane of the ellipse is perpendicular to the surface of the glass substrates a straight black line is observed for the ellipse and passing through it a curved black line is seen for the hyperbola.

a. Parabolic-defects

The parabolic defects[24] are closely related to the focal-conic defects. In many instances these defects are actually incorporated into focal-conic domains. The layers in the smectic phase are stressed. Along one of a pair of parabolic line defects the stress becomes localized. Thus the parabolic defects are formed due to the localisation of the stress in the system. The parabolas are oriented with their planes at right angles to one another and they intersect with each other (see Fig. 11.9). These defects are exhibited by a number of smectic phases and are usually observed at the ends of the arms of the parabolas resting on the surface of substrate. In the presence of focal-conic domains these can occur in the horizontal plane of the sample.

b. The polygonal texture

The polygonal texture is made up of focal-conic domains that are not fully produced, but are half domains where one brush of the hyperbola (above or below the ellipse) is missing. Thus, the domains take on a conical shape, with the ellipse of each cone touching one of the supporting glass surfaces, whereas the other surface is touched by the apex. Since the cones can be oriented in head-to-tail arrangement, a very efficient packing of half domains of the focal-conic defect texture can be produced; in this structure elliptical cross-sections sit on the glass substrates. Smectic A phase exhibits the polygonal texture, where the phase nucleates from the isotropic liquid or the nematic phase close to a surface of the substrates. This texture is usually observed in thick samples.

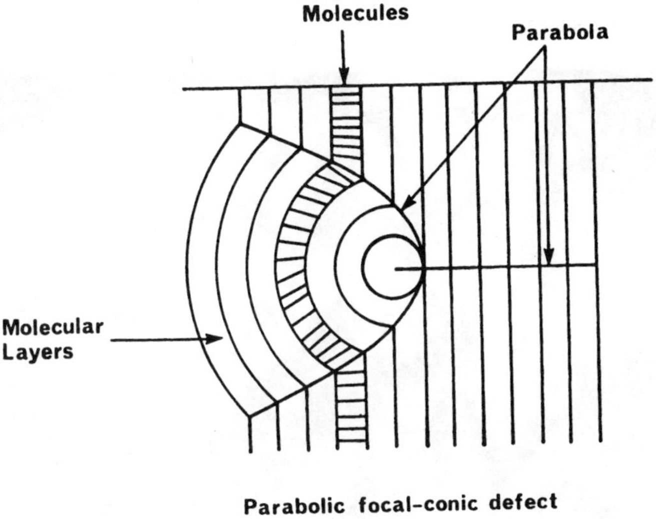

Fig. 11.9. Structure of parabolic focal-conic defect.

11.1.5 Zigzag defects

Almost all ferroelectric mesophases, except those having the 'bookshelf geometry', exhibit the zigzag defects. As shown in Figure 11.10, zigzag defect texture[10,27] shows sharp jagged edges and rounded or hairpin defects between neighbouring domains. The existence of this defect has been shown[28,29] by the X-ray studies. An understanding about the structure of zigzag defects and how they can be eliminated from the cell is of crucial significance for the development of ferroelectric liquid crystal display technology.

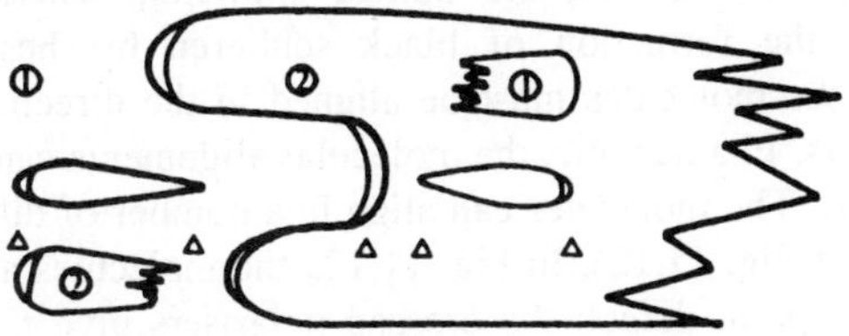

Fig. 11.10. Zigzag and hairpin defects in ferroelectric cells.

11.2 DEFECT TEXTURES IN UNIAXIAL NEMATIC PHASE

In an uniaxial nematic phase point defects and line defects (disclination lines) arise. For the homeotropic preparation the texture is optically extinct and so appears black. However, in case of a quite thick sample there may be small nondescript areas of a birefringent texture with black thread like defect. For the homogeneous alignment close to the clearing point the phase appears most colourful (schlieren texture) and a thin sample looks sharp (Fig. 11.11).

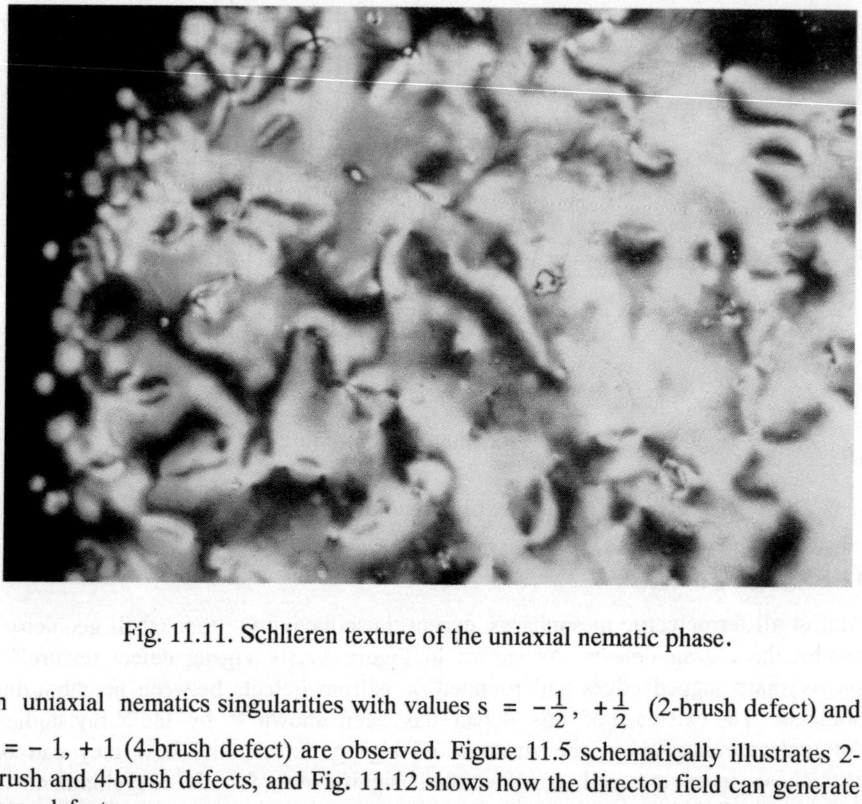

Fig. 11.11. Schlieren texture of the uniaxial nematic phase.

In uniaxial nematics singularities with values s = $-\frac{1}{2}$, $+\frac{1}{2}$ (2-brush defect) and s = − 1, + 1 (4-brush defect) are observed. Figure 11.5 schematically illustrates 2-brush and 4-brush defects, and Fig. 11.12 shows how the director field can generate these defects.

As discussed in sec. 11.1.3, the optical extinction caused by the crossed polarisers leads to the formation of black schlieren brushes. For the optical extinction to occur the molecules must be aligned in the direction of either of the two crossed polarisers. In a nematic, the molecular alignments can often converge to a point (point defect). The molecules can align in a number of different ways (some of these are shown in Fig. 11.12). In Fig. 11.12a the molecules are radially aligned pointing towards the point defect; the crossed polarisers give rise to extinction of light in four regions from the point defect. The molecular alignments shown in Fig.

11.12b also generate a 4-brush schlieren. The molecules arch around the point defect (Fig. 11.12c); in two regions from the point defect the molecules are aligned with either polariser and 2-brush schlieren is generated. Similarly, Fig. 11.12d also shows the formation of 2-brush schlieren. It can be seen that as one moves farther away from the point defect the region of optical extinction increases because more and more molecules align with the polarisers. The situations as shown in Fig. 11.12 only arise when the point defect is seen from the above. When a disclination line connects a point from one surface to the other and is seen parallel to the surface an optically extinct black region is observed in the form of a thread. Figure 11.11 shows the classical schlieren texture of an uniaxial nematic with regions of both 2- and 4-brushes and threads caused by the point and line defects, respectively.

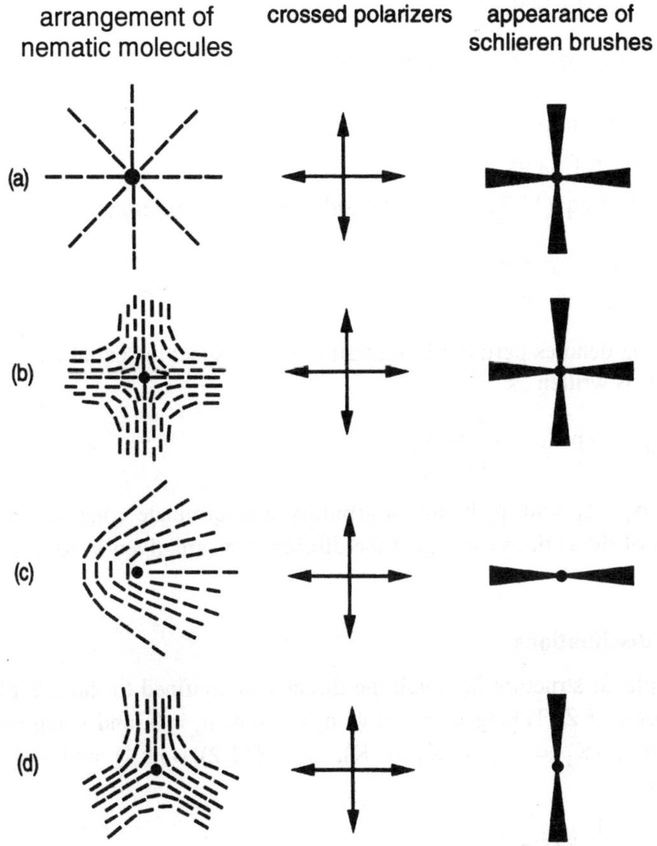

Fig. 11.12. Orientations of the molecules generating Schlieren brushes.

Two disclinations may combine to give rise to what are called as bulk inversion walls[1,10]; here in-between the two sides of the defect the director rotates few degrees. In the nematic texture an inversion wall appears when two schlieren running parallel to one another are joined at two points. A surface inversion wall is observed if a bulk inversion wall is anchored at both surfaces. Where the director is slightly tilted with respect to the supporting surface, tilt inversion wall[30] may also be formed; these regions usually exhibit a maximum degrees of birefringence with respect to the rest of the sample.

11.2.1 Disclinations : Planar Model

The free-energy density of a deformed sample relative to the undeformed one is given by (see eq. (5.14)),

$$f_d = \frac{1}{2}[K_1(\nabla\cdot\hat{\mathbf{n}})^2 + K_2(\hat{\mathbf{n}}\cdot\nabla\times\hat{\mathbf{n}})^2 + K_3(\hat{\mathbf{n}}\times\nabla\times\hat{\mathbf{n}})] \qquad (11.2)$$

The total distortion free-energy of the system is

$$\Delta A_e = \int f_d\, d\mathbf{r} \qquad (11.3)$$

Minimization of eq. (11.3) leads to the following set of differential equations :

$$\left(\frac{\partial f_d}{\partial n_{i,j}}\right)_{,j} - \frac{\partial f_d}{\partial n_i} = 0 \qquad (11.4)$$

where comma denotes partial differentiation with respect to spatial coordinates. The stress tensor is written as[31,32]

$$t_{ji} = -p\,\delta_{ij} - \frac{\partial f_d}{\partial n_{k,j}} n_{k,i} \qquad (11.5)$$

where $p = p_0 - f_d$, with p_0 being an arbitrary indeterminate constant. An elaborate description of the various aspects of the disclinations are given in References 3, 5, 7 and 8.

a. Wedge disclinations

Assume a planar structure in which the director is confined to the XY plane and is not a function of Z. Taking $n_x = \cos\phi$, $n_y = \sin\phi$, $n_z = 0$, and using one constant approximation ($K_1 = K_2 = K_3 = K$), eqs. (11.2), (11.4) and (11.5) reduce, respectively, to

$$f_d = \frac{1}{2}K(\nabla\phi)^2 \qquad (11.6)$$

$$\nabla^2\phi = 0 \qquad (11.7)$$

and

$$t_{ji} = -p\delta_{ij} - K\phi_{,j}\phi_{,i} \tag{11.8}$$

The solution of eq. (11.7) that are discontinuous on the Z-axis and independent of $r(= (x^2+y^2)^{1/2})$ are $\phi = 0$, which is of no interest, and

$$\phi = s\alpha + c \tag{11.9}$$

where c is a constant and $\alpha = \tan^{-1}(y/x)$. At the singular point, ϕ is a linear function of α. Rotation by 2π around the line restores the initial direction of the optical axis; ϕ can change at most by π (apolar medium) or 2π (polar medium). Thus

$$s = \pm\frac{1}{2}, \pm 1, \pm\frac{3}{2}, \cdots, \qquad \text{with } 0 < c < \pi \text{ (apolar medium)}$$

$$s = \pm 1, \pm 2, \pm 3, \cdots, \qquad \text{with } 0 < c < 2\pi \text{ (polar medium)}$$

When the axis of rotation is parallel to the disclination line L, the disclination is referred to as 'wedge disclination'[33], while in a 'twist disclinations'[34] the two are at right angles to each other. If the structure is such that the director is not normal but inclined with respect to Z-axis (e.g. S_C phase), even in case of apolar ordering, half integral values of s are not possible.

When the s values of two neighbouring disclinations are equal and opposite, the brushes joining them are circular. Superposition of the solutions of the type (11.9)

$$\phi = \phi_1 + \phi_2 = s_1\alpha_1 + s_2\alpha_2 + \text{const.}$$

Using $s_1 = -s_2 = s$,

$$\phi = s\beta + \text{const.} \tag{11.10}$$

where $\beta = \alpha_1 - \alpha_2$. Thus, as shown in Fig. 11.13, the curves of constant ϕ will be the arcs of circle passing through the two disclinations. For $s = 1/2$, ϕ changes by $\pi/2$ on going from one side of the chord to the other, but for $s = 1$ it remains unchanged.

In a circular layer of unit thickness and radius R the energy of isolated disclination is given by[34,35]

$$W = \iint f_d \, dx\, dy \tag{11.11}$$

The disclination is assumed to have a core of unknown energy. For this to occur we assume a cut-off radius r_c around the disclination and integrate, for distances greater than r_c, to get

$$W = W_c + \pi K s^2 \ln (R/r_c) \tag{11.12}$$

where W_c is the energy of the central region. As $R \to \infty$, $W \to \infty$ logarithmically. However, in practice, such a situation does not arise because of the existence of disclination pairs of opposite signs. Thus, according to the planar model the energy of a single defect is proportional to s^2; defects with $|s| > \frac{1}{2}$ should be unstable and should dissociate into $|s| = \frac{1}{2}$ defects. The stable defects of $|s| = 1$ (4-brushes) occur very frequently.

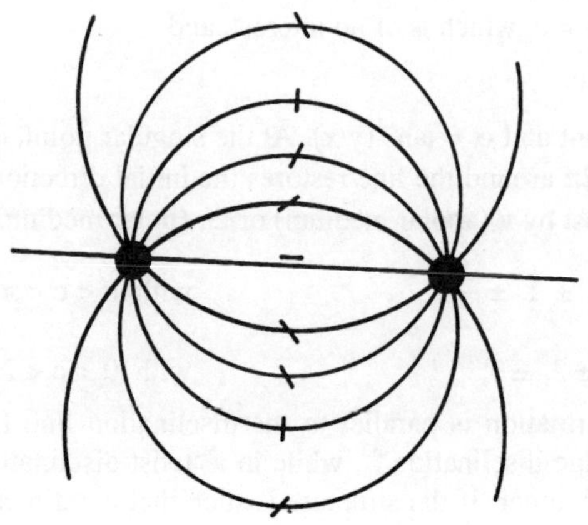

Fig. 11.13. The curves of equal alignment around a pair of singularities of equal and opposite strength for s = 1 and c = 0.

The interaction between disclinations may be evaluated from the superposition of the solutions of the type (11.9)

$$\phi = \sum_i s_i \tan^{-1}\left(\frac{y - y_1}{x - x_1}\right) + \text{const.} \tag{11.13}$$

For a pair of disclination separated by a distance r_{12} ($r_c << r_{12} << R$), one obtains[36]

$$W = \pi K(s_1 + s_2)^2 \ln (R/r_c) - 2\pi K s_1 s_2 \ln (r_{12}/2r_c) \tag{11.14}$$

When $s_1 = -s_2$, W becomes independent of R, and the interaction energy is given by the second term of this equation. Accordingly, the force between two singularities is $2\pi K s_1 s_2/r_{12}$; the disclinations of opposite signs attract and that of like signs repel.
It was shown[37,38] that the singularity at the origin of $|s| = 1$ defect can be avoided by a non-singular continuous structure of lower energy. As confirmed by the optical observations, the director orientation 'escapes' in the Z-direction.
Consider the nematic sample in a capillary of radius r_0. Let the director is tilted towards the Z-axis, and in the region $0 < r < r_0$,

$$n_x = \sin\theta \sin\phi$$
$$n_y = \sin\theta \cos\phi$$
$$n_z = \cos\theta .$$

The deformation energy per unit length is given by

$$W = \frac{1}{2}K\int [(\nabla\theta)^2 + \sin^2\theta(\nabla\phi)^2 + 2\sin\theta\ \cos\theta\{\nabla\phi\times\nabla\theta\}]\,d\mathbf{r} \tag{11.15}$$

The first two terms represent volume integrations, while the third term is a surface integration.

$$\nabla^2\theta - \sin\theta\ \cos\theta\,(\nabla\phi)^2 = 0$$

$$\nabla^2\theta + 2\cos\theta\,(\nabla\theta\cdot\nabla\phi) = 0 \tag{11.16}$$

For a solution with $\phi = \phi(\alpha)$ and $\theta = \theta(r)$, this eq. (11.16) reduces to

$$\frac{1}{r}\frac{\partial}{\partial r}\left(r\frac{\partial\theta}{\partial r}\right) - \sin\theta\ \cos\theta\,(\nabla\phi)^2 = 0 \tag{11.17}$$

$$\frac{\partial^2\phi}{\partial\alpha^2} = 0 \tag{11.18}$$

The solution of eq. (11.18) is

$$\phi = s\alpha + c, \qquad s = \pm 1,$$

For the boundary condition

$$\theta = 0 \qquad \text{at } r = 0$$

$$\theta = \frac{\pi}{2} \qquad \text{at } r = r_0$$

The equation

$$\frac{1}{r}\frac{\partial}{\partial r}\left(r\frac{\partial\theta}{\partial r}\right) - (s^2/r^2)\sin\theta\ \cos\theta = 0$$

gives

$$\theta = 2\tan^{-1}\left(\frac{r}{r_0}\right)^{|s|}, \tag{11.19}$$

This is an escape which involves splay and bend.

The total energy, independent of r_0, is

$$W = 3\pi K, \qquad \text{for } s = +1$$

$$= \pi K, \qquad \text{for } s = -1,$$

while the planar structure for $s = \pm 1$ yields

$$W = W_c + \pi K \ln(r_0 / r_c) \tag{11.20}$$

where r_c is the core radius and W_c the core energy. The planar solution has much higher energy than the continuous structure if the capillary radius r_0 is large enough for the optical observation. For an extremely small values of r_0 (i.e. very large elastic constant) the planar solution energetically may be more favourable.

Figure 11.14 shows the escaped configurations for s = 1, c = $\pi/2$ and s = -1; structures with the nails signify that the director is tilted with respect to the plane of the page.

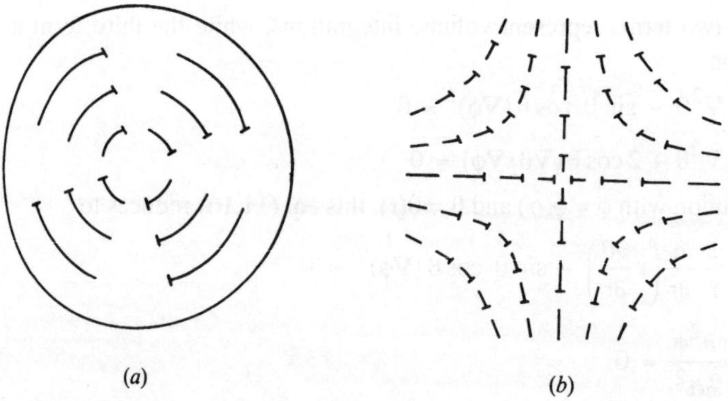

Fig. 11.14. The escaped configurations of (a) s = 1, c = $\pi/2$, and (b) s = -1 disclinations.

b. Twist disclinations

The director is parallel to the XY plane, and ϕ is a function of x and z. The solution of eq. (11.7) is obtained as

$$\phi = s\tan^{-1}(z/x) + c \tag{11.21}$$

The Z-axis (axis of rotation) is at right angles to the Y-axis (the singular line). So it is referred to as the twist disclination. It is observed that there exist in this case all the three basic elastic distortions (splay, twist and bend), while the planar solutions for the wedge disclinations involve only splay and bend deformations.

For elastically isotropic medium, the energies and interactions derived for wedge disclinations are exactly applicable to the case of twist disclinations also. Although, the non-singular solution (11.18) for s = 1 is valid in the present case, but the energy per unit length for the escaped configuration is $2\pi K\,|s|$. The structure of the escaped configuration is more complicated than that for the wedge disclination. From the nature of the director patterns it can be inferred that under the polarizing microscope for the light propagating normal to the film the dark brushes of schlieren cannot be observed. Therefore, the twist disclinations may be expected to be less conspicuous than the wedge disclinations.

Let us consider a twist disclination loop in a twisted nematic. Assume the planar structure with the director parallel to the XY plane and an imposed twist of q per unit length about the Z-axis. Let the disclination loop of radius R is in the XY plane. The net energy of this structure is

$$W = \pi^2 Ks\,[2s\,R\ln(R/r_c) - q\,R^2) \tag{11.22}$$

Here s and q are of the same sign. For the maximum value of W(s = 1/2)

$$R = R_0 \sim \frac{1}{4q} \ln\left(\frac{1}{q\, r_c}\right) \tag{11.23}$$

The energy decreases for the higher and lower values of R. Consequently, large loops ($R > R_0$) may occur, while the smaller loops shrunk (see Fig. 11.15) and eventually disappear. When s and q are of opposite signs, loops may not occur at all.

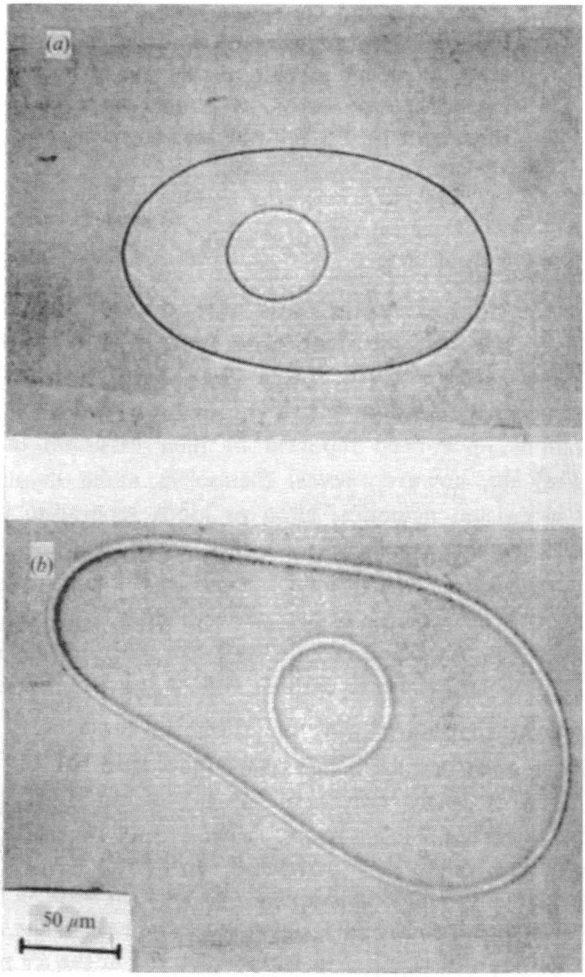

Fig. 11.15. Shrinking of twist disclination loops : (a) thin thread |s| = 1/2, (b) thick thread |s| = 1 with an escaped structure (coreless) (Ref. 7).

c. Surface disclinations

Surface disclinations are line defects associated with the interfacial boundaries[39]. They are much wider than the disclination lines in the bulk. The structure of these lines has been analysed in detail by Vitek and Kleman[40].

Consider a model with the director parallel to the XY plane of a rubbed surface and X-axis is the easy axis (i.e. the rubbed direction). Let the anchoring of the director is not infinitely strong and the deviations θ from the easy axis require an energy

$$\gamma = \frac{1}{2} A \sin^2 \theta$$

For the disclination to be along the Y-axis at z = 0, the total energy is

$$W = \frac{1}{2} K \int_{-\infty}^{\infty} \int_{0}^{\infty} [(\nabla \theta)^2 + \frac{A}{K} \sin^2\theta \; \delta(z)] \, dx \, dz \tag{11.24}$$

where δ(z) is a Dirac delta function.

Minimization of W yields

$$\delta(z) \left[\frac{A}{2K} \sin 2\theta - \frac{\partial \theta}{\partial z} \right] + \nabla^2 \theta = 0$$

The conditions to be satisfied in the bulk and at the interfaces are given, respectively, by

$$\nabla^2 \theta = 0 \tag{11.25}$$

and

$$\frac{\partial \theta}{\partial z} = \frac{A}{2K} \sin 2\theta \tag{11.26}$$

Now one obtains a continuous solution $\hat{\mathbf{n}}$ (cos θ, sin θ, 0), with

$$\theta = \tan^{-1} \left(\frac{z + \ell}{x} \right) \tag{11.27}$$

where

$$\ell = K/A$$

is a certain characteristic length. At z = 0 (plane) and x = ± ∞, the director is along the X-axis. On crossing the disclination line from x = − ∞ to x = + ∞, the orientation of director changes by π. At x = 0, the director is along the Y-axis. On crossing the line the major fraction of the change of π takes place over a distance 2ℓ, known as the 'line width'.

The total energy per unit length of this structure is

$$W = \frac{1}{2} \pi K \left[1 + \ln \left(\frac{R}{2\ell} \right) \right] \tag{11.28}$$

where R is the radius of the sample. The surface disclinations will not occur if the anchoring is very strong.

11.3 DEFECT TEXTURES IN BIAXIAL NEMATIC PHASE

We consider (see sec. 5.1.1.2) that the preferred direction of the orientation of molecules in an orthorhombic nematic phase is described by an orthonormal triad of director vector fields $\hat{\mathbf{n}}$, $\hat{\mathbf{m}}$ and $\hat{\ell}$, and express the elastic free energy by the relation (5.19). It is seen that in going from N_u to N_b phase the deformation modes of splay, twist and bend split each into two modes; and, in addition, six new modes are developed.

The optical textures of N_u and N_b phases are virtually indistinguishable[5]. Let the three mutually perpendicular director vectors, $\hat{\mathbf{n}}$, $\hat{\mathbf{m}}$, $\hat{\ell}$, in the unperturbed state are parallel to the X–, Y– and Z–axes, respectively. Similar to that of N_u phase, in N_b phase also singularities with $s = -\frac{1}{2}, +\frac{1}{2}, -1$ and $+1$ are observed. For $|s| = \frac{1}{2}$ disclination there exist three distinct classes of wedge disclinations, depending on whether the disclination line is parallel to the X, Y or Z axis, and similarly there are three classes of twist disclinations. The $\hat{\mathbf{n}}$ and $\hat{\mathbf{m}}$ director patterns for $s = +\frac{1}{2}$ and $s = -\frac{1}{2}$ wedge disclinations are shown in Fig. 11.16.

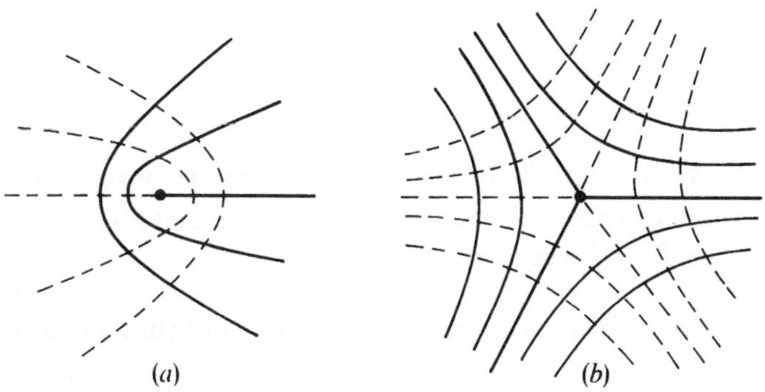

Fig. 11.16. The wedge disclinations in an orthorhombic nematic phase for (a) $s = +\frac{1}{2}$ and (b) $s = -\frac{1}{2}$.

The disclination line is along Z-axis; $\hat{\mathbf{n}}$ and $\hat{\mathbf{m}}$ directors are, respectively, represented by the full and dashed lines. Similar patterns can be drawn for the $\hat{\mathbf{n}} - \hat{\ell}$ and $\hat{\ell} - \hat{\mathbf{n}}$ fields with the disclination lines along the X– and Y–axes, respectively.

For the $s = \pm 1$ disclinations three different types of planar structure can be constructed. Figure 11.17 shows the structure of $s = +1$ wedge disclination with the singular line parallel to the Y-axis. In Fig. 11.17a, the concentric circles represent

$\hat{\mathbf{n}}$ –director and the radial lines the $\hat{\ell}$ –director. When one of the director is allowed to escape in the third dimension, the structure, as shown in Fig. 11.17b results (the dashed line represents the $\hat{\mathbf{m}}$ –director). Topologically all the three structures are equivalent. If the collapse is only over a limited region, at most two of the director fields can have out-of-plane distortions, but the third assumes a strictly planar structure. So the escape mechanism is unable to remove the singularity and it continues to exist at the origin. The energy is lowered by the collapse of $\hat{\ell}$ – director and a stabilizing effect may be caused on the $|s| = 1$ disclinations in the $\hat{\ell} - \hat{\mathbf{n}}$ or $\hat{\ell} - \hat{\mathbf{m}}$ fields.

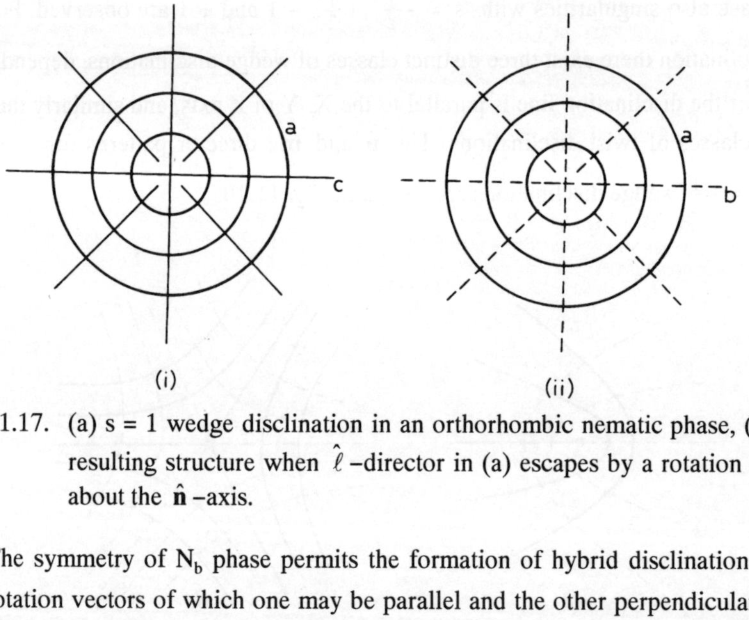

Fig. 11.17. (a) s = 1 wedge disclination in an orthorhombic nematic phase, (b) the resulting structure when $\hat{\ell}$ –director in (a) escapes by a rotation of $\pi/2$ about the $\hat{\mathbf{n}}$ –axis.

The symmetry of N_b phase permits the formation of hybrid disclinations with two rotation vectors of which one may be parallel and the other perpendicular to $\hat{\ell}$ and $\hat{\mathbf{n}}$, i.e., parallel to $\hat{\mathbf{m}}$. Homotopy theory has made remarkable predictions regarding the properties of defects in the N_b phase. The usual law of coalescence of defects breaks down. The combination rule is now non-abelian. There can be entanglement of disclination lines, which in turn leads to the topological rigidity[41]. In the N_u phase the singular points are perfect topological objects. However, such points are absent in the N_b phase.

11.4 TEXTURES AND DEFECTS IN SMECTICS

Tables 11.2 and 11.3 illustrate, respectively, the natural textures and paramorphotic textures exhibited by the smectic mesophases. The S_A and S_B phases exhibit the focal-conic natural textures, whereas the S_C and S_I phases exhibit the broken focal-conic textures. A focal-conic type texture may also be exhibited by the S_F phase and all the crystal phases (S_J, S_G, S_E, S_H, S_K) on cooling from any of the above phases due to paramorphosis. We give a brief discussion about the defect textures found in these phases.

11.4.1 The Smectic A Phase

Two characteristic textures are exhibited by the S_A phase under a polarizing microscope. In case of homeotropic alignment of molecules, the polarized light is extinguished and the texture appears completely black because the polarised light is passing down to optic axis. These black regions often possess birefringent rings because of the disclinations at the edges of air bubbles. When the molecules are not aligned homeotropically, a focal-conic fan texture results (see Fig. 11.18) which arises due to an energetically favourable packing of the lamellar structure to give Dupin cyclides (see Fig. 11.6). The nature (anisotropic, birefringent) of the phase structure now affects the polarized light and a bright texture appears which is the representative of the defects within the structure. In an ideal situation, both the homeotropic texture and the focal-conic texture should appear in the same sample when viewed by the optical polarising microscopy (see Fig. 11.19). However, the formation of focal-conic texture is more complex and more commonly observed.

Fig. 11.18. The focal-conic fan texture of S_A phase.

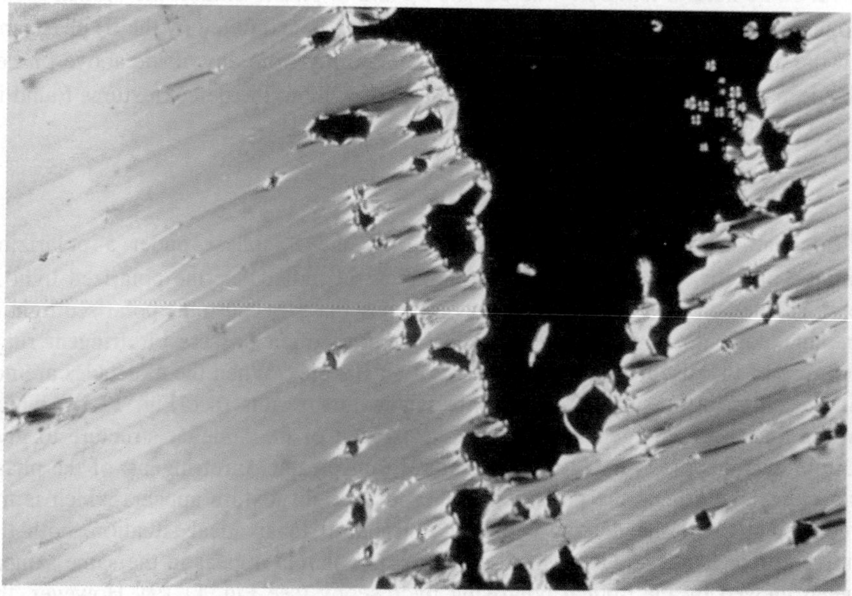

Fig. 11.19. The homeotropic texture and the focal-conic fan texture of S_A phase.

Figure 11.19 illustrates the typical focal-conic fan texture along with some homeotropic black texture. The focal-conic fan texture of an S_A phase develops from a nematic phase or isotropic liquid as bâtonnets that coalesce and eventually generate the focal-conic texture. The dark lines arise from optical discontinuities that are seen in the form of ellipses and hyperbolae and these two lines are called a focal-conic pair. The molecular layering in the S_A phase are such that the layers pack in concentric circles. Two sets of these concentric circles appear to merge into each other in an unsymmetric manner (Dupin cyclides). An additional set of concentric layers completes the 3D structure and confers the ellipsoid base of the focal-conic domain. The ellipses and hyperbolae of the focal-conic texture appear as black lines in the texture.

Figure 11.20 shows the focal-conic and parabolic defect textures of the S_A phase, that have been formed on heating a hexatic S_B phase; parabolic defects are obtained on the backs of some of the focal-conic domains and appear as 'wishbones'.

Fig. 11.20. The focal-conic and parabolic defect textures of the S_A phase.

Figure 11.21 illustrates the polygonal texture of the S_A phase; the elliptical sections are seen in the optical polarising microscope, and show brushes crossing the sections of all of the ellipses in the same direction.

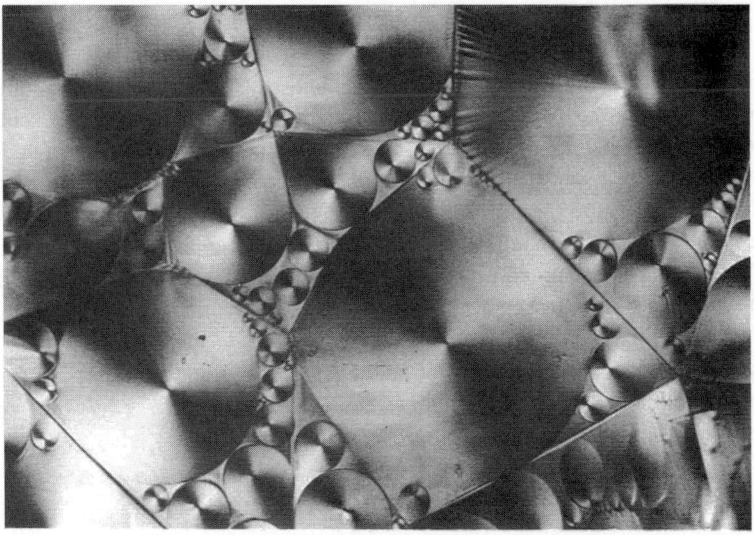

Fig. 11.21. The polygonal texture of the S_A phase.

The edge and screw dislocations in the S_A phase have been discussed in detail by Chandrasekhar and coworkers[5,7]. Topologically the edge dislocation defects are very similar to their counterparts in crystals, but they differ significantly from the point of view of energetics. The screw dislocations are exactly analogous to screw dislocations in crystals; there exists a spiral arrangement of the S_A layers around the dislocation line which is normal to the layers.

11.4.2 The Smectic C Phase

Basically the S_C phase is the tilted analogue of the S_A phase. This structural characteristics is reflected by the optical polarising microscopy. Two types of singularities are possible in the S_C phase : – the disclinations in the $\hat{\mathbf{c}}$ –director field, and the disclinations in the layer ordering. Consequently, the S_C phase exhibits two characteristic textures : the schlieren texture with $s = \pm 1$ (4-brush schlieren defect) disclinations, and the focal-conic texture which is identical to that of the S_A phase except that the molecules are tilted within the layers. When the S_C phase is formed on cooling an S_A phase broken focal-conic texture is generated because the layer spacing is reduced due to the tilted structure. An optically extinct, black homeotropic texture similar to that of the S_A phase cannot appear in S_C phase because due to the tilt of the molecules when the layers are parallel to the glass substrate the light is not directed down the optic axis.

The schlieren texture of the S_C phase continuously moves around as the phase is cooled because of the constantly increasing tilt angle. The schlieren texture in the S_C phase is more commonly generated than the focal-conic one, and is always observed when the S_C phase is formed by cooling the nematic phase. However, when both the textures (schlieren and focal-conic) are present in the same sample, the schlieren texture is accompanied by the broken focal-conic texture (see Fig. 11.22). The focal-conic texture appears broken in the S_C phase because of the disruption to the lamellar packing due to the tilt of the molecules. Both the focal-conic and schlieren textures can occur as the natural texture or paramorphotic texture. When the focal-conic fan texture is generated on cooling the nematic phase or isotropic liquid, the fans are not necessarily broken because no previous layer ordering exists.

The possibility of the occurrence of dislocations in the S_C phase have been discussed by Kléman and Lejček[42] for screw dislocations and by Lejček[43] for edge dislocations. Single edge dislocations have been observed[44] by the optical microscopy at temperatures near the S_CS_A transition.

11.4.3 Other Smectic Phases

The S_B phase can exhibit three textures : the homeotropic texture, the mosaic texture, and the focal-conic fan texture. In the mosaic texture the layers are in planes and not curved as they are in the focal-conic texture. The different levels of colour within the mosaic areas arise from the different angles of the layers. These regions are isotropic and appear black because at the boundaries of the mosaics the optic axis changes abruptly. The paramorphotic focal-conic fan texture of the S_B phase is generated from the focal-conic fan texture of the S_A phase. On cooling the S_A phase into the S_B phase, small changes in birefringence occur and sometimes a rippling effect is observed in the focal-conic fan texture. On the other hand, when the S_B phase is heated to the S_A phase, wishbone or parabolic defects are produced.

Fig. 11.22. The broken focal-conic fan texture and schlieren texture of the S_C phase.

The tilted analogues of the S_B phase are known as the S_I and S_F phases. Both of these phases generate schlieren texture similar to the S_C phase. However, it is almost impossible to distinguish between the S_I and S_F phases if both of these are not exhibited by the same sample.

The mosaic textures are the natural textures exhibited by all the crystal smectic (S_B, S_E, S_H, S_G and S_K) phases, whereas the focal-conic textures found in these phases are paramorphotic textures. In crystal smectic phases, long range order is seen in the form of grain boundaries. These phases do not form a genuine focal-conic fan texture because curved layers are not possible. When a mesophase exhibiting a focal-conic texture is cooled into a crystalline phase, the fans usually assume a more broken and chunky appearance at the phase transition because the long-range order builds up and produces domains with different molecular orientations. The domains are separated by the grain boundaries and walls. As an example, consider the transition from the S_A phase to the $S_{B(cry)}$ phase. When the S_A phase is cooled, the $S_{B(cry)}$ phase begins to build up in the bulk S_A phase and a biphasic region develops which produces a series of dark lines, running parallel to the layers, across the backs of the focal-conic domains. These lines occur only at the phase transition and are called as 'transition bars'. These bars are transitory in nature. Their origin is associated with the build up of the long-range order of the molecules. Upon further cooling the bands expand, meet and eventually coalesce to produce a smooth texture again. Figure 11.23 shows the transition bars produced at the S_A $S_{B(cry)}$ transition. Figure 11.24 illustrates the defect textures of a $S_{B(cry)}$ phase formed on cooling a S_A phase. The homeotropic texture present shows that the molecules are arranged in layers with their long axes perpendicular to the layer planes. The steps are seen in the focal-conic domains which are due to the extensive out-of-plane ordering. It can be seen that in this part of the texture no ellipses or hyperbolae exist. This suggests that the true focal conic domains no longer exist.

Fig. 11.23. Transition bars observed under crossed polars at the S_A $S_{B(cry)}$ transition (X100; Ref. 10).

Fig. 11.24. The homeotropic and homogeneous defect textures of the $S_{B(cry)}$ phase (X100, Ref. 10).

The focal-conic textures can also appear banded; for example, the $S_{E(cry)}$ phase can exhibit lined focal-conic texture (Fig. 11.25). This shows an extensive out-of-plane correlations of the molecules. It can be seen that the bands are parallel which suggest that there exist domains stacked up one on top of another. In the tilted smectic and crystal phases the focal conic domains become crossed in a patchwork pattern. Figure 11.26 shows the patchwork pattern of the focal-conic texture of the S_F phase. This pattern indicates the existence of long-range out-of-plane correlations coupled with domains tilted in different directions to one another.

11.5 TEXTURES AND DEFECTS IN CHIRAL LIQUID CRYSTALS

Chiral systems that are optically active possess the rich diversity of defects and textures.

11.5.1 The Chiral Nematic (Cholesteric) Phase

The chiral nematic (cholesteric) phase, an optically active variant of the achiral nematic phase, may exhibit defects and textures typical of the achiral nematic phase, and some other textures that are not shown by the achiral nematic at all. These additional textures include the focal–conic fan and polygonal textures, oily streaks and various dislocations.

Fig. 11.25. The arced focal-conic texture of $S_{E(cry)}$ phase (X100, Ref. 10).

Fig. 11.26. The patchwork pattern of the focal-conic texture of the S_F phase (X100; Ref. 10).

Based on the ratio L/p (L is the size of the sample and p is the pitch of the helix), there are two complementary approaches to describe the distortions in the chiral nematic phase. For weakly twisted phase $L/p \ll 1$, while for strongly twisted one $L/p \gg 1$. When $L/p \ll 1$, the chiral nematic does not differ much from the achiral nematic phase, and therefore the optical observations may reveal 'thick' (nonsingular) and 'thin' (singular) line defects. For $L/p \gg 1$, the elastic properties of the cholesterics are close to that of lamellar phases. The layered structure of the cholesterics leads to the formation of large scale defects such as focal-conic domains and oily streaks.

11.5.1.1 Disclinations in the chiral nematic phase

The chiral asymmetry of the molecules leads to the transition from a achiral nematic state, with order parameter space carrying all the possible rotations of a unique director, to a chiral nematic state with the order parameter space which carries all the possible rotations of a set of three mutually perpendicular directors. Following Friedel and Kléman[45], these directors are denoted as λ (which represents the local direction defined by the molecule), χ (along the helical axis), and $\tau = \lambda \times \chi$. Thus from a topological point of view, the defects in the chiral nematics is similar to that in the biaxial nematics, and the phenomena of topological entanglement of disclinations and the formation of nonsingular soliton configurations may be observed.

The energy of disclinations strongly depends on how the trihedron λ, χ, τ is distorted. In a uniaxial cholesteric, the three have different physical meaning and different distortion energy. Only λ is a real director, while χ and τ are immaterial directors. Consequently the singularities associated with χ and τ would be generally more costly than belonging to λ. The difference becomes apparent when the disclinations of half-integer strength $s = n + \frac{1}{2}$ (n is an integer) are compared. In the λ disclinations the director λ is not singular, whereas in the τ disclinations the λ is singular. The core of λ disclinations is of radius p ("thick" lines, see Fig. 11.27(a), (b) and (d)), while the core of τ disclinations ("thin" lines, see Fig. 11.27(c)) is of molecular size. When the chiral nematic is unwound into a achiral nematic phase, $p \to \infty$, and λ disclinations vanish. Figure 11.27(b) shows that a λ^- can be annihilated by a collapse with a λ^+. Figure 11.27(d) shows a wedge line with $s = \frac{1}{2}$, i.e., τ^+ disclinations. τ^+ is singular for the χ and λ fields but continuous for the τ field.

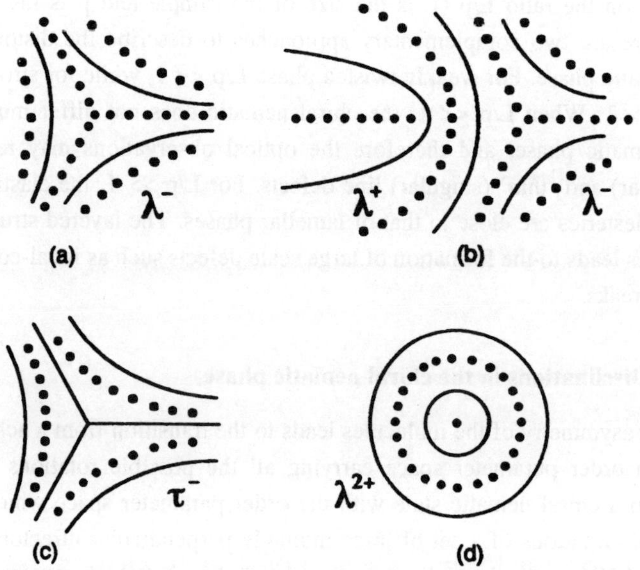

Fig. 11.27. λ and τ disclinations in a chiral nematic phase. + and − superscripts indicate the values of $s = +\frac{1}{2}$ and $-\frac{1}{2}$, respectively.

The disclinations with integer strength are divided into two classes; with odd integer strength s = 2n+1, and with even integer strength s = 2n. The disclinations with s = 2n+1 cannot be eliminated; the region of size ~ p restricts the escape of the director into the third dimension. As shown in Fig. 11.27(d), the λ line with s = 1 cannot be made continuous for the τ and χ fields simultaneously. However, the s = 2n lines do escape in the third dimension.

The λ and τ disclinations are often observed in fingerprint texture (Fig. 11.28). It is expected that the λ defects are more frequent than the τ defects because the line tensions differ by an amount ~ K ln (p/a), a is of the order of few molecular size. On the other hand, the τ defects often occur in pairs with λ defects to replace χ defects.

11.5.1.2 Dislocations

On the symmetry consideration the χ disclinations can be equivalently treated as dislocations[46,47], with the Burgers vector

$$b = -s\,p \tag{11.29}$$

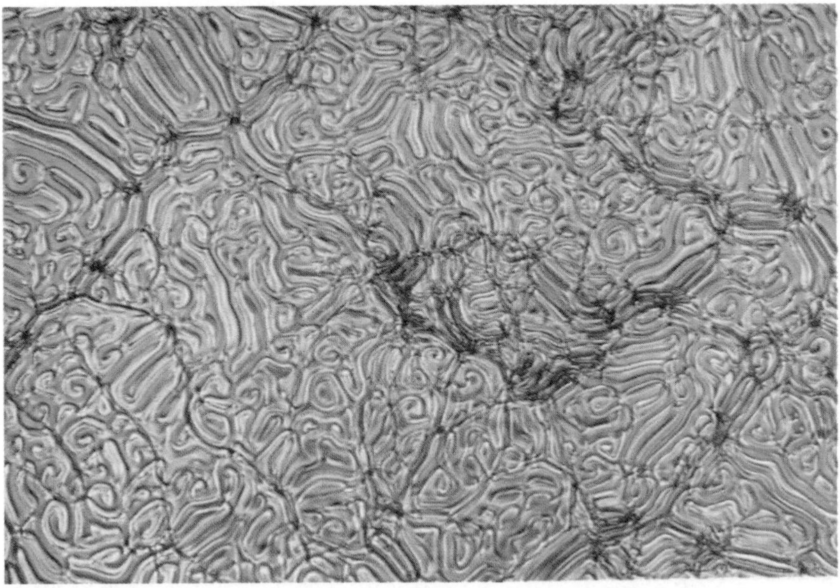

Fig. 11.28. The fingerprint texture of the chiral nematic phase.

Figure 11.29 illustrates a χ^+ wedge disclination (χ is continuous) which can be constructed by performing a Volterra process along the line. The cut surface is opened by an angle π; each cholesteric layer gives a 2D, $s = \frac{1}{2}$ configuration which rotates helically along the line with a pitch p. The χ dislocations can split into combinations of λ and τ disclinations. An example is illustrated in Fig. 11.30; the core splits into a λ^- and τ^+ separated by a distance p/4.

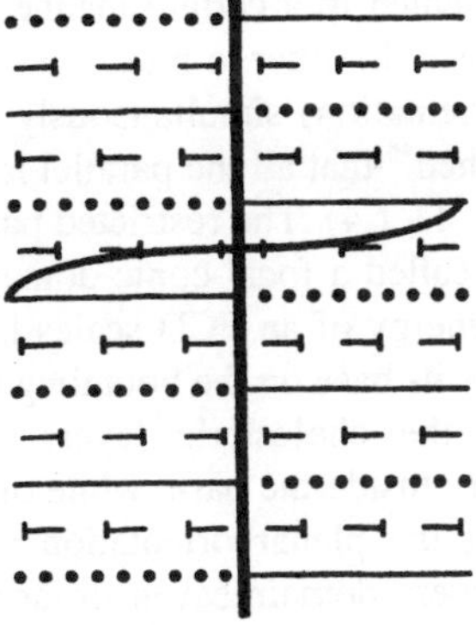

Fig. 11.29. Wedge χ^+ disclination = screw dislocation.

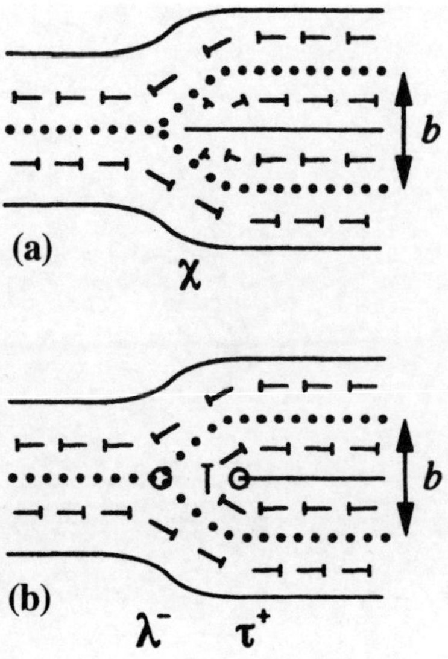

Fig. 11.30. (a) χ–twist disclination = edge dislocation, and (b) splitting of the core of a dislocation into a λ^- and τ^+ disclinations.

11.5.1.3 Effects of the layer structure : Focal-conic domains and oily streaks

The existence of focal-conic domains and oily streaks depends crucially on the layered characteristic of ordering.

For L/p >> 1, the elastic theory considers the medium as a system of equidistant, parallel layers. The dislocations, oily streaks, and focal-conic domains can be described by using the results obtained for the S_A phase.

The liquid crystal samples are always bounded. Due to the surface interactions the cholesteric layers are oriented in a certain manner; most often, the layers align parallel to the bounding surface. In order to meet the two requirements (equidistance of layers and the surface orientation) simultaneously the layers are bent in a very special manner. It is established[48] that all the parallel layers should assume the shape of "Dupin cyclides" (see sec. 11.1.4). The restricted part of space filled with a single family of Dupin cyclides is called a focal-conic domain (FCD). The FCDs involve 3D distortions. The elastic energy of an FCD scales linearly with the characteristic size ~ KL. When an FCD has its base on the bounding surface, it effectively changes the surface orientation of the cholesteric layers. Most often, the layers are perpendicular to the boundary inside the base, while outside they are parallel. When the surface does not favour the planar orientation of layers, FCDs formation is energetically justified, and every domain saves surface energy at the cost of elastic energy.

Liquid crystals and oily streaks were discovered simultaneously, and, as FCDs, are common for many lamellar liquid crystals. Their inner structure is very complicated, and depends on many parameters such as elastic constants and surface anchoring. Figure 11.31 shows the simplest type of oily streaks made of two parallel $s = \frac{1}{2}$ disclinations with a wall between them. The total Burgers vector is zero, so topologically the oily streaks is trivial; it can disappear by pulling the semiround ends together. Their networks coarsen with time because the free-energy per unit length (line tension) is normally positive. At the sample's boundary, the layers within the FCDs and oily streaks, and the layers outside these defects usually have different orientations. Consequently, the problems of surface anchoring and the layers curvature in these defects are strongly connected.

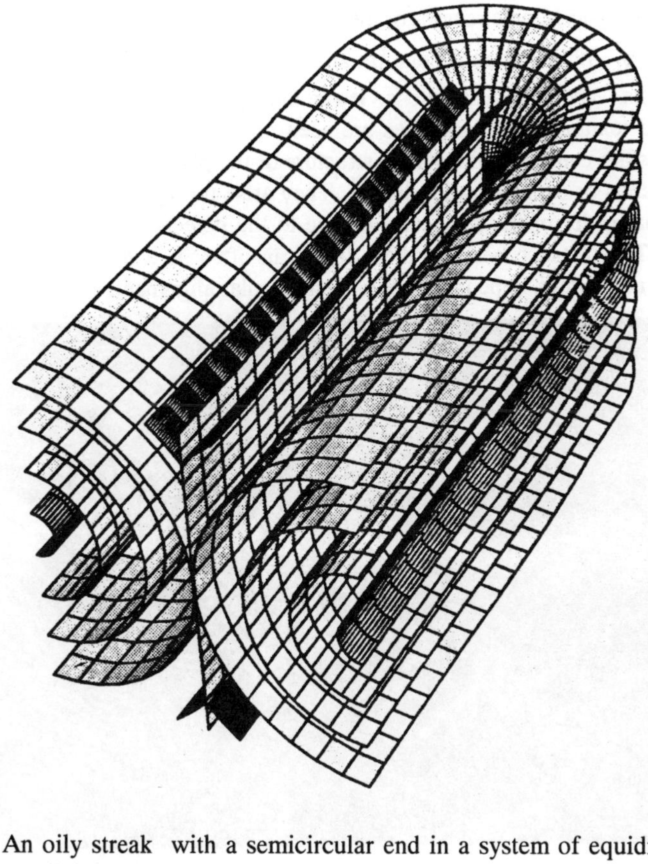

Fig. 11.31. An oily streak with a semicircular end in a system of equidistant and parallel layers.

11.5.2 The Chiral Smectic C (S_C^*) Phase

A homogeneous sample of S_C^* phase obtained on cooling either from an N^* phase or a S_A phase generally exhibits at least two major kinds of defects[49]; the focal-conic and parabolic defects. When the growth of the S_C^* phase from either the nematic phase or the isotropic liquid phase occurs naturally via the formation of bâtonnets a focal-conic defect results. The focal-conic texture formed by this process is the natural texture of the S_C^* phase and so it is the one which is adopted instantaneously at nucleation of this mesophase. A focal-conic defect also results paramorphotically on cooling the S_A phase. Usually a parabolic defect is generated as the growth fronts of the nucleating S_C^* phase meet each other. Sometimes, the parabolic defects result as the stress-induced in the focal-conic regions of the sample.

In the S_C^* phase the molecular tilt and the helical ordering can further complicate these defects. When a thin sample is viewed under optical polarized microscopy, in the focal-conic texture often dark lines are seen on the backs of the fans (Fig. 11.32). These lines are caused by the helical structure of the phase and are known collectively as pitch bands or dechiralization lines[50]. They arise through the continuous change in orientational order of the molecules or due to the defects close to the surface of the substrate. The distance between these lines can often be a measure of half the pitch length but not always, depending upon the molecular alignment.

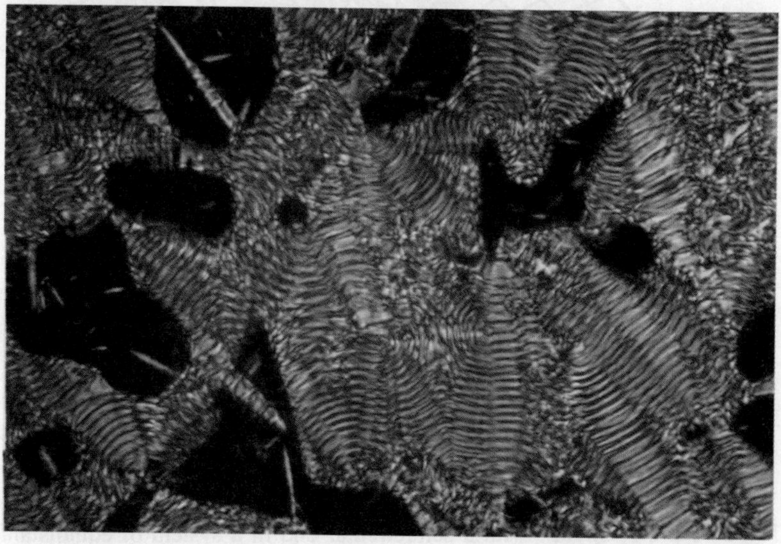

Fig. 11.32. The lined focal-conic fan texture of S_C^* phase.

In the achiral S_C phase, the homeotropic textures do not occur because of the tilt of the molecules. Since in the S_C^* phase the tilt angle spirals to describe a helix and on the average the tilt angle through the phase is zero, the light directed down the average optic axis of the helix is extinguished. When the pitch length of the helix is in the region of visible light, colours are reflected, blue at lower temperatures and red at higher temperatures, and the pseudo-homeotropic texture becomes coloured and is called the petal texture.

11.5.3 The Blue Phases

In the blue phases, as well as in TGB* phases, defects are not only present in the structure, but are essential for the generation and stability of these phases. The structure of blue phases is often referred to as a 'double twist' structure (see sec. 9.2), whereas the chiral nematic phase is considered as a single twist system.

The defects in two of the blue phases, BP I and BP II, occur at every point where the different director of the twist cylinders field meet. These are arranged in a cubic lattice, but unlike in a regular crystal, the constituent molecules move freely and the disclinations occupy the lattice points. In BP I the lattice of defects is a body-centered cubic, whereas in BP II it is a simple cubic[51]. The $s = -\frac{1}{2}$ disclinations have been found in the lattices of BP I, BP II or BP III.

As BP I and BP II have cubic structures, their textures will be based on this cubic arrangement of defects. Accordingly the blue phases exhibit a platelet texture under the polarising microscope (Fig. 11.33). The platelet textures of BP I are striated in nature and appear blue, red and green in colour, whereas that of BP II are cleaner and mostly blue in colour. These platelets have sharp edges because of the cubic structure of the phase.

11.5.4 The Twist Grain Boundary Smectic A* Phase

The structure of TGB_A^* phase is one where blocks of a normal S_A^* phase are rotated with respect to one another to produce a macroscopic helix which occurs in a direction parallel to the planes of the layers. The twisting of blocks is facilitated by the inclusion of screw dislocations between the individual blocks. Along the axis of helix, the screw dislocations occurs periodically. Consequently, the rotation of the layers relative to one another is produced by the rows of screw dislocations between the blocks. Due to relative size of the blocks and optical pitch, selective reflection of light occurs. Accordingly, the defects are too small to see in the microscope. However, the phase exhibits a Grandjean platelet texture similar to the one exhibited by the blue phases.

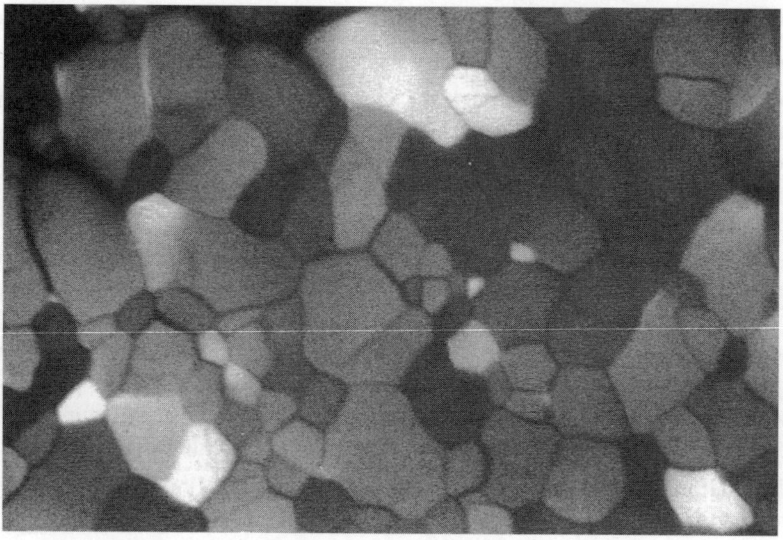

Fig. 11.33. The platelet texture of blue phases (X100, Ref. 10).

When TGB_A^* phase is obtained from a homeotropic S_A^* phase, a filamentary defect texture, as shown in Fig. 11.34, is seen.

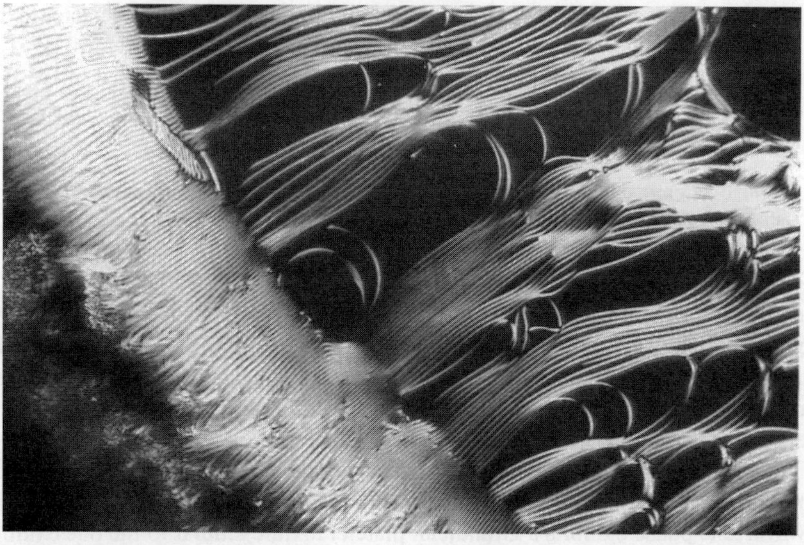

Fig. 11.34. The filamentary defect texture TGB_A^* phase seen in a free-standing film (X100, Ref. 100).

11.6 DEFECTS IN THE COLUMNAR LIQUID CRYSTALS

Defects in hexagonal structure have been investigated in detail by Bouligand[54] and by Kléman and Oswald[55]. The burgers vector **b** of a dislocation in the columnar structure is normal to the columnar axis $\hat{\mathbf{n}}$. Generally

$$\mathbf{b} = \ell\mathbf{a} + \mathrm{m}\mathbf{a}' \qquad (11.30)$$

where ℓ and m are positive or negative integers. When **b** is perpendicular to the dislocation line L, edge dislocations result. Two types of edge dislocations are possible. In longitudinal edge dislocations L is along $\hat{\mathbf{n}}$, while L is normal to $\hat{\mathbf{n}}$ in case of transverse edge dislocations. For the screw dislocation **b** is parallel to the dislocation line. A hybrid of screw and edge dislocations can also results.

In a hexagonal lattice, the longitudinal wedge disclinations are the standard crystal dislocations. Near a positive disclination the lattice gets compressed, whereas it is stretched near a negative disclination. Most of the longitudinal wedge disclinations, in general, possess prohibitively large energies, and so these are seen as unlike pairs at the core of a disclination.

In the transverse wedge disclinations the rotation vector (θ_2 or T_2) is normal to $\hat{\mathbf{n}}$. Figures 11.35(a), (b) and (c) illustrate the $\pm\,\pi$ disclinations. Two transverse wedge disclinations may occur in association (Fig. 11.35d). Such defects have been observed experimentally[56].

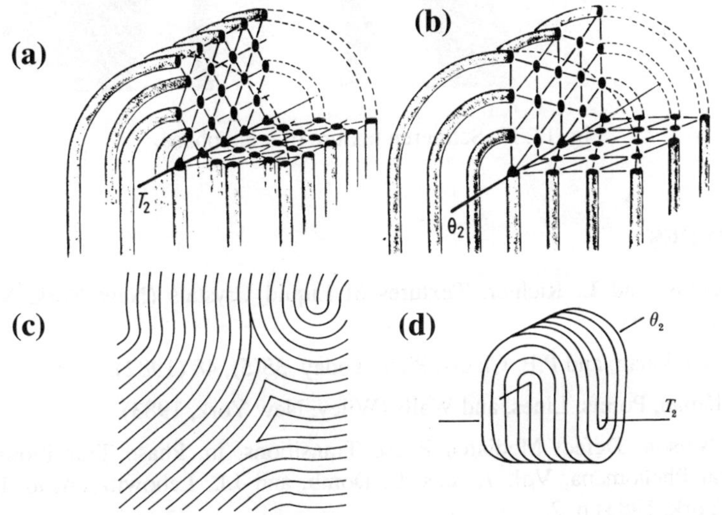

Fig. 11.35. (a) and (b) π transverse wedge disclinations about the binary axes T_2 and θ_2, respectively, (c) $-\pi$ transverse wedge disclination leading to the formation of walls, and (d) two transverse π disclinations, one about T_2 and other about θ_2 (Ref. 54).

The symmetry of uniaxial nematic and discotic nematic phases is the same, and so identical types of defects – the schlieren texture, umbilics, etc., are observed in both cases. Figure 11.36 shows a typical schlieren pattern[57] exhibited by a thin N_D film when viewed under a polarising microscope.

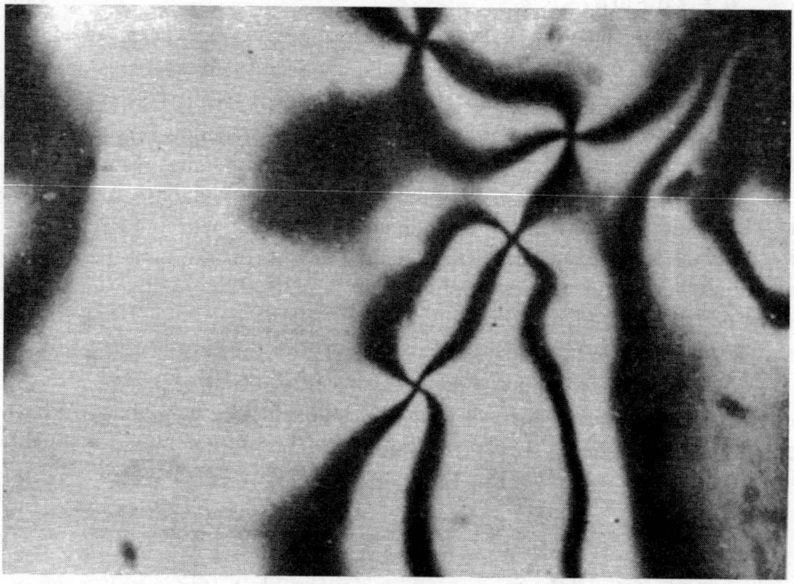

Fig. 11.36. Schlieren texture of a N_D phase.

REFERENCES

1. D. Demus, and L. Richter, Textures of Liquid Crystals (New York, Verlag Chemie, 1978).
2. W.F. Brinkman, and P.E. Cladis, Phys. Today **35(5)**, 48 (1982).
3. M. Kléman, Points, Lines, and Walls (Wiley, New York, 1983).
4. D.R. Nelson, Defect Mediated Phase Transitions, in, Phase Transitions and Critical Phenomena, Vol. 7, eds. C. Domb, and J.L. Lebowitz (Acad. Press, New York, 1983) p. 2.
5. S. Chandrasekhar, and G.S. Ranganath, Adv. Phys. **35**, 507 (1986).
6. M. Kleman, Rep. Prog. Phys. **52**, 555 (1989).
7. S. Chandrasekhar, Liquid Crystals (Cambridge Univ. Press, Cambridge, 1992).

8. P.G. de Gennes, and J. Prost, The Physics of Liquid Crystals (Clarendon Press, Oxford, 1993).
9. P.M. Chaikin, and T.C. Lubensky, Principles of Condensed Matter Physics (Cambridge University Press, 1995); T.C. Lubensky, Solid State Commun. **102**, 187 (1997).
10. A.J. Slaney, K. Takatoh, and J.W. Goodby, Defect Textures in Liquid Crystals, in, The Optics of Thermotropic Liquid Crystals, eds., S. Elston, and R. Sambles (Taylor and Francis Ltd., 1998) Chapt. 13, p. 307.
11. H.S. Kitzerow, and C. Bahr, eds., Chirality in Liquid Crystals (Springer-Verlag, New York, 2001).
12. P.E. Lammert, D.S. Rokhsar, and J. Toner, Phys. Rev. Lett. **70**, 1650 (1993).
13. L. Lam, and J. Prost, eds., Solitons in Liquid Crystals (Springer-Verlag, New York, 1992).
14. C. Kittel, Introduction to Solid State Physics, 5th edn. Frankfurt/Main (Wiley, 1971).
15. G. Toulouse, and M. Kléman, J. Phys. Lett. **37**, L149 (1976).
16. N. Schopohl, and T.J. Sluckin, Phys. Rev. Lett. **59**, 2582 (1987).
17. V. Volterro, Ann. Ecole Normale Super. (Paris) (3) **24**, 400 (1907).
18. G.W. Gray, and J.W. Goodby, Smectic Liquid Crystals, Textures and Structures (Leonard Hill, London, 1984).
19. A.D.L. Chandani, E. Gorecka, Y. Ouchi, H. Takezoe, and A. Fukuda, Jpn. J. Appl. Phys. **28**, L1265 (1989).
20. F.C. Frank, Disc. Faraday Soc. **25**, 19 (1958).
21. Y. Bouligand, J. Microsc. (Paris) **17**, 145 (1973).
22. Y. Bouligand, J. Phys. (Paris) **34**, 603 (1973).
23. N.H. Hartshorne, and A. Stuart, in, Crystals and the Polarising Microscope, 4th edn. (London, Edward Arnold, 1970).
24. C. Rosenblatt, R. Pindak, N.A. Clark, and R.B. Meyer, J. Phys. (Paris) **38**, 1105 (1977);

    J.W. Goodby, and R. Pindack, Mol. Cryst. Liquid Cryst., **75**, 233 (1981).
25. Y. Bouligand, J. Phys. (Paris) **33**, 525 (1972).
26. J.P. Sethna, and M. Kleman, Phys. Rev. **26**, 3037 (1982).
27. M.A. Handschy, N.A. Clark, and S.T. Lagerwall, Phys. Rev. Lett. **51**, 471 (1983).

28. T.P. Rieker, N.A.Clark, G.S. Smith, D.S. Parmar, E.B. Sirota, and C.R. Safinya, Phys. Rev. Lett. **59**, 2658 (1987).
29. Y. Ouchi, J. Lee, H. Takezoe, A. Fukuda, K. Kondo, T. Kitamura, and A. Mukoh, Jpn. J. Appl. Phys. **27**, L725 (1988).
30. C.E. Williams, and M. Kléman, J. Phys. Lett. **35**, L33 (1974).
31. J.L. Ericksen, in, Liquid Crystals and Ordered Fluids, eds., J.F. Johnson, and R.S. Porter (Plenum, New York, 1970) p. 181.
32. F.M. Leslie, Arch. Rational Mech. Anal. **28**, 265 (1968).
33. R. de Wit, in, Fundamental Aspects of Dislocation Theory, eds., J.A. Simmons, R. de Wit, and R. Bullough, Spl. Pub. No., 317 (NBS **1**, 715 (1970)).
34. A. Saupe, Mol. Cryst. Liquid Cryst. **21**, 211 (1973).
35. J. Nehring, and A. Saupe, J. Chem. Soc. Faraday Trans. II **68**, 1 (1972).
36. C.M. Dafermos, Quart. J. Mech. Appl. Math. **23**, S49 (1970).
37. P.E. Cladis, and M. Kléman, J. Phys. (Paris) **33**, 591 (1972).
38. R.B. Meyer, Phil. Mag. **27**, 405 (1973).
39. M. Kléman, and C.E. Williams, Phil. Mag. **28**, 725 (1973).
40. V. Vitek, and M. Kléman, J. Phys. (Paris) **36**, 59 (1975).
41. G. Toulouse, J. Phys. (Paris) Lett. **38**, L-67 (1977);

    V. Poenaru, and G. Toulouse, J. Phys. (Paris) **8**, 887 (1977).
42. M. Kléman, and L. Lejček, Phil. Mag. **42**, 671 (1980).
43. L. Lejček, Czech. J. Phys. **B34**, 563 (1984); Ibid **B35**, 726 (1985).
44. R.B. Meyer, B. Stebler, and S.T. Lagerwall, Phys. Rev. Lett., **41**, 1393 (1978);

    S.T. Lagerwall, and B. Stebler, Phys. Bull. **31**, 349 (1980).
45. J. Friedel, and M. Kléman, in, Fundamental Aspects of Dislocation Theory, eds., J.A. Simmons, R. de Wit and R. Bullough, Spl. Publ. No. 317, (NBS **1**, 607 (1970)).
46. Y. Bouligand, and M. Kléman, J. Phys. (Paris) **31**, 1041 (1970).
47. O.D. Lavrentovich, and M. Kléman, Cholesteric Liquid Crystals : Defects and Topology, in, Chirality in Liquid Crystals, eds., H.S. Kitzerow, and C. Bahr (Springer-Verlag, New York, 2001) Chapt 5, p. 115.
48. G. Friedel, and F. Grandjean, Bull. Soc. Franc. Miner. **33**, 192, 409 (1910); C.R. Hebd. Scan Acad. Sci. **151**, 762 (1910).

49. J.W. Goodby, Properties and Structures of Ferroelectric Liquid Crystals, in, Ferroelectric Liquid Crystals : Principles, Properties and Applications, eds., J.W. Goodby, R. Blinc, N.A. Clark, S.T. Lagerwall, M.A. Osipov, S.A. Pikins, T. Sakurai, K. Yoshino, and B. Žekš, (Gordon and Breach Science Publishers, 1999) Chapt. 9, p. 99.

50. M. Glogarova, L. Lejcek, J. Pavel, U. Janovec, and F. Fousek, Mol. Cryst. Liquid Cryst. **91**, 309 (1983).

51. P.P. Crooker, Liquid Crystals **5**, 751 (1989).

52. A.J. Slaney and J.W. Goodby, J. Mater. Chem. **1**, 5 (1991); Liquid Crystals **9**, 849 (1991).

53. J.M. Gilli, and M.M. Kamaye, Liquid Crystals **12**, 545 (1992).

54. Y. Bouligand, J. Phys. (Paris) **41**, 1297, 1307 (1980).

55. M. Kléman, and P. Oswald, J. Phys. (Paris) **43**, 655 (1982).

56. P. Oswald, J. Phys. (Paris) Lett. **42**, L171 (1981).

57. C. Destrade, M.C. Bernaud, H. Gasparoux, A.M. Levelut, and N.H. Tinh, Proceedings of the International Conference on Liquid Crystals, Bangalore, December, 1979; ed., S. Chandrasekhar (Heyden, London, 1980), p. 29.

Index